INTRODUCTION *to*

Wildlife Management

The Basics

Paul R. Krausman

The University of Arizona

Prentice
Hall

Upper Saddle River, New Jersey 07458

Library of Congress Cataloging-in-Publication Data

Krausman, Paul. R., 1946-
  Introduction to wildlife management: the basics / Paul R. Krausman.
    p. cm.
  Includes bibliographical references (p. 423).
  ISBN 0-13-280850-1
    1. Wildlife management. I. Title

 SK355 .K73 2001
 333.95'4–dc21   2001024583

Executive Editor: Debbie Yarnell
Associate Editor: Kimberly Yehle
Production Editor: Lori Dalberg, Carlisle Publishers Services
Production Liaison: Eileen O'Sullivan
Director of Production & Manufacturing: Bruce Johnson
Managing Editor: Mary Carnis
Manufacturing Buyer: Cathleen Petersen
Senior Design Coordinator: Miguel Ortiz
Cover Designer: Steve Frim
Cover Image: Ted Kerasote, Photo Researchers, Inc.
Marketing Manager: Jimmy Stephens
Interior Design and Composition: Carlisle Communications, Ltd.
Printing and Binding: Phoenix Color Book Tech

Prentice-Hall International (UK) Limited, *London*
Prentice-Hall of Australia Pty. Limited, *Sydney*
Prentice-Hall of Canada Inc., *Toronto*
Prentice-Hall Hispanoamericana, S.A., *Mexico*
Prentice-Hall of India Private Limited, *New Delhi*
Prentice-Hall of Japan, Inc., *Tokyo*
Prentice-Hall Singapore Pte. Ltd.
Editora Prentice-Hall do Brasil, Ltda., *Rio de Janeiro*

10 9 8 7 6 5 4 3 2 1
ISBN 0-13-280850-1

Dedicated to
Martha Irene Grinder
1967–1999
The Indefectible Scholar

Contents

Preface

"What has happened before will happen again. What has been done before will be done again. There is nothing new in the whole world."

Ecclesiastes 1:9

Because of human domination of the earth, wildlife is dependent upon mankind for its existence. Unfortunately, our efforts have not always been successful and we are still struggling to find ways to effectively maintain the habitats to ensure the existence of many animals. It is a complex process and the young field of wildlife management has served as a springboard for a host of yet even younger disciplines including conservation biology, landscape ecology, ecosystem management, and human dimensions. These new fields and their associated jargon are important to conservation and management of wild flora and fauna. However, students in these disciplines should be well founded in the basic principles that underlie the functioning of healthy populations. To that end, those students should have a strong background in ecology and the basic principles of wildlife management. After all, wildlife is the emphasis of the numerous groups, organizations, and federal, state, and international agencies charged with maintaining and enhancing wild and natural places, and places not so wild. I have yet to find The Book on wildlife management and have relied on a variety of media to teach wildlife courses (e.g., textbooks, journals, labs, popular literature, and news). Hands-on laboratories often provide lasting examples of wildlife management practices. However, students also have to be exposed to the underpins of the profession. That is the purpose here. *Introduction to Wildlife Management: The Basics* was written for beginning and advanced wildlife students, and as a reference for professionals who want to brush up on the basics of their profession. All users are expected to have a background in ecology and beginning students would obviously need more lecture time than advanced students. The text was not written to serve as the only source of information, but was designed to include the basic concepts used to manage wildlife and to provide a solid reference source for additional reading.

Many thanks to B. Ballard, J.A. Bissonette, B. Czech, E. de Steiguer, M.L. Morrison, and M. C. Wallace who provided helpful comments about previous drafts; to S. Gillatt, who drafted many of the figures; and to V. Calt and D. Brown, who assisted with production. Thanks also to reviewers Mark Wallace, Texas Tech University; and Ronald M. Case, University of Nebraska, who provided valuable insight and helpful feedback.

Sources of Chapter Opening Quotations

Chapter 16   Psalms 84.

Chapter 17   Leopold, A. 1933. *Game management.* Charles Scribner's Sons, New York, New York, USA.

Chapter 18   Child, G. 1995. *Wildlife and people: The Zimbabwean success.* Wisdom Foundation, Harare, Zimbabwe.

Chapter 19   Leopold, A. 1933. *Game management.* Charles Scribner's Sons, New York, New York, USA.

1

Defining Wildlife Management

"The man should have youth and strength who seeks adventure in the wide, waste spaces of the earth, in the marshes, and among the vast mountain masses, in the northern forests, amid the steaming jungles of the tropics, or on the deserts of sand or of snow. He must long greatly for the lonely winds that blow across the wilderness, and for sunrise and sunset over the rim of the empty world. His heart must thrill for the saddle and not the hearthstone. He must be helmsman and chief, the cragsman, the rifleman, the boat steerer. He must be the wielder of axe and of paddle, the rider of fiery horses, the master of the craft that leaps through white waters. His eye must be true and quick, his hand steady and strong. His heart must never fail nor his head grow bewildered, whether he faces brute and human foes, or the frowning strength of hostile nature, or the awful fear that grips those who are lost in trackless lands. Wearing toil and hardship shall be his; thirst and famine shall he face, and burning fever. Death shall come to greet him with...charging beast or of scaly things that lurk in lake and river; it shall lie in wait for him among untrodden forests, in the swirl of wild waters, and in the blast of snow blizzard of thunder — shattered hurricane."

T. Roosevelt (1916)

INTRODUCTION

Most people have an idea of what wildlife is and there are likely numerous definitions that are used. To discuss the basics of wildlife management it is important to have a common meaning of the subject. To that end this chapter defines wildlife, makes distinctions between active and inactive management, introduces the goals of management, then concludes with classifications of wildlife that have been used.

■ DEFINING WILDLIFE MANAGEMENT

In 1933 Aldo Leopold defined game management as the art of making land produce sustained annual crops of wild game for recreational use. Today, "game" is a term that means whatever the legislature of the state that the animal resides in stipulates as game. The term "game" can have a narrow, precise, legislated meaning that first appeared in British law in an Act of 1389 (McKelvie 1985). The use of the word into British law found its way to North America and has led to many anomalies. For example, mourning doves in parts of the eastern United States are classified as songbirds (i.e., nongame) but they are hunted as game in many western states. In addition, many species are not classified as game but most states require a game license to harvest them. Wildlife does not have a universally defined meaning and it expands and contracts with the viewpoint of the user (Caughly and Sinclair 1994). In general the wildlife profession considers wildlife to be free-living animals of major significance to humans. This definition does not exclude plants or lower animals because wildlife is more than the animal. When wildlife are defined, the habitats that support them also have to be considered. Species and their habitat are interlocked and cannot properly be considered separately. Because wildlife belong to the public (in North America), humans also have to be considered. Giles (1978) presented a holistic view of wildlife that considered the animal, people, and their habitat when considering wildlife and the interactions between each (Figure 1–1). There are numerous definitions of wildlife such as the definition presented by the United States Congress (1973): "The term fish or wildlife means any member of the animal kingdom, including without limitation any mammal, fish, bird (including any migratory, non-migratory, or endangered bird for which protection is also afforded by treaty or

Figure 1–1 The animal, habitat, and human triad that incorporates wildlife and wildlife management (after: *Wildlife Management.* Robert H. Giles, Jr. © 1978 by W.H. Freeman and Company. Used with permission).

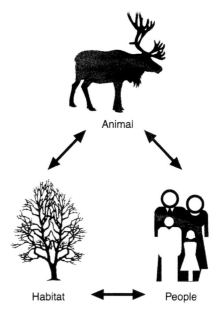

Animal

Habitat ⟷ People

other international agreement), amphibian, reptile, mollusk, crustacean, arthropod or other invertebrate, and includes any part, products, egg, or off-spring thereof, or the dead body parts or parts thereof" (A compilation of federal laws. U.S. Government Printing Office 052-070-02871-4, 1975).

This definition is too descriptive and does not take into account other aspects of wildlife. The Wildlife Society (the professional society for wildlife biologists, researchers, and managers) defines wildlife as "free-living animals of major significance to man." Until the 1970s wildlife was synonymous with animals that were hunted, but in the past three decades most free-living animals have become significant to humans. "Wildlife" is often restricted to terrestrial and aquatic vertebrates other than fish, however, due to a long history.

The Political Discipline

To have a funding base, some of the earliest efforts to manage wildlife were established in the United States Department of Agriculture's Division of Entomology that was funded by the American Ornithologists Union. The Division of Entomology was established to determine the status of bird distributions and their migrations. This group was then transferred to the Division of Economic Ornithology and Mammalogy in 1885. The main function of the Division of Economic Ornithology and Mammalogy was to determine bird distributions and the crop damage they caused. This agriculture base was expanded to deal with the relationship of all wildlife and agriculture and was placed in the Bureau of Biological Survey in 1896. In 1940 a political decision by President F. D. Roosevelt created the Bureau of Wildlife, which combined the Bureau of Biological Survey (birds and mammals) and the Bureau of Fisheries (fish). However, fisheries biologists did not think that the newly created bureau would appropriately represent them under this broad title and the bureau was changed to the Bureau of Sport Fish and Wildlife. It was later changed to the present United States Fish and Wildlife Service. As a result of the bureau's name change the implication was that fish were somehow different than other wildlife and different disciplines and societies have developed; one for fisheries, another for other vertebrates.

In the early 1990s the Secretary of the Interior, Bruce Babbitt, attempted to create a freestanding organization that would combine all of the research conducted in the United States Department of the Interior into a single organization entitled the National Biological Survey. The secretary's reason was that "the purpose of science is not to conquer the land, but to understand the mechanisms of ecosystems and to fit man into the resources he has available on the planet on which he has evolved." From understanding bird distributions (e.g., Division of Entomology) to placing man in ecosystems (e.g., National Biological Survey) is a long stretch over a short period but reflects how public attitudes change and conservation organizations are formed. The latest organization will not be the last. As wildlife populations are dynamic so are the organizations that contain the people that manage them. Regardless of the various definitions, I consider wildlife in this book as *free-ranging* undomesticated animals in natural environments. Animals that are owned by private landowners that keep the animals on their property with a barrier (e.g., most often a fence) are not free-ranging and are not considered wildlife. Animals in captivity are the subject of game ranching, which is developing into a separate discipline akin to animal husbandry. Caughley and Sinclair (1994) argue that the definition may be too restrictive because of the human aspect of the wildlife triad (Fig. 1.1). I agree that the human aspect is important because management of wildlife involves human education, extension, law enforcement, and park administration among many other human related issues. However, to consider each of these would distract from the core of managing wildlife populations, which is to manipulate or protect populations to achieve a goal. The human side of this argument is certainly important, but the core of wildlife management is being able to understand animals and their relationships with their habitats so the decisions that are made to manage wildlife are informed decisions that will achieve the desired goals of humans in a proper way. To manage wildlife populations effectively requires a combination of biological and sociological strategies. This book focuses on the biological strategies related to managing wildlife population. An understanding of these strategies is central to sound wildlife management.

Active vs. Inactive Management

There are numerous ways wildlife is managed and they all imply stewardship. *Active management* does something to the population such as increase or decrease numbers directly through activities like translocations or hunting, respectively. It may also cause similar effects by manipulating an aspect of the habitat that will cause an increase or decrease in the population. If the numbers are dangerously low, other active management may be incorporated such as controlling predators to reduce neonatal mortality or habitat improvement to provide needed cover for neonates from predators. These efforts imply an active approach. Other populations may not need to be actively manipulated such as populations in some national parks. In those situations the only management activities are to minimize external influences on the population and its habitat. Still other populations may be so poorly understood that no action is taken at all because managers do not have enough information to make any informed decisions (i.e., *inactive management*). However, it is important to understand that all populations are managed either actively or inactively. For some, no action at all takes place. Because wildlife belongs to the people, it is managed by biologists even without management because the decision had to be made to do nothing. The latter category is often referred to as *passive* or *nonmanagement*.

The Goals of Management

Caughly (1977) clearly outlines the goals of management as one of four options.

1. Make a population increase.
2. Make a population decrease.
3. Harvest the population for a continuing yield.
4. Do nothing except monitor the population.

These are the options available to wildlife biologists and will vary depending on what the goals are. After goals are established, the manager can determine the appropriate management action and how that action can best be achieved. The establishment of goals is a value judgement, but how the goals are achieved involves technical decisions.

The Wildlife Being Managed

Wildlife has been classified numerous ways. There is a danger to classification schemes because by setting classification criteria some species can fall by the wayside if they do not fit into someone's particular scheme. For example, funds set aside for rare and endangered species cannot be used for other animals (i.e., not endangered regardless of the need). However, classification of flora and fauna is an ongoing process. Estimates of the number of species on the earth are as high as 30 million yet less than 1.5 million have been described: approximately 750,000 insects; 359,000 invertebrates, fungi, algae, and microorganisms; 250,000 plants; and 41,000 vertebrates. Wildlife biologists concentrate their efforts on the ecology of vertebrates. Some of the more familiar classifications were established by Leopold (1933) and others (i.e., rare and endangered, urban, and park) are more recent and include the following:

1. *Farm species.* These are nonmigratory species that can be produced on farms and are a suitable by-product of farming. Typical farm species include bobwhite quail, pheasants, squirrels, rabbits, raccoons, skunks, and opossums.

2. *Forest and range species.* These are also nonmigratory species but those that are compatible with forestry or livestock and are a suitable by-product of such. Typical examples are turkeys, deer, elk, raccoons, fox, grouse, antelope, and collared peccaries.

3. *Wilderness species.* Wilderness species are those that are harmful to or harmed by economic land uses and require special reserves or forests to be preserved as their habitat. Grizzly bears, mountain lions, mountain goats, and bighorn sheep are true wilderness species. Others can include pronghorn, elk, deer, and moose.

4. *Migratory species.* Migratory species are those that always leave the land on which they were raised in the course of their seasonal movements. Ducks, geese, and swans are excellent examples but elk, deer, and other mammals can also fall into this category.

5. *Furbearers.* Furbearers are species that can be produced and marketed due to the commercial value of their pelts. Coyotes, bobcats, muskrats, otters, and beavers are just a few of those typically in this category.

6. *Predators.* Predators are animals that kill other species or are considered dangerous to livestock. Obvious examples are mountain lions, bobcats, coyotes, and feral dogs.

7. *Rare* and *endangered species.* Rare and endangered species are species of native fish and wildlife that are threatened with extinction. They are classified as such

by the United States Secretary of Interior whenever their existence is endangered because:

    a. their habitat is threatened with destruction, drastic modification, or severe curtailment,

    b. over exploitation,

    c. disease,

    d. predation, or

    e. other factors, and their survival requires assistance.

    8. *Urban wildlife.*   More recently emphasis has been placed on species that find habitat in cities, and a new discipline is arising that studies their habitat relationships with humans. These species are defined as urban wildlife. At one time pigeons and rats were the species in cities but deer, raccoons, peregrine falcons, and coyotes are other examples of the urban wildlife class.

    9. *Park wildlife.*   Park wildlife is another classification for those species that exist in parks, which include all those mentioned above plus numerous others. The category has occurred not because of the species but because of the confined management that occurs due to human-created boundaries.

Regardless of how humans classify wildlife the basic goals for management are the same: influence the population to increase or decrease, manage for a harvest, or simply monitor the population.

SUMMARY

Wildlife includes free-ranging undomesticated animals in natural environments. Wildlife is not restricted to the animal but includes their interrelationships with their habitat and humans. Fish are usually considered separately from other animals because of political issues. The goals of management are to make a population increase, decrease, harvest the population for a continuing yield, or simply monitor the population. Management can be active (i.e., influences populations directly) or inactive (i.e., no intentional management). Wildlife is generally placed in different categories: farm, forest, range, wilderness, and migratory species; furbearers; predators; rare and endangered species; and urban and park wildlife.

2

Evolution of Wildlife Management: Our Roots

"I think, as the twentieth century comes to a close, that we are coming to an understanding of animals different from the one that has guided us for the past three hundred years. We have begun to see again, as our primitive ancestors did, that animals are neither imperfect imitations of men nor machines that can be described entirely in terms of endocrine secretions and neutral impulses. Like us, they are genetically variable, and both the species and the individual are capable of unprecedented behavior. They are like us in the sense that we can figuratively talk of them as beings some of whose forms, movements, activities, and social organizations are analogous, but they are no more literally like us than are trees... they move in another universe, as complete as we are, both of us caught at a moment in mid evolution."

B. Lopez (1979)

INTRODUCTION

All too often we get caught up in the moment and fail to consider how contemporary wildlife management was established. It is important to have an idea of the "early days" of wildlife management to understand why we manage the way we do. This chapter is a brief exploration of the evolution of wildlife management with specific examples from Britain and the United States. This history began without concern for any form of wildlife conservation, then waxed and waned over the years until large mammals in North America were severely depleted. It was at this point that the "wildlife management experiment" began in North America.

▒ EVOLUTION OF WILDLIFE MANAGEMENT

When Europeans first began inhabiting the North American continent they found a bounty of renewable natural resources. The resources were used but their conservation was seldom considered. As game became scarce in the east, settlers simply moved farther west. By the late 1800s, the wildlife in North America had been explored, exploited, and many populations were reduced or eliminated. The problem may have been recognized but conservation efforts were not established or practiced. This history has taught an important lesson: if wildlife is to exist with humans, management is critical for survival.

Humans tend to be shortsighted when thinking of wildlife and consider changes in terms of the present, or at best, in relation to their lifetime. To understand the magnitude of wildlife and its relationship with humans we need a long-term perspective because wildlife has evolved over thousands of years but humans are new in geological and evolutionary time.

Instead of starting with reductions of wildlife in North America in the 1800s, it is informative to take a bottom-up view beginning with the development of Earth approximately 4.5 billion years ago. The first animal life appeared somewhere around a billion years ago but humanoids have been around for only 4 to 2 million years, a snap of the fingers in geological time (Figure 2–1). Although humanoids have been here for a short time their technological powers have had tremendous influences on wildlife, the products of thousands of years of evolution. The human population is driving all ecosystems because of our unprecedented rate of increase (Table 2–1) and use of natural resources. This runaway increase is detrimental and strongly influences wildlife populations. At a minimum we should attempt to understand the importance of wildlife and potential implications of these changes on wildlife populations and the quality of our lives. Of equal importance is to continue to learn how to mitigate or avoid the kind of effects on wildlife and wildlife habitat that are unacceptable.

Evolution of Wildlife Management in Europe and Asia

The evolution of game management goes back to the beginning of mankind. It was practiced as taboos in the early stages of social evolution, and the survival of tribes depended on their competition for and management of game. Quite simply, those more successful survived and from here the roots of game management began.

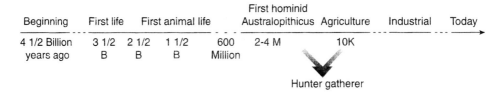

| | | | | First hominid | | | |
Beginning	First life	First animal life		Australopithicus	Agriculture	Industrial	Today
4 1/2 Billion years ago	3 1/2 B	2 1/2 B	1 1/2 B	600 Million	2-4 M	10K	

Hunter gatherer

Figure 2–1   Time line from the creation of earth to contemporary society.

Table 2-1 Doubling time of the human population.

Years ago	Date	Population	Doubling time (years)
10,000		5 million	1,500
300	1700	500 million	200
100	1900	1 billion	80
50	1950	2 billion	45
20	1980	5 billion	37
2	2000	6 billion	10–20

The first written law about wild animals was recorded in the Bible (Deuteronomy 22:6) as Moses decrees that the breeding stock should be saved. Early Egyptians were keen sportsmen but their record reveals no worries over the conservation of sport. However, early philosophers Solon and Xenophon questioned the role of sport hunting to society — a question still debated today. The Greeks and Romans had some game laws but they were to prevent negligence of the mechanical arts, not for conservation. However, their "training for war" (e.g., forcing soldiers to sleep on hard surfaces, making them alert, knowledge in arms proficiency, and knowledge about the outdoors) did promote an awareness of nature.

The first clear record of game management comes from Marco Polo as he described the game of laws of Kublai "The Great Khan" (A.D. 1215–1294) in the Mongol Empire. The Great Khan initiated at least four "management activities": harvest restrictions to allow species to increase and multiply, establishment of food plots, winter feeding, and cover control.

In Europe (1350 A.D.) food and cover were not controlled but regulation of hunting for the ruling class was developed as customs rather than laws; hunting became the sport of kings. Laws for closed seasons began in the 1300s and others followed that controlled weapons, species to be harvested, and areas where hunting could occur. Many of the laws were to enhance the hunter, not as a conservation effort. And most hunters were the ruling elite. This method of management has serious drawbacks (Geist 1996) as wildlife in private ownership turns into the very symbol of the hated elite and becomes an object of diversion and persecution by the public when the opportunity arises. For example, consider the Robin Hood poacher that took from the rich. Landowners responded to poaching with vigilante action. The last herd of ibex in Austria was exterminated on the order of a local bishop to stop the bloodshed between foresters protecting them for royalty and local villagers. Other examples are reported by Threlfall (1995) and Geist (1988).

In England, King Canute (1016) prevented people from hunting on his land with the threat of death. Emperor Sigmund passed a *Mandat* in 1425 that anyone caught snaring a hare shall forfeit his right thumb. William of Hessia in 1567 commanded that all commoners caught in the act of poaching or pursuing deer were to be hanged on the spot and their children and wives would be treated as slaves. Earlier, 1517, Ulrich von Württemberg ordered that anyone caught with a crossbow or gun in hand that was not authorized should have his eyes gouged out. Baron Galeasso Sforza caught a poacher who killed a hare and made him devour it on the spot — intestines, fur, and all. And, Vitold von Leptau had poachers sewn in green hides and fed to the dogs. These are extreme punishments and Threlfall (1995) suggests that fines or imprisonment were the norm. However, a point is reached where locals rebel against such outrageous acts.

Martin Luther rallied against the abuses of peasants by nobility. The "wildlife of nobility" destroyed fields and gardens, and hunting parties galloped through growing grain and flattened vineyards. When the peasant revolts in German (1524–1525) occurred they were in good part a revolt against wildlife and hunting. Royal hunting parties of up to 200 horses and dogs for three to four days, chasing a single deer, created substantial damage (Geist 1988). This attitude created a loath view of wildlife and peasants took it out on wildlife.

It happened in feudal times.

It happened after the eighteenth-century revolutions.

It happened after the nineteenth-century revolutions.

It happened with the ascent of the Nazis and Communists.

It happened during Desert Storm and is probably occuring in Iran and elsewhere today.

Owning wildlife by customs is one of the dangers of letting wildlife slip into the hands of the ruling elite (Geist 1988).

Wildlife management in Europe has changed significantly as conservation incorporates planned and wise use of resources so that the quality of habitats is maintained. This process is dynamic and strongly influenced by attitudes about wildlife and its management in the past. In Europe, Britain provides an excellent example of this (Threlfall 1995) because the entire island (only 230,000 km^2) is inundated by 54,000,000 humans (Ratcliffe 1984) and their history has a strong impact on current wildlife management in North America.

Wildlife Management In Great Britain

Today, all British land is influenced by humans (Edlin 1972, Rackham 1976, Woodell 1985, Threlfall 1995) although over thousands of years, climate and succession imposed their influences. Prior to 8000 B.C. humans depended on hunting, fishing, and food gathering but probably had minimal influence on the resources (Hawkes and Hawkes 1958). Even through the Mesolithic times (8000–3250 B.C.) there was little evidence of permanent human settlements but the first signs of woodland clearances were reported (Hawkes and Hawkes 1958). Forest clearings became more common in the Neolithic period (3250–1700 B.C.) for farming, usually on high ground (Dimbleby 1984). Changes in vegetation were also influenced by the introduction of sheep that reduced the subsistence use of red deer.

Threlfall (1995) summarizes the influence of humans on the landscape in England over the centuries into several periods.

The Bronze Age (1750–500 B.C.). During the Bronze Age metallurgical skills were developed that enhanced forest cleaning, and humans moved to lower ground. The first significant forest clearances probably began in the middle or late Bronze Age leading to heathland (Tubbs 1986).

The Iron Age (500 B.C.–A.D. 43).   Farms and forts were settled, soil was moved and degraded, land was cleared for cattle and sheep, and forest clearing continued.

The Roman Invasion and Occupation (A.D. 43–410).   Towns and roads were developed (Hodder 1978), forests continued to dwindle to make way for arable farming, houses, and bathhouses. Bathhouses also required wood for heat, which further depleted forests.

The Dark Ages (A.D. 410–1066).   England was dominated by Germanic tribes (Jones 1978) and more towns were developed. Additional movement from higher ground to the lowlands (perhaps due to the loss of soil fertility in higher areas due to years of cultivation) caused more environmental alteration.

The Norman Conquest (A.D. 1066).   Construction of castles, villages, roads, and monasteries (Darby 1973, Dodgshon 1978) changed the countryside. Monasteries wielded power and controlled more and more land. The word "forest" acquired a legal meaning with the ascension of William the Conqueror to mean large, unfenced tracks of land on which deer were protected by special bylaws. Only the king or those designated by him could hunt deer in the forest. Over the next hundred years or so Norman kings continued to add to the repressive forest laws and spread Royal Forests throughout the island. Habitat alteration continued, the human population continued to increase (to 4.5 million by A.D. 1300), and most farmland was developed. When bubonic plague occurred, combined with weather-related famines, up to 50% of the human population died (between 1348 and 1350) and over 2,000 villages were eliminated (Bowsky 1971).

After The Norman Conquest.   Over the next three centuries the landscape was changed dramatically as fields were enclosed (Chapman 1987, Turner 1984, Threlfall 1995), the influence of the church increased, monasteries were reduced, and the land reverted to royalty, yielding country houses and landscaped parklands. In addition people ate more game (Girouard 1987) and industry increased, thus demanding more fuel and further reducing woodlands.

As farmland became more and more scarce, more remote areas, including the richest area in England, the Fenland, were exploited. The Romans started in 1630 but the job was not completed until 1984 when the Fenland had been reduced from 3,380 km^2 to 10 km^2 (Threlfall 1995).

Rivers were modified to provide power and canals substituted for transportation. Vegetation and other natural resources continued to be exploited but few efforts were directed toward renovation of uplands. Those that were, were unsuccessful (Threlfall 1995).

The Industrial Revolution (1700s).   With the advent of industry, society changed from agrarian and handcraft to one dominated by urban life and industry that continued to alter landscapes. To support this lifestyle coal mining also increased, which contributed to altered landscapes.

Contemporary Britain.   All of these previous activities had serious impacts on the native flora; most of it was destroyed by 1700 and today agriculture affects 80% of Britain's surface. Related pollution, recreational pressure, and urban growth contribute to the decline. Only 1.3% (300,000 ha) of ancient and semi-natural broadleaf woodland remain. The hedgerow, which became common as a means to cause enclosure of lands, has created new wildlife habitat but is also being destroyed for agriculture. In 1946–1947 there were 800,000 km of hedgerow but only 576,000 km in 1974. Since 1974, 4,160 to 6,400 km of hedgerow per year are destroyed (Muir and Muir 1987).

The human changes due to agriculture have been obvious, but habitat alteration from pollution (Mellanby 1967), recreation (Bracey 1972), and draining lands (Stewart and Lance 1983) have made serious contributions to habitat destruction. The battle between agriculture and conservation continues in Britain as it does worldwide but conservation efforts for vegetation are being made. New forests are being planted (Watkins 1987) and Peterken (1983) predicts that by 2050 woodlands will double from 9 to 18%. Britain is developing a National Park system also, so extensive areas of natural beauty are available for the public.

Britain's Wildlife Resources

When the last glaciation ended about 20,000 B.P. in Britain, it was connected to Europe. About 5000 B.C. Britain was separated, leaving most species that exist in Britain today. In 1974 there were 56 mammals (14 had been introduced) (Threlfall 1995). Overall, carnivores have been severely reduced or exterminated (i.e., wolf, brown bear) and some introduced species (i.e., muskrat) have been exterminated because of the damage they caused.

Avifauna has fluctuated with the changing landscapes but with proper management the original avifauna will be able to adapt to the changing forest cover (Williamson 1972).

Game Laws in the United Kingdom

Game laws have always been in some state of confusion and little has been done to simplify matters. Animals have been hunted since Paleolithic times and anything that was caught or killed belonged to the hunter regardless of where it was killed. This philosophy did not last and although little evidence suggests that active game management occurred prior to the 1600s, there were abundant and often confusing regulations controlling the use of game (Threlfall 1995). Political upheaval was exacerbated because the people were poor, in need of food, and unhappy with the virtually exclusive privilege of the rich to hunt. As a result poaching was common and numerous laws were established to keep hunting in the hands of those with money and land. Gamekeepers were the front line of defense against poachers for more than 200 years and they represented the "qualified" sportsmen (rich landowners) that made up less than 0.5% of the population. Gamekeepers could "arrest" suspect poachers and when found guilty they were penalized by being whipped, imprisoned for two years, or sent to the armed forces. In 1803 it became a capital offence to forcibly resist arrest. Still poaching provided an expanding trade in the early nineteenth century (Kirby 1941). London was the center of the game traffic where it was openly sold as normal fare, and the market was often glutted due to the vast amounts of game being poached. Over time wildlife became the property of the owner of the land, not its occupier, causing problems for tenant farmers who could not kill animal pests. This was changed in 1880 when the Ground Game Act gave tenant farmers the right to shoot hares and rabbits that were destroying their crops. This was not a popular law but it did reduce sympathy for the poacher (Emsley 1987).

Threlfall (1995) reports that game laws are still in a state of confusion in Britain and there has been little effort to rationalize or consolidate them. Poaching is still covered by earlier law but additions have been made in 1960, 1963, and 1968 with the Game Laws (amendment) Act, Deer Act, and Firearms Act, respectively. "Laws against poaching are concerned with criminal trespass and possessing the equipment for poaching, not theft. Although sporting rights are conceived as property, the owner of land or sporting right does not own wild creatures until they are dead or in his/her power."

Hunting was relatively uncontrolled in Britain; with the exception of protecting animals in the months in which they were breeding, just about anything was legal. Crossbows, dogs, falcons, horses, and anything to give the hunter an edge were acceptable. Blackmore (1971) reported the Marquis of Ripon from 1867 to his death in 1923 shot 97,503 red grouse, 124,193 grey partridge, 241,234 pheasants, 31,934 hares, and 40,138 rabbits. This was obviously extreme but illustrates the uncontrolled harvests that lead to overexploitation. At the beginning of the twentieth century game was scarce because of overexploitation, changing agricultural practices, and a rising human population (increased from 6 million in 1756 to 42 million in 1911). Some recognized the importance of preserving game and killing "vermin" but the public was largely unconcerned. With the First World War came an end to hunting as it had occurred for the past few decades as horses were used by the cavalry and packs of hounds were destroyed. Shooting no longer was for "sport"; it became a necessity for food. Attempts to preserve game ceased and once again poaching became rampant. This caused a hiatus between rural and urban people, and vocal antihunting groups in urban areas were formed. Antihunters were counted with groups such as the British Field Sports Society (1930) to "...represent and safeguard field sports interests," and to "...counter the increasing calls for legislation against field sports" (Rogers et al. 1985). These changes continued after World War Two and the number of guns owned by the public increased. Many people now hunt in Britain and several organizations such as The British Association for Shooting and Conservation control the behavior of hunters. Numerous field sports are available due to the redistribution of wealth since the First World War and the commercialization of many estates (Threlfall 1995). In addition, institutions are purchasing sporting rights as an investment or fringe benefits for their employees (Blackell and Gilg 1981).

This brief history of using wildlife resources is important to understand the current management of wildlife in Britain and to some extent the attitudes present in North America. The historic and current management in Britain have certain similarities.

Managing Wildlife In Britain

Munsche (1981) suggests that historic preservation consisted of breeding game, destroying vermin, and creating protected areas for game. As early as the Middle Ages, rabbits had been protected (Shaeil 1971) and deer were afforded shelter in restricted parks.

There were numerous attempts to rear and release pheasants and grey partridges but they were too tame to be used for sport hunting. In the 1750s and 1760s eggs were obtained from the wild and hatched in the pheasantries, and in the 1770s crops were grown to attract and keep wild birds (Threlfall 1995).

The hunting seasons of today began in 1762 and were modified through 1776. The same basic laws have been in use for over 200 years.

Numerous predators including owls, corvids, foxes, badgers, stoats, weasels, squirrels, rats, and hedgehogs were controlled through the 1950s (Vesey-Fitzgerald 1946, MacIntyre 1952, Neale 1986, Creswell et al. 1989). Gamekeepers became important figures to manage estates and produce game, and their numbers and duties have changed over the years. In 1910 there were over 22,000 gamekeepers, but only 5,000 in 1971 (Threlfall 1995). In addition, modern gamekeepers have more duties than in the past but many only work part time. To be successful, however, modern gamekeepers need to be able to shoot game and kill vermin, raise pheasants, and train dogs. The remuneration received varies from poor to adequate.

Actual management is shared between the government (e.g., Nature Conservancy) and nongovernmental organizations (e.g., Royal Society for the Protection of Birds, Royal Society for Nature Conservation). This relationship has evolved through four phases: the natural history/humanitarian period (1830–1890), the preservation period (1890–1940), the scientific period (1910–1970), and the popular/political period (1960–present) (Lowe 1983). Britain may be in a new period now, the management era (Blunden and Curry 1985, Threlfall 1995). The latter era may have serious implications on all wildlife in Britain, as instead of the historic management of a small number of selected species (i.e., pheasants, red grouse, grey partridges, hares, and deer), management for biodiversity will increase (Westman 1990). With more interest in biodiversity and nonconsumptive use of wildlife, more sections of the public need to be considered in wildlife management decisions (Gilbert and Dodds 1987). There will need to be better integration between agriculture, forestry, and the public in developing game policies. The public interest in wildlife resources and their habitats will have to continue for healthy wildlife resources to coexist with humans (Threlfall 1995).

■ BRIEF HISTORY OF WILDLIFE MANAGEMENT IN NORTH AMERICA

It is important to have some knowledge of our historical roots so we know where we have been in order to determine where we are going. With scanty data it has been a challenge to reconstruct man and wildlife's placement in the Americas and it may be something we can only imagine.

Placing Man and Wildlife in America

Trefethen's (1975) account of the beginnings of wildlife management in North America is a classic. Try to imagine a succession of caribou, bison, mammoths, and other herbivores followed by wolves, saber-toothed cats, bears, and other predators followed in turn by two-legged hunters. This was a time, at the end of the last ice age (approximately 10,000 years ago), of kilometer-thick glaciers where the way of life followed chlorophyll in vegetation being consumed by grazers, which in turn were consumed by predators. This ice age lowered the oceans enough to create a land bridge greater than 1,600 km wide between Siberia and Alaska that was lush with vegetation.

The two-way traffic on the land bridge enriched the wildlife population of both continents just as a tropical land bridge to the south (i.e., Isthmus of Panama) enabled movement between North and South America. To be able to appreciate the relationship between primitive man, flora, and fauna at that time compared to today's society is questionable. It would require an understanding of people living off the land (i.e., *ecosystem people*) versus a society more aware of the biosphere (i.e., *biosphere people*). The normal ice melt filled streams and riparian areas were lush with willows, pines, cedars, birch, cottonwood, deer, moose, elk, beavers, and other flora and fauna. Ducks, geese, and swans were common in lakes and sturgeon and trout were in most waterways. Whales even moved up big rivers like the Hudson. North America was clearly a land of plenty.

After humans colonized America and after the mammoths and other megafauna disappeared, North America was left with the present assemblage of large mammals (Table 2–2) (Martin and Szuter 1999). Megafauna extinction has been related to changes in climate, vegetation, and exploitation. Although causes of the great die-off continue to be debated, historical ecologists recognize at least two catastrophes that shaped the biogeography of large mammals in the new world. The first was the megafuna extinction late in the Quaternary (approximately 13,000 years ago) when the Americas lost nearly 66% of their large mammals including elephants, horses, and ground sloths (Table 2–2) (Martin and Klein 1984, Martin 1996). In the past 10,000 years the Americas lost a species of bison but no other animal larger than 44 kg (as an adult) (Martin 1996).

The second event occurred 500 years ago when European explorers unknowingly introduced pandemics to the Americas. Native Americans were highly vulnerable to the diseases, like smallpox, and human populations were severely reduced. The introduction of guns, the horse, and new markets for fur transformed the New World, and the extinctions of the late Quarternary reduced large mammals. The second event reducing human populations reduced hunting pressures on big game (Martin 1996).

When native Americans were at war in North America, William Clark saw large numbers of game in the patches between the warring parties (Kay 1990). Native Americans had a profound influence on the distribution and abundance of animals, and the species Lewis and Clark observed as they surveyed the Louisiana Purchase was not wild natural America, but America that had already been shaped by humans (Martin 1996); shaped to the point that the distribution of large mammals was largely confined to areas that were not inhabited by humans. The controversy is sure to continue, but there are at least three early groups of humans that had a strong influence on wildlife in America: the Spanish, other Europeans, and Russians.

Table 2–2 Mammals (>40 kg) of the late Quaternary, western United States and northern Mexico. After Martin and Szuter (1999).

Extant Species		Extinct Species	
Scientific name	Common name	Scientific name	Common name
Alces alces	Moose	*Arctodus simus*	Giant short-faced bear
Antilocapra americana	Pronghorn	*Bison priscus*	Steppe bison
Bison bison	Bison	*Bootherium bombifrons*	Bonnet-headed musk ox
Canis lupus	Timber wolf	*Camelops hesternus*	Western camel
Cervus elaphus	Elk	*Canis dirus*	Dire wolf
Odocoileus hemionus	Mule deer	*Equus conversidens*	Mexican horse
Odocoileus virginianus	White-tailed deer	*Equus occidentalis*	Western horse
Oreamnos americanus	Mountain goat	*Equus spp.*	Horses, asses
Ovis canadensis	Bighorn sheep	*Euceratherium collinum*	Shrub ox
Panthera onca	Jaguar	*Glossotherium harlani*	Big-tongued ground sloth
Puma concolor	Mountain lion	*Glyptotherium floridanum*	Glyptodont
Rangifer tarandus	Woodland caribou	*Hemiauchenia acrocephala*	Long-legged llama
Ursus americanus	Black bear	*Mammut americanum*	American mastodon
Ursus arctos	Grizzly bear	*Mammuthus columbi*	Columbian mammoth
		Mammuthus jeffersonii	Jefferson's mammoth
		Mammuthus primigenius	Wooly mammoth

The Spanish (1500s)

The Spanish were initially looking for a route from Europe to the East for tea and spices. The American land mass was viewed more as a nuisance than as a discovery with potential riches. The Spanish later discovered that Central America and Mexico harbored gold and silver that served as a motive for exploitation. The Spaniards also had the religious desire to Christianize the natives.

It is interesting that even though there was considerable exploration and settlement, the Spanish were largely blinded by their quest for gold and they made very little notice of the flora and fauna in North America. About the only detailed descriptions they made were of armadillo and bison. Even though they were oblivious

to wildlife they had a significant impact on wildlife. The impacts were indirect but they changed the lifestyle of the native Americans and subsequently influenced wildlife in several ways.

Introduction of the Horse to North America. By bringing the horse back to North America, the Spanish were able to convert stationary lifestyles of Native Americans to mobile life styles.

Introduction of Domestic Livestock. This introduction was more pervasive than the horse and long lasting in terms of an influence on wildlife and its habitat. Livestock also created lifestyle changes by changing hunters to shepherds. The impacts of livestock on the land and habitat in the Southwest transformed the land, severely depleting wildlife. It began when Spanish missionaries introduced Texas longhorns that lived in the wild, and naturally competed with native grazers. Domestic sheep were also affecting native grazers and they transmited diseases that caused the reduction of native sheep.

Ultimately the overgrazing that began at this early time resulted in major changes in vegetation and soils. It took more than 1,000 years to deplete the top soil in Mediterranean Africa. In the southwestern United States, humans may be able to complete the same task in less than 200 years with increases in grazing and abuses in land management.

Other Europeans (1500–1600s)

Other Europeans were also looking for routes to the East, and in retrospect it is amazing how little interest they showed in learning about this new land they discovered that appeared to be blocking their discovery of a route to the spices and treasures of India. From 1496 to 1620, 124 years, only one expedition penetrated the mainland (by land) for more than 8.0 km from the coast!! This lack of exploration occurred even with considerable activity along the coast and a growing number of temporary settlements. For this entire period the land and its wildlife resources remained *terra incognito*. The marine resources, however, could not be ignored. The early travelers could not help but notice the tremendous abundance of fish on the grand banks of Newfoundland where they were "...able to collect codfish by baskets." It was not long before meat-hungry fishermen from every European port came to fish. In so doing they also exploited sea birds for their eggs and marine mammals for their fur and meat. Some birds, like the colonial nesting great auk, were extremely vulnerable and in spite of their abundance were exterminated (by 1840).

Others, like some whales, held out a little longer. To process fish and whales efficiently, temporary settlements were established but abandoned each winter. Some terrestrial wildlife was exploited: deer for meat and beaver for fur. However, this exploitation only became important after colonization began in 1620.

Russians (1700s)

Russians are another group of visitors to North America that deserves to be mentioned. They arrived in the 1700s and discovered the incredible richness of marine mammals in the Pacific and Alaskan islands. One expedition in 1741 returned with

$1,000,000 worth of sea otter skins and started a stampede of fur hunters. After the Russians took the most the Americans moved in and decimated the rest. The effects caused drastic reductions and even extinctions of marine animals. The Stellar's sea cow was so large and vulnerable that it was exterminated before naturalists even had a complete skeleton. And the southern sea otter (believed extinct until the late 1950s) and northern seals were nearly eliminated.

Adding exotic organisms (e.g., humans, horses, livestock, and associated diseases) and exploiting others for profit certainly had an influence on the wildlife resources of North America. The impacts continued with the arrival of the colonists.

The Colonists

Humans in America had been living in temporary settlements along the coast where their impacts could be reversed after they left within 5 to 20 years. When the colonists arrived they created permanent settlements and with the settlements arrived the first real interest in terrestrial wildlife. The colonists came to the New World to escape conditions in Europe, but the European countries that "sponsored" the travel of many of the colonists viewed their trip to America as a business proposition. Europeans wanted colonies to facilitate exploitation and because the ocean voyages had been paid by Europeans, the new Americans had to repay them. The first payment was a shipment of beaver skins. The primary exports were fish, deer, beaver skins, and lumber.

To understand the role of wildlife in the lives and economics of the colonies it is necessary to understand why people came to America. Religious freedom was one reason but they also came to escape the European system in which wild animals belonged to the very few landed gentry (Threlfall 1995). They wanted access to land and they wanted freedom to exploit the earth's resources. One petition for emigration to the government read that the petitioner wanted "...to settle in America for freedom to worship and catch fish." In a religious sense they believed that man's role on earth was to multiply and subdue and the wealth gained from exploiting the earth's resources was an indication of God's favor for the hardworking and ingenious. The colonists found such an abundance and diversity of flora and fauna that they readily adopted a belief in the "myth of superabundance." There was so much wildlife it was incomprehensible. There were 2,100 species of vertebrates, 380 species of mammals, 650 species of birds (compared with 622 in all of Russia and only 420 in all of Europe). And all were in as pristine condition as could be imagined and in phenomenal numbers. There may have been billions of passenger pigeons, and bison represented the largest biomass of any existing species that inhabited the land with 60,000,000 individuals. With this type of natural capital, conservation did not seem necessary leading to the Era of Exploitation, which included the fur industry and market hunting. These events were the precursors to modern wildlife management in North America.

The Era of Exploitation

It is not surprising that early Americans adopted some of their ideas about wildlife to create the myth of superabundance. The resources they encountered in the New

World were boundless and these resources led to a philosophy of expansion for the next 250 years. This philosophy lead to a legacy by the end of the 1800s where North America was left with a vastly reduced wildlife component.

Wildlife is viewed so differently today compared to the colonial period that it may be difficult to even think how it was used. In reality it was exploited. Wildlife was thought of as a commodity like coal or oil; there were no aesthetic or recreational values, there was no ecology or management, and certainly no discipline related to population management. The use of wildlife was very different from today. Anything that could be used was! People ate songbirds, shorebirds, beaver flesh, predators, and skins were sold including swan skins, baleen was used for corsets, feathers for garments, and even bones were used for fertilizer. Out of this use came two major industries that exploited wildlife: fur, and market hunting.

Fur Industry

When one discusses wildlife today it is only one of many components of modern society and is certainly not the foremost concern of our entire economy and society. But for a long time in our history it was, and furs were our most valuable export surpassing even fish as the primary export. All types of fur were used but the most precious were from the weasel family (i.e., Mustelidae: weasel/ermine, otter, mink, skunk, fisher, marten). However, the most significant fur was that of the largest rodent in the United States, the beaver.

Beaver inhabited all waterways throughout North America and the 24 species constructed dams in streams everywhere. These impoundants created a cycle of vegetational change from plant succession creating ponds that were a focal point of wildlife. Over time, silting occurred leading to other successional changes from pondweeds to willows to water grasses to meadows to forests. These seres created habitat for a broad array of wildlife.

When beaver were exploited it was rapid and fierce. In 1638 the King of England demanded that beaver be the compulsory ingredient in hats and North America was the sole source of beaver for Europe. The Hudson Bay Company kept records of what was sold and provides the most information about the fur trade.

By 1880 beaver were almost extinct east of the Mississippi and by 1930 they were almost extinct in North America. They were saved only by the whims of fashion with the acceptance of silk hats in the 1830s.

With the overharvesting of beavers and the subsequent dewatering of North American landscape, wildlife was as drastically changed as any other single change since the discovery of America. Beaver created habitat for wildlife and when they were removed an important part of America left with them. Their beneficial water conservation, stabilization of stream flow, raised water levels, erosion control, water purification, and mechanical filter were also lost. Part of this history was saved by Lewis Caroll who coined the term "mad hatter" in his book *Alice in Wonderland*. The term came from people that made beaver hats because excessive exposure to the mercury compounds used in tanning caused brain damage. Beaver were only one example of the exploitative fur market. Anything with fur was exploited.

Market Hunting

The same attitudes prevailed with market hunting as did with the fur market. With market hunting the real pressure was applied at the beginning of the 1800s because man then had the percussion rifle, which made killing effective, and the railroad provided a mechanism to get the meat to market. When these two technologies developed, market hunting became very efficient in reducing wildlife. Little was left alone and nearly everything that could be harvested was consumed (Figure 2–2). Do not get market hunting confused with hunting today. Market hunting was uncontrolled exploitation that in part led to modern wildlife management. This wanton use of wildlife resources clearly impacted wildlife, but two species more than any others triggered the recognization that the myth of superabundance was just that, a myth. Those species and their wasteful harvest include the bison and passenger pigeon. These were two of the most spectacular and numerous species in parts of North America. They were also two of the most valuable species in terms of human uses. They were also two of the most rapidly and spectacularly declining species.

Bison

The bison was obviously attractive for human use. It had excellent meat, was large (less than 1,000 kg), easily killed with rifles, provided an excellent robe, hide, and the hair was useful for insulation. Buffalo were also abundant and they guaranteed a source of these components for settlers. Their decimation proceeded rather slowly at first but the real destruction occurred around 1830 and was complete in 50 years. Their range incorporated about 30% of the United States (Figure 2–3). The gradual slaughter of their limited range east of the Mississippi River led to all but a few eliminated east of the river by 1800. However, the large herds on the plains were still largely untouched until a variety of factors led to the sudden explosion in use. Western settlement was gradual until the gold rush of 1849 followed by homestead laws of 1862. The kiss of death was the incursion by railroads crossing the plains beginning in 1860 that created a large scale commercial market in fresh and dried meat, tallow, marrow, tongues, robes, hides, chips, and sinews. However, the overwhelming market was for hides. Less than 1/1000 of the buffalo was used (Hornaday 1889). Hornaday (1889: 464–465) summarized the reduction of buffalo:

> "The causes which led to the practical extinction (in a wild state, at least) of the most economically valuable wild animal that ever inhabited the American continent, are by no means obscure. It is well that we should know precisely what they were, and by the sad fate of the buffalo be warned in time against allowing similar causes to produce the same results with our elk, antelope, deer, moose, caribou, mountain sheep, mountain goat, walrus, and other animals. It will be doubly deplorable if the remorseless slaughter we have witnessed during the last twenty years carries with it no lessons for the future. A continuation of the record we have lately made as wholesale game butchers will justify posterity in dating us back with the mound-builders and cave-dwellers, when man's only known function was to slay and eat.
>
> The primary cause of the buffalo's extermination, and the one which embraced all others, was the descent of civilization, with all its elements of destructiveness,

PROCESSION OF GAME

Soup

Venison (Hunter Style)   Game Broth

Fish
Broiled Trout, Shrimp Sauce
Baked Black Bass, Claret Sauce

Boiled
Leg of Mountain Sheep, Ham of Bear
Venison Tongue, Buffalo Tongue

Roast
Loin of Buffalo, Mountain Sheep, Wild Goose, Quail,
Redhead Duck, Jack Rabbit, Blacktail Deer, Coon,
Canvasback Duck, English Hare, Bluewing Teal,
Partridge, Widgeon, Brant, Saddle of Vension, Pheasant,
Mallard Duck, Prairie Chicken, Wild Turkey, Spotted
Grouse, Black Bear, Opossum, Leg of Elk, Wood Duck,
Sandhill Crane, Ruffled Grouse, Cinnamon Bear

Broiled
Bluewing Teal, Jacksnipe, Blackbirds, Reed Birds,
Partridges, Pheasants, Quails, Butterballs, Ducks,
English Snipe, Rice Birds, Red-Wing Starling, Marsh
Birds, Plover, Gray Squirrel, Buffalo Steaks, Rabbits,
Venison Steak

Entrees
Antelope Steak, Mushroom Sauce; Rabbit Braise, Cream
Sauce; Fillet of Grouse with Truffles; Venison Cutlet,
Jelly Sauce; Ragout of Bear, Hunter Style; Oyster Pie

Salads
Shrimp, Prairie Chicken, Celery

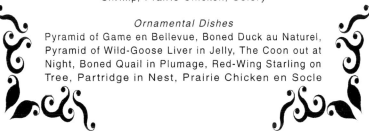

Ornamental Dishes
Pyramid of Game en Bellevue, Boned Duck au Naturel,
Pyramid of Wild-Goose Liver in Jelly, The Coon out at
Night, Boned Quail in Plumage, Red-Wing Starling on
Tree, Partridge in Nest, Prairie Chicken en Socle

Figure 2–2 Thanksgiving menu at a Chicago hotel, 22 November
1886 (after *Wildlife in America* by Peter Matthiessen, line drawings
by Bob Hines, copyright © 1959, revised and renewed 1987 by Peter
Matthiessen. Copyright © 1959 by Bob Hines, line drawings. Used by
permission of Viking Penguin, a division of Penguin Putnam).

Figure 2–3 The extermination of the American bison (after Hornaday 1889).

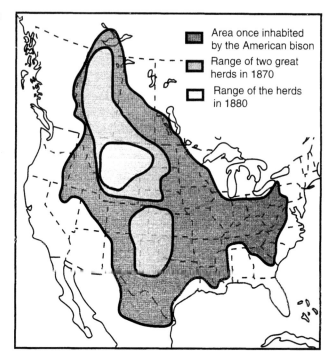

Area once inhabited by the American bison

Range of two great herds in 1870

Range of the herds in 1880

upon the whole of the country inhabited by that animal. From the Great Slave Lake to the Rio Grande the home of the buffalo was everywhere overrun by the man with a gun; and, as has ever been the case, the wild creatures were gradually swept away, the largest and most conspicuous forms being the first to go.

The secondary causes of the extermination of the buffalo may be catalogued as follows:

(1) Man's reckless greed, his wanton destructiveness, and improvidence in not husbanding such resources as come to him from the land of nature ready made.

(2) The total and utterly inexcusable absence of protective measures and agencies on the part of the National Government and of the Western States and Territories.

(3) The fatal preference on the part of hunters generally, both white and red, for the robe and flesh of the cow over that furnished by the bull.

(4) The phenomenal stupidity of the animals themselves, and their indifference to man.

(5) The perfection of modern breech-loading rifles and other sporting firearms in general.

Each of these causes acted against the buffalo with its full force, to offset which there was *not even one* restraining or preserving influence, and it is not to be wondered at that the species went down before them. Had any one of these conditions been eliminated the result would have been reached far less quickly. Had the buffalo, for example, possessed one-half the fighting qualities of the grizzly bear he would have fared very differently, but his inoffensiveness and lack of courage almost leads one to doubt the wisdom of the economy of nature so far as it relates to him."

The extermination of buffalo from the eastern part of their range occurred from 1720 to 1830 and the systematic slaughter throughout their range then began. Buffalo hunting expeditions began in 1820 using Indians and more than 500 carts. In 1840 a single expedition included 1,210 carts, 620 Indians and hunters plus women and children. When the railroads came they advertised the game, split the herds, and began the rapid elimination. Hunters lined up along rivers to shoot the surviving buffalo that came to drink. By 1875 the Southern herd was nearly gone.

From 1872 to 1874 white hunters killed 3,158,730, Indians about 390,000, and settlers killed 150,000 buffalo. The Northern herd lasted a little longer but by 1882 the end was near. Five thousand hunters and skinners shipped about 200,000 hides in 1882 but they went broke in 1883 and were only able to ship 300 hides.

By 1889 Hornaday predicted there were 85 scattered remnants and 200 buffalo in Yellowstone. There was also a wild herd in Canada so remote it was not wiped out (in the Great Slave Lake area).

There were calls for the protection of buffalo "... to prevent the useless slaughter of buffalo within the territories of the U.S.," and Congress passed protection laws in 1872 but they were not signed by the president. President Grant saw the extermination of bison as a means to exterminate the troublesome plains Indians.

When the Cree Indians and white hunters moved up the Cannon Ball River in North Dakota to attack the last large herd of bison, they ended a chapter of American history started only a few years before.

The Passenger Pigeon

The fate of the passenger pigeon is similar to that of the buffalo except humans were more efficient and caused total extinction. This is impressive for several reasons. The passenger pigeon was perhaps the most numerous bird ever to exist on earth and there may have been as many as 5,000,000,000 at one time. They were also good to eat and very easily captured.

The vast concentrations of passenger pigeons are almost beyond belief, except for the reports from Canada to the Gulf of Mexico. Trefethen (1975) relays the observations of two naturalists: Alexander Wilson, the father of American ornithology, and John J. Audubon. Wilson reported one flight of birds in Kentucky in 1806 that was more than 1.6 km wide, 64 km long and contained 2,230,072,000 birds! Audubon (in litt.) observed a flight of passenger pigeons so dense "the light of the noonday sun was obscured as by an eclipse." He estimated that he saw 1,115,136,000 birds in one day! Nesting areas looked like they were covered with snow, and so many birds descended in trees that they broke branches as the weight increased. One nesting colony in Wisconsin was over 160 km long with hundreds of thousands of nests. The characteristics of this bird made it a market hunter's dream. They were in vast numbers, could be easily located, mass killings were easy, and they tasted good.

These characteristics caused the demise of this bird, and as civilization moved west the decline of passenger pigeons increased. The last great nesting concentration was in New York in 1868 and a large roost was found in 1875. Once found, the slaughter began by well-organized "pigeoners" (Trefethen 1975:64). They used the telegraph to keep each other informed about bird movements and locations, and when new sites were located they were descended upon *en masse*. Pigeoners netted the adults and felled trees for the squabs. One warehouse used to store birds for shipment was 7 × 20 meters and was covered with squabs to a depth of up to 1.3 meters. The man in charge of the operation estimated that his total shipments during the nesting season were between 40 and 50 tons of squabs (Trefethen 1975:64). Trefethen (1975:64–65) described the last great nesting colony:

> "In 1878, the last great nesting colony was discovered near Petoskey, Michigan. Twenty-five hundred netters descended upon it. Within the space of a few weeks they marketed 1,107,800,066 pigeons... This was the end of the passenger pigeon as a viable species. A few scattered flocks persisted until after the turn of the century, but they dwindled in numbers each year. The last known survivor of the species died in the Cincinnati Zoological Garden on September 1, 1914."

The pigeoners contributed to the demise but their efforts ended when netting was no longer profitable. However, in the absence of effective game laws, those remaining were shot whenever they came within range. More important, however, was the destruction of the virgin eastern forest that was necessary for forage and nesting sites. As trees were felled pigeon habitat went with them. At the turn of the century America was not the place for agriculture, growing villages and farms, or passenger pigeons.

Other Exploited Species

Buffalo and passenger pigeons were not the only species that were exploited — everything was exploited if money was to be made: Canada goose, trumpeter swan, wild turkey, white-tailed deer, Rocky Mountain elk, and pronghorn. This was the beginning of the largest wildlife management experiment in the world. At the turn of the century Americans had turned their wildlife heritage into a wasteland. However, Americans learned one very valuable lesson: market hunting and disregard for habitat destroy wildlife. Other species that slipped into extinction from a lack of any management were the Labrador duck, heath hen, Carolina parakeet, and the great auk. It was only in the early 1900s, when the United States and Canada outlawed market hunting, that wildlife began to return in abundance and diversity. This was not an overnight process as it took decades to reeducate Americans not to see curlews, sand pipers, and plovers as food or to regard elk and bears as dollars on the hoof or paws.

During the height of exploitation (1870 – 1915) the United States Cavalry was forced to protect wildlife in Yellowstone National Park from 1886 to 1918 (Hampton 1971). When the Army replaced the superintendent of Yellowstone and took over the administration of the park, that was the beginning of the end of the tragedy that decimated North America's wildlife and was the beginning of modern North American wildlife management (Geist 1995:77). Aldo Leopold, however, was the one that foraged modern-day wildlife management in the United States with the publication of *Game Management* (Leopold 1933, Flader 1974, McCabe 1987). With modern management, the experiment of returning wildlife to America was a success and wildlife was returned to North America so there are now more than 30 million big game animals north of Mexico (Schmidt and Gilbert 1978). Geist (1995) pointed out that this happened "...despite the ready availability of firearms, which today outnumber big game 8 to 1, despite a human population that outnumbers big game 9 to 1, and despite an aggressive agricultural industry whose livestock outnumbers big game at least 120 to 1" (Pimentel et al. 1980).

Our current system of wildlife conservation may have been rooted in the destruction of wildlife by settlers and traders, markets, and a growing human population, but once the results of the glut were observed (and people could actually see the demise of wildlife in their lifetime), national policy was implemented that had begun in the 1600s in the New World (but was not enforced). Hampton (1971) outlines the role General Sheriden had in protecting wildlife by sending the cavalry into Yellowstone Park. The national parks in North America are a cornerstone of conservation of wildlife but even Yellowstone was under threat of unmanaged abuses. They were saved by the Army. Leaders like Theodore Roosevelt began and implemented the conservation movement in North America that is used as a model around the world.

SUMMARY

When Europeans arrived on the shores of North America, wildlife was so abundant it was used without any regard for any type of conservation and management. Many of the early Europeans arrived in America to escape the restrictions that had been imposed on their use of wildlife in Europe. The abundance that awaited them led to the "myth of superabundance," which was shattered as wildlife populations dwindled with increasing anthropogenic influences. When the buffalo and passenger pigeon were nearly eliminated and exterminated, respectively, leaders recognized that something had to be done or the fate of other species would follow that of the passenger pigeon. Thus began the "wildlife management experiment."

3

The Conservation Idea

"There are two spiritual dangers in not owning a farm. One is the danger of supposing that breakfast comes from the grocery, and the other that heat comes from the furnace."

A. Leopold (1949)

INTRODUCTION

Conservation in North America has been practiced since the early 1900s. This chapter introduces the conservation concept, leaders that made conservation a national issue, and important laws that influence why wildlife management is practiced as it is. The legal basis of wildlife management is an important and basic concept of why and how some management occurs as it does.

▨ THE CONSERVATION IDEA

In Europe the objectives derived for wildlife were to improve hunting for and by private landowners. In America the objectives surrounding wildlife at the turn of the century were to save and perpetuate wildlife rather than to improve or create hunting. Remember that Americans observed the reduction of big game in their lifetimes and the virtual elimination of buffalo. Numerous individuals were responsible for the beginnings of conservation in North America after market hunting and exploitation left North America a wildlife wasteland at the turn of the nineteenth century.

The first steps towards conservation involved restrictions on harvesting animals, or law enforcement. Wildlife conservation followed this concept until Theodore Roosevelt, the 26th United States President, brought the idea of "conservation through wise use" to the American public. He recognized that renewable natural resources could last forever if harvested scientifically and not faster than reproduced. Theodore

Roosevelt introduced "conservation" to the public (something they had not heard of before) and determined the subsequent history of American game management in three ways:

1. Roosevelt recognized the landscapes, waters, vegetation, and animals as entire ecosystems.
2. He recognized conservation through wise use.
3. He demanded that science be the cornerstone of conservation.

These were new concepts in the early 1900s. Most of the public maintained provincial attitudes and had the opinion that the "forces" that put animals on the earth would be the same that would remove them. A change of attitude was clearly necessary. Naturalists were aware of wildlife but not of their ecology (i.e., the relationships between animals and their habitats). Progress was being made in the identification of flora and fauna, and hunting seasons had been established; concerned citizens knew why species had to be managed. However, nobody was dealing with how wildlife should be managed. The "how" of wildlife management was the beginning of true wildlife conservation in North America. Deliberate and purposeful manipulation of the environment was needed for conservation. This manipulation could only be accomplished with the help of the landowners, and landowners needed some incentive for being involved in the conservation process. These were the rudimentary steps during the 1930s. Since then, with the guidance of Aldo Leopold, the art and science of wildlife management has flourished. Leopold recognized that management of natural resources was necessary and that management began with landowners. Leopold is recognized as the father of modern wildlife management in North America. Many of the contemporary philosophies in the arena follow his early practices for which Theodore Roosevelt, the grandfather of wildlife management, paved the way.

Conservation Movements in America

In the United States there have been essentially two conservation movements with a third emerging at the end of the twentieth century. The first began in the late 1920s and early 1930s when Americans realized 1) the buffalo was not coming back; 2) the passenger pigeon was gone; 3) there were only 500,000 white-tailed deer left in the nation; 4) there were less than 25,000 pronghorn in the United States; and 5) elk numbers had been reduced from millions to less than 50,000 scattered throughout the western states. Furthermore, numbers of trumpeter swans, Canada geese, turkeys, and other wildlife were low. Other species were extinct (i.e., great auk, Steller's sea cow, eastern sea mink, birds) and yet others were so drastically depleted that their trend toward extinction was irreversible and followed in the early 1900s (i.e., heath hen, Carolina parakeet). These were population declines the public observed within their lifetime and they realized that if measures were not taken, America would lose an important part of its natural heritage. Between 1885 and 1910 the big game resources of the United States had been reduced by more than 80% as a result of absolutely no control on use. The breeding stock was available and with sound management, visionary leaders recognized that the numbers could be returned. Thus, the wildlife management experiment began.

The second major conservation movement began in the 1960s and is still present due to a similar set of problems: loss of habitat for wildlife, pollution, controversy over predators, and increased demand for land use. The third conservation movement is essentially a reawakening of the second at the end of the twentieth century. North Americans are concerned about wildlife and their role in its management is recognized. It is not that there was absolutely no recognition of the need for sound management of natural resources 100 years ago, there were only a few farsighted individuals that called for moderation in use. However, they did not receive the public support needed to be effective until the late 1800s and early 1900s, when populations of wildlife began to dwindle so fast that people could actually see changes before their eyes. Then the ideas of protection and later conservation began to catch on.

The Beginnings of Conservation

From the 1880s to 1930s there was widespread concern for wildlife, but until the 1930s Americans were at a loss as how to best benefit wildlife. Most efforts were simply protection—not active management or habitat improvement. For example, by 1800 all of the colonies had restrictions on deer hunting, in 1844 the New York Sportsman Club was formed to protect and preserve game (Trefethen 1975), laws were established against spring shooting of birds (Belanger 1988), prairie chickens were protected in Wisconsin, and California protected elk and deer (Table 3–1). Nongame birds were protected in Vermont in 1851, Massachusetts in 1855, and in ten other states by 1864 (Trefethen 1975). Maine employed salaried game wardens in 1852 (Trefethen 1975), licenses were required as early as 1864 in New York (Leopold 1933), national parks were established (Yellowstone) in 1872 (Adams 1993), and market hunting was first outlawed by Arkansas (Leopold 1933) (Table 3–1). Laws restricting hunting continued to develop until 1880 when all states had some game enforcement (Leopold 1933, Belanger 1988). Many of the laws established had little effect because of lack of enforcement. That was drastically changed with the emergence of the federal government into wildlife law enforcement with the Lacey Act of 1900. The Lacey Act strengthened and supplemented state wildlife conservation and laws, prohibited importation of wild vertebrates and other animals injurious to humans, agriculture, and wildlife resources, and prohibited the shipment, transport, sale, or purchase of wildlife and their products taken or possessed in violation of federal, state, or foreign laws. The Lacey Act essentially eliminated "market hunting."

Just four years earlier (1896) the Supreme Court case of Greer vs. Connecticut (161 U.S.C.) determined that game is property of the state unless otherwise stated (Adams 1993). Thus, by the turn of the century a general public consciousness about natural resources was developing, especially in the eastern United States (the West was still frontier oriented). People (like Hornaday) were speaking out but there was no real uniting force and no real leadership. The time was right for a

---

Table 3-1   A chronology of representative legislation affecting wildlife in North America

---

The first written law about wild animals comes from the Bible (Deuteronomy 22:6) where Moses decrees that the breeding stock should be saved.

1300s. First clear written record of a well-rounded system of game management. Marco Polo describes the game laws of Kublai Khan (1215–1294) in the Mongol Empire:

1. Harvest was prohibited between March and October to allow species to reproduce.

2. Food patches were established.

3. Winter feeding programs were initiated.

4. Cover control was initiated.

Important dates and regulations influencing wildlife in the United States:

1581. English statute passed to preserve pheasant and partridge; prohibited any person from taking birds in the night and outlawed hawking of birds in cornfields (Favre 1983).

1616. Bermuda government protects cahow (Matthiessen 1959).

1620. Bermuda government protects green turtle (Matthiessen 1959).

1629. The West India Company grants hunting rights to persons planting colonies in the New Netherlands (Matthiessen 1959).

1630. First New World bounty system: Massachusetts Bay Company authorizes payment of one penny per wolf (Trefethen 1975).

1646. First closed season on a mammal (i.e., deer), Rhode Island (Trefethen 1975).

1677. Connecticut prohibited exporting game.

1700. Most of the deer shot out in the colonies; deer harvesting laws were made but not enforced.

1708. First closed season on birds: ruffed grouse, quail, turkey, and heath hens in certain New York counties (Matthiessen 1959).

1710. Massachussets prohibits use of camouflaged canoes or boats equipped with sails in pursuit of waterfowl (Matthiessen 1959).

1718. Massachusetts closed deer season for three years (Leopold 1933).

1739. First warden (deer "reeves") system in Massachussets (Trefethen 1975).

1776. First federal game law required closed season on deer in all colonies except Georgia (Belanger 1988).

1782. Bald eagle recognized as the national emblem (Belanger 1988).

1790. First exotics from the Old World; Hungarian partridges were introduced to New Jersey (Matthiessen 1959).

1800. Enforcement of laws was a problem even though all colonies had restrictions on deer hunting.

1844. New York Sportsmen's Club formed for the protection and preservation of game, and in 1846 proposed a law protecting game (Trefethen 1975).

1846. First law against spring shooting (wood duck, black duck, woodcock, snipe); later repealed (Rhode Island) (Belanger 1988).

1849. U.S. Department of the Interior established (Belanger 1988).

1850–1860. Wisconsin protected prairie chicken; California (1852) protected the elk and pronghorn.

1850. First protection for nongame birds in Connecticut and New Jersey (screech owl and insectivorous birds) (Trefethen 1975).

1851. Nongame bird protection established in Vermont; subsequently in Massachussets (1855) and 10 other states by 1864 (Trefethen 1975).

Table 3–1 A chronology of representative legislation affecting wildlife in North America. *Continued*

1852. Maine first to employ salaried game wardens (Trefethen 1975).

1860–1870. Required collectors to have permits. Ban on spring shooting over most of United States. Movement of humans west doomed the bison.

1864. First protection of bison, in Idaho; closed season (1 February–1 July) on bison, deer, elk, pronghorn, goat, and sheep (Matthiessen 1959).

1864. A hunting license was first required by hunters in New York (Leopold 1933).

1865. First game department—Massachussets Commission of Fisheries and Game (Trefethen 1975).

1871. Establishment of Federal Commission of Fish and Fisheries (later Bureau of Fisheries). Establishment of $30 bounty on Adirondack wolves in New York (Matthiessen 1959).

1872. First national park—Yellowstone Park Act (Adams 1993).

1873. New Jersey establishes first hunting licenses for nonresidents (Belanger 1988). Maine established the first bag limit on deer (3 per year) (Trefethen 1975).

1874. A bill "to prevent the useless slaughter of buffaloes within the Territories of the United States" is passed by Congress but is pigeonholed by President Grant (Allen 1954).

1875. Market hunting first outlawed by Arkansas (Leopold 1933).

1877. Florida passes a plume-bird law prohibiting wanton destruction of eggs and young (Matthiessen 1959).

1878. First upland-game bag limit law passed in Iowa limiting harvest of prairie chickens to 25 per day (Belanger 1988).

1880. All states had game laws (Leopold 1933, Belanger 1988).

1880–1890. William T. Hornaday made first buffalo count.

1883. American Ornithological Union established (Matthiessen 1959).

1885. Banff was established as the first Canadian national park (Matthiessen 1959).

1885. Establishment of Federal Division of Economic Ornithology and Mammalogy (in 1896 becomes the Division of Biological Survey and in 1940 the U. S. Fish and Wildlife Service) (Dunlap 1988).

1886. American Ornithological Union "Model Law" for bird protection first promoted (however the English sparrow was exempt from protection) (Matthiessen 1959).

1886. Audubon Society formed by George Bird Grinnell to combat bird plume hunters (Matthiessen 1959).

1887. Boone and Crocket Club formed by Theodore Roosevelt and G. B. Grinnell (Matthiessen 1959).

1891 Forest Reserve Act created that permitted the President to set aside forest resources (later national forests) on the public domain (Belanger 1988).

1892. Sierra Club formed by John Muir (Trefethen 1975).

1894. Yellowstone Park Protection Act provides protection for wildlife, timber, and minerals within the park (Adams 1993).

1894. George Bird Grinnell first warned of dangers of lead poisoning to waterfowl from ingesting spent shot pellets (Belanger 1988).

1895. Modern system of resident and nonresident hunting licenses established in several states (Matthiessen 1959).

1896. The Supreme Court in Geer vs. Connecticut (161 U.S. 519) decrees that game is the property of the state unless otherwise stated (Adams 1993).

1898. League of American Sportsmen formed (Belanger 1988).

Table 3–1 *Continued*

1899. Refuse Act. Designed to regulate discharge of pollutants into navigable waters. Now used to require polluters to obtain permit from the Army Corps of Engineers before discharging pollutants (Adams 1993).

1900–1910. Wildlife at its lowest ebb in this country. First conservation movement began. Roosevelt doctrine—laws protecting game. Predator control began as a means of increasing wildlife. Early attempts to control environmental factors.

1900. Lacey Act forbids importation of foreign wildlife without a permit. Prohibited interstate commerce in illegal game (Adams 1993).

1902. First wildlife managers' meeting held in Yellowstone; attended by eight game wardens from six states; formed an association that would later become the International Association of Fish and Wildlife Agencies (Belanger 1988).

1903. First federal wildlife refuge established at Pelican Island, Florida (Belanger 1988).

1905. Creation of U. S. Forest Service and under leadership of Gifford Pinchot, with President Roosevelt, created many of the national forests in the United States (Trefethen 1975).

1906. Congress refuses appropriations for further investigations by President Roosevelt's Natural Resources Committee (Matthiessen 1959).

1908. Conference of Governors at the White House that established the National Conservation Commission under Gifford Pinchot. "Conservation" first used by Pinchot in 1907, in establishing the theme that states must join the federal government in inventories, management, and husbandry of their respective natural resources (Matthiessen 1959, Trefethen 1975).

1909. Pronghorn removed from list of game animals everywhere in its range except Arizona (which did so in 1911) and Canada (Matthiessen 1959).

1911. Seal Treaty controlling the take of fur seals and sea otters (the latter still receive complete protection) signed by United States, Great Britain, Russia, and Japan (Trefethen 1975, Bean 1983).

1913. Federal Tariff Act of 1913. Prohibited the possession or sale of plumes, feathers, quills, wings, or any skin or parts of skins of wild birds, either raw or manufactured (Matthiessen 1959).

1913. Weeks-McLean Migratory Bird Act. Declared all migratory game and insectivorous birds to be "within the custody and protection of the government of the United States" through the Department of Agriculture (Belanger 1988).

1914. Last passenger pigeon died in captivity (Belanger 1988).

1914. Last Carolina parakeet died in captivity (Matthiessen 1959).

1915. Act of Congress authorizes funds for the control of wolves, prairie dogs, and other injurious animals (Reed and Drabelle 1984).

1918. Migratory Bird Treaty Act—Implements treaty with Great Britain (for Canada) for protection of migratory birds, whose welfare is a federal responsibility. Regulates taking, selling, transporting, and importing migratory birds and provides penalties for violations ($500 and/or six months in jail for each violation) (Adams 1993).

1920–1930. Enforcement of laws in most states become more effective. Period of neat farming. Very little cover left on farms. First attempts to control hoof-and-mouth disease.

1924. Cooperative bobwhite quail investigation by H. L. Stoddard. This bobwhite study was one of the best studies on a bird (Leopold 1933).

1926. Black Bass Act prohibits interstate shipment of black bass, or purchase of such fish, contrary to the laws of any affected state. Amended in 1947 to apply to other game fish, and in 1952 to apply to all fish. The act was repealed in 1981 (Adams 1993).

Table 3–1 A chronology of representative legislation affecting wildlife in North America.
Continued

1929. Migratory Bird Conservation Act. Provided for further acquisition of migratory bird refuge land (Belanger 1988).

1930. Tariff Act. Import duties on certain wild birds and on bird feathers and skins; prohibits importation of wild mammals or birds or parts in violation of laws of country from which exported (Committee on Merchant Marine and Fisheries 1975).

1931. Animal Damage Control Act authorized the "control or eradication of mountain lions, wolves, coyotes, bobcats, prairie dogs, gophers, ground squirrels, jack rabbits, and other animals, and birds...." Sanctioned cooperative relationships between federal agencies and nonfederal entities in executing field operations, and it authorized control activities on nonfederal lands (Bean 1983).

1933. *Game Management* by Aldo Leopold published. Father of game management in the United States.

1934. Fish and Wildlife Coordination Act provides assistance to states from the federal government for wildlife protection, development stocking, and rearing of fish and wildlife and controlling losses (Adams 1993).

1934. Migratory Bird Hunting Stamp Act of 1934 requires every person over 16 years old hunting migratory waterfowl to purchase a duck stamp, the proceeds from which are deposited in a special migratory bird conservation fund for acquiring and managing lands as migratory bird refuges (Adams 1993).

1934. Taylor Grazing Act directed the Secretary of the Interior to "provide for the orderly use, improvement, and development of the range," and to cooperate with "official state agencies engaged in conservation or propogation of wildlife interested in the use of grazing districts" (Adams 1993).

1935. Wilderness Society formed by Aldo Leopold, Bob Marshall, and others (Adams 1993).

1936. First National Wildlife Conference called by President Franklin D. Roosevelt. Now called the North American Wildlife and Natural Resources Conference. Sponsored by Wildlife Research Units and the American Wildlife Institute (later the Wildlife Management Institute) (Belanger 1988).

1936. Migratory Bird Treaty Act of 1918 amended to include treaty with Mexico for protection of migratory birds and mammals (Adams 1993).

1937. Federal Aid in Wildlife Restoration Act (Pittman-Robertson Act) establishes 11% excise tax on sporting arms and ammunition that are provided to states on a matching basis ($3 by Federal government for each $1 of state money) for land acquisition, research, and management projects. Money must be used each year, or it reverts back to the federal government. Each state allotted a certain amount determined by license sales and geography (size of state) (Adams 1993).

1937. Ducks Unlimited formed. Private organization in United States and Canada that contributes money to improve waterfowl habitat.

1937. The Wildlife Society organized.

1946. Wildlife Management Institute formed.

1940. Bald Eagle Protection Act provides for protection of the bald eagle. Amended in 1962 to extend the protection to golden eagles and name changed to the Eagle Protection Act (Anderson 1991).

1946. International Convention for the Regulation of Whaling. International Whaling Commission (IWC) was formed to regulate whaling (Bean 1983).

1948. Federal Water Pollution Control Act, included financial and technical assistance to states and communities willing to act on sewage and industrial pollution problems (Adams 1993).

Table 3–1 *Continued*

1950. Federal Aid in Fish Restoration Act (Dingell-Johnson Act) does the same for fishing as Pittman-Robertson Act did for hunting (Adams 1993).

1950. Nature Conservancy formed (Dunlap 1988).

1954. Congress authorized the Army Corps of Engineers to construct watershed projects, stream channelization and management, under section 205 of the Flood Control Act of 1948 (e.g., "205 projects"), and authorized the Soil Conservation Service to engage in similar activities under the Watershed Protection and Flood Protection Act of 1954, Public Law No. 566 ("566 projects") (Adams 1993).

1956. Fish and Wildlife Act of 1956. Established a comprehensive national fish and wildlife policy; established present U.S. Fish and Wildlife Service (Adams 1993).

1960. Sikes Act. Provides for cooperation by the Departments of the Interior and Defense with state agencies in planning, development, and maintenance of fish and wildlife resources on military reservations through the United States (Adams 1993).

1960. Cooperative Research and Training Units Act authorizes the Secretary of the Interior to enter into cooperative agreements with colleges and universities, state fish and game agencies, with nonprofit organizations for the purpose of: 1) training fish and wildlife workers, 2) research on fish and wildlife species, and 3) educational work with the cooperative extension service (Adams 1993).

1964. Wilderness Act of 1964 directs the Secretary of the Interior, within 10 years, to "review every roadless area of 5,000 contiguous acres or more in the...national park system and every such area of...the national wildlife refuges and game ranges... [and to] report to the President his recommendation as to the suitability...for reservation as wilderness" (Adams 1993).

1966. Endangered Species Preservation Act. Authorized the secretary "to carry out a program in the United States of conserving, protecting, restoring and propagating selected species of native fish and wildlife found to be threatened with extinction." It also directed the Secretary to publish in the Federal Register a list of native fish and wildlife species threatened with extinction. Provided protection for only native species but it was a beginning. Repealed in 1973 with the passing of the Endangered Species Act (Bean 1983).

1968. Wild and Scenic Rivers Act of 1968. Establishes a national wild and scenic river system.

1969. National Environmental Policy Act of 1969 requires all federal agencies to consult with each other and to employ systematic and interdisciplinary techniques in planning. All actions significantly affecting the quality of the human environment require a detailed statement on the environmental impact, any adverse environmental effects, and alternatives. Also established the Council on Environmental Quality (Adams 1993).

1969. Endangered Species Conservation Act. Directed the Secretary of Interior to provide a list of fish and wildlife species and subspecies threatened with worldwide extinction. The act made it a crime to import said listed species, and encouraged international agreements for the protection, conservation, and propagation of fish and wildlife. Toward that end the act seeks the convening of an international meeting for the purpose of signing an international treaty on the conservation of endangered species. That meeting and future treaty was the Convention on International Trade in Endangered Species of Wild Fauna and Flora (CITES) (Adams 1993).

1971. Wild Free-Roaming Horses and Burro Act. The stated purpose of this act was to "require protection, management, and control of wild free-roaming horses and burros on public lands" because they were declared to be "living symbols of the historic and pioneer spirit of the West" (Bean 1983).

1972. Marine Mammal Protection Act. This act establishes a moratorium on the taking and importation of marine mammals and products made from them. Exceptions include Indian taking for consumption or use in native crafts; and taking incidental to commercial fishing activities. Establishes a Marine Mammal Commission to coordinate research and take part in the regulatory processes (Adams 1993).

Table 3–1   A chronology of representative legislation affecting wildlife in North America. *Continued*

1972. Migratory Bird Treaty Act of 1918 was amended to include Japan (Adams 1993).

1973. Endangered Species Act of 1973 repeals the Endangered Species Preservation Act of 1966. The act provides for the conservation of threatened and endangered species of fish, wildlife, or plants by federal action and by encouraging the establishment of state programs (Adams 1993).

1974. CITES (see 1969). Provided the international forum and treaty powers justification for the United States' endangered species program. In 1975 it was ratified by 10 countries, in 1976, 30 countries, in 1982, 77 countries, and as of August 1996, 147 countries have signed (Adams 1993).

1974. Forest and Rangeland Renewable Resources Planning Act. This act is designed to gear development and use of forest lands and rangelands for long-term benefits and goals without degradation (Adams 1993).

1976. Federal Land Policy and Management Act. This act was designed to establish land policy by providing guidelines for administration, management, protection, development, and enhancement of public lands (Adams 1993).

1976. National Forest Management Act provided for multiple-use planning and management of timber on federal lands by the Forest Service. Public participation was incorporated into decision making. Amended in 1981 to include an interdisciplinary approach to land and resource planning and fish and wildlife must be considered in any activity planned by the Forest Service (Adams 1993, Anderson 1991).

1976. Migratory Bird Treaty Act of 1918 was amended to include the Soviet Union (Adams 1993).

1980. Alaska National Interest Lands Conservation Act protects 42,086,900 hectares including 22,662,200 hectares in the National Wilderness Preservation System. This act made the National Park Service the major custodian of wilderness (Adams 1993).

1980. Fish and Wildlife Conservation Act provided federal funds, by reimbursement, to support nongame species programs. The Pittman-Robertson and Dingell-Johnson Acts and this act provided the foundation of state wildlife and freshwater fish research and management activities (Adams 1993).

1981. Lacey Act Amendments of 1981 combined the features of the Lacey Act of 1900 and the Black Bass Act of 1926. It is a federal offense to cross state boundaries in the commission of any act involving any animal or plant when that act would be unlawful in either state (Adams 1993).

1984. Wallup-Breaux Act supplanted the Dingell-Johnson program for fish restoration and management and increased its income by adding 3% tax on boats and motors (Adams 1993).

1986. United States and Canada signed the North American Waterfowl Management Plan to restore declining waterfowl populations. Formalized long-range cooperative plans for managing international waterfowl resources and their habitats. This plan was implemented at the flyway, national, provincial, territorial, state, and private levels (Belanger 1988).

---

prominent figure to assume a leadership role. In 1901 when President McKinley was assassinated, Theodore Roosevelt became president. Roosevelt was always interested in nature, developed into a devoted sportsman, and was very progressive and idealistic. He filled that leadership role.

Roosevelt's chief advisor in forestry, Gifford Pinchot, championed the idea of sustained yield forestry and advocated wise use rather than total preservation and he coined the term "conservation" to form a middle ground between preservation and total controlled exploitation.

Roosevelt created the first official wildlife refuge at Pelican Island, Florida (Belanger 1988) to protect animals from plumage hunters and fisherman. By 1904 Roosevelt and his administration created 51 more refuges. Today the national wildlife refuge system is the backbone of the United States Fish and Wildlife Service and includes thousands of hectares. In 1908 Western interests succeeded in obtaining a rider attached to an appropriations bill to abolish the Forest Reserve Act. Roosevelt and Pinchot spent an entire day with maps of the western United States and withdrew thousands of hectares in one day—and then signed the bill. During Roosevelt's term, the forest reserve went from 17,000,000 to 69,605,390 hectares.

Other federal laws and treaties relating to wildlife were instrumental in shaping contemporary wildlife management (Table 3–1).

The Migratory Bird Treaty Act of 1918, as amended (16 U.S.C., 703-711).   The Migratory Bird Treaty Act implemented treaties with Great Britain for Canada ratified in 1916, and Mexico ratified in 1936, for the protection of migratory birds (managed by the federal government) and provided for regulations to control taking, selling, transporting, and importing migratory birds. The act was amended in 1974 to extend provision to the Convention between the United States and Japan. Migratory species are not the sole property of any country. This act was an important step in the development of international law.

Black Bass Act of 1926, as amended in 1969 (16 U.S.C., 851-856).   The terminology of the Lacey Act dealt with game birds and fur-bearing mammals. The Black Bass Act provided protection for black bass and all game fish and prohibited importation or transportation in interstate or foreign commerce of black bass and other fish in violation of foreign or state laws. Both the Lacey Act and the Black Bass Act protect animals and parts of animals including eggs.

Migratory Bird Conservation Act of 1929 (16 U.S.C. 715-715s).   This act provided a mechanism to acquire additional bird refuge habitat by authorizing acquisition, development, and maintenance of refuges. It also authorized research and publications on North America birds.

Tariff Act of 1930 (19 U.S.C. 1202, schedule 1: and sec. 1527).   The Tariff Act established import duties on certain wild birds (and parts) and prohibited importation of wild mammals and birds (and parts) in violation of regulations of the country from which they were exported.

Animal Damage Control Act of 1931 (7 U.S.C., 426-426b).   The Animal Damage Control Act provided broad authority to investigate and control mammalian predators, rodents, and birds. It also authorized cooperative agreements with other agencies to conduct similar activities.

Fish and Wildlife Coordination Act of 1934 (16 U.S.C., 661-661c).   The Fish and Wildlife Coordination Act authorized federal assistance for the protection, rearing, stocking, and censusing of fish and wildlife by federal, state, and other agencies. It also authorized some federal agencies to obtain lands in connection with water-use projects specifically for the conservation of fish and wildlife. An important aspect of the law required consultation with the United States Fish and Wildlife Service and state wildlife agencies that have waters that are proposed or authorized to be influenced by

Table 3–2   Wildlife restoration account receipts (National Survey of Fishing, Hunting, and Wildlife Associated With Recreation, 1996)

Revenue source	Year				Total
	1939–1993	1994	1995	1996	
Sporting arms and ammunition	$2,082,943,582	$144,470,129	$132,107,023	$121,790,516	$2,481,311,250
Pistols and revolvers (since 1971)	$ 550,661,166	$ 67,903,169	$ 54,411,412	$ 39,402,257	$ 712,378,004
Archery equipment (since 1975)	$ 182,938,782	$ 11,278,372	$ 30,186,051	$ 18,278,953	$ 242,682,158
Total	$2,816,543,530	$223,651,670	$216,704,486	$179,471,726	$3,436,371,412

any federal agency, or to prevent loss or damage to wildlife resources in connection with the water resources.

Migratory Bird Hunting Stamp Act of 1934 (16 U.S.C., 718-718h).   The Migratory Bird Hunting Stamp Act, also known as the Duck Stamp Act, required waterfowl hunters 16 years of age and older to purchase duck stamps. Funds generated are used to purchase waterfowl habitats.

Taylor Grazing Act of 1934.   The Taylor Grazing Act provided for the orderly use, improvements, and development of range resources. The act also established cooperation with official state agencies charged with the conservation and management of wildlife and rangelands.

Migratory Bird Treaty Act of 1918 amended in 1936 (50 statute 1311).   The 1918 act was amended to include the United Mexican States for the conservation, protection, and management of migratory birds and game animals.

Federal Aid in Wildlife Restoration Act of 1937 (16 U.S.C., 669-669i).   The Federal Aid in Restoration Act is also known as the Pittman-Robertson or PR Act and provided federal aid to states for wildlife restoration work. Funds are obtained from an excise tax on sporting arms (e.g., rifles, pistols, revolvers, bows) and ammunition and allocated to states on a matching basis for land acquisition and development, research, management, and hunter safety programs (Table 3–2).

Bald Eagle Act of 1940 (16 U.S.C., 668–668d).   The bald eagle is America's national emblem and was provided protection under this act along with the golden eagle.

Federal Aid in Fish Restoration Act of 1950 (16 U.S.C. 777-777k).   The Federal Aid in Fish Restoration is also known as the Dingell-Johnson Act or D.J. Act, and provided federal aid to states for sport fish restoration work. Funds are obtained from an excise tax on some sport fishing tackle and allocated to states on a matching basis for land acquisition, research, development, and management (Table 3–3).

Fish and Wildlife Act of 1956 (16 U.S.C. 742a-742j).   The Fish and Wildlife Act established a comprehensive national fish and wildlife policy, which resulted in the U.S. Fish and Wildlife Service. The U.S. Fish and Wildlife Service is a cornerstone of wildlife management that is instrumental in research extension, de-

Table 3–3   Sport fish restoration account receipts (Maharaj and Carpenter 1996)

Revenue source	Year				
	1952–1993	1994	1995	1996	Total
Fishing tackle and equipment	$1,151,224,870	$ 93,432,000	$ 95,689,000	$ 98,253,000	$1,438,598,870
Trolling motors and fish finders	$ 194,312,028	$ 2,090,000	$ 2,566,000	$ 2,573,000	$ 26,680,028
Motorboat fuels	$ 529,297,998	$ 93,079,133	$ 95,029,320	$127,199,085	$ 242,682,158
Small engine fuels[a]	$ 140,413,000	$ 49,531,000	$ 50,734,000	$ 53,330,000	$ 294,008,000
Import duties[b]	$ 232,087,045	$ 24,853,449	$ 27,199,391	$ 28,103,356	$ 312,243,241
Interest on Investments	$ 213,511,169	$ 14,455,154	$ 33,318,995	$ 40,813,652	$ 302,098,970
Total	$2,816,543,530	$223,651,670	$216,704,486	$179,471,726	$3,218,214,465

[a]Authorized by Coastal Westlands Planning, Protection, and Restoration Act.

[b]These figures represent U.S. Treasury estimates deposited into the Sports Fish Restoration Account.

velopment, management, protection, and conservation of fisheries and wildlife resources.

Sikes Act of 1960 (16 U.S.C. 670a, et seq.).   The Sikes Act provided for planning, development, and maintenance of fish and wildlife resources on Department of Defense lands throughout the United States via cooperation with the state agencies, Department of the Interior, and Department of Defense.

Cooperative Research and Training Units Act of 1960 (74 Statute 733; 16 U.S.C. 753a-753b).   The Cooperative Research and Training Units Act developed adequate, coordinated cooperative research and training programs about fish and wildlife resources through the Secretary of the Interior. The Secretary enters into cooperative agreements with colleges, universities, state game and fish agencies, and certain nonprofit organizations.

The Wilderness Act of 1964 (16 U.S.C., 1131-1136).   The Wilderness Act established "wilderness" in the United States. The secretary of the Interior is responsible for reviewing roadless areas of 2,023 hectares or more, and all roadless lands within national wildlife and game refuges. The Secretary recommends to the President the suitability of each area for formal preservation as wilderness under special Acts of Congress.

National Endowment Policy Act of 1969 (Public Law 91-190; 83 statute; 42 U.S.C., 4321-4347).   When management of federal lands occurs, all federal agencies involved consult with each other to employ systematic and interdisciplinary techniques in planning and decision making. All actions that significantly influence the quality of the human environment require a detailed statement on the environmental impact, any adverse environmental effects, and alternatives. The environmental impact statements are required to be available to the public. The Council on Environmental Quality was also established under this act.

Marine Mammal Protection Act of 1972 (Public Law 92-522, 86 statute 1027; 16 U.S.C. B61, 1362, 1371-1384, 1401-1407).   Marine mammals are protected under this act by a moratorium on the taking and importation of marine mammals or products from them. Under certain conditions taking incidental to commercial fishing activities and Indian taking for consumption or use in native craft industries are acceptable.

Endangered Species Act of 1973 (Public Law 92–516; 16 U.S.C., 1531-1543, 87 statute 884).   The Endangered Species Act is an important law because it provided for the conservation and management of endangered or threatened flora and fauna via federal action. The act authorized the determination and listing of endangered and threatened flora and fauna and their habitat, prohibits unauthorized taking, possession, sale or transport of an endangered species, authorized habitat acquisition, authorized the establishment of cooperative agreements with states that have active programs for endangered and threatened species, and provided civil and criminal penalties for violations of the act.

Convention on International Trade in Endangered Species of Wild Fauna and Flora (1974).   The Convention on International Trade in Endangered Species of Wild Fauna and Flora (CITES) provided protection for endangered species by establishing an import/export procedure. The Marine Mammal Protection Act, Endangered Species Act, and CITES constitute the legal basis of the endangered species program in the United States.

Forest and Rangeland Renewable Resources Planning Act of 1974 (Public Law 93-378, 88 statute 467).   The Forest and Rangeland Renewable Resources Planning Act called for the preparation of land management plans for the use of forest and rangelands that would result in long-term benefits and goals without degradation.

Federal Land Policy and Management Act of 1976.   The Federal Land Policy and Management Act was designed to establish land policy by providing guidelines to state and federal administration, management, protection, development, and enhancement of public lands. It specifically mandated that the Bureau of Land Management develop land use plans.

National Forest Management Act of 1976.   The National Forest Management Act stipulated that national forest management plans comply with the National Environmental Policy Act via multiple-use planning and management of timber on federal lands by the U.S. Forest Service. Management would maintain viable populations of native vertebrates on national forests.

The evolution of modern game management (Leopold 1933) is apparent from an examination of these and earlier laws related to wildlife (Table 3–1). The first laws involved restrictions on taking wildlife then turned to reducing predators to protect certain species. Those alone were not enough to protect the habitat of wildlife, so laws that set aside land were enacted followed by laws allocating land and dictating how it would be managed. The evolution of wildlife in North America thus follows an evolutionary path of:

1. restriction of hunting,
2. predator control,
3. reservation of land,

4. artificial replenishment of animals,
5. environmental control (Leopold 1933) or more specifically,
6. habitat control and management.

None of these steps operates in a vacuum and the effective management of wildlife resources involves the synergetic influences of each. However, the ability to enforce the laws and the involvement of people in the management process have been key to the success North American conservation has had.

Personalities that Shaped Conservation in the United States

Wildlife conservation and management are new and dynamic. However, basic concepts are well founded in some of the earliest "conservationists". Below are a few of the personalities that shaped the philosophies of conservation in the United States:

John J. Audubon (1785–1851). Naturalist, author, and artist. He obtained international fame with *The Birds of America*, which emphasized the spectacular drama of wildlife in the New World.

George Perkins Marsh (1801–1882). Author concerned with humans and their impact on nature. In his lifetime he saw the United States transformed from an infant to a large consumer of natural resources. His books *Man and Nature* (1905) and *The Earth as Modified by Human Action* (1907) were the first comprehensive overviews of man's influence on nature.

Henry David Thoreau (1817–1862). Author that devoted himself to nature and wilderness and was concerned with the man/nature ethic. He was an early pioneer that called for the conservation of open space. He is considered a forefather of the environmental movement (Atkinson 1965).

Frederick Law Olmstead (1822–1903). Father of Landscape Architecture whose plans were used to design Central Park in New York City, Boston parks, the arboretum at Amherst College, Stanford University Campus, and the U.S. Zoological Park in Washington, D.C.

John Wesley Powell (1834–1902). Author, naturalist, and explorer best known for his exploration of the Colorado River and the Grand Canyon (Stegner 1953).

John Burroughs (1837–1921). Naturalist and author who stimulated interest in wildlife on a national scale (Teal 1952).

John Muir (1838–1922). Naturalist, explorer, and author. He was the founder of the Sierra Club and instrumental in the development of the U.S. National Park Service. He called for a halt of unregulated use of natural capital (Wolfe 1951, Stetson 1994).

George Bird Grinnell (1849–1938). Naturalist, author, explorer, and conservationist. He fought against wanton hunting, founded the Audubon Society, and is often called the Father of American Conservation.

Bernhard Eduard Fernaw (1851–1923). Brought scientific management and sustained use of forests to the United States when they were being exploited. He was America's first professional forester (Rodgers 1951).

Table 3–4 The lifespans of some of the important individuals leading up to modern wildlife management in the United States

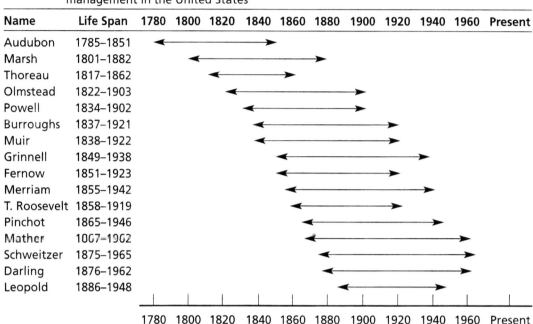

Name	Life Span	1780	1800	1820	1840	1860	1880	1900	1920	1940	1960	Present
Audubon	1785–1851											
Marsh	1801–1882											
Thoreau	1817–1862											
Olmstead	1822–1903											
Powell	1834–1902											
Burroughs	1837–1921											
Muir	1838–1922											
Grinnell	1849–1938											
Fernow	1851–1923											
Merriam	1855–1942											
T. Roosevelt	1858–1919											
Pinchot	1865–1946											
Mather	1867–1962											
Schweitzer	1875–1965											
Darling	1876–1962											
Leopold	1886–1948											

Clinton Hart Merriam (1855–1942). Naturalist, taxonomist, and author of the Life Zone concept

Theodore Roosevelt (1858–1919). Grandfather of wildlife management. As a naturalist, explorer, author, and student of big game, he created a passion for wildlife. As President of the United States, he was able to bring conservation to the American people from a position of authority (Lewis 1919) and created the U.S. Forest Service wildlife reserves and was a founder of the Boone and Crocket Club.

Gifford Pinchot (1865–1946). Worked with Theodore Roosevelt in the conservation movement and presented a utilitarian definition of conservation. Conservation was 1) the development of natural resources to their fullest uses for present and 2) future generations and implied 3) the prevention of waste. Roosevelt provided the authority for many of Pinchot's ideas (McGeary 1960).

Stephen T. Mather (1867–1946). First Director of the National Park Service.

Albert Schweitzer (1875–1965). Musician, theologian, medical doctor, and author of *Reverence for Life* (Schweitzer 1966). His attitudes about killing have been used as a rallying cry by antihunters.

Jay Norwood "Ding" Darling (1876–1962). Cartoonist, naturalist, artist. He conceived the Cooperative Wildlife Research Unit Program, was Father of the National Wildlife Federation, and was a leading conservationist instrumental in wildlife restoration efforts (See *Journal of Wildlife Management* 1963:489–502).

Aldo Leopold (1886–1948). The Father of modern wildlife management. He wrote *Game Management,* which established the basic principles of wildlife management in North America. Earlier writings of Leopold were published after his death, giving Americans a land ethic model to follow (Leopold 1949).

Since the work of Leopold, the field of wildlife management has flourished with scientists involved with the scientific study of wildlife and the relationships to their habitat, and the American consciousness has been aroused so that wildlife management is ingrained in our daily lives. As these early pioneers interacted with each other (Table 3–4) they set the stage for a philosophy of sustained use of wildlife resources that would allow wildlife resources to always be a part of American culture.

SUMMARY

The first conservation movement in the United States began in the early 1900s when Americans realized some changes had to be made to preserve their natural heritage. Subsequent conservation movements occurred in the 1960s and the end of the twentieth century in response to habitat alteration, controversy over predators, competing uses for landscapes, and pollution.

The evolution of the science of wildlife management can be traced through laws that follow a pattern of hunting restrictions, predator control, land reservation, translocations, environmental control, and habitat control and management.

The philosophy of contemporary wildlife management was, in part, developed by pioneers like J.J. Audubon, G.P. Marsh, J.W. Powell, J. Burrough, J. Muir, G. Grinnell, B.E. Fernow, and C.H. Merriam. T. Roosevelt and Aldo Leopold forged the philosophy into policy and active management. Who will be the leaders into the twenty-first century?

4

The Professional Wildlife Biologist

"I have read many definitions of what is a conservationist, and written not a few myself, but I suspect that the best one is written, not with a pen, but with an axe. It is a matter of what a man thinks about while chopping, or while deciding what to chop. A conservationist is one who is humbly aware that with each stroke he is writing his signature on the face of his land. Signatures of course differ, whether written with axe or pen, and this is as it should be."

A. Leopold (1949)

INTRODUCTION

A professional wildlife biologist has a complex role in contemporary society. This chapter outlines the general role of wildlife biologists and the educational requirements for entry level positions, and presents an array of job opportunities. Because The Wildlife Society is a professional society for wildlife biologists, I also present their code of ethics and standards for professional conduct.

■ THE PROFESSIONAL WILDLIFE BIOLOGIST

The complexity of issues involving wildlife and the habitats they use boggle down the meaning of wildlife today so it is often difficult to know what a "wildlife biologist" is. In defining the wildlife biologist of the new century, several concepts need to be kept in mind. First of all, wildlife science is one of the newest sciences. Unlike astronomy, physics, or chemistry, which have a long history hundreds of years old, wildlife science is relatively new and was only established in 1933 with the efforts of Aldo

Leopold (1933). The field continues to grow and diversify and emphasis often changes. Humans have managed wildlife for hundreds of years (Wisterskov 1957); however, in North America where the lack of any management resulted in the demise of the nation's wildlife resources, serious concerns for America's wildlife developed in the late 1900s. Citizens' organizations called for a halt to abuse and reduction of wildlife, which led to federal, state, provincial, and private organizations to conserve and manage wildlife. By the 1920s it was evident that proper skills and knowledge were necessary to properly manage wildlife (Everden, no date).

Second, wildlife had to be managed. North America had just witnessed the results of unregulated use and recognized that without management wildlife would not be able to coexist with humans. Third, there are four basic objectives that wildlife scientists are primarily engaged in: increasing populations, decreasing populations, maintaining populations, or simply keeping an eye on populations (Caughley and Sinclair 1994). Fourth, although in its early stages wildlife management cut its teeth on game animals, the field has developed and broadened to include all wildlife and the habitats it uses, whether it is harvested or not. This broadened view has led to federal, state, and provincial laws that protect game, nongame, and endangered species and their habitats (Jensen and Krausman 1993).

Finally, one needs to keep in mind that in the United States, wildlife belongs to the public. Each of these concepts places a different type of responsibility on wildlife biologists.

One way to understand the professional wildlife biologist is to examine the education he or she receives. Educational requirements vary for biologists at different universities, but there are core classes in which all biologists should be well versed.

Required Education

Because of the broad aspects of wildlife biology and the triad that makes up wildlife management (i.e., animal, habitat, and people) (Giles 1978), disciplines need to be incorporated into a basic education. In general, the field of wildlife management borrows from many other sciences to form the core academics (Figure 4–1), and professional wildlife biologists have developed from many disciplines. The diverse array of disciplines has been needed to combine into a team effort to solve management problems. In the development stages of the profession, many that entered had strong roots to the natural world and had grown up on ranches, farms, and forest settings, and had experienced a life of sustained use of our natural capital. As urbanization increased throughout the 1900s and into this century, more and more individuals did not develop interest in wildlife issues until their late teens or even later in life. Regardless of when one decides on a professional career in wildlife, certain classes have to be included to incorporate the field. In other words, wildlife biologists do not only work with animals and their habitats; they have to work with people.

There are over 100 institutions in North America that offer curriculum specializing in wildlife biology and most have a science base (Appendix 4A). Some universities may emphasize one wildlife specialty over others because of location, expertise of the faculty, emphasis of the institution (i.e., research or teaching), or available

Figure 4–1 Wildlife management borrows from numerous disciplines to formulate the curriculum for the profession.

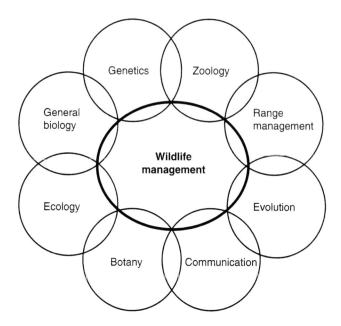

equipment. However, some basic level of education should be required for all wildlife biologists. The Wildlife Society, the professional society for wildlife biologists, provides guidelines which many of the institutions that confer degrees follow. The Wildlife Society recommends that the course of study for wildlife biologists should be from a college or a university leading to a Bachelor of Science (BS) or Bachelor of Arts (BA), or equivalent or higher degree and should include:

1. 36 (semester) hours in biological sciences that include:
   a. 6 hours of courses emphasizing the principles and practices of wildlife management
   b. 6 hours of courses in the biology and behavior of birds, mammals, reptiles, or amphibians to include at least one course concerning birds or mammals
   c. 3 hours of courses in general plant or animal ecology
   d. 9 hours of courses in general zoology, genetics, physiology, anatomy, invertebrate zoology, or taxonomy
   e. 9 hours of courses in general botany, plant genetics, plant morphology, plant physiology, or plant taxonomy
   f. 3 additional hours in a-e above
2. 9 hours in physical sciences (e.g., chemistry, physics, geology, agronomy) with at least two disciplines represented
3. 9 hours in quantitative research that includes:
   a. 3 hours in basic statistics
   b. 6 hours in calculus, biometry, advanced algebra, systems analysis, mathematical modeling, sampling, computer science, or other quantitative science
4. 9 hours in humanities and social sciences (e.g., economics, sociology, psychology, political science, government, history, literature, foreign language)

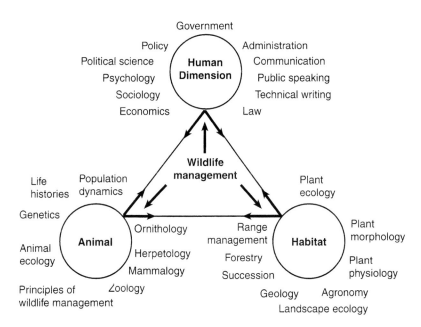

Figure 4–2 Some of the disciplines necessary to understand the arena of wildlife management. Notice the interaction between each point of the triangle.

5. 12 hours in classes designed to improve communication skills (e.g., English composition, technical writing, journalism, public speaking, uses of mass media)
6. 6 hours in policy, administration, and law (e.g., resource policy, resource administration, environmental or wildlife law, or natural resource or land use planning)

The B.A. or B.S. degree that includes these courses is the basic education required. However, to be certified as a professional wildlife biologist (by The Wildlife Society), one would also need a minimum of five years of professional experience gained within the 10 years prior to applying for certification. Because the field is so popular, everyone who pursues a career does not become a wildlife biologist. The most recent review of wildlife graduate placement indicates that 34% of those with bachelor's degrees obtain wildlife related employment, compared to 66% and 87% with master's and Ph.D. degrees, respectively (Hodgdon 1988). Although many jobs in the profession require only the bachelor's degree, the master's degree appears to be required to be truly competitive.

Applying the relevant course work required to be a professional to Giles' (1978) triad results in a more complex triad (Figure 4–2). Each one of the disciplines listed (Figure 4–2) is a specific field. One of the aspects of a wildlife biologist is working in the interactive environment where all of the disciplines come together for a common cause. Thus, a working definition of a professional wildlife biologist would be one with demonstrated expertise in the art and science of applying the principles of ecology to the sound stewardship and management of the wildlife resources and their environment (The Wildlife Society 1999). One can obtain a better understanding of a wildlife professional by examining the Code of Ethics and Standards for Professional Conduct prescribed by The Wildlife Society.

Code of Ethics:

"Associate and Certified Wildlife Biologists have a responsibility for contributing to an understanding of mankind's proper relationship with natural resources, and in particular for determining the role of wildlife in satisfying human needs. Certified individuals will strive to meet this obligation through the following professional goals: They will subscribe to the highest standards of integrity and conduct. They will recognize research and scientific management of wildlife and their environments as primary goals. They will disseminate information to promote understanding of, and appreciation for, values of wildlife and their habitats. They will strive to increase knowledge and skills to advance the practice of wildlife management. They will promote competence in the field of wildlife management by supporting high standards of education, employment, and performance. They will encourage the use of sound biological information in management decisions. They will support fair and uniform standards of employment and treatment of those professionally engaged in the practice of wildlife management."

Standards for Professional Conduct:

"The following tenets express the intent of the Code of Ethics as prescribed by The Wildlife Society and traditional norms for professional service. **Wildlife biologists shall at all times:**

1. Recognize and inform prospective clients or employers of their *prime* responsibility to the public interest, conservation of the wildlife resources, and the environment. They shall act with the authority of professional judgment, and avoid actions or omissions that may compromise these broad responsibilities. They shall respect the competence, judgement, and authority of the professional community.

2. Avoid performing professional services for any client or employer when such service is judged to be contrary to the Code of Ethics or Standards for Professional Conduct detrimental to the well-being of the wildlife resource and its environment.

3. Provide maximum possible effort in the best interest of each client/employer accepted, regardless of the degree of remuneration. They shall be mindful of their responsibility to society, and seek to meet the needs of the disadvantaged for advice in wildlife related matters. They should studiously avoid discrimination in any form, or the abuse of professional authority for personal satisfaction.

4. Accept employment to perform professional services only in areas of their own competence, and consistent with the Code of Ethics and Standards for Professional Conduct described herein. They shall seek to refer clients or employers to other natural resource professionals when the expertise of such professionals shall best serve the interests of the public, wildlife, and the client/employer. They shall cooperate fully with other professionals in the best interest of the wildlife resource.

5. Maintain a confidential professional-client/employer relationship except when specifically authorized by the client/employer or required by due process of law or this Code of Ethics and Standards to disclose pertinent information. They shall not use such confidence to their personal advantage or to the advantage of other parties, nor shall they permit personal interests or other client/employer relationships to interfere with their professional judgment.

6. Refrain from advertising in a self-laudatory manner, beyond statements intended to inform prospective clients/employers of qualifications, or in a manner detrimental to fellow professionals and the wildlife resource.

7. Refuse compensation or rewards of any kind intended to influence their professional judgment or advice. They shall not permit a person who recommends or employs them, directly or indirectly, to regulate their professional judgement. They shall not accept compensation for the same professional services from any source other than the client/employer without the prior consent of all the clients or employers involved. Similarly, they shall not offer a reward of any kind or promise of service in order to secure a recommendation, a client, or preferential treatment from public officials.

8. Uphold the dignity and integrity of the wildlife profession. They shall endeavor to avoid even the suspicion of dishonesty, fraud, deceit, misrepresentation, or unprofessional demeanor."

As can be seen from the required courses, code of ethics, and standards for professional conduct, a wildlife professional is complex. Those entering the field of wildlife management "...in addition to being intelligent, well educated, professionally skilled, diligent, and patient, must like people, be communicative, be willing to work with a diversity of human personalities, and be dedicated to working toward a secure future for wildlife resources" (Eveden, no date).

EMPLOYMENT OPPORTUNITIES

Because the jobs conducted by wildlife biologists are so diverse, so are their employment opportunities. A good part of the education previously emphasized concentrated on classroom education. However, a lot of practical outdoor skills are often required to enhance one's opportunity for employment. As an example, examine these questions that were part of an oral interview to prospective wildlife managers in Arizona in the 1980s:

1. What is the purpose of the uniform?
2. What procedure would you use to get a truck out of a sandtrap or mudhole?
3. What equipment would you take on a foot survey for deer in desert habitat?
4. During a patrol, you find that the deputy sheriff has illegally killed a deer. What would you do?
5. What equipment would you take on a boat patrol?
6. Can you swim?
7. Describe your most recent hunt.
8. What job, of these, would you prefer: boat officer, investigator, district manager.
9. During a white-tailed deer season, you find a hunter that shot a mule deer spike on sympatric range. What would you do?
10. What type of equipment would you carry in your department vehicle when traveling in remote areas?
11. You are operating an outboard motor and it develops a shimmy. What would you do?

Most of these questions are related to outdoor skills and safety issues. The review of applications revealed the necessary academics. During the interview, the department wanted to determine the outdoor skills and outdoor sense of the applicants. Consider one more set of questions for a wildlife technician that was advertised in December 1999.

"The position for which you are interviewing is a Wildlife Technician. This is a permanent, full-time position. The salary starts at $22,151 annually. The position includes insurance benefits and vacation and sick leave.

The successful candidate for this position will work in all aspects of a research project. The candidate will trap and mark wildlife species; maintain complex data bases both on field forms and the computer, be responsible for cleaning data and preliminary analysis of data from projects, and writes (or assists) reports based on field data collection. The successful candidate will work independently under general supervision of the project biologist, and will communicate progress and/or suggest modifications needed to improve data collection and analysis. The candidate will also support the project through conducting literature searches and reviews, maintaining inventory of project equipment, and keeping equipment in proper condition, and scheduling volunteers. Special training in the use of chemical restraint and aerial radio telemetry will be required.

The Technician is primarily a field-oriented position, requiring extensive travel and work on weekends. During some periods of fieldwork, a 10 on 4 off schedule may be worked, other times 8 on, 6 days off. Either way, the position will require long absences from home. A medical/physical evaluation is required prior to appointment. Wildlife Technician Interview:

1. List the experience you have with the following software and computer functions:
    1) Microsoft Office including Word, Excel, Access, and Presentations,
    2) Statistical packages such as SPSS, Statistica, or any other statistics software, and with retrieving information off of the internet.

2. While stationed at a remote field camp, you find the technicians working for other specialists (or even other branches) are disruptive (lack of cleanliness, staying up late, playing loud music). If your immediate supervisor were on vacation for the next 3 weeks, explain what steps you would take to help alleviate problems.

3. Tell us your experience with aerial telemetry, be sure to include what type of antennae system used, and possible experience with mapping coordinates, GPS and data loggers. While flying a deer telemetry flight, you find that a buck is not in the area where you typically find him, go through the procedures you would do (including instructions to the pilot) to help find the buck.

4. In developing a manuscript for publication, what is the information that should be included in 1) the Introduction, 2) Methods and Materials, 3) Study Area, 4) Results, and 5) Discussion. Be sure to include key parts of any paper including literature review, tables, figures, comparisons with other similar studies, objectives etc.

5. Please give us your background and training in capturing and handling both large and small animals. (Be sure to include types of organisms, types of capture [pitfalls, leg-hold snares, helicopter, etc.], and experience with using drugs.

6. You have a large bear (> 150 kg) captured in a snare, that you successfully shot with a dart 30 minutes ago with the recommended amount of Telazol (for an animal of that size). Although he is sluggish, he shows no signs of going down. What would be your next series of steps?

7. Give us your background with horses and mules. Be sure and include possible experiences with packing. Please describe the following equipment used with horses and mules: 1) difference between a Decker and Sawbuck saddle, 2) lash rope, 3) halter, 4) latigo, 5) cantle and swells, and 6) a "britchen".

8. You are entering a known female's occupied bear den with a jab stick in one hand, a loaded pistol in a holster, and a flashlight in the other. Your goal is to determine reproduction and replace the collar. Upon entrance you notice that both the female and a yearling occupy the den. However, the female is wide awake, acting very agitated, consistently bluff charging, and acting more nervous than any other bear whose den you have entered in the past. What would you do?

9. Please give us your experience with ground radio telemetry. You are radio-tracking a desert bighorn and the signal leads you to believe she is in a steep, narrow canyon. Upon entering the canyon, you have a hard time determining which side of the canyon to glass. Give 2 or 3 things you could do to help you determine which side of the canyon the ewe is on. Give your feelings on visual versus triangulation in this kind of situation. If this was a bear on a day bed instead of a bighorn, how would you go about triangulating the location?

10. Give the general use of the following statistical tests, with the assumptions that go with each: t-test, ANOVA, Mann-Whitney U test, Spearman rank correlation Kruskal-Wallis test, and linear regression.

11. Identify the following group of organisms by their genus:

		Odocoileus spp.
Canis spp.	*Dipodomys* spp.	*Zenaida* spp.
Arctostaphylos spp.	*Crotalus* spp.	*Cercocarpus* spp.
		Quercus spp.

12. Do you have a valid driver's license? If offered the position, when would you be able to start?

13. Given the duties described above, would you be able to perform the essential functions of this job with or without reasonable accommodations?"

The 20 years between the development of these two sets of questions reflect how technology and standard data analysis have become everyday tools, but they also reflect the importance of common field savvy in being competitive as a wildlife biologist or technician. However, there are numerous positions in the field: wildlife manager, researcher, extension specialist, educator, professor, law enforcement officer, statistician, computer analyst, political liaison, land use planner, naturalist, land acquisition, public relations, administration, consultant, and others. These are each highly specialized, but a broad and diverse background and rigorous education will prepare the wildlifer to be successful. Below are some of the common types of jobs where wildlife biologists are required.

Wildlife Managers

This is the backbone position for most organizations and agencies working with wildlife. Wildlife managers are often called "jacks of all trades" because of the diversity of their jobs. The manager has an array of jobs from law enforcement to making hunt recommendations and surveying populations. The wildlife manager is usually the most recognized employee in an organization because he or she usually wears some type of uniform, drives an agency vehicle, works among the public, and in general, serves as an ambassador for the agency. They not only work with animals, but also with animals' habitats. Thus, their duties are broadened so they interact with federal agencies that often manage habitats and private landowners. In many organizations the wildlife manager is the entry-level position that one could develop over a lifetime or depart from to concentrate more in a specific area instead of being spread out over many aspects of the wildlife management triad.

Wildlife Law Enforcement

Some organizations combine wildlife law enforcement with wildlife management; others have personnel that specialize in each. Enforcement officers require additional law enforcement training after college, as their jobs involve the enforcement of laws and regulations that enhance wildlife populations. As with the wildlife manager, the law enforcement officer is highly visible to the public.

Wildlife Research

Wildlife research obtains facts through the scientific methods that are useful in the management of wildlife. Wildlife researchers should have a master's degree or Ph.D. with the educational emphasis on how research should be conducted. There is no limit to the areas researchers work in but the areas are usually related to animal ecology, physiology, population dynamics, disease, predation, habitat, and anthropocentric influences on animals and their habitats. The wildlife research scientist needs to be objective and efficiently obtain information that can assist the wildlife manager. They must also be able to clearly communicate their findings to the scientific community, managers, and the general public. A good wildlife manager keeps abreast of the latest research on contemporary wildlife management and depends on new information gathered, especially with the additional demands placed on wildlife habitats.

Public Relations and Extension

Wildlife belongs to the people and they have a desire to know how it is managed. Those in public relations and extension present research and management efforts to the public to promote understanding so the public knows why certain practices are being, or not being, conducted. An important part of this job is working effectively with people, so communication skills are extremely important. One may be cooperating with the press, radio, or television on one issue, but then need to write an agency report or provide photographs on training and education on another issue.

Extension specialists are usually associated with universities where technical expertise is supplied to individuals or organizations to solve problems. They also explain the results of research to those that need up-to-date information, and on many occasions extension specialists conduct their own research to obtain the information. Those in extension and public relations should have excellent communication skills and truly enjoy working with people.

Wildlife Education

The traditional positions for the education of professional biologists have been in colleges and universities. However, since the 1970s there has been a growing interest in teaching the basics of wildlife ecology at elementary and secondary schools, and international training. Education for elementary and secondary schoolteachers is shared among universities, game and fish agencies, and other organizations. Universities also provide education for international wildlife management, but the U.S. Fish and Wildlife Service and nongovernmental organizations also provide educational opportunities.

Colleges and universities that offer a wildlife curriculum hire wildlife professionals for their teaching and research. Professors usually teach, conduct research with students, provide public service, and serve professional organizations. University professors in wildlife programs have diverse backgrounds, but should be well versed in ecology, management, and administration of wildlife resources. A Ph.D. degree is essential.

Wildlife Administration

Solid leadership is essential also for effective wildlife management. Wildlife administrators should have an understanding of the various positions that meet the organization's function. As a result administrators often come from the professional ranks and are able to work effectively with people.

Administration is a specialized job because it requires technical knowledge, business management, communication, and personnel evaluation. A good administrator is one who has the best interests of wildlife in mind and succeeds by making those working with him or her successful.

Private Organizations

Because of our laws that specify how wildlife and their habitats can be used (e.g., National Environmental Policy Act), wildlife employment in the private sector has increased. Opportunities exist with private companies (e.g, ranching, mining, and timber), professional organizations (e.g., Rocky Mountain Elk Foundation), game ranching, sportsman clubs, and other organizations. Artists, writers, and photographers specialize in wildlife also. Each year new organizations emerge that hire wildlife professionals in the private sector for the management of wildlife (e.g., The Rob and Bessie Welder Wildlife Foundation, the Turner Endangered Species Fund).

Consultants

When the National Environmental Policy Act dictated that federal, state, and private agencies and groups (e.g., developers) proposing major actions that would have an impact on the environment or public lands, they had to report the likely effects of the proposed action on the environment. Wildlife biologists were often those that did the "reporting" and this opened a new field that employs trained professionals. Many consultants have other wildlife jobs but serve as part-time consultants (for a host of wildlife issues) on a job-by-job basis. Others have successfully created their own consulting firms that hire full-time employees.

Table 4–1   Employment tracking on regular members of The Wildlife Society for 1999. (Compiled by H. Hodgdon, Executive Director, The Wildlife Society.)

Employer	Regular Members
State/province	1806
Federal	1690
University	874
Consulting firm	533
Corporation	284
Nonprofit organization	385
Other	366
No data	549
Total	6487

Public Agencies

Most wildlife professionals work with state, provincial, or federal agencies (Table 4–1) and involve wildlife management, research, education, administration, extension, conservation, and law enforcement. Most public agencies require their wildlife professionals to have a bachelor's degree. State or provincial agencies include departments of fish and game, conservation, natural resources, wildlife, forestry, parks and recreation, and other related tittles. U.S. agencies include the U.S. Fish and Wildlife Service, U.S.D.A. Forest Service, the National Park Service, Bureau of Land Management, U. S. Bureau of Land Reclamation, Bureau of Indian Affairs, Department of Defense, the Environmental Protection Agency, Natural Resources Conservation Service, and the U.S. Geologic Survey

Regardless of the job that wildlife biologists are involved in, communication skills are important for success. To that end I have included the guidelines for preparing scientific articles to be published by The Wildlife Society (Appendix 4B). These guidelines are only one way to present information and can certainly be modified. However, they do not present the basic information that should be included in scientific writing.

SUMMARY

The entry level for a wildlife biologist is a B.S. or B.A. degree that includes specific courses. There are numerous academic institutions that educate wildlife biologists. During a biologist's career, which could be in an array of positions (i.e., wildlife manager, law enforcement, research, public relations and extension, education, administration, private organization, state or federal agencies, consultant), his or her activities should be guided by a code of ethics. Regardless of the specific career track, wildlife biologists should excel in communication.

APPENDIX 4A

Universities and colleges offering curricula in wildlife conservation in the United States and Canada, 1996 (From the Wildlife Society 1996)

This roster includes those North American campuses that have indicated they have special curricula related to the fields of wildlife conservation and management. The Wildlife Society believes strongly that graduates from wildlife curricula should meet certain minimum educational requirements. A recommended curriculum for a bachelor's degree in wildlife was first adopted by the Society in 1965 and revised and strengthened in 1971 and 1977. The 1977 recommended minimum educational requirements subsequently became the initial minimum standard of education for The Wildlife Society's Certified Wildlife Biologist Program. These standards are revised periodically. The current educational requirements for certification appear at the end of this roster, and prospective students in wildlife are encouraged to incorporate them into their higher education plans.

Colleges and universities use a variety of approaches in educating students in wildlife conservation and management; some offer named degrees in wildlife, others have a wildlife major option, and still others have neither named degrees nor a major option in wildlife, but offer wildlife courses. To assist you in identifying campuses of potential interest, the list is organized by state and province and each entry is followed by a code to provide you with a general idea about the institution and its wildlife program. An interpretive "key" to the coding system is provided below. Additional information may be obtained by writing directly to the institution at the address provided. Each institution provided the information on its wildlife program.

State or Province	University	Address	Degree Code
Alabama	Auburn University	Department of Zoology And Wildlife Science, Auburn, Alabama 36949-5414	(B1,MD2j,4)S
Alaska	University of Alaska-Fairbanks	Department of Biology and Wildlife,Fairbanks, Alaska 99775-6100	(BMD1,4)S
Arizona	University of Arizona	School of Renewable Natural Resources, Tucson, Arizona 85721	(BMD1,4)S
	Arizona State University	Wildlife Professor, Department of Zoology, Tempe, Arizona 85287-1501	(B1,MD3j,5)S
	Northern Arizona University	School of Forestry, Flagstaff, Arizona 86011-5018	(MD2g,5)S
Arkansas	University of Arkansas	Wildlife Professor, School of Forest Resources, Monticello, Arkansas 71655	(B1,5)S

State or Province	University	Address	Degree Code
	Arkansas State University	Head, Department of Biological Science, State University, Arkansas 72467	(B1,4,M2c)
	Arkansas Tech	Head, Department of Fish and Wildlife Biology, Russellville, Arkansas 72801	(B1,5)S
California	University of California	Wildlife Professor, Department of Environmental Science, Policy, and Management Berkeley, California 74720-3114	(BMD2k,5)
	University of California-Davis	Chairman, Department of Wildlife, Fish, and Conservation Biology, Davis, California 95616	(B1, MD2d,5)S
	California Polytechnic State University	Wildlife Professor, Biological Biological Sciences Department;	(BM2cdfj,5)
		Wildlife Professor, Natural Resource Management Department, San Luis Obispo, California 93407	(BM2gk5)
	California State University	Wildlife Professor, Department of Biological Science, Sacramento, California 95819-6077	(BM2c,5)
	Humboldt State University	Chairman, Department of Wildlife, Arcata, California 95521	(B1,M2k,5)S
	San Diego State University	Chairman, Department of Biology, San Diego, California 92182	(BM3cdfj,D3d)
	San Jose State	Chairman, Wildlife, Zoology, Curriculum,Biological Science Department, San Jose, California 95192.	(B2j,M2c,5)
Colorado	Colorado State University	Head, Department of Fish and Wildlife Biology, Fort Collins, Colorado 80523	(BMD1,5)S
Connecticut	University of Connecticut	Wildlife Professor, Natural Resources Management and Engineering, Storrs, Connecticut 06269	(B3,M2k,5)S
	Yale University	Wildlife Professor, School of Forestry and Environmental Studies, New Haven, Connecticut 06511	(MD2dgk)

State or Province	University	Address	Degree Code
Florida	University of Florida	Chairman, Department of Wildlife Ecology and Conservation, Gainesville, Florida 32611-0430	(BMD1,5)S
Georgia	University of Georgia	Wildlife Professor, School of Forest Resources, Athens, Georgia 30602	(BMD1,4)S
Idaho	University of Idaho	Head, Department of Fisheries and Wildlife Resources, Moscow, Idaho 83843	(BMD1,5)S
	Idaho State University	Chairman, Department of Biological Sciences, Pocatello, Idaho 83209	(BMD2,5)
Illinois	University of Illinois	Director, Center for Wildlife Ecology, Illinois Natural History Survey, Champaign, Illinois 61801	(BMD3bcdg,5)
	Southern Illinois University	Wildlife Professor, Department of Zoology, Carbondale, Illinois 62901	(BMD2j,5)S
	Western Illinois	Wildlife Professor, Department of Biological Science, Macomb, Illinois 61455	(BM3c,5)S
Indiana	Ball State University	Department of Biology, Muncie, Indiana 47306	(BM2c,5)S
	Purdue University	Wildlife Professor, Department of Forestry and Natural Resources, West Lafayette, Indiana 47907	(BMD1,2g,4)S
Iowa	Iowa State University	Department of Animal Ecology, Ames, Iowa 50011	(BMD1,5)S
Kansas	Kansas State University	Wildlife Professor, Division of Biology, Manhattan, Kansas 66502-4901	(BMD2c,5)S
Kentucky	Eastern Kentucky University	Wildlife Professor, Department of Biological Science, Richmond, Kentucky 40475	(B1M,5)S
	University of Kentucky	Wildlife Professor, Department of Forestry, Lexington, Kentucky 40546-0073	(B3,M2g,5)S

State or Province	University	Address	Degree Code
	Murray State University	Wildlife Professor, Department of Biological Science, Murray, Kentucky 42071-0009	(BMD1,5)S
Louisiana	Louisiana State	Director, School of Forestry, Wildlife and Fisheries, Baton Rouge, Louisiana 70803	(B1,4,MD1,5)S
	Louisiana Tech	Head, Department of Biological Science, Ruston, Louisiana 71270	(B2Mc,5)S
	McNeese State	Wildlife Professor, Agriculture Department, Lake Charles, Louisiana 70609	(B2a,5)
	Northeast Louisiana University	Head, Department of Biology, Monroe, Louisiana 71209	(B3,M2c,5,WP)
	Northwestern State University	Wildlife Professor, Department of Life Science, Section of Wildlife Management, Natchitoches, Louisiana 71497	(B2c,WP,5)
	Southwestern Louisiana University	Wildlife Professor, Department of Biology, Lafayette, Louisiana 70504	(B2f,M3,D3,5)
Maine	University of Maine	Department of Wildlife Ecology, 210 Nutting Hall, Orono, Maine 04469-5755	(B1,4;MD1,5)S
Maryland	Frostburg State University	Wildlife Professor, Biological Department, College Park Campus and AEL, Frostburg, Maryland 21532	(B1,4,M1,5)S (D2,5)
	Maryland University	Wildlife Professor, Agriculture and Extension Education, College Park, Maryland 20742	(B2k,5)
Massachusetts	Framingham State College	Chairman, Department of Biology, Framingham, Massachusetts 10701	(B2c,5)
	University of Massachusetts	Head, Department of Wildlife Management, Amherst, Massachusetts 01002-4210	(BMD1,5)S
Michigan	Central Michigan University	Wildlife Professor, Department of Biology, Mount Pleasant, Michigan 48859	(BM1cdfk,5)

State or Province	University	Address	Degree Code
	Lake Superior State University	School of Science and Natural Resources, Department of Fish and Wildlife Biology, Sault Ste. Marie, Michigan 49783	(B1,5)
	University of Michigan	Chairman, School of Science and Natural Resources and Environmental Resource Ecology and Management Concentration, Ann Arbor, Michigan 48109-1115	(BMD2k)S
	Michigan State University	Chairman, Department of Fisheries and Wildlife, East Lansing, Michigan 48824	(BMD1,5)S
	Michigan Tech University	Dean, School of Forestry, Houghton, Michigan 49931	(BMD2g,5)S
	Northern Michigan University	Wildlife Professor, Department of Biology, Marquette, Michigan 49855	(BM2c,5)S
Minnesota	University of Minnesota	Department of Fish and Wildlife, St. Paul, Minnesota 55108	(BMD1,5)S
Mississippi	Mississippi State University	Head, Department of Wildlife and Fisheries, Mississippi State Mississippi 39762	(B1,MD1,4)S
Missouri	University of Missouri	Wildlife Professor, Fish and Wildlife Program, Columbia, Missouri 65211	(BMD1,5)S
Montana	University of Montana	Wildlife Professor, Wildlife Biology Program, School of Forestry, Missoula, Montana 59812	(BM1,D2gc)S
	Montana State University	Wildlife Professor, Department of Biology, Bozeman, Montana 59717	(B2c,M1,D2c,5)
Nebraska	University of Nebraska at Kearney	Wildlife Professor, Department of Biology, Kearney, Nebraska 68847	(B2c,M2e,4)S
	University of Nebraska	Wildlife Professor, Department of Forestry, Fish and Wildlife, Lincoln, Nebraska 68583-0814	(B2k,M1,D2,g,5)
Nevada	University of Nevada	Wildlife Professor, Department of Environmental and Resource Sciences, Reno, Nevada 89512	(BM2k,D2d,5)

State or Province	University	Address	Degree Code
New Hampshire	University of New Hampshire	Wildlife Professor, Forestry Resource Department, Durham, New Hampshire 03824	(B,M2k,D2k,5)S
New Jersey	Rutgers University	Wildlife Professor, Department of Environmental Resources, New Brunswick, New Jersey 08903	(B2k,MD2k,5)S
New Mexico	New Mexico State University	Head, Department of Fish and Wildlife Science, Las Cruces, New Mexico 88003-0003	(B2a,M1,5)S
New York	Cornell University	Chairman, Natural Resources Ithaca, New York 14853-3001	BMD(2k,5)S
	State University of New York	College of Environmental Science and Forestry, Chairman, Department of Environmental and Forest Biology, Syracuse, New York 13210	(BMD2f,5)S
	State University of New York - Cobleskill	Wildlife Professor, College of Agriculture and Technology, Cobleskill, New York 12043-9986	(B2b,5)
North Carolina	North Carolina State University	Coordinator, Fish and Wildlife Science, Raleigh, North Carolina 27695	(Bm1,D2gj,5)
North Dakota	University of North Dakota	Wildlife Professor, Department of Biology, Grand Forks, North Dakota 58202	(B1,MD2c,5)S
	North Dakota State University	Chairman, Department of Zoology, Fargo, North Dakota 58105	(BMD2j,5)S
Ohio	Ohio State University	Director, School of Natural Resources; Chairman, Department of Zoology, Columbus, Ohio 43210	(B2k,MD2, d,e, g,k,4) (MD2dj,5)S
Oklahoma	Northeastern State University	Head, Department of Biological Science, Talequah, Oklahoma 74464	(BMD1,5)S
	Oklahoma State University	Head, Department of Zoology, Stillwater, Oklahoma 74078	(BMD1,5)S
	Southeastern Oklahoma State University	Chairman, Department of Biological Science, Durant, Oklahoma 74701	(B1,5)

State or Province	University	Address	Degree Code
Oregon	Oregon State University	Head, Department of Fisheries and Wildlife, Corvallis, Oregon 97331-3803	(BMD1,5)S
Pennsylvania	California University of Pennsylvania	Chairman, Department of Biological and Environmental Science, California, Pennsylvania, 15419	(B2c,M3c,5)
	Pennsylvania State University	Director, School of Forest Resources, University Park, Pennsylvania 16801	(BMD1)S
Rhode Island	University of Rhode Island	Chairman, Department of Natural Resource Science, Kingston, Rhode Island 02881	(B2k,M2k,4)S
South Carolina	Clemson University	Chair, Department of Aquaculture, Fish and Wildlife, Clemson, South Carolina 29634	(B1,M1,D2c,5)S
South Dakota	South Dakota State University	Head, Department of Wildlife and Fishery Science, Brookings, South Dakota 57007	(B1,M1,D2c,5)S
Tennessee	Lincoln Memorial University	Chairman, Department of Math and Natural Science, Harrogate, Tennessee 37752	(B1,5)
	University of Tennessee	Head, Department of Forestry, Wildlife and Fisheries, Knoxville, Tennessee 37901	(B1,M1&2D, D2d, 4)S
	University of Tennessee	Dean, School of Agriculture and Home Economics, Martin, Tennessee 38238	(B1,M2c,4)S
	Tennessee Tech University	Wildlife Professor, Department of Biology, Cookeville, Tennessee 38505	(B1,M2c,4)S
Texas	Southwest Texas State University	Wildlife Professor, Department of Biology, San Marcos, Texas 78666	(BM2c,5)S
	Stephen F. Austin State University	Wildlife Professor, College of Forestry, Nacogdoches, Texas 75962	(BM2cg,D2g,5)S
	Sul Ross State University	Wildlife Professor, Division of Range Animal Science, Alpine, Texas 79832	(B2k,M1,5)S

State or Province	University	Address	Degree Code
	Texas A&M University	Head, Wildlife and Fishery Science, College Station, Texas 77843-2258	(BMD1,4)S
	Texas A&M University - Kingsville	Chairman, Department of Animal and Wildlife Science, Kingsville, Texas 78363	(BMD1,4)
	Texas Tech University	Chairman, Department of Range, Wildlife and Fishery Management, Lubbock, Texas 79409	(BMD,4)
Utah	Brigham Young University	Coordinator, Division of Wildlife and Range Resources, Department of Zoology, Provo, Utah 84602	(B2ij,MD1,4)S
	Utah State University	Head, Department of Fisheries and Wildlife, Logan, Utah 84322	(BMD1,5)S
Vermont	University of Vermont	Chairman, Wildlife and Fishery Biology Program, Burlington, Vermont 05405	(B1,M1,D2k,5)S
Virginia	Ferrum College	Wildlife Professor, Life Science Division, Ferrum, Virginia 24088	(B3,5)
	Virginia Polytechnic Institute and State University	Head, Department of Fish and Wildlife Science, Blacksburg, Virginia 24061-0321	(B2g,MD1,4)S
Washington	University of Washington	Wildlife Professor, Division of Ecosystem and Conservation, College of Forest Resources, Seattle, Washington 98195	(BMD2g,5)S
	Washington State University	Chairman, Department of Natural Resource Science, Pullman, Washington 99164-6410	(BMD2k,5)S
West Virginia	University of West Virginia	Wildlife Professor, Division of Forestry, P. O. Box 6125, Morgantown, West Virginia 26506-6125	(BMD1,4)S
Wisconsin	University of Wisconsin	Chairman, Department of Wildlife Ecology, Madison, Wisconsin 53706	(BMD2k,5)
	University of Wisconsin - Stevens Point	Dean, College of Natural Resources, Stevens Point, Wisconsin 54481	(BM14,)S

State or Province	University	Address	Degree Code
Wyoming	University of Wyoming	Head, Department of Zoology and Physiology, Laramie, Wyoming 82071-3166 CANADA	(B1,MD2j,5)S
Alberta	University of Alberta	Head, Department of Zoology Wildlife Professor, Faculty of Agriculture and Forestry Environment and Conservation Science,	(BMD3cj,5); (BMD2agk,5);
		Wildlife Professor, Chair Department of Biological Science, Edmonton T6G 2H1	(BMD3,cdfj,5)
British Columbia	The University of British Columbia	Wildlife Professor, Faculty of Forestry, Vancouver V6T 1Z4	(BMD3jk,5)
Manitoba	University of Manitoba	Director, Natural Resources Institute, Winnipeg R3T 2N2	(M3k,2)
New Brunswick	University of New Brunswick	Wildlife Professor, Departments of Biology And Forest Resources, Fredericton E3B 6C2	(BM2g,4,MD2c, 5)S
Nova Scotia	University of Arcadia	Nova Scotia Center for Wildlife and Conservation Biology, Department of Biology, Wolfville B0P 1X0	(BM3,5)
Ontario	University of Guelph	Chairman, Department of Zoology, Guelph N1G 2W1	(BMD1,5)
	University of Western Ontario	Wildlife Professor, Department of Zoology,Ecology, and Evolution Program, London N6A 5B7	(BD2dj,5)
Quebec	McGill University (MacDonald Campus)	Wildlife Professor, Department of Natural Resources, Ste Anne de Bellevue H9X 3V9	(BMD2a,5)S

Key:

B = Bachelor's degree
M = Master's degree
D = Doctorate
WP = Wildlife Professor within nonwildlife department
S = Student chapter of The Wildlife Society at the institution
1 = Institution grants a named degree in wildlife
2 = Institution grants a wildlife major option with a degree in:
a = agriculture; b = animal science; c = biology; d = ecology; e = education; f = field biology; g = forestry; h = parks; i = range; j = zoology; k = natural resources
3 = Institution grants no degree in wildlife, but has curriculum or courses in wildlife
4 = Institution requires, as part of a wildlife degree or wildlife option, all courses necessary to meet the educational requirements for certification by The Wildlife Society
5 = Institution offers, but does not require, all courses necessary to meet educational requirements for certification by The Wildlife Society

APPENDIX 4B

Writing guidelines for the Journal of Wildlife Management

1 September 1997
John T. Ratti
Department of Fish and Wildlife Resources
University of Idaho
Moscow, ID 83844-1136
208-885-7741; FAX 208-885-9080; E-mail jratti@uidaho.edu

RUE: *JWM Manuscript Guidelines* • *Ratti and Smith*

MANUSCRIPT GUIDELINES FOR THE *JOURNAL OF WILDLIFE MANAGEMENT*

JOHN T. RATTI, [1,2] Department of Fish and Wildlife Resources, University of Idaho, Moscow, ID 83843, USA

LOREN M. SMITH, Department of Range, Wildlife, and Fisheries Management, Mail Stop 2125, Texas Tech University, Lubbock, TX 79409, USA

Abstract: This guide provides information for preparing manuscripts submitted to the *Journal of Wildlife Management (JWM)* for publication consideration. Authors should submit manuscripts in the format and style presented in these guidelines, i.e., your manuscript format should be identical to this example. Proper preparation increases the probability and speed of acceptance.

JOURNAL OF WILDLIFE MANAGEMENT 00(0):000-000

Key words: author, format, guidelines, instructions, manuscript, style, *Journal of Wildlife Management.*

These guidelines update Gill and Healy (1980), Ratti and Ratti (1988), and those on the back cover of some issues of *JWM*. This update was prepared to make the guidelines more available to authors, to include basic format and style changes, and to provide additional examples. Authors should review a recent issue of the *JWM* but should understand there are differences between articles in final printed form and correct format of submitted manuscripts (e.g., key words, text columns, placement of tables and figures, line spacing). Check recent *JWM* issues for instructions that my supersede these guidelines, and for the name and address of the current Editor in Chief. Papers that clearly deviate from *JWM* format and style may be returned for correction before review.

---

[1]Present address: (Use this format to give present address of an author if it differs from the address during the time research was conducted).

[2]E-mail: jratti@uidaho.edu

HIGHLIGHTS OF GUIDELINES CHANGES

For those authors with experience and knowledge of *JWM* Guidelines, it may be helpful to identify and review significant changes in this manuscript. Fundamental changes include (1) most abbreviations have been eliminated from the LITERATURE CITED section; (2) spell out country names at the end of author and publisher addresses, except for United States use "USA;" (3) no use of underlined words to indicate italic type, i.e., use italic fonts where appropriate; and (4) ACKNOWLEDGMENTS are a separate section preceding LITERATURE CITED. Please review this document for additional changes.

POLICY

Referees and editors judge each submitted manuscript on data originality, concepts, interpretations, accuracy, conciseness, clarity, appropriate subject matter, and contribution to existing literature. Prior publication or concurrent submission to other refereed journals precludes review or publication in *JWM* (additional information in section on Transmittal Letter and Submission). The *JWM, Wildlife Society Bulletin,* and *Wildlife Monographs* have similar quality standards. Fisheries manuscripts are discouraged unless information is a part of an account that mainly concerns terrestrial vertebrates.

PAGE CHARGES AND COPYRIGHTS

Current policies regarding page charges offer alternatives and are explained to authors after manuscripts are submitted, and when they are accepted for publication. Page charges may change annually; for members of The Wildlife Society in 1997, they were $65/page for the first 8 pages plus $125 for each succeeding page (for nonmembers the rate was $125/page for all pages). Authors pay for alterations to page proofs (in 1997, $3.25/reset line), except for typesetting errors and editorial errors. If a manuscript not in the public domain is accepted for publication, authors or their employers must transfer copyright to The Wildlife Society. Publications authored by federal-government employees are in the public domain. Manuscript submission implies entrusting copyright (or equivalent trust in public-domain work) to the Editor in Chief until the manuscript is either rejected, withdrawn, or accepted for publication. If accepted, The Wildlife Society retains copyright.

COPY

Use quality white paper, 215 × 280 mm (8.5 × 11 inches) or metric size A4. Do not hyphenate words at the right margin, and do not right-justify text. Manuscripts produced on dot matrix printers are not acceptable.

Margins should be 3 cm (1 3/16 inches) on all sides. Do not violate margin boundaries to begin a new paragraph or the LITERATURE CITED at the top of a new page; i.e., do not leave >3 cm of space at the bottom of a page (except to prevent a widow heading). Type the senior author's last name (upper left) and page numbers (upper right) on pages 2 through the LITERATURE CITED, on tables and figure title pages, but not on the first page, figures, or illustrations. Do not underline words or use bold or italic font in the text to indicate emphasis.

Scientific names should be in italic font. Keep the original manuscript and submit 4 quality copies. Submit a transmittal letter (see below) with your manuscript.

RUNNING HEAD, TITLE, AND AUTHORS

Page 1 of the manuscript should begin with the date (update with each revision), corresponding author's name, address, and telephone, FAX, and E-mail numbers (if available), single-spaced in the upper left corner. Thereafter, all text is double-spaced, including tables.

The running head (RH) is the first line following the correspondent's address. The RH is limited to 45 characters, left-justified, and typed in upper- and lower-case letters followed by a dot (or raised period) and the last name(s) of ≤2 authors. For ≥3 authors, use the name of the first author followed by "et al."

Type the author's name(s) in italic font. The RH is used in final printed form as an abbreviated title at the top of each page following the title page.

The title follows the RH, is also left-justified in bold font, all upper-case letters, should not include abbreviations, acronyms, punctuations, and should not exceed 10 words (unless doing so forces awkward construction). In such cases, use ≤13 words. The title identifies manuscript content. Do not use scientific names in the title except for organisms that do not have, or are easily confused by, common names. Do not use numbers in titles or the RH.

Author's names are left-justified in upper-case letters followed by affiliation and address in upper- and lower-case letters (usually where the author was employed during the study). The second and third lines of the author's address are indented 5 spaces. Use available U.S. Postal Service (USPS) abbreviations (Appendix A), zip codes, and the country abbreviation (e.g., USA), in each address. Write out words like Street, Avenue, and Boulevard but abbreviate directions (e.g., N. and N.W.). For multiple authors with the same address, repeat the address after each author's name.

FOOTNOTES

Footnotes appear at the bottom of the first page to reference present address of an author when it differs from the by-line address, and for E-mail address of the corresponding author. Footnotes also may be used to indicate a deceased author. The footnote appears immediately below a left-justified solid line of 10 characters, and each footnote is indented 5 spaces and starts with a numerical superscript; subsequent lines are left-justified. The footnote origin corresponds to the superscript number following the author's name. Endorsement disclaimers and pesticide warnings should be incorporated in the text. For table footnotes, see the TABLES section.

ABSTRACT

Begin with the word "Abstract" in italic and bold fonts followed by a colon, and left-justified. The Abstract text begins after the colon on the same line, and should be a single paragraph not exceeding 1 line/page of text, including LITERATURE CITED. The Abstract should include:

Problem Studied or Hypothesis Tested—Identify the problem or hypothesis and explain why it was important. Indicate new data, concepts, or interpretations directly or indirectly used to manage wildlife.

Results—Emphasize the most important results, positive or negative, but keep the methods brief unless a new or much-improved method is reported.

Utility of Results—Explain how, when, where, and by whom data or interpretations can be applied to wildlife problems or contribute to knowledge of wildlife science.

On the line following the Abstract, type *"JOURNAL OF WILDLIFE MANAGEMENT 00(0):000-000"* right-justified and in capital letters, bold font, and italics (see page 1 of this manuscript).

KEY WORDS

Key words follow the Abstract. The phrase "Key words" is typed in italic and bold fonts followed by a colon, left-justified, and followed by 10-12 key words in alphabetical order. Include some words from the title and others that identify (1) common and scientific names of principal organisms in the manuscript; (2) geographic area, usually the state, province, or equivalent, or region if its name is well known; (3) phenomena and entities studied (e.g., behavior, populations, radiotelemetry, habitat, nutrition, density estimation, reproduction); (4) methods—only if the manuscript describes a new or improved method; and (5) other words not covered above but useful for indexing. Type a solid line from the left to the right margin beneath the key words; begin the text below this line.

HEADINGS AND MAJOR SECTIONS

Headings

Three levels of headings may be used and examples of each appear in this manuscript. First-level headings are in upper-case letters, are left-justified, and in bold type. Second-level headings also are bold type and left-justified, but only the first letter of each word (except articles, conjunctions, and prepositions) is upper-case. Third-level headings have the first letter of each word upper-case, but are indented 5 spaces, italicized, and followed by a period and 3 hyphens. Although short papers (≤4 pages) may not require any headings, most require at least first-level headings. Under a first-level heading, use only third-level headings if all subsections are short (≤2 paragraphs; e.g., see Abstract section of this manuscript). Avoid repeating exact wording of the heading with second- and third-level headings, do not leave first- or second-level headings standing alone on the last line of a page (i.e., as a "widow line"), and avoid 1-sentence paragraphs.

Major Sections

The introduction (no heading) starts below the line under key words and is a concise synthesis of literature specific to the manuscript's main topic. The latter part of this section states objectives or hypotheses tested.

Most *JWM* manuscripts have 8 major sections: introduction, STUDY AREA, METHODS, RESULTS, DISCUSSION, MANAGMENT IMPLICATIONS, ACKNOWLEDGMENTS, and LITERATURE CITED. It is permissible to combine STUDY AREA and METHODS, but do not combine RESULTS and DISCUSSION. Merging these sections so that results can be interpreted when first presented leads to superfluous wording, unnecessary discussion, and confusion.

Most study-area descriptions should be presented in past tense; e.g., "average annual precipitation was 46 cm," "habitat was primarily grass." Exceptions include geological formations that have been present for centuries. Methods should be brief and include dates, sampling schemes, duration, research or experimental design, and data analysis. Previously published methods should be cited without explanation. New or modified methods should be identified as such and explained in detail. Many research projects require animal-welfare protocols, and these should be cited here. If an approval number for the protocol was necessary, list it parenthetically following the statement,

Present results in a clear, simple, concise, and organized fashion. Avoid overlapping text with information in tables and figures; do not explain analyses that should be presented in the METHODS section. Results should be presented in past tense (e.g., body-mass loss occurred during winter). Reserve interpretation comments for the DISCUSSION section.

The discussion provides an opportunity for interpreting data and making literature comparisons. Reasonable speculation and new hypotheses to be tested may be included in the DISCUSSION. Do not repeat results and comment only on the most important findings. Systematic discussion of every aspect of the research leads to unnecessarily long manuscripts.

The MANAGEMENT IMPLICATIONS section should be short and direct, but explain issues important to conservation. This section may include speculation, but should address specific management opportunities or problems.

STYLE AND USAGE

Manuscripts with publishable data may be rejected because of poor writing style (e.g., long and complex sentences, superfluous words [Table 1], unnecessary information, and poor organization). Most editors are patient with this problem and are willing to offer helpful suggestions. However, referees are less tolerant of poor writing, and this problem may lead to negative reviews. Many of these problems can be corrected by having your manuscript critically reviewed by colleagues before submission for publication. Authors are urged to review Chapters 3 and 4 in the "CBE Style Manual" (CBE Style Manual Committee 1994) and "Writing with Precision, Clarity, and Economy" by Mack (1986). Manuscripts should be direct and concise. Many common problems may be avoided by use of a carefully prepared outline to guide manuscript writing. Other helpful suggestions are presented by Strunk and White (1979), Day (1983), and Batzli (1986). Use first person and active voice whenever appropriate to avoid superfluous wording. Review the list of commonly misused words (Table 2) before preparing your manuscript (e.g., use the word "mass" rather than "weight" to conform to international standards).

Numbers and Unit Names.—Use digits for numbers (e.g., 7 and 45) unless the number is the first word of a sentence, where it is spelled out. Use symbols or abbreviations (e.g., % and kg) for measurement units that follow a number unless the number is indefinite (thousands of hectares), is a "0" (zero) standing alone, or is the first word in a sentence. In such cases spell out the number and unit name or recast the sentence. Avoid using introductory phrases such as "A total of. . . ." Spell out numbers used as pronouns (i.e., one) or adverbs and ordinal numbers (e.g., first and second). However, use digits for cases such as 3-fold and 2-way. Convert fractions (1/4, 1/3, etc.) to decimals except where they misrepresent precision.

Hyphenate number-unit phrases used as adjectives (e.g., 3-m^2 plots and 3-year-old males), but not those used as predicate adjectives (e.g., plots were 3 m^2). Insert commas in numbers $\geq 1{,}000$ (except for pages in books, clock time, or year dates). Do not insert a comma or hyphen between consecutive, separate numbers in a phrase (28 3-m^2 plots). Do not use naked decimals; i.e., use 0.05, not .05.

Time and Dates.—Use the 24-hr system: 0001 through 2400 hr (midnight). Date sequence is day month year, without punctuation. Do not use an apostrophe for plural dates (e.g., 1970s). Spell out months except in parentheses, tables, and figures, in which 3-letter abbreviations are used with no period (e.g., 31 Mar 1947, Appendix B).

Mathematics and Statistics.—Use italic font for Roman letters used as symbols for quantities (e.g., n, $\bar{x}$, F, t, Z, P, and X). Do not underline or italicize numbers, Greek letters, names of trigonometric and transcendental functions, or certain statistical terms (e.g., in, e, exp, max, min, lim, SD, SE, CV, and df). Use bold font for items that should be set in boldface type.

Insert a space on both sides of symbols used as conjunctions (e.g., $P > 0.05$), but close the space when used as adjectives (e.g., >20 observations). Where possible, report exact probabilities ($P = 0.057$, not $P > 0.05$). A subscript precedes a superscript (X_z^3) unless the subscript includes >3 characters. Break long equations for column-width printing (67 mm) if they appear in the main body of the manuscript; long equations and matrices can be printed page-width (138 mm) in appendices. Swanson (1974) or the CBE Style Manual Committee (1994: 206–218) should be followed for general guidance, and Macinnes (1978) for advice on presentation of statistics. Authors are urged to read Tacha et al. (1982) and Wang (1986) for reviews of common statistical errors. Authors should consider statistical power when judging their results (*JWM* 59:196–198).

Abbreviations and Acronyms.—Metric units, their appropriate prefixes, and abbreviations identified by an asterisk in Appendix B may be used in the text. All other abbreviations or acronyms (except DNA) used in the Abstract or text must be defined the first time used; e.g., Bureau of Land Management (BLM). Acronyms established in the Abstract should not be reestablished in the text. Do not start sentences with acronyms; do not use an apostrophe with plural acronyms (e.g., ANOVAs). All abbreviations in Appendices A and B may be used within parentheses.

Punctuation.—Use a comma after the next-to-last item in a series of ≥ 3 items (e.g., red, black, blue). Do not hyphenate prefixes, suffixes, or combining forms unless necessary to avoid confusion. Common hyphenation errors occur in 3 cases: (1) a phrase containing a participle or an adjective is hyphenated as a compound when it precedes the word modified, and is written without a hyphen when it follows the word modified (e.g., a small-bird study vs. a study of small birds); (2) a modifier containing a number is usually hyphenated (e.g., a 6-year-old mammal); and (3) a 2-word modifier containing an adverb ending in *ly* is not hyphenated (e.g., a carefully preserved specimen).

Closing quotation marks are placed after periods and commas, but may be placed either before or after other punctuation (CBE Style Manual Committee 1994: 177-181). Fences must appear in pairs, but the sequence varies. Use ([]) in ordinary sentences, use {[()]} in mathematical sentences, and use (()) only in special cases such as chemical names. Brackets are used to enclose something not in the original work being quoted (e.g., insertion into a quotation or a translated title [CBE Style Manual Committee 1994:58–59]).

Enumerating Series of Items.—When enumerating series, a colon must precede the numbered items unless preceded by a verb or preposition. Place numbers within parentheses for presentation of a simple series (e.g., Key words section of this manuscript). When enumerating lengthy or complexly punctuated series, place the numbers at the left margin, with periods but no parentheses, and indent run-on lines (see example in tables subsection below).

COMMON AND SCIENTIFIC NAMES

Do not capitalize common names of species except words that are proper names (e.g., Canada goose [*Branta canadensis*], Swainson's hawk [*Buteo swainsoni*], white-tailed deer [*Odocoileus virginianus*]). Scientific names should follow the first mention of a common name, except in the title. If a scientific name is given in the Abstract, do not repeat it in the text or tables. Scientific names following common names should be in italic font in parentheses with the first letter of the genus upper-case and the species name in lower-case letters. Abbreviate genus names with the first letter when they are repeated within a few paragraphs, provided the meaning is clear and cannot be confused with another genus mentioned in the manuscript with the same first letter; e.g., we studied snow geese (*Chen caerulescens*) and Ross's geese (*C. rossii*).

Do not use subspecies names unless essential and omit taxonomic authors names. Use "sp " (not italicized) to indicate unknown species. Use "spp." for multiple species; e.g., the field was bordered by willow (*Salix* spp.). Use the most widely accepted nomenclature where disagreement occurs. Use the most current edition of The American Ornithologists' Union Check-list (e.g., 1997) and periodic supplements published in *Auk* as general references for North American birds. For mammals, use Nowak (1991) or Whitaker (1996). There is no single reference for North America plants; we recommend citing the most widely accepted regional flora reference (e.g., in northwestern states, Hitchcock and Cronquist 1973). Omit scientific names of domesticated animals or cultivated plants unless a plant is endemic or widely escaped from cultivation, or is a variety that is not described adequately by its common name.

MEASUREMENT UNITS

Use Systeme International d'Unites (SI) units and symbols. Use English units (or another type of scientific unit) in parentheses following a converted metric unit only in cases that may misrepresent (1) the statistical precision of the original measurement or (2) the correct interpretation of the results. However, these non-SI units are permitted:

area—hectare (ha) in lieu of $14^4 m^2$;
energy—calorie (cal) in lieu of Joule (J);
temperature—Celsius (C°) in lieu of Kelvin (K);
time—minute (min), hour (hr), day, etc. in lieu of seconds (sec);
volume—liter (L) in lieu of dm^3.

The CBE Style Manual Committee (1994:200–205) provided definitions of SI units and prefixes. The American Society of Testing Materials (1979) included many conversion factors.

CITING LITERATURE IN TEXT

In most cases, reference citations parenthetically at the end of a sentence; e.g., mallard-brood survival was higher in the wettest years (Rotella 1992). Published literature is cited by author and year; e.g., Jones (1980), Jones and White (1981). With ≥3 authors use "et al."; e.g., (Jones et al. 1982). Do not separate the author and date by a comma, but use a comma to separate a series of citations and put these in chronological order; e.g., (Jones 1980, Hanson 1986). If citations in a series have >1 reference for the same author(s) in the same year, designate the years alphabetically (in italics) and separate citations with semicolons;

e.g., (Jones 1980a,b; Hanson 1981; White 1985, 1986). For citations in a series with the same year, use alphabetical order within chronological order; e.g., (Brown 1991, Monda 1991, Rotella 1991, Allen 1995). Do not give more than 6 citations in the text to reference a specific issue or scientific finding. For a quotation or paraphrase, cite author, year, colon, and page number(s); e.g., we used Neyman allocation to minimize variance (Krebs 1989:216). Use the same style for a book or other lengthy publication unless the reference is to the entire publication; e.g., Odum (1971:223).

Cite documents that are cataloged in major libraries, including theses and dissertations, as published literature. These citations include symposia proceedings and U.S. Government reports that have been widely distributed. However, cite such references as unpublished if they are not easily available. Cite unpublished information in the following forms: (J. G. Jones, National Park Service, personal communication), (D. F. Brown, Arizona Game and Fish Department, unpublished data), (D. E. Timm. 1977. Annual Waterfowl Report, Alaska Department of Fish and Game, Juneau, Alaska, USA).

A manuscript accepted for publication is cited as a published manuscript in the text using the anticipated publication year. In the LITERATURE CITED, show the year after the name(s) of the author(s) and "In Press" after the volume number (see below). Do not cite manuscripts that are in review; use the unpublished style.

LITERATURE CITED STYLE

Type the citations double-spaced immediately following the text, not necessarily on a new page. Spell out all words in cited literature, i.e., do not use abbreviations. However, the following 3 exceptions are allowed: Washington D.C.; "U.S.," e.g., U.S. Department of Agriculture, and "USA" in author and publisher addresses. Alphabetize by author's surname(s), regardless of the number of multiple authors for the same publication. Within alphabetical order the sequence is chronological. Use upper- and lower-case letters (typing all capital letters complicates editing names such as DeGraaf and vanDruff). Use 2 initials (where appropriate) with 1 space between each initial. For multiple citations with the same author(s), use a 5-spaced line to replace the author's name(s) after the first citation. For serial publications, show the issue number only if the pages of each issue are numbered separately. As in the text, spell out ordinal numbers (e.g., Third edition). Use the word Thesis to denote Master of Science (M.S.) or Master of Arts (M.A.), and Dissertation for Doctor of Philosophy (Ph.D.). Do not write the total page number of books at the end of citations. Omit unnecessary words, but do not remove a conjunction if the meaning may be changed (e.g., Game and Fish vs. Game Fish). For publishers, do not include words like Company, Incorporated, Limited, or Publishing (e.g., Macmillan, not Macmillan Publishing Company). Please review the following examples.

Book—More than 1 Edition
Smith, R. L. 1974. Ecology and field biology. Second edition. Harper & Row, New York, New York, USA.

Book—More than 1 volume
Palmer, R. S. 1976. Handbook of North American birds. Volume 2. Yale University Press, New Haven, Connecticut, USA.

Book—Editor as Author
Temple, S. A., editor. 1978. Endangered birds: management techniques for preserving threatened species. University of Wisconsin Press, Madison, Wisconsin, USA.

Chapter Within Book
Zeleny, L. 1978. Nesting box programs for bluebirds and other passerines. Pages 55–60 in S. A. Temple, editor. Endangered birds: management techniques for preserving threatened species. University of Wisconsin Press, Madison, Wisconsin, USA.

Theses or Dissertations
Tacha, T. C. 1981. Behavior and taxonomy of sandhill cranes from mid-continental North America. Dissertation, Oklahoma State University, Stillwater, Oklahoma, USA.

Journals—General Format
Miller, M. R. 1986. Molt chronology of northern pintails in California. Journal of Wildlife Management 50:57–64.

Journals in Press—Year and Volume Known
Zelenak, J. R., and J. J. Rotella. 1997. Nest success and productivity of ferruginous hawks in northern Montana. Canadian Journal of Zoology 75:in press.

Journals in Press—Year and Volume Unknown
Giudice, J. H., and J. T. Ratti. In Press. Biodiversity of wetland ecosystems: review of status and knowledge gaps. BioScience.

Symposia and Proceedings—Complete Volume
DeGraaff, R. M., technical coordinator. 1978. Proceedings of workshop on management of southern forests for nongame birds. U.S. Forest Service General Technical Report SE-14.

Symposia and Proceedings—Individual Article
Dickson, J. G. 1978. Forest bird communities of the bottomland hardwoods. Pages 66-73 *in* R. M. DeGraaf, technical coordinator. Proceedings of workshop on management of southern forests for nongame birds. U.S. Forest Service General Technical Report SE-14.

Symposia and Proceedings—Part of a Numbered Series
Palmer, T. K. 1976. Pest bird control in cattle feedlots: the integrated system approach. Proceedings of the Vertebrate Pest Conference 7:17-21.

Multiple Citations of the Same Author(s)
Peek, J. M. 1963. Appraisal of a moose range in southwestern Montana. Journal of Range Management 16:227–231.
_____. 1986. A review of wildlife management. Prentice-Hall, Englewood Cliffs, New Jersey, USA.
_____, and A. L. Lovaas. 1968. Differential distribution of elk by sex and age on the Gallatin winter range, Montana. Journal of Wildlife Management 32:553–557.
_____, _____, and R. A. Rouse. 1967. Population changes within the Gallatin elk herd, 1932-1965. Journal of Wildlife Management 31:304–316.
_____, and R. A. Rouse. 1966. Preliminary report on population changes within the Gallatin elk herd. Wildlife Science 82:1298–1316. (Note: fictitious citation used for example only.)

Government Publication
Lull, H. W. 1968. A forest atlas of the Northeast. U.S. Forest Service, Northeastern Forest Experiment Station, Upper Darby, Pennsylvania, USA.

Government Publication—Part of a Numbered Series
Anderson, D. R. 1975. Population ecology of the mallard: V. Temporal and geographic estimates of survival, recovery, and harvest rates. U.S. Fish and Wildlife Service Resource Publication 125.

Government Publication—Agency as Author
National Research Council. 1977. Nutrient requirements of poultry. Seventh edition. National Academy of Science, Washington, D.C., USA.

Note: Cite in text as National Research Council (1977). For additional examples, see the LITERATURE CITED section of this manuscript.

TABLES AND FIGURES

Submit only essential tables and figures. Often tables overlap with presentation in the text, or the information can be easily printed in the text with less journal space. Do not present the same data in a table and a figure. Number tables and figures independently. In the text limit reference of tabular data to highlights of the most important information. Reference tables and figures parenthetically, and avoid statements such as "The results are shown in Tables 1-4." Prepare line drawings only for data that cannot be presented as clearly in a table. For general guidance follow CBE Style Manual Committee (1994: 677-693).

Tables and figures should be able to stand alone (e.g., self-explanatory). Avoid reference to the text, and be sure the title includes the species or subject of the data, and where and when data were collected. In rare cases, titles or footnotes of tables and figures may be cross-referenced to avoid repeating long footnotes for the same data. However, this violates the "self-explanatory" rule and should be avoided.

Tables

Do not prepare tables for small data sets, those containing many blank spaces, zeros, repetitions of the same number, or those with few or no significant data. Put such data or a summary in the text. Day (1983) presents a practical discussion of tables.

For data that must be shown in a table, items that provide the most important comparisons usually read vertically, not horizontally. Construct tables for column-width (67 mm) printing. If the table will not fit in 1 column width, construct it for page-width printing not wider than 23 cm (9 inches). Some extra-wide tables can be printed vertically (e.g., *JWM* 50:192, 51:461), but such tables usually waste space. Extra-long and extra-wide tables require justification from the author.

Table titles may vary, but we recommend this sequence: (1) name of the characteristic that was measured (e.g., mass, age, density), (2) measurement unit or units in parentheses (e.g., cm. No./ha, M: 100F, or %), (3) name of organism or other entity measured (e.g., "of Canada geese"), and (4) location and date. Each part of the sequence can include >1 item (e.g., "Carcass and liver fat [%] and adrenal and kidney weight [mg] of white-tailed deer in Ohio and Michigan, 1975)."

Avoid beginning the title with superfluous words (e.g., The , Summary of, and Comparisons between) and words that can be presented parenthetically as symbols or abbreviations (e.g., %). Symbols such as *n* and % in the title seldom need repetition in table headings. Do not use abbreviations in table title, except within parentheses. However, use standard abbreviations and symbols (Appendix B) in the table body and in footnotes.

The lines printed in tables are called "rules," and *JWM* standards are

1. None drawn vertically within the table.
2. Three rules across the entire table: below the title, below the column headings, and at the bottom. Type each as a single, continuous line.
3. Use rules that straddle subheadings within the column heading (e.g., *JWM* 50:48).
4. None to show summation; use "Total" or equivalent in the row heading.
5. Do not use rules to join the means in multiple-range tests. Use Roman upper-case letters instead of rules (e.g., 12.3A, 16.2A, 19.5B) where the superscript "a" references a footnote such as "[a]Means with the same letters are not different ($P > 0.10$)" (e.g., *JWM* 50:22). Upper-case letters may be used in a similar fashion to reference the relationship of data among columns (e.g., *JWM* 50:371).

In column headings use straddle rules liberally to join related columns and reduce wordage (e.g., *JWM* 50:31). Label columns to avoid unnecessary print in the data field. For example, instead of "$\bar{x} \pm SE$," label $\bar{x}$ and SE separately so that $\pm$ need not be printed. Similarly, label sample size columns "*n*" instead of using numbers in parentheses in the data field.

Keep column- and row-heading words out of the data field. Type main headings flush left, and indent their subheadings (e.g., *JWM* 50:86). In the data field, do not use dashes (often misused to mean "no information") or zeros unless the item was measured, and 0, 0.0, or 0.00 correctly reports the precision. Similarly, respect digit significance in all numbers, particularly percentages. Do not use percentage where n is <26, except for 1 or 2 samples among several others where n is >25. Where the number of significant digits varies among data in a column, show each datum at its precision level; i.e., do not exaggerate precision. For P values only use 3 digits past the decimal and do not list $P = 0.000$; the correct form is $P \leq 0.001$.

For footnote superscripts use asterisks for probability levels and lower-case Roman (not italic) letters for other footnotes. Use this sequence for placing letters alphabetically: in the title, then left-to-right, and then down. Make certain that each footnote character in the title and table matches an explanation that is indented below the table. Left justify run-on lines of footnotes. Footnotes may be used to reduce cluttering the title and table with details. The most common errors in tables are single spacing, incomplete titles, naked decimal points, and ambiguous or unnecessary characters in the data field.

Figures

Most figures are either line (or computer) drawings or pictures ("picture" is used to distinguish scene or object photographs from photos of drawings). If possible, photographic prints should not exceed 20 x 25 cm. Submit 4 prints of a picture; for drawings submit either 4 prints or 1 print and 3 photographic copies. Retain original drawings to guard against loss or damage. Consult Allen (1977), Day (1983), and the CBE Style Manual Committee (1994:693–699) for additional guidance.

Type all figure captions on 1 (or more) page(s). On the back of each figure lightly print (in soft pencil) the senior author's name, figure number, and "Top." Figure titles tend to be longer than table titles because figures are not footnoted. The title may be several sentences and include brief suggestions for interpreting the figure content.

Pictures.—Few pictures are accepted. They must have sharp focus, high tonal contrast, a reference scale if size is important, a glossy finish, and must be unmounted. Letters, scales, or pointers can be drawn on the prints, but they must be of professional quality. Sets of 2-4 related pictures can be mounted as 1 figure if prints are the same width and will fit in a space $67 \times < 170$ mm when reduced for printing. Label prints A, B, C, D or use "Top," etc., for reference in the figure title. Cropping improves composition of most pictures, but do not put crop marks on prints. Instead, put them on xerographic copies or sketches. Do not submit color prints unless you are able to pay for printing at approximately $1,200/plate (as of 1997).

Line Drawings.—Consider whether a drawing can be printed column width (67 mm) or is so detailed that it must be printed page width (138 mm). The difference depends mainly on size of characters and lengths of legends drawn on the figure. If page width is necessary, consider omitting some of the detail and look for ways to shorten legends. Column-width figures are preferred (e.g., *JWM* 50:145).

Before revising the first sketch, determine the minimum height for letters, numbers, and other characters, which must be ≥1.5 mm tall after reduction for printing. Determine width in millimeters for the revised sketch. To determine the minimum height (mm) for characters, multiply the width by 0.0224 for column-width printing or 0.0109 for page-width printing. If in doubt as to printed width, use the column-width multiplier. The product is the minimum height in millimeters. Plan to use at least the next larger character height available. Hand-drawn lines and lettering and typewriter characters are not acceptable. We recommend professionally prepared line drawings. Lettering from modern personal computer graphics software and printers is acceptable.

For axis labels, use lower-case or italic letters where they are essential to the meaning, as in mathematical terms and most metric units (see subsection on Mathematics and Statistics and Appendix B). Otherwise use upper- and lower-case letters, which are more legible when reduced. Identify arbitrary symbols

Table 1 Common expressions with superfluous words[a]

Superfluous wording	Suggested substitute
the purpose of this study was to test the hypothesis	I (or we) hypothesized
in this study we assessed	we assessed
we demonstrated that there was a direct	we demonstrated direct
were responsible for	caused
played the role of	were
on the basis of evidence available to date	consequently
in order to provide a basis for comparing	to compare
as a result of	through, by
for the following reasons	because
during the course of this experiment	during the experiment
during the process of	during
during periods when	when
for the duration of the study	during the study
the nature of	(omit by rearrangement)
a large (or small or limited) number of	many (or few)
conspicuous number of	many
substantial quantities	much
a majority	most
a single	one
an individual taxon	a taxon
seedlings, irrespective of species	all seedlings
all of the species	all species
various lines of evidence	evidence
they do not themselves possess	they lack
were still present	persisted, survived
the analysis presented in this paper	our analysis
indicating the presence of	indicating
despite the presence of	despite
checked for the presence of	checked for
in the absence of	without
a series of observations	observations
may be the mechanism responsible for	may have caused
it is reasonable to assume that where light is not limiting	with light not limiting
in a single period of a few hours	in a few hours
occur in areas of North America	are in North America
adjacent transects were separated by at least 20 m	≥20 m apart
in the vicinity	nearby
separated by a maximum distance of 10 m and a minimum distance of 3 m	3-10 m apart
the present-day population	the population
their subsequent fate	their fate
whether or not	whether
summer months	summer
are not uncommon	may be
due to the fact that	(omit by rearrangement)
showed a tendency toward higher survival	had higher survival
devastated with drought-induced desiccation	killed by drought

[a]Mack (1986:33). Reprinted with permission from the Ecological Society of America.

Table 2 Words that commonly need correction in *Journal of Wildlife Management* manuscripts[a]

Word and proper usage

Accuracy (see precision): extent of correctness of a measurement or statement.

affect (see effect): verb, to cause a change or an effect; to influence.

among (see between): use in comparing >2 things.

between (see among): use in comparing only 2 things.

cf.: compare

circadian: approximately 24 hours.

continual: going on in time with no, or with brief, interruption.

continuous: going on in time or space without interruption.

diurnal: recurring every 24 hours; occurring in daylight hours.

effect (see affect): usually a noun, the result of an action; as an adverb (rare), to bring about or cause to exist, or to perform.

e.g. (see i.e.): for example.

enable (see permit): to supply with means, knowledge, or opportunity; to make possible.

ensure (see insure): to make certain or guarantee.

farther: more distant in space, time, or relation.

further: going beyond what exists, to move forward.

i.e. (see e.g.): that is.

incidence (see prevalence): number of cases developing per unit of population per unit of time.

insure (see ensure): to assure against loss.

livetrap: verb.

live trap: noun.

logistic: symbolic logic.

logistics: operational details of a project or activity.

mass (see weight): proper international use for measures of mass.

ovendry: adjective.

oven-dry: verb.

percent: adjective, adverb, or noun. Spell out only when the value is spelled out or when used as an adjective. Use "%" with numerals.

percentage: noun, part of a whole expressed in hundredths; often misused as an adjective, e.g., percent error, not percentage error.

permit (see enable): to allow, to give formal consent.

precision (see accuracy): degre of refinement with which a measurement is made or stated; e.g., the number 3.43 shows more precision than 3.4, but is not necessarily more accurate.

prevalence (see incidence): number of cases existing per unit of population at a given time.

sensu: as understood or defined by; used in taxonomic reference.

since: from some past time until present; not a synonym for "because" or "as."

presently: in the future, not synonymous with "at present" or "currently."

that (see which): pronoun introducing a restrictive clause (seldom preceded by a comma).

usage: firmly established and generally accepted practice or procedure.

utilization, utilize: avoid by using "use" instead.

various: of different kinds.

varying: changing or causing to change. Do not use for different.

very: a vague qualitative term; avoid in scientific writing.

weight (see mass): should seldom be used.

viz: namely.

which (see that): pronoun introducing a nonrestrictive clause (often preceded by a comma or preposition [for, in, or of which]); the word most often misused in *JWM* manuscripts.

while: during the time that. Use for time relations but not as synonym for "whereas," "although," and "similarly," which do not imply time.

[a]Adapted in part from CBE Style Manual Committee (1994:123-125); also see Day 1983:123-125).

Appendix A   Abbreviations for United States and Canadian political units. Spell out geographic locations given parenthetically in the text or in the LITERATURE CITED, but use ANSI abbreviations in tables, figures, and footnotes. Use U.S. Postal Service (USPS) abbreviations only in addresses with zip codes (e.g., author addresses). A blank means do not abbreviate.

Unit	ANSI	USPS	Unit	ANSI	USPS
U.S. and territories			U.S. and territories (continued)		
Alabama	Ala.	AL	Oklahoma	Okla.	OK
Alaska	Alas.	AK	Oregon	Oreg.	OR
American Samoa	Am. Samoa	AS	Pennsylvania	Pa.	PA
Arizona	Ariz.	AZ	Puerto Rico	P.R.	PR
Arkansas	Ark.	AR	Rhode Island	R.I.	RI
California	Calif.	CA	South Carolina	S.C.	SC
Canal Zone		CZ	South Dakota	S.D.	SD
Colorado	Colo.	CO	Tennessee	Tenn.	TN
Connecticut	Conn.	CT	Texas	Tex.	TX
Delaware	Del.	DE	Trust Territory	Trust Territ.	TT
District of Columbia	D.C.	DC	Utah	Ut.	UT
Florida	Fla.	FL	Vermont	Vt.	VT
Georgia	Ga.	GA	Virginia	Va.	VA
Guam		GU	Virgin Islands	V.I.	VI
Hawaii	Haw.	HI	Washington	Wash.	WA
Idaho	Id.	ID	West Virginia	W.Va.	WV
Illinois	Ill.	IL	Wisconsin	Wis.	WI
Indiana	Ind.	IN	Wyoming	Wyo.	WY
Iowa	Ia.	IA			
Kansas	Kans.	KS	Canadian provinces and territories		
Kentucky	Ky.	KY	Alberta	Atla.	AB
Louisiana	La.	LA	British Columbia	B.C.	BC
Maine	Me.	ME	Manitoba	Manit.	MB
Maryland	Md.	MD	New Brunswick	N.B.	NB
Massachusetts	Mass.	MA	Newfoundland	Newf.	NF
Michigan	Mich.	MI	Northwest	Northwest	
Minnesota	Minn.	MN	Territories	Territ.	NT
Mississippi	Miss.	MS	Nova Scotia	N.S.	NS
Missouri	Mo.	MO	Ontario	Ont.	ON
Montana	Mont.	MT	Prince Edward	Prince Edward	
Nebraska	Nebr.	NE	Island	Isl.	PE
Nevada	Nev.	NV	Quebec	Que.	PQ
New Hampshire	N.H.	NH	Saskatchewan	Sask.	Sk
New Jersey	N.J.	NJ	Yukon Territory	Yukon Territ.	YT
New Mexico	N.M.	NM			
New York	N.Y.	NY	Other		
North Carolina	N.C.	NC	United States	USA	
North Dakota	N.D.	ND	New Zealand	N.Z.	
Ohio	Oh.	OH	United Kingdom	U.K.	

Appendix B Abbreviations commonly used in *Journal of Wildlife Management* tables, figures, and parenthetic expressions. Only those metric units and their appropriate prefixes (CBE Style Manual Committee 1994:202–205, 206–218) identified with an asterisk may be abbreviated in the text. A blank means do not abbreviate.

Term	Abbreviation or symbol	Term	Abbreviation or symbol
Adult	ad	Logarithm, base e	*in or log
Amount	amt	Logarithm, base 10	*$\log_{10}$
Approximately	approx	Maie	M
Average	$\bar{x}$	Maximum	
Calorie	*cal	Meter	*m
Celsius	*C°	Metric Ton	t
Chi-squared	x^2	Minimum	
Coefficient	coeff	Minute	*min
Coefficient of		Month	
correlation, simple	r	Month names	Jan, Feb, etc.
multiple	R	More than	*>
determiniation, simple	r^2	Number (of items)	No.
multiple	R^2	Observed	obs.
variation	CV	Outside diameter	o.d.
Confidence interval	CI, a $< \bar{x} \leq$ a	Parts per billion	*ppb
	or $\bar{x} \pm$ a	Part per million	*ppm
Day		Percent	*%
Degrees of freedom	df	Population size	N
Diameter	diam	Probability	P
Diameter, breast height	dbh	Range	
Equations(s)	eq(s)	Sample size	n
Expected	Exp	Second	*sec
Experiment	exp.	Spearman rank correlation	r_s
Female	F	Square	sq
F ratio	F	Standard deviation(s)	SD
Gram	*g	Standard error(s,)	SE
Gravity	g	Student's t	t
Hectare	*ha	Temperature	temp
Height	ht	Trace*	tr
Hotelling's T^2	T^2	Versus	vs.
Hour(s)	*hr	Volt	*V
Inside diameter	i.d.	Volume: liquid, book	vol, Vol.
Joule	J	Watt	*W
Juvenile	juv	Week	
Kilocalorie	*kcal	Weight	wt
Lethal dose, median	LD_{50}	Wilcoxon test	T
Less than	*<	Year	yr
Limit	lim	z-statistic	z
Liter	*L		

*Define in a footnote (e.g., tr = <1%).

INDEX

Note: The Index and Appendices (above) have been single spaced to save printing space. However, all text and tables submitted for publication should be double-spaced.

5

Basic Concepts of Population Dynamics

". . . . no mathematics, however sophisticated, can remedy inadequate observation."

D. W. Macdonald, F. G. Bull, N. G. Haugh

INTRODUCTION

Population dynamics are the heart and soul of wildlife populations. In this chapter I discuss the early attempts to explain how animal numbers fluctuate and some of the early pioneers that set the stage for the field of population studies. The concept of a population is then discussed, followed by a description of population parameters and the use of life tables which is one method to obtain a snapshot of the population.

▉ SOME HISTORICAL ROOTS

The study of animal populations is a young science but its roots extend to ancient times (Egerton 1968). The few historical accounts about populations present observations on population phenomena with some interpretations, but for the most part population increases were accepted as supernatural punishment by plagues or of divine providence in the creation (Egerton 1968). Even philosophers that had an interest in biology and reproduction (e.g., Aristotle) left no works that recognized the importance of animal numbers. However, early recorded accounts were influential upon later naturalists involved with population, and it is important to see where the field has been in order to understand why we are at the current stage and where the future lies.

The earliest recorded accounts of animals' influences on humans came from the Bible and historical writings of Hebrews, Greeks, and Romans. Most of these early accounts describe plagues or other events that affected human society. For example, the early Egyptians believed that one of the leaders (Sennacherib) was defeated in 701 B.C. because of divine intervention that caused rodents to consume the quiver and bowstrings of his army. There are also the plagues suffered by the Egyptians (e.g, rodents, locust, frogs) (Exodus 7:14-20). These and other examples were accepted as the relationship that humans had with their mortality (e.g., the plagues of disease and rodents that struck the Philistines when they stole the Ark of God [I Samuel 5-6]).

For the most part, the plagues were sporadic and unexpected and most men regarded them as too mysterious to be understood. There are, however, a few attempts at explaining ancient outbreaks summarized by Egerton (1968).

Aristotle (384–322 B.C.) explained the rapid increase and subsequent decrease of vole and locust populations were related to the environment. When they were superabundant (e.g., consume entire corn crops), only rain succeeded in reducing numbers, even though predators (i.e., foxes and ferrets) killed many. How numbers increased was less decisive, but Aristotle reported, "In a certain district of Persia when a female mouse was dissected, the female embryos appear to be pregnant. Some people assert, and positively assert, that a female mouse by licking salt can become pregnant without the intervention of the male (Ross 1964)."

Pliny (23–79 A.D.), armed with more information, concluded, as did Aristotle, that reproduction and climate were important in understanding animal population fluctuations. Pliny added that means of controlling the species would be enhanced by understanding life histories. Pliny's accounts contained gross exaggerations (e.g., locusts in India are a meter long, with legs and thighs that can be dried and used as saws), but he may have stimulated others to study plagues.

Historically, nature was viewed very differently than it is today. Greek science was built upon the assumption that nature was constant and harmonious, leading to providential ecology. The basic observations that providential ecology is based on are:

a. a differential reproduction,
b. mutualism,
c. niche concept, and
d. survival of all species.

Many of these ideas came from Plato's (427–347 B.C.) ramblings, which were never meant to form theory, but because of his reputation, providential ecology was accepted as fact and the harmony of nature (i.e., everything has a purpose) was generally accepted.

However, Cicero (106–43 B.C.) supported providential ecology but emphasized mutualism and minimized competition, whereas Plotinus (205–270 A.D.) attempted to show that competition and predation were not at odds with providential ecology. Plotinus' writings emphasized population interactions more than the writings of others.

Another contribution of Cicero came in the form of a discussion he has on the importance of diseases as a critical factor influencing population growth. Diseases were often discounted because they are difficult to account for as a factor in population dynamics, because the basis to understand them was not present until the nineteenth century, and because they are not a regular phenomenon (Egerton 1968).

Providential ecology, therefore, was the guiding light until students of natural history and human ecology began to focus the ideas and concepts of ecology to provide an analytical framework from which to work, which took place nearly 2000 years later.

Pioneers of Population Dynamics

J. Grant (1620–1674). Grant was called the Father of Demography because he was the first to recognize the importance of measuring quantitatively the birth rate, death rate, sex ratio, and age structures of the human population in England. He constructed the first table of mortality and estimated the potential growth rate.

G.L.L. Buffon (1707–1788). Buffon was the author of *Natural History*, in which he recognized that populations of humans, flora, and fauna were subjected to the same processes. Buffon discussed how fertility of species was counterbalanced by numerous destructive agents. He argued that population numbers were controlled by many biological agents (e.g, disease, starvation). Buffon dealt with many problems of population regulation that are still debated today.

Thomas R. Malthus (1766–1834). Malthus was an English economist and authored *Essays on Population* (1798), in which he criticized England for supporting large families. He calculated that although the numbers of organisms can increase geometrically (i.e., 1, 2, 4, 8, 16, 32...), their food supply may never increase faster than arithmetically (i.e., 1, 2, 3, 4, 5...). He was criticized by society for his first monograph so he prepared a second in 1803 that had numerous empirical examples to support his theories. However, the thrust of his ideas was negative: what prevents populations from reaching the bare subsistence level that his theories predicted? What checks operate against the tendency toward a geometric rate of increase? We still ask these questions. Malthus was the one who put the problems before the public.

L.A. J. Quetelet (1796–1874). Quetelet was a Belgian statistician who suggested in 1835 that the ability of a population to grow geometrically was balanced by a resistance to population growth.

P. Verhulst (1804–1849). Verhulst was a student of Quetelet who derived an equation describing the cause of population growth over time. The S-shaped curve he developed is also called the logistic curve (Figure 5–1).

W. Farr (1807–1883). Farr was one of the earliest demographers who was concerned with mortality. He discovered (in England) that there was a relationship between the density of the population and the death rate, such that mortality increased as the sixth root of density. Farr also pointed out that in the United States, Malthus'

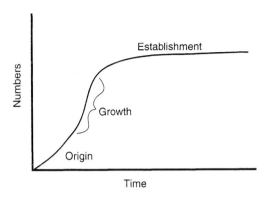

Figure 5–1 The logistic or S-shaped curve.

postulate was wrong, as food production increased geometrically at a rate even greater than the human population.

During the 1700s and 1800s the philosophical background relating to the animal numbers was providential in design (i.e, differential reproduction, mutualism, niche, survival of all species). However, late in the 1800s and early in the 1900s, two ideas emerged that undermined the concept of a "balance of nature": many species became extinct, and naturalists recognized that competition caused by population pressure is important in nature. At this point providential ecology and the balance of nature were replaced by natural selection and the struggle for existence.

Raymond Pearl (1879–1940). Pearl introduced life tables to ecologists. He also worked with the logistic curve and tried unsuccessfully to determine the number of animals in the entire world.

G. F. Gause (1910–1986). Gause conducted basic studies in predator-prey relationships and made direct applications to the logistic curve. His work yielded the concept that two species with the same requirements for survival cannot exist together in the same place.

These are some of the roots that support and have led to the development of modern population theory. There are many others (i.e., L.C. Cole, D. Lack, H.G. Andrewortha, L.C. Birch, V.C. Wynne-Edwards, C.J. Krebs, G. Caughley), and because they were first ecologists their concepts have been applied broadly to the field.

The Population Concept

A *population* can be defined as a group of organisms of the same species that occupies a given area over a specific time period. That is easy enough to define but becomes more complex when the concept is applied to the management of wildlife. Slobodkin (1980) tried to get one to visualize a population by simply placing a fence around a specified area and listing all of the organisms within the borders of the fence. This is another concept that is easy to visualize. However, think of the challenges that are presented. There are thousands of types of animals; the number of species is high in some areas but low in others; how can you know how many animals are present,

Figure 5–2 Similarities of a population and lake dynamics. Water flows into the lake (i.e., births), flows out (i.e., deaths), and rain adds to the lake volume (i.e., immigration), whereas evaporation (i.e., emigration) reduces the volume.

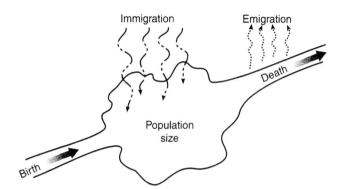

and how many pass through the fence, stay in the area all year, or inhabit the area for part of the time? For example, biologists have been working with desert mule deer and collared peccaries that are within an enclosed area (300 hectares). In the latest studies we have been counting animals annually, but from time to time animals escape and others get in the enclosure. The enclosure is a fairly simple system, but we still have to make estimates when counting animals.

Farner (1955) compared the concept of a population to the dynamics of a lake (Figure 5–2). Water flows into a lake (i.e, birth), out of the lake (i.e., death), and there is interplay between inflow and outflow (i.e, animal numbers). Precipitation into the lake contributes to the lake volume (i.e., immigration) and evaporation from the lake removes water (i.e., emigration). These simple analogies would break down if elaborated further but they do represent the basic properties that populations have in common. Populations are influenced by birth, death, emigration, and immigration (Figure 5–3). However, most populations do not have a fence around them, and even management units or study areas are artificial and at best fuzzy (Lancia et al. 2000).

To be more precise in discussing populations, biologists should be able to discuss the birth rate, death rate, sex ratio, and age structure of a population. More specifically, biologists should have information on survival by age class, fecundity by age, frequency distribution of ages, sex ratio, and numbers or density. These are difficult data to obtain for most populations and are often not available because of limited time and resources. As a result, seat-of-the-pants management is common. However, there is too much at risk in not having these basic data because managers must be able to defend their management plans and activities to their peers and public with scientific authority (Lancia et al. 2000). As a result it is important to select the correct parameters when defining populations.

Parameters are necessary for the description of most things. For example, a square can be easily described with a single measurement of one side, a circle has radius, a rectangle has a length and width, and so on. The biologist's decision when working with populations is to decide on which parameter to use. Because populations are complex, the reasons certain parameters are selected is because of their usefulness and their relationship to the objectives that have been established. As

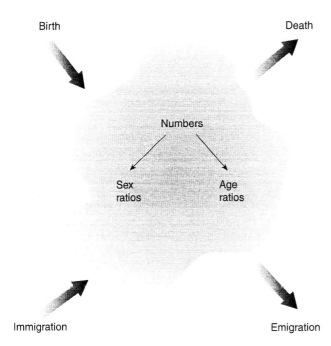

Birth

Death

Numbers

Sex ratios

Age ratios

Immigration

Emigration

Figure 5–3 The numbers within a population are difficult to determine because of the fuzzy borders of the population and fuzzy calculation of population parameters (after Lancia et al. 2000).

pointed out by Caughley (1977:6), the usefulness of population parameters depends on at least five conditions.

1. Ease of estimation is one consideration. Obviously, if a parameter is impossible to obtain, it will be of little use. For example, pellet groups can be used to obtain an estimate of the number of deer in an area, but it would be impossible to always have an exact count of all wild deer in a population.

2. How well do the selected parameters describe significant population properties? Using the same example as above, it is common to estimate the number of deer with pellet groups, but a total enumeration has yet to be obtained.

3. Selection of population parameters is of more value when one has the ability to extrapolate beyond the data from which they were calculated and collected. For example, after you obtain a parameter (such as birth rate), how much information do you have? If birth rate is all that can be obtained, it is not as meaningful as if you could extrapolate from the birth rate to be able to make other statements relating to productivity or the health of the population.

4. What is the directness of the parameter to the population process? For example, the birth weight of neonates is important but does not tell one as much about the population as does an accurate measure of productivity.

5. To what extent can the population parameter be applied to other populations? If your data are unique to a specific population, they will not be as useful to other scientists because their applicability would be limited. Caughley (1977) applied this concept to mammalian breeding systems in populations by describing two general types of systems from which he generated population models: *birth flow*

and *birth pulse*. Birth flow pertains to populations whose rate of breeding is constant throughout the year (e.g., humans, mountain lions, deer in southern latitudes). A birth pulse model implies that birth occurs at one time of the year (e.g., insects, most mammals and birds) ± 30 days, year after year. These two concepts can apply to a wide variety of populations, but it would be difficult to find a population that fit either model 100% of the time. As a result one should select whichever concept approximates the population. That would be more efficient than having a specific model for every population, which would hinder progress.

■ BASIC POPULATION PARAMETERS

Because the *density* (i.e., number of animals per area) in a population is influenced by births, deaths, immigration, and emigration, each will be discussed in more detail.

Birth Rate or Natality

Natality is the number of individuals produced per unit of time and is often referred to as the natality rate. Natality is an important parameter in population dynamics. Mean litter size is commonly used to compare natality between populations (McLellan 1989). If the number of breeding females is known, one can discuss a *specific natality rate,* which is the number of individuals produced per unit of time per breeding female. Dasmann (1981) used quail as an example. Quail breed seasonally so the unit of times is a year. If 640 quail are produced, the natality rate is 640 quail per year. If 80 breeding females produced the 640 quail, the specific *natality rate* would be eight young per year per breeding female (640 ÷ 80).

With big game, natality rates are expressed as the number of young per 100 breeding females per year [e.g., 80 fawns/ 100 female deer or 50 calves/100 female elk (Box 5–1)]. Standardizing these ratios at 100 females allows comparison between populations.

Natality is one of the most important characteristics for a biologist to be able to measure because of the information it yields (Dasmann 1981). Natality is an important parameter because it provides information about potential yield of the population, which relays information about population health, and thus, the relationship of the population with its habitat. However, natality is difficult to measure because of the variables that influence it.

Dasmann (1981) outlined five limitations to measuring natality: size of clutch or litter produced, number of clutches or litters produced per year, minimum and maximum breeding age for individuals, sex ratio and mating habits, and population density.

Size of a Clutch or Litter Produced.   Species vary in the number of young produced in a single breeding cycle. Quail can produce up to 15 eggs but the California condor produces only one. African ungulates generally produce one young per year but deer in North America generally produce two fawns as adults. Collared peccaries can

Box 5–1

HOW TO COMPUTE SEX RATIOS

For big game natality rates are usually expressed as number of young/100 females/year and male: female ratios are expressed as males: 100 females. If you observed 1063 males and 2784 females during a survey of adult elk, the adult ratio should be expressed as 38 males/100 females:

$$\left(\frac{1063}{2084} = \frac{x}{100}\right) \text{ and } x = 38$$

If you are presented a ratio of 60:100, you can calculate the percent of males and females with simple math.

$$\left(\frac{100\%}{60 \text{ males} + 100 \text{ females}}\right) = 0.625 \times 100 = 62.5\% \text{ females}$$

In the example above with the ratio of 38:100, 72.4% are females:

$$[(100 \div 38 + 100) \times 100]$$

The math is easy to check:

$$1063 + 2784 = 3847$$

$$2784 \div 3847 = 72.4\%$$

have many young, but normally have twins, and armadillos have the unusual condition of polyembryony, in which the fertilized ovum divides four ways resulting in same-sex quadruples. For the most part the size of the clutch is a relatively fixed characteristic of a population with little variance from the mean. However, clutch or litter size is influenced by the environment. Lack (1954) demonstrated differences in the clutch size of robins (Figure 5–4) due to differences in availability of food. Populations with more forage resources produced larger clutches. Allen (1954) argued that when you had good soil, you would have successful crops. This analogy applies to wildlife as well; populations with adequate resources are generally successful.

Determining natality is also difficult in the field. Natality studies of birds often involve locating nests and counting eggs, but with mammals determining natality is more difficult because locating young is not easy and obtaining adequate sample sizes is problematic. As a result other indices are used when possible, such as counting the number of corpus lutea in the ovary (Box 5–2 and 5–3).

Length of Breeding Season and Young Produced per Year. The number of litters (or clutches) a female produces each year will influence the population structure, which

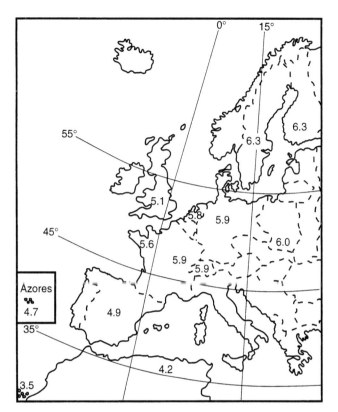

Figure 5–4   The average clutch size of robins in different countries (from Lack 1954).

Box
5–2

USE OF OVARIES TO ESTIMATE PRODUCTIVITY IN WHITE-TAILED DEER (CHEATUM 1949, HAUGEN AND TRAUGER 1962)

White-tailed deer breed in autumn and have a gestation period of approximately 200 days before parturition. During this process the ovaries undergo various changes that are useful in determining productivity.

In autumn, immature Graafian follicles in the ovary grow, then mature, and rupture during estrus. After discharge of the ovum from the ovary, several changes in the ovulated follicle occur, forming a *corpus luteum* that is a grayish to yellowish body 6 – 9 mm in diameter.

If the female is not bred, the *corpus luteum* of cycle degenerates in 14–15 days and another crop of follicles begins to develop. Another estrus occurs 28 days following the previous estrus.

If the female is bred, the *corpus luteum* of pregnancy persists and remains as a small red or brown spot, 1–2 mm in diameter, that persists for 6–8 months and is now called a *corpus luteum of pregnancy*. Thus, the *corpus luteum of pregnancy* is visible in ovaries that are collected in autumn hunting seasons.

Box
5–3

USE OF OVARIES TO ESTIMATE PRODUCTIVITY IN WHITE-TAILED DEER (CHEATUM 1949, HAUGEN AND TRAUGER 1962)

How are these ovarian structures used?

1. The *corpus lutea* counted in the ovaries collected in the hunting season give an indication of the health of the females because yearlings in poor condition may fail to ovulate, or ovulation may be reduced in other females in poor condition.

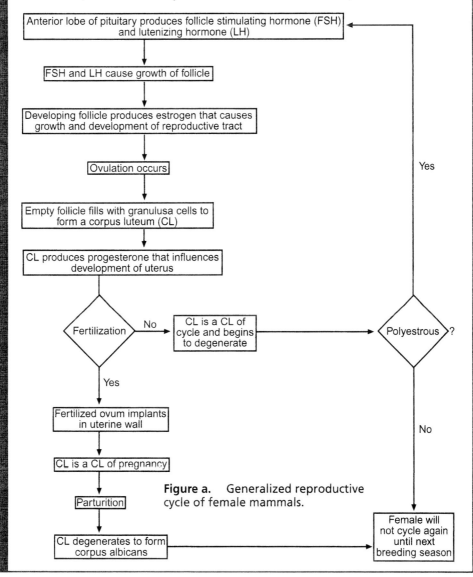

Figure a. Generalized reproductive cycle of female mammals.

(Continued)

Box 5-3

USE OF OVARIES TO ESTIMATE PRODUCTIVITY IN WHITE-TAILED DEER (CHEATUM 1949, HAUGEN AND TRAUGER 1962) *CONTINUED*

2. The *corpus lutea of pregnancy* from females collected in late winter show the number of ova that were shed when the female was bred. The number of fetuses should be similar to the number of *corpus lutea of pregnancy* but not identical (Figure b). Thus biologists can obtain an index of prenatal loss with the following formula:

$$\bar{\chi} \text{ corpus albicans/female} - \bar{\chi} \text{ fetuses/female} =$$
prenatal loss of fawns (from death of ova,
unfertilized ova, death of embryos).

Figure b.   Mean number of wolverine fetuses observed and number of corpus lutea, Yukon 1982–1985. Bars represent standard deviations and sample sizes and indicated (from Banci and Harestad 1988).

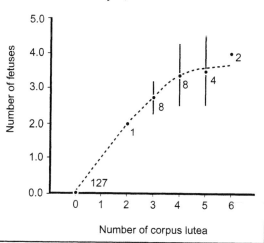

in turn is influenced by at least four events: the length of the breeding season, the gestation period, the nature of the sexual cycle, and the fate of the preceding neonates.

The length of the breeding season varies considerably in animals, which influences population dynamics. For example, a meadow vole has a 21-day gestation period and can breed immediately after parturition. In temperate zones, meadow voles could produce up to 10 litters per year with six young per litter. At the other extreme are elephants, with a 22-month gestation period, which at best could only produce a calf every two years if bred immediately after parturition. The interval, however, is considerably longer and the nature of the sexual cycle allocates energy to raising the calf instead of breeding more frequently.

Raising young is an expensive life history strategy. If a neonate is successfully raised one year, subsequent successes may be reduced. For example, Krausman et al. (1989) documented that female bighorn sheep in extreme environments that successfully raised a lamb to live more than six months in one year were less likely to raise a lamb to live six months in the following year. Natality is generally lower when females face environmental stress (Sadleir 1969, Caughley 1977, Clutton-Brock et al. 1982).

Also, some birds will not usually breed twice in a year if the first clutch is successful. If, however, the first clutch is not successful early in the breeding season, renesting is common.

Breeding Age. The reproductive success in mammals is often lowest among the youngest and the oldest females in the population (Sherman 1976, Hrdy 1977). However, the probability of successfully weaning young generally continues to increase with age (Reiter et al. 1981, Riedman et al. 1994). When a female first gives birth it will drastically affect the numbers in a population. Female black rhinoceros, for example, breed for the first time when they are 4–10 years of age (Hall-Martin 1986, Smith and Read 1992). Their social structure is like humans as there will be many nonbreeders in a population at the young and older age classes. On the other hand some ungulates (i.e., common duiker) breed and produce their first young by one year of age, and they rarely live past the breeding age. Therefore the common duiker populations will consist of many younger animals that are all breeding.

The influence of breeding age can also be seen in contrasting bighorn sheep and deer. Both *can* breed by one year of age; however, deer generally do not breed until they are at least one year old and bighorn sheep generally do not breed until they are two years old or older. In addition, deer generally produce two fawns per year, but sheep produce only a single lamb.

The influence of breeding age can also be seen in the human populations. Those countries that begin producing children when reproductively able have a considerably higher growth rate than many western cultures that delay childbirth until the females are in their late 20s, 30s, and older.

Sex Ratios and Mating Habits. The natality of a population is influenced by the sex ratios and mating habits of its members. For example, quail are monogamous, so a 1:1 sex ratio would favor a higher production of young than an unequal sex ratio. Polygamous species, on the other hand, such as deer or sheep, could still reach maximum productivity with a sex ratio that favored females because males are breeding as many females as they can.

Density and Breeding. If numbers are too low in a population, males and females will not be able to find each other to breed. This was the case with the white rhinoceros in southeast Africa, and biologists captured males to place with females (Player 1972). This is obviously a critical situation where interference by humans was necessary to maintain the species. On the other hand when densities exceed some level, productivity can drop because of competition for forage and related stress that inhibit breeding or do not enhance the survival of young. This was documented for colonies of rats (Calhoun 1952) and has subsequently been described for deer (Teer et al. 1965) and other species.

Biologists have to understand what is involved in natality. Many of the parameters listed by Dasmann (1981) may be difficult to measure, but to understand how natality operates in populations, measurements are required.

Births and birth rates are important to know in wildlife management. Births are the total number of young born or recruited into the population and *birth rates* are expressed as births per female per year (or other appropriate time interval) (Lancia et al. 2000). Births are the product of birth rate times the number of females in a population and potential rates can be determined by examining the reproductive tracts of females (Box 5–2). Lancia et al. (2000) defined *realized births* as the number of young at a particular point in the annual cycle, which often is measured immediately following the peak of births in a birth pulse population. Realized births are often calculated by direct counts of the number of young per female. For these observations to be reasonable the observabilities of young and females should be equal. If young: female ratios are used to estimate total births, the total number of females must be known to obtain an accurate estimate of the number of births (Lancia et al. 2000). Another important measurement of natality is fecundity.

Fecundity and Fertility

Fecundity is the actual number of animals born, whereas *fertility* is the potential number of animals that could be born. For example, under ideal conditions mule deer can produce two fawns per adult female per year for the reproductive life of the female (i.e., 1–8 years) resulting in realized fertility. Under realistic conditions mule deer produce less than two fawns per year per adult female for their reproductive life, which represents fecundity. Caughley (1977) has developed fecundity tables for species to provide more information about natality in populations (Table 5–1).

The fecundity pattern of a population is tabulated as a *fecundity table* or a fecundity schedule that lists the mean fecundity in each age class. Two tables could be constructed (i.e., one for males and one for females), but in population analysis the female segment is often more interesting so the male segment is often ignored. As a further simplification only female offspring are recorded. Most fecundity tables used in population analysis list the mean number of female offspring produced per female for each interval of age (Table 5–1).

Data required to construct the fecundity table include the mean litter size, mean number of litters per year, and the sex ratio at birth. For most species these three parameters have been established. The mean litter size can be obtained from field observations or from the literature, and the mean number of litters per year or frequency of births can also be determined from field observations or the literature. The sex ratio at birth for most vertebrates is close enough to 1:1 so the slight bias of males or females can be ignored. If this information is not available, Caughley (1977) offers techniques to determine the season of birth (Box 5–4) and the frequency of birth (Box 5–5).

Calculating the Season of Birth.   Because the frequency of births climbs rapidly to a peak and then declines more slowly, the seasons of birth in most mammals are

Table 5–1 A fecundity schedule calculated for chamois (from Caughley 1970)

Age in years (x)	Sampled number (f_x)	Number of pregnant or lactating (B_x)	Female births/female ($B_x/2f_x$) m_x
0	—	—	0.000
1	60	2	0.0167
2	36	14	0.194
3	70	52	0.371
4	48	45	0.469
5	26	19	0.365
6	19	16	0.421
7	6	5	0.417
>7	10	7	0.350

Box 5–4

CALCULATING THE SEASON OF BIRTH (FROM CAUGHLEY 1977)

Period (15.2 days)	$\frac{1}{2}$ Month	No. births (f)		Class range	
0	Oct[a]	0		0 – 0.499	i = interval
1	Oct[b]	5		0.5-1.499	
2	Nov[a]	22	2	1.5–2.499	
3	Nov[b]	39	3	2.5–3.499	
4	Dec[a]	17	2	3.5–4.499	
5	Dec[b]	17	1	4.5–5.499	
6	Jan[a]	2		5.5–6.499	
7	Jan[b]	1		6.5–7.499	
8	Feb[a]	0		7.5–8.499	
		= 103			Median = 52

$$\text{Median date of birth (M)} = L = \; + \; \frac{(n - 1/2)i}{f} \; \text{where}$$

L = true lower limit of the class in which the sample median lies
n = the number of desired observations within the class to obtain the median
i = class interval
f = absolute frequency of median class

$$M = 2.5 + \frac{(25 - 1/2)1}{39} = 2.5 + \frac{24.5}{39} = 3.12$$

3.12 periods $\times$ 15.2 days/period = 47 days
Add to midpoint of first period = 7 October
7 October + 47 days = 23 November

Box 5–5

CALCULATION OF THE FREQUENCY OF BIRTH (FROM CAUGHLEY 1977)

The mean number of litters produced per female over a period of time (when more than one litter per year is produced) is estimated from the following method.

1. Estimate the mean number of pregnancies over that portion of the year in which pregnancies can be detected. The *prevalance of pregnancy* ($\bar{P}$) should be measured on several occasions during the period when pregnant females can be detected. These estimates are averaged to calculate $\bar{P}$

For example, if pregnant females can be found over five months of the year and a sample is collected each month, ratios of pregnant females: total females can be recorded.

Month	2	3	4	5	6
No. pregnant	3	24	12	12	7
Total females	30	48	30	40	70

Each season must be properly represented in the sample.

$$\bar{P} = \frac{\dfrac{3}{30} + \dfrac{24}{48} + \dfrac{12}{30} + \dfrac{12}{40} + \dfrac{7}{70}}{5 \text{ months}} = 0.28$$

2. Calculate the mean incidence of pregnancy ($\bar{I}$), which is the number of times an average breeding female becomes pregnant during a time period (usually one year).

$\bar{I} = \dfrac{\bar{D}}{\bar{P}}$ where $\bar{D}$ is the mean duration of visible pregnancy and is measured as a fraction of the length of the season over which pregnancies were sought. In this example pregnancies were sought for over 5 months, so:

$$\bar{D} = 1/5 = 0.2$$

$\bar{I} = \dfrac{\bar{D}}{\bar{P}} = \dfrac{0.28}{0.20} = 1.4$ indicating that the average female has 1.4 pregnancies per year.

skewed in the positive direction. When this occurs (Figure 5–5), the median date of birth is a better indication of the average birthday of all animals in the population than the mean date of birth. Caughley's example (Box 5–4) is straightforward. One hundred and three births occurred from October through January and the median was 5.2. Each month was divided into two 15.2-day periods and births were recorded for each period. Use of the formula allows the calculation of the median date of birth.

Calculation of Frequency of Birth. The frequency of birth for most animals is available to biologists, but assumptions are often made that may or may not be accurate (i.e., the frequency of birth in one area is known so that must be applicable in other

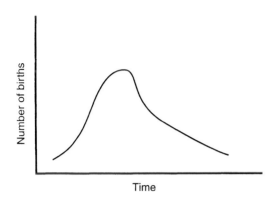

Figure 5–5 Number of births rising rapidly then gradually declining.

areas). The mean number of litters produced per female over a period of time (usually one year, when more than one litter per year is produced) can be estimated from the calculations presented by Caughley (1977) (Box 5–4). Although the incidence of pregnancy provides information on the mean number of pregnancies per female per year, the limitations of the calculation need to be recognized.

First of all, the duration of visible pregnancy should be the length of the entire pregnancy. For most mammals pregnancy can be established only after physical visual conditions occur (e.g., swelling of the flanks, uterus, or teat development). Even under the best field conditions these visual conditions can be difficult to obtain. However, pregnancy actually occurs earlier than cues can be detected, so the formula records the duration of visible pregnancy rather than the true pregnancy. As a result the incidence of pregnancy refers to the incidence of visible pregnancy rather than the incidence of the total gestation period.

Secondly, there may be the loss of entire litters during pregnancy that would create another limitation. The incidence of pregnancy refers to the midpoint in time of visible pregnancies. Considering these two restrictions, the incidence of pregnancy takes on a new meaning. The incidence of pregnancy then will overestimate the number of live litters produced due to intrauterine mortality or abortion and it only rates to visible pregnancy that is always going to be a of a shorter duration than the actual gestation period.

Why then, even use the incidence of pregnancy? There are at least two reasons. First of all, the incidence of pregnancy can be used to contrast within species in different habitats or situations. Second of all, the incidence of pregnancy for each age class of mothers is needed in fecundity tables. The duration of visible pregnancy will be independent of age (the animal can become pregnant only during certain times of the year). The prevalence of pregnancy, however, will vary with age.

With the mean litter size, mean number of litters per year, and sex ratio at birth, a fecundity table can be developed. In population biology the fecundity rate of a given female is not as important as the mean fecundity rate for each female age class. A complete schedule of these values, covering all ages from birth to the oldest observed age, is a fecundity table (Table 5–1). The importance of the fecundity table is

Figure 5–6 Expanding, declining, and stable age pyramids are useful in visualizing the sex and age structure of populations.

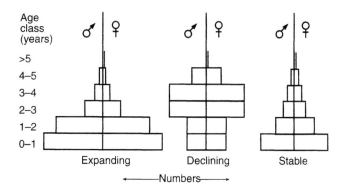

that it shows where breeding is taking place in the population, and comparisons can be made between species and habitat. Production can be in one short interval or over a longer period. Most vertebrates exhibit a birth pulse model of reproduction with an estrous cycle of 5–60 days (Caughley 1977). In Caughley's example, 275 females were shot during the season of birth, aged, and classified as to their reproductive condition. The sex ratio at birth was 1:1 and the litter size was one.

Age Pyramids

Most ecology texts discuss age pyramids when describing population structure. The numerical relationship between the sexes and ages of individuals in a population can be expressed with pyramids if a complete set of data is available to develop them. The information needed to develop age pyramids includes the number of animals in a population, the ages of the animals, and the sex ratio. By placing these data in a pyramid, one can obtain a visual picture of the population structure. An expanding population would have a pyramid that is broad at the base because of the numerous young being produced. A declining population, on the other hand, would have a narrow base, suggesting fewer young but numerous older age class individuals. A stable population would have a more symmetrical form, indicating a balanced sex and age ratio (Figure 5–6). Age pyramids are difficult to accurately construct for most wildlife populations because accurate data about numbers, age classes, and sex ratios are difficult to obtain even under the best circumstances. If the data can be obtained or approximated, the models can help to understand the history of the population and predict the probable future development. Age pyramids are determined as much by mortality as they are by natality.

Mortality

Because biologists work with animals, a common question they are often asked is, "How long does this species or that species live?" Just as a population is influenced by natality, it is influenced by mortality. If animals are not killed by predators, disease, accidents, or anthropogenic factors, they will die from plain physiological breakdown. To answer the question with "it depends", however, is not complete.

Many of the reported ages for animals come from zoo records (Table 5–2). However, some of the estimates are unreliable and most are overestimates of longevity. For example, the percent of birds that reach the flying stage from an egg is generally not more than 80% (Krebs 1978). Most species live for a year or less and die before they are able to breed (Caughley 1966, Ricklefs 1979). Understanding the mechanisms of death, rather than how old an animal can live, is the question of interest to biologists. However, obtaining accurate data is difficult. Even with the human population, where communication is highly sophisticated, data are erroneous because illiterate and semi-illiterate people provide false data, older citizens tend to add years (e.g., if you are 88, why not just round off your age to 90?), whereas younger folks often take off a year or two (e.g., if you are 30, why not stay in your late 20s?).

To compound problems in studying mortality, most mortality that seriously alters population structure is noticed after any management action could alter the situation. Even when declining populations are noticed, it is often difficult to determine the decimating factor. The carcasses of larger mammals can be found, examined, and used to help determine causes of death. Also, scavengers (e.g., vultures, magpies, ravens, crows) congregate around carcasses, assisting in locating large animals that recently died. During a visit to Tiger Reserves in India, my colleagues and I located seven of eight kills made by tigers and leopards from scavenging activity of various species of vultures (Figure 5–7). Kills made by mountain lions and other predators have also been found from the activity of scavenging avifauna (Hornocker 1970, Krausman and Ables 1981). However, obtaining accurate information on death of smaller mammals and birds is very difficult unless they have been marked so they can be located.

One important and meaningful way of examining the ages of animals and understanding the role age plays in a population is to visualize a population of 1000 animals of all age classes and determine the mean life expectancy or the number

Figure 5–7 Nilgai that was killed by a tiger in one of India's Tiger Reserves, 1998. This carcass and others were located from observing the scavenging activity of vultures (photo by P. R. Krausman).

Table 5–2 Reported longevity of selected animals

Animal	Average age in wild[a]	Maximum in wild	Maximum in captivity	Comments	Source(s)
American robin		13.9	12.8		Klimkiewicz et al. (1983), Mitchell (1911)
Bison	<20.0	>40.0			McHugh (1958) Meagher (1973)
Black bear	3–5 (M) 5–8 (F)	32	>25.0		MacDonald (1984) Pelton (2000)
Box turtle			123.00		Missouri Conservationist
Caribou	14–15 11–13(M) 15–16(F)		18–20 (F)	Without predators	Bergerud (2000) Tyler (1987) Miller (1974) Miller (1974)
Condor			50(M)		Hediger (1950)
Deer mule	10–12(F) 8(M)	19(M) 20(F)			Robinette et al. (1977) Ross (1934) Mackie et al. (1998) Mackie et al. (1998)
Columbian black tailed		22(F)		Semiwild	Cowan (1956)
white tailed	<10	20			Halls (1978)
Elephant		<14F			Ozoga (1969), Tullar (1983) Fowler (1947)
Elk		15(F)	157[b]	Unhunted	Bubenik (1982), Hines and Lemos (1979)

Species			Reference
Chimney Swift		<10(M)	Wisdom and Cook (2000)
Grizzly and polar bear	14.0		Bodick and Peffer (1993)
	20–30	30	Pasitschniak-Arts and Messier (2000)
Gorilla		54	Arizona Daily Star
House sparrow	13.3		Klimkiewicz et al. (1983)
Lion and big cats			Hediger (1950)
Moose	18(F)	20	Franzmann (2000)
	14(M)		Franzmann (2000)
Panda		37	Arizona Daily Star
Penguin	13.0		Scolario (1990)
Pronghorn	9–15		Hepworth (1965)
			Kerwin and Mitchell (1971)
Royal Albatross	58.0	55	del Hoyo et al. (1992)
Salamander, giant			Hediger (1950)
Swan		102	Missouri Conservationist
Tortoise		>39	Slobodkin 1980
Tortoise, giant		152	Missouri Conservationist
Vulture		117[c]	Flower (1938)
Wolf		16	Young (1944)
	14		Ballard and Gipson (2000)

[a]When sex was reported, it was indicated as male (M) or female (F).

[b]This elephant was born as an African elephant but died as an Asian elephant. These types of records, even from zoos, make many longevity records questionable (Fowler 1947).

[c]Questionable (Fowler 1938).

of years of life that will be lived by an individual in one of the age classes. This snapshot picture of a population can be obtained with a life table.

Life Table

A *life table* is a clear and systematic picture of mortality and survival and is a convenient format for describing the mortality schedule of a population (Deevey 1947). Life tables were developed by human demographers, especially those working for life insurance companies that have to know how long any one person can be expected to live (Krebs 1972) (Table 5–3). Although life tables related to human populations fill the literature (Krebs 1972), little is known about the life tables of animal populations. Life tables were first introduced to ecologists by Pearl and Miner (1935) and were useful because of the mortality pattern (Table 5–4). In most life tables the authors assume all animals were born on the same day and the numbers surviving to each subsequent birthday were recorded. For example, Caughley (1966) constructed a life table for female Himalayan thar, which began with 205 animals (Table 5–4).

The columns of the life table are standard in ecology and are summarized by letters. The table consists of several columns.

x = age interval. The age interval for Himalayan thar is one year but could be longer (e.g, 5 years for humans) or shorter (e.g., 1 month for rodents). With shorter intervals the detail of the life table increases.

f_x = the actual number observed or still surviving (f_x) at one, two, three, four, and so on years after birth of the original number born.

l_x = the number of survivors (survivorship at the start of age interval x) The probability at birth of living to birth is obviously unity but it is often multiplied by 1000 as are the other values in the series, so survivorship is viewed as the number of animals in a cohort of 1000. It is easier to visualize the cohort of 1000 than age-specific probabilities. However, it is the probabilities that are used in analysis (e.g., 1.0 versus 1000). Converting numbers to probabilities is not complex.

d_x = the number dying during the age interval x to $x + 1$. d_x can be calculated from the l_x series by

$$d_x = l_x - l_{x+1}$$

For our example, d_x at $x = 10$ would equal

$$73 - 47 = 26$$

Because l_x starts at 1000 (or unity), by definition the d_x schedule totals 1000 (or unity).

q_x = the mortality rate during the age interval x to $x + 1$. The mortality rate is calculated as d_x/l_x. In our example q_x for $x = 2$ would equal

$$28 \div 461 = 0.013$$

Table 5–3 Actuarial table for humans

				Life Expectancy in Years				
Age	Male	Female	Age	Male	Female	Age	Male	Female
0	70.83	76.83	34	39.54	43.91	68	12.14	15.10
1	70.13	75.04	35	38.61	42.98	69	11.54	14.38
2	69.20	74.11	36	37.69	42.05	70	10.96	13.67
3	68.27	73.17	37	36.78	41.12	71	10.30	12.97
4	67.34	72.23	38	35.87	40.20	72	9.84	12.26
5	66.40	71.28	39	34.96	39.28	73	9.30	11.60
6	65.46	70.34	40	34.05	38.36	74	8.79	10.95
7	64.52	69.39	41	33.16	37.46	75	8.31	10.32
8	63.57	68.44	42	32.26	36.55	76	7.84	9.71
9	62.62	67.48	43	31.38	35.66	77	7.40	9.12
10	61.66	66.53	44	30.50	34.77	78	6.91	8.55
11	60.71	65.58	45	29.62	33.88	79	6.57	8.01
12	59.75	64.62	46	28.76	33.00	80	6.18	7.48
13	58.80	63.67	47	27.90	32.12	81	5.80	6.98
14	57.86	62.71	48	27.04	31.25	82	5.44	6.49
15	56.93	61.76	49	26.20	30.39	83	5.09	6.03
16	56.00	60.82	50	25.36	29.53	84	4.71	5.59
17	55.09	59.87	51	24.52	28.67	85	4.46	5.18
18	54.18	58.93	52	23.70	27.82	86	4.18	4.80
19	53.27	57.98	53	22.89	26.98	87	3.91	4.43
20	52.37	57.04	54	22.08	26.14	88	3.66	4.09
21	51.47	56.10	55	21.29	25.31	89	3.41	3.77
22	50.57	55.16	56	20.51	24.49	90	3.13	3.45
23	49.66	54.22	57	19.74	23.67	91	2.94	3.15
24	48.75	53.28	58	18.99	22.86	92	2.70	2.85
25	47.84	52.34	59	18.24	22.05	93	2.44	2.55
26	46.93	51.40	60	17.51	21.25	94	2.17	2.24
27	46.01	50.46	61	16.79	20.44	95	1.87	1.91
28	45.09	49.52	62	16.08	19.65	96	1.54	1.56
29	44.16	48.59	63	15.38	18.86	97	1.20	1.21
30	43.24	47.65	64	14.70	18.08	98	0.84	0.84
31	42.31	46.71	65	14.04	17.32	99	0.50	0.50
32	41.38	45.78	66	13.39	16.57			
33	40.46	44.84	67	12.76	15.83			

1980 Commissioners Standard Ordinary Mortality Table

Table 5–4 Life table for female Himalayan thar (from Caughley 1966)

Age in years (x)	No. observed alive each year (fₓ)	No. surviving at the start of age interval (lₓ)	No. dying within age interval x to x + 1 (dₓ)	Rate of mortality, $\frac{d_x}{l_x}$ (qₓ)	Rate of survival, 1-q (pₓ)
0	205	1000	533	0.533	0.4670
1	96	467	6	0.013	0.987
2	94	461	28	0.061	0.939
3	89	433	46	0.106	0.894
4	79	387	56	0.145	0.855
5	68	331	62	0.187	0.813
6	55	269	60	0.223	0.777
7	43	209	54	0.258	0.742
8	32	155	46	0.297	0.703
9	22	109	36	0.330	0.670
10	15	73	26	0.356	0.644
11	10	47	18	0.382	0.618
12	6	29	29	1.000	0.000

p_x = the survival rate during the age interval x to $x + 1$. The survival rate is calculated as $1 - q_x$.

Some life tables include another column, the expectation of further life (e_x) for each age class. This column is a carryover from the insurance industry and has often been applied uncritically to nonhuman populations. This statistic adds little to an understanding of population processes with the exception of the life expectancy at birth (Caughley 1977).

To calculate e_x two other columns need to be generated, L_x and T_x (Table 5–5).

L_x = the number of individuals alive on average during the age interval x to $x +1$. L_x is calculated as $l_x + l_{x + 1} \div 2$.

In the example (Table 5–5), L_3 (i.e, the average number alive in age interval 3 to 4 years)

$$\frac{l_3 + l_4}{2} = \frac{433 + 387}{2} = 410$$

T_x = summation of L_x from the bottom of the life table to the top

$$or \ \sum_{x}^{\infty} L_x$$

For the thar data (Table 5–5) $T_{10} =$

$$L_{12} + L_{11} + L_{10} = 14.5 + 38.0 + 60.0 = 112.5 \text{ thar years}$$

Table 5–5 Life expectancy for female Himalayan thar (from Caughley 1966)

Age in years (x)	No. surviving at the start of age interval (l_x)	No. individuals alive on the average during the age interval x to x + 1 (L_x)	Summation of L_x (T_x)	$\dfrac{T_x}{l_x}$ (e_x)
0	1000	733.5	3,470.0	3.5
1	467	464.0	2,736.5	5.9
2	461	447.0	2,272.5	4.9
3	433	410.0	1,825.5	4.2
4	387	359.0	1,415.5	3.7
5	331	300.0	1,056.5	3.2
6	269	239.0	756.5	2.8
7	209	182.0	517.5	2.5
8	155	132.0	335.5	2.2
9	109	91.00	203.5	1.9
10	73	60.00	112.5	1.5
11	47	38.00	52.5	1.1
12	29	14.50	14.5	0.5

The statistic e_x can be obtained by dividing T_x by l_x:

$$e_x = \frac{T_x}{l_x}$$

In our example, the life expectancy at birth would be:

$$e_x = \frac{T_x}{l_x} = \frac{3470}{1000} = 3.5 \text{ years (rounded off).}$$

Given any one of the f_x, l_x, d_x, or q_x columns of the life table, the others can be calculated. There is nothing new in each column; they are simply different ways of summarizing a data set and conversion of one column to another in a simple arithmetic exercise. Each column is presented with life tables so the table can be used for the different views of a population (Caughley 1977). The l_x column is in most population dynamic equations. The study of genetic and evolutionary consequences of mortality uses the d_x schedule. The q_x schedule is a good projection of the mortality pattern and is useful to compare within and between species. Finally, p_x is useful in harvesting calculations and population simulation.

Life Table Data

Life tables are often developed from data collected in conjunction with other data collected in field studies. To simply collect data for a life table would be time consuming, tedious, and expensive. If a life table is needed to solve a problem, the biologists should decide on the most effective way of collecting accurate data. There are several common types of data used to generate a life table (Caughley 1966, 1977): direct and approximate methods.

Direct Methods. There are two direct methods and both require information from intensively studied natural populations or captive populations. The first involves recording the ages at death of a large number of animals born at the same time to obtain d_x information. A second way to obtain data would be to record the number of animals in the original cohort that are still alive at regular intervals (e.g, one year for large mammals) to obtain l_x information.

Approximate Methods. There are four approximate methods. The first is similar to the first direct method except that the ages at deaths are collected from animals marked at birth but not born at the same time. The data are still used to calculate d_x information. All three of the methods (i.e., the two direct and the first approximate method) are generally used in studies of small mammals. The remaining methods are more commonly used for large mammals and are valid only when the data are collected from a *stationary age distribution* (i.e., an age distribution that results when a population does not change in size and where the age structure of the population is constant with time).

The second approximate method requires the recording of ages at death of a representative sample by aging carcasses. These data are also used to generate d_x information.

The third approximate method requires a sample of ages at death from a population where the animals were killed so that an unbiased sample of ages in a living population could be recorded as d_x data. This method is valid only if all those killed are representative of the entire population. It is not valid when the frequencies of ages at death are obtained only on one range (i.e., summer or winter) of a population, or the data reflect the mortality pattern over only part of the year. This method can not be used for d_x information if the sampled deaths resulted from a rare event (e.g., fire, flood, avalanche) that removed a sample of the population's age distribution during life. The method is also invalid if the perishibility or conspicuousness of the carcasses varies with age. For example, the mortality in the first year of life is usually underestimated by a sample of picked-up skulls because they deteriorate faster than mature skulls. Finally, this method would not be appropriate if one mixed natural deaths with mortality from catastrophes or shootings because the combination would confound d_x and l_x effects and could not be converted to either schedule. Mixing natural mortality and hunting mortality could only be appropriately combined in the same sample if the mortality from each were sampled in proportion to their relative effects on the population, which is difficult or impossible.

The final approximate method would be with data collected from a sample of females at the season of birth. The resultant frequencies are used to develop l_x information as long as the sample is representative of the population. It would not be representative if all age classes were not represented or if hunters selected for one age class over others (e.g., larger versus smaller animals).

Number and Type of Sample. There is no absolute number of samples required for the development of a life table, but all age classes in the population should be

Table 5–6 Dynamic or cohort life table for red deer, Isle of Rhum 1957 (from Lowe 1969)

Age in years (x)	Survivors at beginning of age class x (l_x)	Deaths (d_x)	Further expectation of life, years (e_x)	1,000 q_x (mortality rate/1000) (e_x)
Males				
1	1000	84	4.76	84.0
2	916	19	4.15	20.7
3	897	0	3.25	0.0
4	897	150	2.23	167.2
5	747	321	1.58	430.0
6	426	218	1.39	512.0
7	208	58	1.31	278.8
8	150	130	0.63	866.5
9	20	20	0.50	1,000.0
Females				
1	1000	0	4.35	0.0
2	1000	61	3.35	61.0
3	939	185	2.53	197.0
4	754	249	2.03	330.2
5	505	200	1.79	396.0
6	305	119	1.63	390.1
7	186	54	1.35	290.3
8	132	107	0.70	810.5
9	25	25	0.50	1,000.0

adequately represented. Life tables that do not adequately represent the entire age distribution and are based on less than 50 ages at death or 150 ages of living animals likely are not adequate (Caughley 1966, 1977). Likewise, life tables that have been developed with probable sampling bias, where death and emigration have been confounded, where l_x is recorded as d_x data, or where aging is not adequate, will be of little use (Caughley 1966).

General Types of Life Tables

There are two general types of life tables: dynamic or cohort and time specific. The *dynamic life table* is constructed from records of the fate of a group of animals all born at the same time. Obtaining these data involves following the population throughout its life and recording the date of death for each individual. Lowe (1969) developed a dynamic life table for red deer on the Isle of Rhum off Scotland (Table 5–6). The table is based on deer born in 1957 and samples of deer that died from 1957–1966. The table is constructed from the number that die in each age class over a lifetime and the rest of the table is derived from d_x values.

Table 5–7   Time-specific life table for red deer, Isle of Rhum, 1957 (from Lowe 1969)

Age in years (x)	Survivors at beginning of age class x (l_x)	Deaths (d_x)	Further expectation of life, years (e_x)	1,000 q_x (mortality rate/1000) (e_x)
Males				
1	1000	282	5.81	282.0
2	718	7	6.89	9.8
3	711	7	5.95	9.8
4	704	7	5.01	9.9
5	697	7	4.05	10.0
6	690	7	3.09	10.1
7	684	182	2.11	266.0
8	502	253	1.70	504.0
9	249	157	1.91	630.6
10	92	14	3.31	152.1
11	78	14	2.81	179.4
12	64	14	2.31	218.7
13	50	14	1.82	279.9
14	36	14	1.33	388.9
15	22	14	0.86	636.3
16	8	8	0.50	1000.0
Females				
1	1000	137	5.19	137.0
2	863	85	4.94	97.3
3	778	84	4.42	107.8
4	694	84	3.89	120.8
5	610	84	3.36	137.4
6	526	84	2.82	159.3
7	442	85	2.26	189.5
8	357	176	1.67	501.6
9	181	122	1.82	672.7
10	59	8	3.54	141.2
11	51	9	3.00	164.6
12	42	8	2.55	197.5
13	34	9	2.03	246.8
14	25	8	1.56	328.8
15	17	8	1.06	492.4
16	9	9	0.50	1000.0

The *time-specific life table* is developed from mortality in each age class of a given population that is recorded at a point in time or over a period of time (e.g., one year). This type of sample yields a set of mortality rates (q_x) from which the rest of the table is derived (Table 5–7).

The dynamic and time-specific life tables will not be identical unless the environment does not change from year to year, allowing the exponential rate of increase to equal zero, or the population is stable. In other words, the survival and fecundity schedules have been constant for some time.

The Value of Life Tables

With all of the limitations and assumptions outlined for the proper construction of a life table, an obvious question is, "What is the value of a life table?" Caughley (1977) outlines at least four uses of the life table.

1. A life table shows how the birth rate and death rate relate to each other and provides the ecologist with insight about a population strategy for survival.

2. A life table is useful in identifying the weakest link in the population, which is important in trying to decrease or stimulate an increase in the population. This weak link can then be attached if a population reduction is desired or enhanced if the goal is to increase the population. The weakest link of a population is usually that age class that shows the greatest between-year variability in mortality, or fecundity rates, or their interaction. In short, the ecologist can use life tables to understand how the population pattern of mortality changes over the short run in response to the environment.

3. Life tables are useful in modeling a population's dynamics.

4. Life tables can also be used with fecundity tables to estimate the intrinsic rate of increase (i.e, the rate at which a population increases when no resource is limiting) for questions of ecology, genetics, evolution, and sustained yield harvesting (Caughley 1977).

An accurately constructed life table can provide additional information about the dynamics of a population with basic curves: mortality curves and survivorship curves. Both can reflect the general life history strategy for a population.

Mortality Curves.   Mortality curves are based on the rate of mortality (q_x) from life tables. Mortality curves plot the mortality rates directly as 1000 q_x against age. The mortality curve developed from the Himalayan thar life table (Table 5–4) is typical of mortality pattern for mammals (Figure 5–8) because it can be divided into two parts: a juvenile phase that is indicative of a high rate of mortality and a post-juvenile phase, in which the rate of mortality is initially low but then rises at a relatively constant rate with age. Mortality curves often look like the letter "J" and are referred to as J-shaped or fishhook curves. Other mammals (Figure 5–9), fish, and reptiles display similar mortality patterns even though the agents of mortality are quite diverse.

Survivorship Curves.   Survivorship curves are based on the l_x column of life tables. As with mortality curves, survivorship curves depend on the validity of the life table. Life tables and survivorship curves are not typical of all populations, but do depict the nature of a population at different places and under different environmental conditions. Therefore, survivorship curves are useful for comparing the population from one area at one time or with one sex to another population (Figure 5–10).

Figure 5–8 Mortality curve for Himalayan thar (from Caughley 1966). The juvenile phase is characterized by a high rate of mortality, followed by a postjuvenile phase in which the rate of mortality is initially low but generally rises at a relatively constant rate with age.

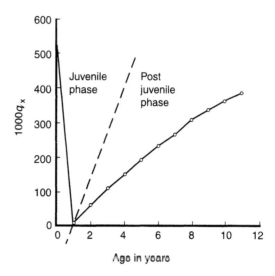

Figure 5–9 Mortality curve for male and female Dall's sheep (from Murie 1944).

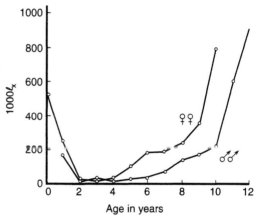

Figure 5–10 The survivorship of boars in Taiwan from two different periods (from McCullough 1974).

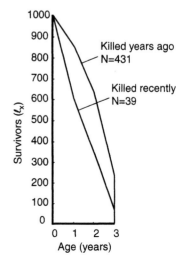

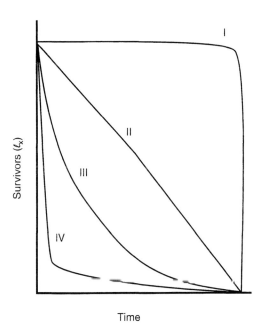

Figure 5–11 Four general types of survivorship curves. Type I represents mortality concentrated on old age animals (e.g., elephants). Type II represents constant mortality over time (e.g., hydra). Type III represents death nearly constant with age. Type IV represents mortality at young ages (e.g., most vertebrates) (from Slobodkin 1980).

There are several general types of survivorship curves that can be used to model most populations. None of them are exact fits but are useful in discussing life history characteristics (Figure 5–11). Curve I is representative of individuals that live out their physiological life span. There is a high degree of survival throughout life with high mortality at the end of the species life span. Curve I is typical of a long-lived species (e.g., humans, elephants, mountain sheep). Curve II is rare as it depicts a constant rate of mortality throughout life (e.g., hydra). Because so few species fit this curve, it is usually combined with Curve III. Curve III represents a constant fraction of animals alive that die in each age. It is nearly linear and is characteristic of birds and rodents after they have reached the adult stage. Curve IV is typical of most fauna. It indicates high mortality rates early in life with fewer individuals living into the postjuvenile stage. Although Curve IV is most common in nature, natural populations do not fit any of the curves exactly.

Different age classes are susceptible to different sources of mortality, causing survivorship curves to be a mixture of the various hypothetical curves. For example, some fish at birth have a mean life expectancy of 12 hours at birth and the curve they would fit is Curve IV. However, when a month old, they have high probability to live three to four more years and realistically fit Curve III. Most animals exhibit Curve IV at birth but those that survive to adulthood have a more constant rate of mortality resulting in a survivorship pattern similar to Curve III.

On the other hand, if the rate of mortality in a population is sufficiently low or causes of mortality are low, most death may occur from physiological breakdown of the organism (i.e., Curve I). However, certain populations of humans have primitive public health practices, resulting in survivorship curves that more closely represent Curve III.

Table 5–8 Life equation for quail (from Leopold 1933).

Date	Variable	Gain	Loss	Population
1 Nov	Sample n			100
1 Jan	Kill = 33%		33	66
	Crippling = 33 × 18%		7	
	Σ hunting mortality		40	60
1 Mar	Winter loss = 60 × 30%		18	42
1 May	42 × 47.5% F = 20 pairs			
	Eggs = 20 × 14	280		322
1 Aug	Nest mortality = 280 × 60%		168	154
	F mortality = 20 × 14%		3	151
	Juvenile mortality ≈			
	151−100 = 51 in 8 surviving			
	nests @ 6/nest		51	100

When constructing survivorship curves, several assumptions need to be made:

1. The population size is constant (i.e., there is constant annual production).
2. The death rate is constant from year to year.
3. The population is free of immigration and emigration, or immigration and emigration are equally balanced in age and number of animals.

The Life Equation

Another version of a life table is a life equation that allows wildlife managers to develop a mental picture of the factors that influence a population (Leopold 1933). A life equation allows for a holistic approach to populations before one tries to isolate single influencing factors. Another way to view the life equation is as a model of the population because numerous variables will have to be estimated until better data are obtained. Leopold (1933:183) presented a simple life equation for quail using some local data plus information from other studies (Table 5–8). In this example the population is hunted, so some estimate of the harvest (33%) and crippling loss (18%) of the harvest is made, leaving 60 birds by 1 January. Because data are not yet available on this population, additional data from elsewhere are used for winter mortality (30%), which leaves 42 birds on 1 March. The sex ratio of local populations of quail averages 52.5 males to 47.5 females that would yield 20 pairs. The 20 pairs are assumed to produce 14 eggs per pair for a gain of 280. That increase plus the parents yields a population of 322.

Another common cause of mortality is nest mortality. Again, in the absence of data for this population, other reasonable data are applied and nest mortality is estimated at 60%. Hen mortality was estimated at 14% of the 20 paired females, or three, leaving 151 animals alive at the time of hatching. The last variable in the model is juvenile mortality that is estimated as six for each of the eight remaining nests, yielding a loss of 51.

Table 5–9 Life equation for deer in Middle Park, Colorado (from Freddy and Gill 1977)

Population data	Males	Females	Male fawns	Female fawns	Total	Percent yearling males–females
Winter population	3000	4000	1650	1350	10000	
(January census)	(30.0%)	(40.0%)	(16.5%)	(13.5%)	(100%)	
Winter mortality	218	213	203	166	800	
	(27.3%)	(26.6%)	(25.4%)	(10.7%)	(100%)	
Spring population	2782	3787	1447	1184	9200	
	(30.2%)	(41.2%)	(15.7%)	(12.9%)	(100%)	
Prefawning	4229	4971			9200	34.2–23.8
population	(46.0%)	(54.0%)			(100%)	
Fawn production			2187	1790	3977	
			(55.0%)	(45.0%)	(100%)	
Prehunt population	4229	4971	2187	1790	13,177	
	(32.1%)	(37.7%)	(16.6%)	(13.6%)	(100%)	
Harvest–early	50	40	6	4	100	
seasons	(50.0%)	(40.0%)	(6.0%)	(4.0%)	(100%)	
Wounding loss	10	8	1	1	20	
(20% of harvest)	(50.0%)	(40.0%)	(6.0%)	(4.0%)	(100%)	
Harvest–regular	780	300	84	36	1200	
season	(65.0%)	(25.0%)	(7.0%)	(3.0%)	(100%)	
Wounding loss	156	60	17	7	240	
(20% of harvest)	(65.0%)	(25.0%)	(7.0%)	(3.0%)	(100%)	
December post-	3233	4563	2079	1742	11617	
season population	(27.8%)	(39.3%)	(17.9%)	(15.0%)	(100%)	

This model obviously has limitations, but it also provides a basis from which to operate and indicates where better data are needed. Furthermore, the manager can use the life equation to examine potential mortality from various causes in relation to potential or assumed natality. Leopold's model assumes the fall population remains constant from year to year; however, with better data over time a more accurate image will emerge that will assist in the management of the quail.

Another example is provided for mule deer from Colorado (Freddy and Gill 1977) (Table 5–9), where the life equation is used to set harvest objectives, seasons, and estimating the number of permits to issue. Life equations can provide useful prehunt data about the population size and composition to make management decisions functional.

The winter population estimate was based on a helicopter survey the previous December and numbers were allocated as though there were 10,000 deer in the population. Fawn composition was based on the sex of fawns harvested during regular hunting seasons.

Winter mortality was estimated from field surveys that were projected to the entire winter range. The total winter mortality was 8.0% of the winter population but varied from 0.7 to 27.3% for the various sex and age classes. The winter mortality is subtracted from the winter population to obtain the spring population estimate.

The prefawning population incorporates the fawns into the adult population as yearlings and fawn production estimates net fawn production (i.e., after mortality on fawns during summer and the hunting season is accounted). In this example, fawn production was measured after the fawning period and the ratio of fawns: 100 females measured in December was used to measure the net fawn crop ($4971 \times 0.80 = 3977$). The prehunting season population size and composition are then determined by allocating the 55% of the fawn production to males and 45% to females and adding those values to the prefawning population for a total production of 13,177 deer.

The harvest was based on results of hunter surveys and wounding loss was set at 20% of the legal harvest. Thus, the postseason population can be estimated by subtracting the harvest and wounding loss from the prehunting population (i.e., 13,177 −1560 = 11,617)

As with the life equation for quail, the initial estimates used in the model will be imprecise. However, after several years of obtaining information about the population, reasonable estimates of mortality and reproduction can be generated that will fine-tune management of the population in question.

Emigration and Immigration

Emigration and immigration are two other basic components of populations and are usually discussed in conjunction with dispersal. *Dispersal* (i.e., movement away from natural areas or into unfamiliar areas) is a constant characteristic of populations but more pronounced when populations are high. When animals leave a population and do not return, they are *emigrants. Immigration* is one-way movement into another area or population. Emigrants from one region are immigrants into another region (Smith 1977). Managers often assume that the effects of emigration and immigration cancel each other out. This is not always the case. In stable populations, those at or near carrying capacity, animal movements may not influence population dynamics. However, emigrations caused by overpopulation and forage shortages that reduce populations may influence the age structure and reproductive rate of the remaining population. On the other hand, immigration into populations can increase the rate of growth by adding new members and by increasing the birthrate (Smith 1977).

Emigration involves density-independent (or innate) or density-dependent dispersal. Innate dispersers are predisposed at birth to move beyond the confines of their parental home. Dispersers leave their habitat and often move into unfavorable areas (Howard 1960). This type of movement may enable a species to expand its range and gene flow between populations (Wynne-Edwards 1965).

Density-dependent dispersal is a response to altered habitats (e.g., deteriorating environmental conditions, crowding) that is most likely to occur when a surplus of individuals must die or emigrate due to population pressures and declining carrying capacity (Errington 1956). This forced emigration increases the chances that

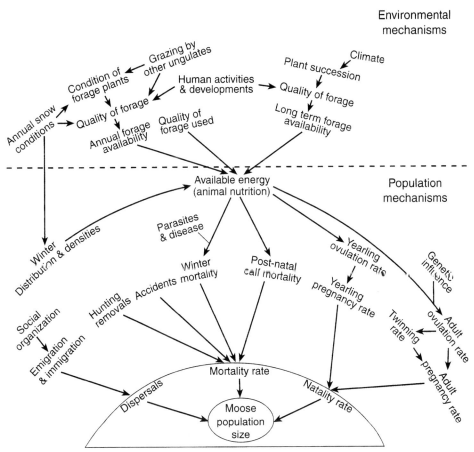

Figure 5–12 Environmental and population mechanisms that influence the size of moose populations in Jackson Hole, Wyoming. The growth of the population can be negative, positive, or zero, depending on the interaction of all factors and the shifts in the values assigned to the components that control growth (from Houston 1968).

those individuals will not be in population crashes; however, survival into the new areas is low. There are advantages from the immigrants: colonization into previously unoccupied areas and genetic exchange from the successful individuals (Smith 1977).

These basic components of a population (i.e., natality, mortality, emigration, immigration) do not operate independently. Houston (1968) illustrated how environmental mechanisms influenced the population mechanisms for a moose population in Jackson Hole, Wyoming (Figure 5–12). Each of these mechanisms needs to be examined to obtain a solid understanding of the complexity of populations. At this point the analogy of a population as a lake (Figure 5–2) clearly breaks down and the dynamics of population are exposed.

SUMMARY

Populations are influenced by birth, death, immigration, and emigration. One way to incorporate these basic parameters into a snapshot of the population is to create a life table or life equation.

6

The Rate of Increase and Population Growth

"Rate of increase reveals much more about a population than the speed with which it grows. It measures a population's general well-being, describing the average reaction of all members of the population to the collective action of all environmental influences. No other statistic summarizes so concisely the demographic vigour of a population."

G. Caughley (1977)

INTRODUCTION

Knowledge of the rate of increase (r) may be the single most important statistic a biologist can obtain from a population because it measures a population's general health by describing the average reaction of all members in the population to the array of environmental influences (Fig. 5–11) that shape the population (Caughley 1977). Life tables summarize the mortality schedule of a population. Biologists also need to consider the reproductive rate of a population and techniques that can combine reproduction and mortality estimates to determine net population changes. This is especially true for managers that want to document the response of a population (i.e., increase, decrease, remain the same) to some management activity.

White (2000) compares this documentation to bookkeeping where the biologist can keep track of the four components of population change (i.e., births, deaths, immigration, and emigration) with the model:

$$N_{t+} = N_t + B_t - D_t + I_t - E_t$$

N = population size at time t + 1 is equal to the population size at time t plus births (B) minus deaths (D), plus immigrants (I) minus emigrants (E). By knowing

115

the net change in the population from year to year, you know if the population remains the same, increases, or decreases. The rate of increase for a population changes constantly and is strongly influenced by environmental conditions. To obtain insight into population changes, an examination of the rate of increase is important. To facilitate that understanding, a review of the concept of rates and examples of how the rate of increase is measured are helpful.

■ WHAT ARE RATES?

A rate is a numerical proportion between two sets of things of interest. For example, the number of caribou with a disease might be measured as six per 100 animals in the population or 6%. If 30 deer out of 1500 deer were killed by vehicles as they crossed an interstate, the road-killed rate would be 2%. In population studies a rate is expressed with a standard time base. If eight quail chicks out of 12 die within one year, the mortality rate would be 66.6% per year. If the population grows from 100 to 150 within one month, the rate of population increase will be multiplied by 1.5 per month (150 ÷ 100) or increase 50% per month [(1.5 − 1) 100]. Wildlife biologists usually think in terms of finite rates, which are simple expressions of observed values. For example, the annual survival rate $= \frac{N_t}{N_0}$ where N_t equals the number alive at the end of the year and N_0 equals the number alive at the start of the year.

Rates can also be expressed as instantaneous or finite. The difference between the instantaneous and finite rate is that with the instantaneous rate, the time base becomes very short rather than a month or a year.

The finite rate $= e^{\text{instantaneous rate}}$ where

instantaneous rate $= \log_e$ finite rate,

where $e = 2.71828$, which is the base of natural logs (i.e., Naperian).

Some examples will illustrate this relationship.

Given a population size of $N = 100$, increasing at a finite rate of 10% per year, the population size at the end of year 1 will be:

$$100\left(1 + \frac{1}{10}\right) = 110$$

At the end of year 2, it would be

$$110\left(1 + \frac{1}{10}\right) = 121$$

At the end of year 3,

$$121\left(1 + \frac{1}{10}\right) = 133.1$$

To repeat these calculations for a finite rate of 5% per 6 months:
at 6 months,

$$100\left(1 + \frac{1}{20}\right) = 105$$

and at 1 year

$$105\left(1 + \frac{1}{20}\right) = 110.25$$

Compare this rate (i.e., 110.25) to the higher increase rate per year (i.e., 110.0). It is easy to see that a 5% increase per 6 months provides a bigger yield than 10% per year.

Because biological systems often operate on a time schedule of hours and days, it is realistic to use rates that are instantaneous, rates that break up the year into many short time periods. For example, if the instantaneous rate of increase = 10% per year, we can divide the year into 1000 short time periods with each having a rate of increase of $0.1 \div 1000 = 0.0001$. Then r = 0.0001 for each of the 1000 periods. For the first 1000th of the year, 100 (1 + 0.0001) = 100.01, until the last where 100 $(1 + 0.0001)^{1000} = 110.5$.

Looking now at rates of increase where

N_o = starting population size and,

N_t = population size at a later date (usually for a
time period of a week, month, or year).

By definition:

(e^r) finite rate of increase $= \dfrac{N_t}{N_o} = e^{rt}$ and

$\log_e$ (finite rate of increase) = instantaneous rate of increase

The rate of increase can be measured by either the finite rate (e^r) or the instantaneous rate. However, the instantaneous rate is more useful in population studies for at least three reasons.

1. Instantaneous rates are easier to deal with mathematically. The instantaneous rate is centered at zero. If the rate of increase of one population was measured as a finite rate, $e^r = 1.649$ and a second as $e^r = 0.607$, a direct comparison of the rates does not indicate that the rate of increase of the first population equals the rate of decrease of the second. However, if the rates were expressed as instantaneous rates, then the rate of increase of the first population is 0.5 and the second is –0.5 and the relationship is obvious (Table 6–1).
2. Instantaneous rates convert easily from one unit of time to another. When

$$r \,/\, \text{year} = \chi, \text{r/day} = \frac{\chi}{365}$$

This conversion is not possible with finite rates.

Table 6-1   The relationship between finite and instantaneous rates of increase. The relationship shows how both rates diverge as they become large. It also illustrates that the change in the size of a hypothetical population that starts with 100 members increases or decreases at the specific rate for one time period. Note that finite rates are always positive or zero but instantaneous rates range from $-\infty$ when decreasing to $+\infty$ when increasing (from Krebs 1978)

% Change	Finite rate	Instantaneous rate	Hypothetical population at end of one time period
−75	0.25	−1.386	25
−50	0.50	−0.693	50
−25	0.75	−0.287	75
−10	0.90	−0.105	90
− 5	0.95	−0.051	95
0	1.00	0.000	100
5	1.05	0.049	105
10	1.10	0.095	110
25	1.25	0.223	125
50	1.50	0.405	150
75	1.75	0.560	175
100	2.00	0.693	200
200	3.00	1.099	300
400	5.00	1.609	500
900	10.00	2.303	1000

	decreases		increases	
	0	1.00	$+\infty$	Finite rates
	$-\infty$	0.0	$+\infty$	Instantaneous rates

3. Instantaneous rates are a useful indication of the rate at which a population grows by providing doubling time. The doubling time can be calculated from r by dividing 0.6931 by r (Caughley and Sinclair 1994).

(If $\dfrac{N_t}{N_0}$ = 2, then e^{rt} = 2, rt = $\log_n 2$ = 0.6931, thus doubling time = 0.6931 ÷ r; Caughley and Sinclair 1994)

If r = 0.405 for a population, then it will take
0.6931 ÷ 0.405 = 1.7 years to double.

If r = 0.23, it takes 3 years to double.

If r = − 0.23, it halves every 3 years.

To apply these rates to a population, imagine you have data from a deer population with a high coyote density and poor habitat conditions. If during fawning the fawn mortality in the first week is 50% and 90% in the second week, how could you combine these mortalities? If expressed as finite mortality rates, you cannot add them because 50% plus 90% is not 140%, but only 95% mortality (e.g., 50% + 90% of 50% = 95%). If, however, mortality is expressed as an instantaneous rate, the mortality could be added directly.

$$\text{First week mortality} = 50\% = \frac{N_t}{N_o} = \frac{50}{100} = 0.5$$

$$\log_e (0.5) = -0.693$$

$$\text{Second week mortality} = 90\% = \frac{N_t}{N_o} = \frac{5}{50} = 0.1$$

$$\log_e (0.1) = -2.303$$

$$\text{Combined loss} = (-0.693) + (-2.303) = -2.996$$

To convert back to a finite rate,

$$e^{-2.996} = 0.05$$

Another example would be a population that increases from 73 to 97 in one year. This increase can be expressed as a finite rate of population growth:

$$\frac{N_t}{N_o} = \frac{97}{73} = 1.329, \text{ meaning the population was multiplied by 1.329 or}$$

$(1.329 - 1)(100) = 32.9\%$, meaning the population increased 32.9% in one year.

The increase can also be expressed as an instantaneous rate of population growth as $\log_e (\text{finite rate}) = \log_e (1.329) = 0.284$.

If a population decreases from 67 to 48 in one month, the finite rate of population growth is

$$\frac{N_t}{N_o} = \frac{48}{67} = 0.7164, \text{ meaning the population was multiplied by 0.7164 or}$$

$$(0.7164 - 1)(100) = -28.4\%, \text{ meaning the population decreased 28.4\% over the month.}$$

The instantaneous rate $= \log_e (\text{finite rate}) = \log_e (0.7164) = -0.334$.

■ MEASURING THE RATE OF INCREASE

Obtaining a measure of the instantaneous rate of increase is easier to describe than calculate because of the factors that influence it. As a result, parameters are established around rates so they can be discussed in a universal manner. The intrinsic rate of increase (r_m) or biotic potential is the exponential rate at which a population with a stable age distribution grows when no resource (i.e., food, water, cover, special factor) is in short supply. The need for a stable age distribution when calculating r_m is obvious because age influences reproduction and mortality rates. The intrinsic rate of increase has also been called the "innate capacity for increase" and the Malthusian parameter (Andrawantha and Birch 1954) and is actually the potential rate of increase of a population that is in an optimum situation. It is a theoretical concept because the optimum rarely occurs. This does not make r_m useless from the point of view of understanding nature. This rate of increase provides a model with which biologists can compare the actual observed rates of increase in nature.

Models are often downplayed because they are based on incomplete information but they are useful to: 1) predict results of management activity, 2) conceptualize the dynamics of a population in mathematical notation, and 3) test hypotheses about population dynamics from observed data (White 2000). Each of these is important to better understand the resources that biologists manage. An exact measure of r_m is difficult, but any population in a particular environment will have a:

1. mean longevity or survival rate,
2. mean birth rate, and
3. mean growth rate of individuals

Each of these values will be determined by the environment and the innate quality of the organisms themselves. Because environments constantly vary (e.g., winter versus summer, wet versus dry) and organisms are dynamic, the rate of increase is not constant (Figure 6–1). In nature biologists often assume that negative influences on populations are compensated for by positive influences, resulting in an average population rate of increase around zero. In other words, over time the negative rates of increase are balanced by the positive rates of increase, so in the long term the rate of increase is zero.

The environment varies and is never consistently favorable (to a population) or unfavorable. Instead the environment fluctuates between favorable and unfavorable. When conditions are favorable, the population's capacity for increase is positive and numbers increase and vice versa (Fig. 6–1). As a result, the actual rate of increase constantly varies from positive to negative in response to changes within the population in

a. age distribution,
b. social structure,
c. genetic composition, and
d. response to the environmental (Krebs 1978).

Krebs (1978) further defines different ways of examining rates of increase .

Figure 6–1 Measuring the rate of increase for a population over time is difficult because of the negative and positive influences on populations.

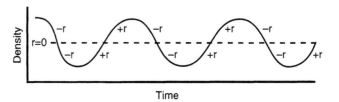

Maximal Rate of Increase (r_m)

The r_m, or the innate capacity for increase, is obtained under controlled conditions (e.g., particular combination of temperature, humidity, quality of forage, without negative influences). In other words, there are two types of environmental factors when determining r_m: optimal and not necessarily optimal but controlled and specified (Andrawantha and Birch 1954). Optimal conditions include the ideal density, space, and resources. Specified conditions would be where forage, temperature, space, and other resources are controlled. Thus, r_m is arbitrarily defined in relation to a specific set of conditions. In free-ranging populations r_m will depend on the birth rate, longevity, and speed of development, which are measured by birth rates and death rates. When the birth rate exceeds the death rate, the population will increase. However, the birth rate and death rate of a population vary with age so life tables can be useful. The l_x column, the proportion of the population surviving to age x, can be used to calculate r_m. Furthermore, the birth rate of a population is expressed as an age schedule of births or a fecundity table (Caughley 1977).

Given this information, one can obtain the net reproductive rate (R_o) and mean length of a generation (G) that can be used to calculate r_m.

$$\text{First calculate } R_o \text{ as: } \frac{\text{number of daughters born in generation } t + 1}{\text{number of daughters born in generation } t}$$

R_o is the multiplication rate per generation (i.e., mean period elapsing between the birth of parents and the birth of offspring) and is obtained by multiplying together the l_x and m_x schedules and summing over all age groups:

$$R_O = \sum_0^\infty l_\chi m_\chi = \sum_0^\infty V_\chi \text{ where } m_\chi = \text{female live births/female (Table 6–2)}$$

Table 6–2 Calculation of the net reproductive rate (R_o) of thar (from Caughley 1966)

Age group (x)	Proportion surviving To age (l_x)	No. female live births per female at age x (m_x)	Product of l_x and m_x (V_x)
0	1.000	0.000	0.000
1	0.467	0.005	0.002335
2	0.461	0.135	0.062235
3	0.433	0.440	0.190520
4	0.387	0.420	0.162540
5	0.331	0.465	0.153915
6	0.269	0.425	0.114325
7	0.209	0.460	0.096140
8	0.155	0.485	0.075175
9	0.109	0.500	0.054500
10	0.073	0.500	0.036500
11	0.047	0.470	0.022090
12	0.029	0.470	0.013630
> 12	0.000	0.000	0.000
		$R_o = \Sigma l_x m_x =$	0.983905

The derived value of 0.98 (Table 6–2) for R_o means the population would multiply 0.98 times in each generation. If $r_m = 1.0$, the population is exactly replacing itself. If $r_m < 1.0$, it is decreasing, and if > 1.0, the population is increasing.

The next step is to determine the mean length of a generation. Because young are born over a period of time and not all at once, the mean generation length is defined as (Dublin and Lotka 1925):

$$G = \frac{\sum l_x m_x X}{\sum l_x m_x} = \frac{\sum l_x m_x X}{R_o}$$

Applying this to the thar population (Table 6–2)

$$G = \frac{\sum l_x m_x X}{R_o} = \frac{5.340480}{0.983905} = 5.4 \text{ years}$$

Finally, with R_o and G, r_m can be calculated as an instantaneous rate:

$$r_m = \log_e (R_o) \div G$$

For the thar population,

$$r_m = \frac{\log_e (0.983905)}{5.4} = 0.003$$

This example provides an approximate estimate of r_m because G is an approximation when generations overlap.

Lotka's Equation

The r_m can be determined more accurately by using Lotka's (1907) equation, which is the basic equation of population dynamics from which many statistics of population analysis are derived:

$$\sum l_x e^{-rx} m_x = 1.0$$

It is obvious that l_x and m_x are directly related to r by Lotka's equation. This formula expresses the dynamics of the female segment of a population having fixed schedules of survival (l_x) and fecundity (m_x) in relation to the rate of increase (r_m). The derivation of the equation is explained by Caughley (1977) and Krebs (1978) with a hypothetical animal used as an example.

Assume that r_m generated from R_o and G was 0.826 from the following data:

x	L_x	m_x	$V_x = l_x m_x$	$x(l_x)(m_x)$
0	1.0	0.0	0.0	0.0
1	1.0	2.0	2.0	2.0
2	1.0	1.0	1.0	2.0
3	1.0	0.0	0.0	0.0
4	0.0	—	—	—

Then e^{-rx} for each age class x would be:

x	$e^{-0.826x}$		$l_x e^{-rx} m_x$
0	1.00		0.00
1	0.44		0.88
2	0.19	and	0.19
3	0.08		0.00
4	0.04		0.00
	Therefore, $\Sigma l_x e^{-rx} m_x = 1.07$		

This value is >1.0 as defined by the equation, so r_m of 0.826 is slightly low. By trying different values of r_m (i.e., 0.881) the results would be as follows:

x	$e^{-0.881x}$		$l_x e^{-rx} m_x$
0	1.00		0.00
1	0.41		0.82
2	0.17	and	0.17
3	0.07		0.00
4	0.03		0.00
	Therefore, $\Sigma l_x e^{-rx} m_x = 0.99$		

This is close enough to 1.0 as directed by Lotka's equation and is a better approximation for r_m so $r_m = 0.881$ and not 0.826.

The variables in this equation have been defined in relation to the female segment but could also have been expressed as male survival and male births per male. As long as both male and female schedules of survival and fecundity are fixed, the value of the rate of increase appropriate to the male segment is the same as the rate of increase for the female segment (Caughley 1977).

Lotka's formula is an abstraction from nature because survival and fecundity schedules do not remain constant. However, by comparing complex natural populations to the basic equation and evaluating the deviations from the simple system, more complex systems can be examined.

Calculation of r from Population Estimates

Caughley (1977) presents a formula to calculate an observed exponential rate of increase ($\bar{r}$). This calculation can be made when the popualtion exhibits a birth pulse, population estimates are accurate, and population estimates are completed at the same time each year. First, convert the indices of population size to natural logs (N). Then, set yearly time limits (t) so the first $= 1.0$. Then determine $\bar{r}$ over the period of observation with the following formula:

$$\bar{r} = \frac{\Sigma Nt - \dfrac{(\Sigma N)(\Sigma t)}{n}}{\Sigma t^2 - \dfrac{(\Sigma t)^2}{n}}, \text{ where}$$

$$N = \text{natural log of the population}$$

$$t = \text{time, and}$$

$$n = \text{number of estimates.}$$

Assuming that the census technique used is accurate the calculation or $\bar{r}$ is useful in many wildife studies. For example, given the following:

Number of elk in a population	(Number)	1500	1550	1600	1575
Number converted to natural logs	(N)	7.313	7.346	7.378	7.362
Yearly Census	(t)	1	2	3	4

$$\bar{r} = \frac{7.313(1) + 7.346(2) + 7.378(3) + 7.362(4) - (7.313 + 7.346 + 7.378 + 7.362)(1 + 2 + 3 + 4) \div 4}{1^2 + 2^2 + 3^2 + 4^2 - (1 + 2 + 3 + 4)^2 \div 4} =$$

$$\frac{73.587 - 73.4975}{30 - 25} = \frac{0.0895}{5} = 0.0179$$

Values of $\bar{r}$ are often used in ungulate management to understand how much populations fluctuated from year to year. Several yearly censuses can provide these data.

To understand rates of increase, information on mortality and fecundity are important. Generally r_m will be less than 0.5 for large mammals and no more than 5.0 for small mammals and birds (Caughley 1977).

■ THE INFLUENCE OF AGE DISTRIBUTIONS ON r_M

When survival and fecundity schedules are fixed, the age distribution of males and females converge to a stable form. When this occurs, the rate of increase becomes constant (Lotka 1907, Sharpe and Lotka 1911). This effect will occur whether or not rates of mortality are constant with age, whether or not survival and fecundity schedules of males and females are the same, and regardless of whether the rate of increase intrinsic to these schedules is positive, negative, or zero (Caughley 1977). Unless a population consists of seasonal breeders with nonoverlapping generations (i.e., annual plants, some insects), it will be characterized by a certain age structure that will have considerable influence on r_m. When l_x and m_x are fixed for a population, each segment of the population (e.g., males, females, age classes, breeders, nonbreeders, juveniles and adults, or any combination or subdivision) increases at the same rate when a stable rate of increase is achieved. Furthermore, if any segment of the population (except a post-reproductive stage) shifts to a new rate of increase and holds to the rate, the rate of increase of all other segments will converge to it (Caughley 1977).

This concept has serious implications in the management of populations. By manipulating a single segment of the population (e.g., juveniles, adult males), the rate of increase for the population can be altered because the rate of increase will change in a single segment as discussed by Caughley (1977). For example, if the objective of management were to reduce or stabilize the numbers of a growing population, the result could be obtained by stabilizing numbers in a single segment because the size of all other segments of the population would then stabilize because their rates of increase converge (i.e., r = 0) to that of the manipulated segment.

Populations can usually be divided into three ecological age periods: prereproductive, reproductive, and postreproductive. The relative length of each obviously depends on the organism. For example, elephants have a long prereproductive period (i.e., they are not sexually active until at least 10 years of age) and a long postreproductive period. Other large species like deer begin to breed by the second year of life and do not have a postreproductive period. Among annual species the length of the prereproductive period has only minor influences on the rate of increase. In longer lived plants and animals the length of the prereproductive period has a greater influence on the rate of growth. Organisms with a short prereproductive period often increase rapidly with a short span between generations. Organisms with long prereproductive periods often increase more slowly and have a longer span of time between generations.

Cole (1954) demonstrated how humans respond to the age at first reproduction. If all women first reproduced at 20, they would have 3.0 children for $r_m = 0.02$. If the first birth is delayed to 30, 3.5 children would be required to provide $r_m = 0.02$. To obtain $r_m = 0.04$, women could start at 13 and produce 3.5 children or start at 25 and have 6.0 children. The age at first reproduction is obviously important to determine growth. Two other factors have a strong influence on r_m: increase in litter size and increase in the number of litters (i.e., increased longevity) (Krebs 1978).

Changes in the age class distribution reflect changes in the production of young, their survival to maturity, and the period of life when most mortality occurs. Any influence that causes age ratios to shift because of changes in age-specific death rates affects the population birth rate. If a population has life expectancy for the oldest age classes reduced, a higher proportion will fall into the reproductive class and the birth rate will increase. On the other hand, if life is extended, a larger portion falls into the post-reproductive class and the birth rate will decrease.

Theoretically, continuously breeding populations move toward a stable age distribution where the population has a constant schedule of birth rates and death rates and will maintain this stable age distribution indefinitely (Lokta 1922). When the proportion of age classes in a population is constant, mortality = natality. If this stable age distribution is disrupted by anything (e.g., humans, nature, disease), the age composition will tend to return to predisturbance distribution when the disruption is eliminated as discussed earlier.

Populations that have reached a constant size in which the birth rate equals the death rate assume a fixed age distribution (i.e., stationary age distribution) (Krebs 1978), which is a special case of the stable age distribution (Caughley and Sinclair 1994). This stationary age distribution and stable age distribution are the only two situations in which a constant age structure is maintained in a population. Populations, however, do not assume a constant age structure because they change over time and in natural populations age structure is constantly changing. Furthermore, natural populations rarely have a stable age structure or stationary age distribution because populations do not

Figure 6-2 Relationship between factors and rates on population increase or decrease (after Houston 1968, Krebs 1978).

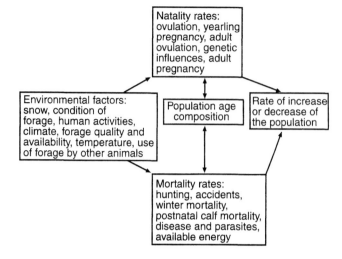

increase for long in an unlimited fashion, and are rarely in a stationary phase for long (Krebs 1978). The complexity of these relationships (Figure 6–2) is similar to the mechanisms that influence moose populations (Figure 5–11).

Although Krebs (1978) presents stable age distributions as completely theoretical, Caughley and Sinclair (1994) argue that the stable age distribution is a robust configuration that occurs rapidly after mortality and fecundity patterns stabilize. When populations of large mammals have mortality and fecundity schedules that do not fluctuate much from year to year, they will have an age distribution close to the stable age distribution. This would not be the case for large mammals living in a fluctuating environment or small mammals, fish, and birds whose age-specific fecundity and mortality is much more variable (Caughley and Sinclair 1994).

■ THE GROWTH OF A POPULATION

How populations grow is dependent on the four fundamental demographic processes: birth, death, immigration, and emigration (Johnson 1994). As a result, growth forms vary from population to population and even between environments for the same population. Because each of the four demographic processes is influenced by environmental and physical factors (Figure 5–11), growth is determined by the availability of resources, something an animal needs to survive and reproduce (Caughley and Sinclair 1994). For example, the most important resource for a population of desert bighorn sheep may be freestanding water. Water may limit distribution and abundance. However, another population may have abundant water but only limited habitat for lamb rearing. Yet another population of bighorn may simply be limited by a shortage of space. As an example, desert bighorn sheep near Tucson, Arizona were surrounded by human development, essentially fenced in, and movement to other areas was terminated (Krausman 1996). In this case, space was a required resource just as important as others.

Figure 6–3 The mechanisms that control population growth during period b should be similar to the mechanisms that control growth during period c. The growth curve during period c is essentially a segment of the initial population growth curve (a and b).

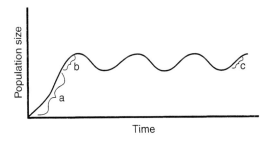

The goals of wildlife management (i.e., increase, decrease, or keep numbers stable) often require the modification of population growth so it is important to understand what kind of process is being modified. The utility of understanding the mechanisms that control population growth (or decline) does not need to be restricted only to situations where populations have grown from the lowest viable population size to a high population size. For example, the mechanisms that control population growth during one point in time should be similar to the mechanisms that control it at another (Figure 6–3). The change in population size (i.e., lambda, λ) between time intervals is the ratio of the size of the population next year (t+1) to the size this year (t):

$$\lambda = \frac{N_{t+1}}{N_t} \text{ or the finite rate of population growth}$$

For convenience, λ is often expressed as an exponent or $\lambda = e^r$ as discussed earlier and referred to as exponential growth.

Exponential growth occurs when no resources are limiting the population and it describes a population that grows at a constant rate. This rate is determined by the innate physiology interacting with elements such as temperature, topography, and other components of the environment that are not resources. As a result, the rate of increase is not a constant for species (Caughley and Sinclair 1994). Collared peccaries released in northern Arizona with all the resources they need will not increase at the same rate as if they were in the southern part of their range because the temperature is too cold for them to reproduce in northern Arizona. The intrinsic rate of increase varies for a species according to the physical environment (Caughley and Sinclair 1994).

When no resources are limiting, the population increases (or decreases) at the rate of increase for that species in that area (i.e., environment) so that

$$N_{t+1} = N_t e^{rt}$$

The exponential growth will be a steepening curve (Figure 6–4) when population size is graphed against time. New individuals are continuously added to the principal that continuously provides an expanded base of reproductive individuals. Exponential growth is abruptly terminated when the number of animals the environment can support is surpassed (e.g., due to a shortage of space). At this point population growth levels off and population growth shifts from one state to another (Figure 6–5).

Figure 6–4 Exponential growth that describes a population growing at a constant rate.

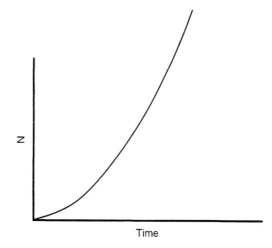

Figure 6–5 Population growth when limited by a nonconsumable resource (after Caughley and Sinclair 1994).

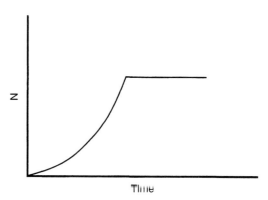

If a resource is eliminated by the species, the population will crash. First, the population will increase while the resource is present. As the resource is reduced, the growth rate will reduce until the population trajectory reverses and the population declines (Caughley and Sinclair 1994). For example, caribou were introduced to Saint Matthew Island, Alaska, where forage (i.e., lichens) were abundant. However, caribou consumed the lichen faster than the forage renewal rate. When forage was available, the caribou population increased but then decreased when the forage was exhausted (Figure 6–6).

These types of curves (i.e., exponential) are generally characteristics of rapidly reproducing organisms (e.g., maturing annual plants, seasonal insect flushes, some rodents) and because exponential growth cannot continue indefinitely (because of the influence of growth-depressing factors, external and internal to the population), an S-shaped curve will result. The S-shaped curve (also called the logistic curve on the Malthusian growth curve) (Figure 6–7) is a specialized case and applicable only in the unusual circumstances when the limiting resource for a population is produced at a rate independent of the animals using it. Essentially, animals are us-

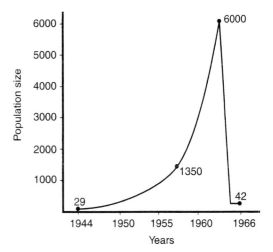

Figure 6–6 Assumed population growth of the reindeer herd on St. Matthew Island, Alaska (from Klein 1968).

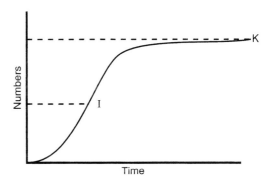

Figure 6 7 The S-shaped or logistic curve resulting from population growth following the logistic equation (Odum 1971). Carrying capacity (K) represents an equilibrium level at which limiting factors prevent further growth. The inflection point (I) corresponds to the population density at which maximum production by the population occurs (Mentis 1977).

ing the current growth without influencing the parent from which the growth was generated. Environmental resistance acts on a growing population and growth is progressively inhibited until the population reaches some asymptotic level (i.e., carrying capacity) that represents the maximum number of individuals that can be supported in a given habitat. Because natural systems rarely fit the specialized case, why is the logistic equation used so commonly in wildlife management?

Caughley and Sinclair (1994) explain that it is so commonly used because when presented as a formula:

$$dN/dt = r_mN(- N/K)$$

there is no information about the resource. When the population (N) is at carrying capacity (K), the logistic curve flattens out to zero, so dn/dt = 0 is only a function of N, r_m, and K, and implies nothing about resources or that the population's relationship to them is necessary. This is clearly unrealistic, as the logistic curve is an oversimplification of nature. There are several assumptions that must be valid before a population follows the logistic growth curve (Slobodkin 1980, Krebs 1978).

1. All animals in the population have identical ecological properties. In other words, all members are equally likely to die, eat, be eaten, or give birth. This would eliminate the influence of age structure.
2. All animals in the population respond instantly and in the same way to changes in the environment. There would be no time lags. In reality it takes time for the body to respond to environmental changes (e.g., sickness) and bodies do not all operate in an identical manner.
3. There is some constant upper limit to population size in any particular situation, and the rate of increase of the population at any particular time is linearly proportional to the difference between the population size at that time and this upper limit.
4. Density is measured properly.
5. The population initially has a stable age distribution. The logistic model assumes that a population beginning growth increases at a rate nearly equal to rN.

These assumptions can be reduced to;

$$\frac{dN}{dt} = rN\left(\frac{K - N}{K}\right)$$

Because all experimental populations can be made to start at a relatively small size and terminate at a larger size, the logistic curve can be fitted fairly well to at least the initial portions of population growth curves.

Even though the logistic curve is an abstraction and should be considered cautiously, it is useful to wildlife management in several situations (Caughley and Sinclair 1994).

1. It is a convenient and intuitively satisfactory mathematical description of a frequently observed pattern of population growth. As a result, it is used in modeling even when not entirely appropriate. In many management situations animals are added to or removed from populations. Prior to the action, managers often ask what the influence will be on other animals. Answers will depend on the population growth, so the logistic equation is often used as a representative example of a model where growth is limited by a resource shortage.
2. When used as a model of population growth, it can provide insight into the factors that affect population growth. In the past a lot of emphasis was placed on logistic growth curves. Its main use was to demonstrate the mathematical techniques to ecology and to serve as a catalyst for conceptual thinking. However, it is not useful as a model to test theories because judgement has to be subjective and there are always deviations from the best-fit curve and statistical methods for evaluating them are lacking. Also, a variety of theoretical curves can be accepted as a satisfactory fit to a heterogeneous set of ecological conditions, and if the fit is nice, it is usually accepted without questioning the faults.
3. The logistic curve has been and still remains a useful demographic tool (if not overextrapolated from insufficient data). When a population has an intrinsic rate of increase greater than 0.1 on a yearly basis, the population growth tends

Figure 6–8 Most animal populations do not meet the assumptions of the logistic curve and grow in nonsymmetrical fashion and not in a symmetrical manner.

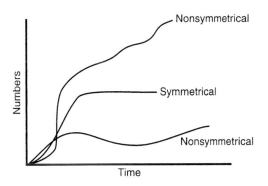

to an asymptote to some degree. Thus, the logistic curve provides a reasonable approximation to the actual growth curve.

4. Plant growth follows the logistic curve because plants do not influence the rate of renewal of their resources (i.e., water, sunlight) and cannot influence the availability of these resources to the next generation. The growth of many plant populations is close to the logistic (Noy-Meir 1975)

5. Finally, the logistic curve is used when showing how factors influence wildlife populations because the relationship is close to validating logistic assumptions. Houston (1982) described an elk population in Yellowstone National Park that consumed hayed grass in winter. As a result, the consumption of grass by elk has no influence on the amount of grass available at the beginning of the next winter and the elk population growth approximates a logistic growth.

In reality, animal populations do not grow in the logistic form (Figure 6–7). The assumptions for logistic growth cannot be met and populations tend to increase more in nonsymmetrical fashion (Figure 6–8). Nonsymmetrical growth is any type of S-shaped curve that can have any length through time or numbers (Figure 6–9).

Most animal populations have an influence on the resources they rely on and the parent resources that produced them, and there is no distinction between them. The relationship is interactive (Caughley and Sinclair 1994) and is common in nature (i.e., predator, prey, plant-herbivore). The interactive relationship yields two outcomes (in a constant environment): a point equilibrium or a stable limit cycle. A point equilibrium is reached when the animal population and the resources are balanced at fixed densities. A stable limit cycle results in fluctuation by the animal population and the resource. Both are the result of the same processes, and which occurs depends on the constants of the equations describing the interaction. Both are theoretical because the stable limit cycle is rare in nature (i.e., one has not been identified) and the point equilibrium implies that populations and their consumable resources are locked into a fixed ration (which they are not). Although theoretical, they are useful in wildlife management because they are unifying (i.e., the density of the animals and the levels of the resource will interact to move towards the equilibrium point). As the animal or resource is displaced from the equilibrium, there is a stronger pull back towards the point. This is referred to as negative feedback

(Caughley and Sinclair 1994). Negative feedback is the process through which wildlife managers manage populations. Caughley (1976) provides a detailed treatment of the rate of increase and population growth.

SUMMARY

Rates of increase are important to know so populations can be better understood. Rates (i.e., a numerical proportion between two sets of things of interest), or more precisely the rate of increase for a population, can be measured from Lotka's equation or from population estimates. Age distribution has an important influence on the rate of increase. Population growth ultimately depends on birth, death, immigration, and emigration and how they interact with resource availability.

7

The Concept of Carrying Capacity

"I did not discover in order to learn; I learned in order to discover."

E. O. Wilson

INTRODUCTION

Carrying capacity (K) is central to the management of wildlife populations. Most biologists have a fairly good idea of what K means, but it is used so often in so many ways that the meaning is often obscured. For example, Macnab (1985) included K with overpopulation, overharvesting, and overgrazing in his list of resource management shibboleths. Because the use of K varies, it is important for wildlife biologists to understand the biological and sociological meanings (Hall et al. 1997) and the methods used to estimate K. Miller and Wentworth (2000) provided an excellent discussion of K from which this chapter follows.

■ DIFFERENT VIEWS OF CARRYING CAPACITY

Leopold (1933:50–55) was the first to differentiate between the maximum number of animals of the same species a wide area may support and the maximum density that a particular but less perfect range is capable of supporting. The latter limit is literally saturation for that particular range and is variable between several ranges. The true K capacity occurs when numerous widely separated optimum ranges exhibit the same K. Saturation points are a property of the species but K is a property of a unit of range (Leopold 1933:51). Since 1933 an array of terms have been applied to K as a result of the diverse fields where K is applied (i.e., ecology, wildlife management, range management, wildland recreation). Unfortunately, many of the definitions have little value (Edwards and Fowle 1955) and are theoretical. For example, Dasmann (1981) listed three ways the term K is used in the literature. First, K is the number of animals of a given species that are actually supported by a habitat, over a period of years. This is used mainly for bird populations where the population levels off at about the same point each year. This level was reached by a population growth in a habitat and is usually used in connection with the logistic curve. Second, K is the upper limit where natality equals mortality and is the balance between biotic potential and environmental resistance. The third situation involves the number of animals that a habitat can maintain in a healthy vigorous condition and was adapted from range management and animal husbandry. This use essentially requires that no resources are limiting and all animals (i.e., livestock) are in the best possible condition (Stoddart et al. 1975). With the use of the last definition there is a marked distinction between Leopold's (1933) ecological K and the concept of economic K, which takes nonwildlife into consideration.

Further confusion results from opinions that K only has meaning with reference to forage quality or that K should be expanded to include other factors (e.g., predation, water, disease). For example, Caughley (1976, 1979) defined K as an equilibrium between animals and vegetation. However, other factors influence animal abundance (e.g., cover, social tolerance, distribution). For some species these views of K are just as important as forage resources and need to be considered (Dasmann 1981). Since Leopold (1933) introduced the concept of K, numerous authors have attempted to synthesize the various meanings and to present a standard terminology (Caughley 1976, 1979; Dasmann 1981; Bailey 1984; Macnab 1985; McCullough 1992). They have been categorized as ecologically based and culturally based K (Table 7–1).

Ecologically Based Carrying Capacities

All definitions of K should be ecologically based. Some definitions, however, are based only on the response of populations to their environments without considering anthropogenic influences (i.e., ecologically based K). Each ecologically based K describes populations at a stable equilibrium with some environmental influence (i.e., forage, water, cover).

K-Carrying Capacity

K-carrying capacity is the maximum number of animals of a given population that can be supported by the available resources (McCullough 1992). This is the most commonly used definition of K (Miller and Wentworth 2000) and has been called *subsistence density* (Dasmann 1981), *potential K* (Riney 1982), and *ecological K* (Caughley 1976). Although some authors include limitations based on space, cover, or other resources in their definition of K-carrying capacity, Miller and Wentworth (2000) treated definitions based on limitations other than forage as unique cases because removal or manipulation of these other limitations may allow a population to grow until it becomes limited by the availability of forage.

Table 7–1 Defintions of carrying capacity (K) (adapted from McCullough 1992)

Types of K	Definition	Synonyms
Ecologically-based		
K-carrying capacity (KCC)	Maximum number of animals of a given population supportable by the resources of a specified area (i.e., K of the logistics curve).	Subsistence density (Dasmann 1981, Bailey 1984) Ecological carrying capacity (Caughley 1976) Potential carrying capacity (Riney 1982)
Behavorial K (BCC)	Maximum density of a population limited by intraspecific behavior such as territoriality.	Saturation point (Leopold 1933) Tolerance density (Dasmann 1981)
Refugium K (RCC)	Maximum population density below which animals are relatively invulnerable to predation.	Threshold of security (Errington 1934, 1956) Security density (Dasmann 1981)
Equilibrium K (ECC)	Density of a population at a consistent equilibrium, but the limiting variables are unknown (May include KCC, BCC, or RCC).	
Culturally-based		
I-carrying capacity (ICC)	Population density that yields maximum sustained yield (i.e., inflection point of the logistic growth curve).	Optimum density (Dasmann 1981) Economic carrying capacity (Caughley 1976) Maximum harvest density (Bailey 1984)
Optimum carrying capacity (OCC)	Population density that best satisfies human expectations for it (may be equivalent to any of the other defintions of K).	Relative deer density (deCalesta and Stout 1997)
Minimum-impact carrying capacity (MCC)	Population density that minimizes impact on other wildlife, vegetation, or humans without eliminating the population (Bailey 1984).	Wildlife acceptance capacity (Decker and Purdy 1988)

K-carrying capacity is the basis for most models of population growth and harvest management, and it corresponds to the upper asymptote of the logistic growth curve (Figures 6–7, 7–1). An equilibrium is essentially reached between the animal and its food supply so that the rate of increase is zero. Mortality will increase and natality will decrease due to increased juvenile or adult mortality and

Figure 7–1 Locations of different views of carrying capacity associated with the logistic curve (from Miller 2000).

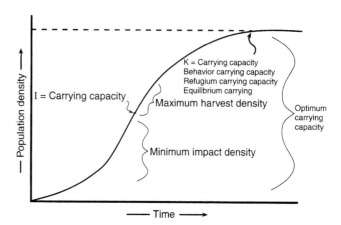

depressed reproductive rates and advanced age of sexual maturity, respectively. The result is the number of births minus the number of deaths equals zero.

Competition for forage is the factor that causes reduced natality and increased mortality at K-carrying capacity, resulting in relatively poor habitat conditions that yield relatively poor animal conditions. This differs from Dasmann's definition (Dasmann 1945) of K, which indicates animal condition is considered in defining K because K is the maximum number of animals that can be maintained in good condition without harming the range. Miller and Wentworth (2000) argue that animal condition not be included in definitions of K because the animal condition automatically declines (due to reduced resources) as the inflection point is passed (Figure 6–7).

In natural systems, there are natural fluctuations in environmental condition and populations fluctuate around K. When populations are at K, they fluctuate more than those that have been reduced (e.g., hunting) and the large fluctuations are more likely to damage vegetation that will reduce K (McCullough 1979). Also, in highly variable environments, populations at K fluctuate more widely than populations in stable environments. Thus, theoretically, damage to the vegetation from unhunted populations is not inevitable because a constant equilibrium may be maintained in stable environments (McCullough 1987). Also, in extreme environments (e.g., deserts), the populations may be intermittently reduced, so it rarely, if ever, reaches K.

Because K-carrying capacity is the most commonly used definition, guidelines for the management of ungulates are based on understanding of K-carrying capacity and habitat (McCullough 1987).

1. Relatively stable environments exhibit a narrow distribution of good and bad years (Figure 7–2a). In these stable environments, competition for resources drives populations, and the population responses to environmental variations are small. Thus, management can be precise. Even when environmental variation is greater but still small (Figure 7–2b), management is less precise but still predictable.

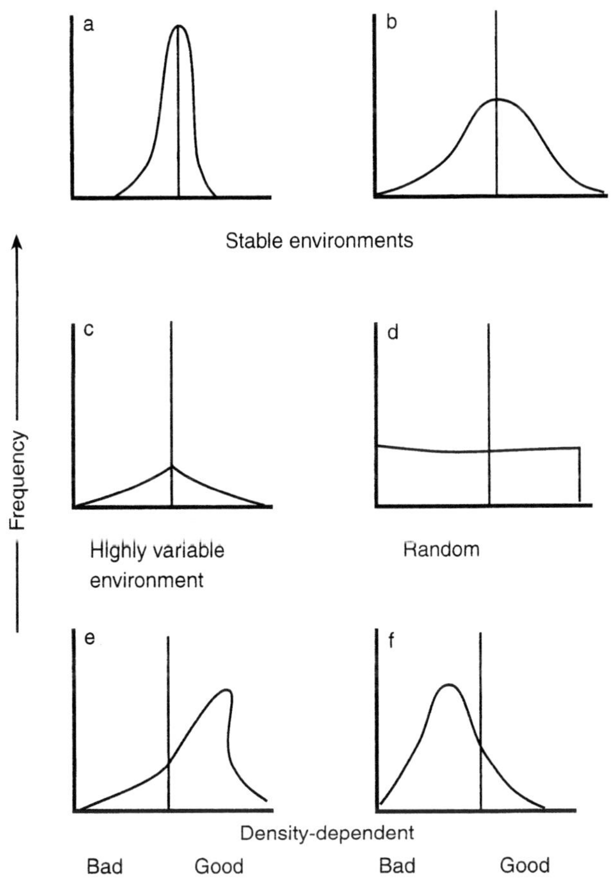

Figure 7–2 Frequency distribution of the occurrence of good and bad years in different environments (from McCullough 1987).

2. In highly variable environments (Figure 7–2c) or environments with random distributions (Figure 7–2d), environmental variations will overshadow density, and management must respond to the uncontrollable environmental variable on an *ad hoc* basis. In other words, there is little that can be done by managers if most of the population is manipulated by the environment. In areas that have many good years (Figure 7–2e) management can be precise, but modifications would be needed for the poor years. Management can also be relatively precise and reflect density-dependent responses in areas where most years are poor with few good years (Figure 7–2f).

3. K-carrying capacity (for many species of ungulates) is influenced by the successionary stage of the habitat. Carrying capacity can increase or decrease depending on the successional stage and the habitat of the species. Alterations in the successional stages can be natural or anthropogenic and can shift plant communities to alternate seres resulting in different K. Also, ungulates can drive successional patterns (Schmitz and Sinclair 1997) in two ways. First, if herbivore numbers approach K during early succession, deer may determine the structure of the plant community by preventing regeneration. Second, low herbivore numbers during

early succession may allow succession to proceed, resulting in a very different vegetation community and K.

4. The dynamics of plant and ungulate interactions are very complex, and herbivores can influence vegetation and the future of herbivore populations (Tilghman 1989, Miller et al. 1992, Waller and Alverson 1997). There are at least three mechanisms where herbivores can create different stable states in vegetation communities: fire, clear-cutting after browsing, and sustained long-term suppression of regeneration (Stromayer and Warren 1997).

Refugium Carrying Capacity

Refugium K is the number of animals a habitat will support when the necessary welfare factors (e.g., escape cover, habitat interspersion) to alleviate predation are present in the proper amount (Errington 1956, Dasmann 1964, Bailey 1984). When the necessary factors are present in the proper amount, animals are secure from predators as long as the population is below the refugium K. Errington (1934, 1956) demonstrated how bobwhite quail and muskrats fit this model. However, the refugium K is not applicable to large mammal populations.

Behavioral Carrying Capacity

Behavioral K is the maximum population size a given area will support when intrinsic behavioral or physiological mechanisms (e.g., territoriality) are the primary forces controlling animal populations. This concept is synonymous with *tolerance density* (Dasmann 1964) and *saturation point density* (Leopold 1933).

Behavioral K occurs at the asymptote of the logistic curve (as with K-carrying capacity). The difference between them is the mechanisms through which population growth rate declines and ultimately stops. As behavioral K is approached, declines in net recruitment may be due to increased dispersal of juvenile or subordinate individuals or to reduced per capita reproductive rates because the percentage of animals breeding declines as population increases. Those animals that have territories are in good condition because the animals and the resources are not limiting.

Equilibrium Carrying Capacity

Equilibrium K is a relatively new view of K where the density of a population is at a constant equilibrium but the limiting variables are unknown (McCullough 1992). Equilibrium K encompasses one of the other ecologically based Ks (e.g., K-carrying capacity, behavioral K, refugium K). Equilibrium K is the most ambiguous of the various meanings of K.

Culturally Based Carrying Capacities

The ecologically based Ks represent populations that are essentially free of human-induced factors. However, in our society laws have dictated that multiple use of many landscapes (e.g., livestock grazing, recreation, timber harvesting) is called for. As a result, populations maintained at a steady density will necessarily be below ecological Ks. Because anthropogenic influences are responsible for the additional use of resources, these Ks are culturally based. They may be determined from direct exploitation of the resource (e.g., sustained yield) or use of some aspect of an animal's habitat that could influence density (e.g., reduction of forage by livestock grazing). In these situations K is influenced more by humans than the ecological K, and K is the result of an objective that has to consider the ecology of the species and the values placed on that animal's habitat.

I-Carrying Capacity

I-carrying capacity is the population density at the inflection point of the logistic curve (Figures 6–7, 7–1). This density is the maximum sustained yield (McCullough 1992) because at this point the population growth is at a maximum.

I-carrying capacity is how K is viewed in range management because animals will be in good health and their habitat will be in good condition. It has also been called *optimum density, economic carrying capacity,* and *maximum harvest density* (Dasmann 1964, Caughley 1976, Bailey 1984, Caughley 1979).

Growing populations go through the inflection point but to document that time in natural populations has been rare (McCullough 1987, McCullough 1979). At the inflection point the rate of increase begins to decrease because of an imbalance between the animals and their resources. To manage at this level would require maximal removal to keep numbers down (and in balance with resources) but would yield a young age structure and a high turnover rate. As a result, there would be few older individuals.

I-carrying capacity is theoretically 50% of K-carrying capacity. However, I-carrying capacity (when viewed as a residual population relative to K-carrying capacity) varies according to the reproductive potential of the species, including the age of sexual maturity (McCullough 1987). McCullough (1987) demonstrated that maximum sustained yield for white-tailed deer occurs at residual (i.e., postharvest) populations that are 56% of K-carrying capacity and for mule deer at 63%. If I-carrying capacity is a larger percentage of K (than predicted), an opportunity for overexploitation exists.

Minimum-impact Carrying Capacity

Minimum-impact density is the population density that minimizes impact on other wildlife or vegetation without eliminating the population (Bailey 1984). In this context, greater value is placed on the forage resource or on some other wildlife species dependent on that resource. For example, rare and endangered cacti at Buenos Aires National Wildlife Refuge are given priority over ungulate populations. If the herbivores had an impact on cactus, they may be managed to lower densities to prevent browsing impacts. As a result, there would be fewer animals, they would be in good condition, and the range would be in excellent condition; thus the plant in question would receive only minor use. However, the reproductive rates of the depressed population would also be high so it would involve intensive management to keep numbers down. Decker and Purdy (1988) refer to minimum-impact K as *wildlife acceptance capacity* or the *maximum population level* in an area that is acceptable to humans. This may or may not be biologically based, depending on human values (e.g., tolerance for damage caused by a species, competition with other species of more interest, disease transmission, or other economic, aesthetic, ecological, educational, scientific, or intrinsic values). In other words, wildlife carrying capacity is related to human values, will vary from place to place, and may change as values change. This concept has been called *optimum K* by McCullough (1992). Because wildlife managers have to consider the animal, the habitat, and human relations with them, determining a wildlife acceptance capacity or optimum K is an important role. As a result, carrying capacities are defined by human expectations and not by the potential of the habitat (Miller and Wentworth 2000). This has resulted in controversial management when managing for a human-conceived notion of ecological balance. Management for a culturally based K is easily justified. However, maintaining ecological balance where natural processes are valued (Warren 1991, Porter 1992) can be complex. If population fluctuations are a result of human activity, then management may be necessary. However, if fluctuations are part of the natural population process, management to maintain ecological balance is not defensible (Warren, 1991, McCullough 1997).

The concept of carrying capacity can be viewed as a continuum along the logistic curve (Figure 7–1), depending on one's objective. When discussing K, it is important to understand how the concept is being used and how it is functional in meeting set objectives.

■ MEASURING K-CARRYING CAPACITY

Wildlife biologists have measured K for mule deer (Wallmo et al. 1977, Albert and Krausman 1993), white-tailed deer (Potvin and Huot 1983, McCall et al. 1997), moose (Crete 1989), elk (Hobbs et al. 1982), and mountain sheep (Mazaika et al. 1992), based on the assumption that forage quantity or quality determines how many animals an area can support. Most measurements require information on diet, forage availability, forage quality, and nutritional requirements. Carrying capacity is then determined from the relationship between available forage resources and the specific nutritional requirements of the species in the following forage quality or biomass model (Harlow 1984, Hobbs 1988).

$$K = \frac{A}{B \times days}$$

where:

K = carrying capacity
A = usable forage (kg/ha)
B = mean daily dry-matter intake (kg/day), and
days = days of use.

The formula is simple but complex. Estimates of the usable forage are crude at best and usually only the dominant food items are considered. Models have been refined by incorporating measures of the nutritional content of the forages to create nutritional-based models as follows (Hobbs et al. 1982).

$$K = \frac{(B_i \times F_i)}{(R_q \times days) - E_n}$$

Where:

K = the number of animals the range can support,
B_i = consumable biomass of principle of forage species i,
F_i = nutrient content of principle forage species,
R_q = individual animal requirements, metabolic requirements for daily maintenance,
Days = number of days animals occupy the range, and
E_n = endogenous reserves of nutrients.

Nutritional-based models require more data but may be more reflective of K when forage quality is limiting. However, these models make a number of assumptions (e.g., consumable biomass is known) and the estimates can vary greatly with relatively small changes in the model assumptions (Hobbs et al. 1982, Potvin and Huot 1983). Refinements to these models will increase as we learn more of the nutritional requirements of animals and animal-forage interactions. They will also need to incorporate environmental variables on carrying capacity and environmental stochasticaticy (McLeod 1997) to be accurate. These latter two will not likely ever be precise. Also, because data need to be species- and site specific, forage-based estimates of carrying capacity will limit their use (Miller and Wentworth 2000).

SUMMARY

Carrying capacity is the maximum number of animals of a given population that can be supported by the available resources. However, the term has been used in numerous ways, resulting in an array of confusion. Accurate measures of carrying capacity have not been developed, but introductory models have made attempts based on resource availability.

8

Leopold's Population Model

"The form of the growth curve sketched by a population increasing from low numbers is determined by the relationship between the population and the dynamics of its resources."

G. Caughley and A. R. E. Sinclair

INTRODUCTION

It is not a mystery that wildlife populations naturally fluctuate (although it is often a mystery as to why). The increases or decreases are viewed as positive or negative, depending on how the species in question is viewed. For example, predators have always been considered too abundant by many ranchers, and attempts to translocate predators into areas they have been extirpated from have created considerable controversy. It may give you pleasure to see rabbits in your yard unless you are trying to grow a garden. Likewise, deer and elk are always enjoyable to view unless they are creating hazards or causing damage to crops and property. Furthermore, hunters desire a huntable supply of game, so a huntable surplus is often managed for. Each of these situations demands some acceptable population level of the species in question. Some ranchers will not tolerate any predator that threatens their livestock, but those trying to reestablish predators (e.g., the wolf in the Yellowstone ecosystem and the Southwest) want at least a minimum viable population. Rabbits, deer, and elk will be acceptable as long as they do not compete with human pursuits. Unfortunately, rabbits, deer, and elk cannot survive as a species to be viewed without consuming the crops humans set before them! These conflicting views (and hundreds of others relating to wildlife) represent an important question central to wildlife management: at what population level should various species be maintained? The answer is easy: somewhere above a minimum viable population so they are not endangered but below a population level where they interfere with human values. That still leaves a wide range, and deciding on a level is more difficult than obtaining the desired level.

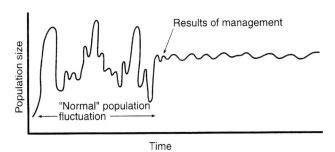

Figure 8-1 Without management, populations will fluctuate and may reach unacceptable highs and lows.

Wildlife could be left alone to fluctuate naturally. If this is the choice that is made, the side effects must be anticipated and deemed acceptable. Thus, all populations are managed. Managerial inaction is itself a decision and all species are under the control of someone (e.g., a homeowner who controls or does not control household insects, or a state or federal agency). Remember, the role of the wildlife manager is to be able to manipulate populations or their habitat to achieve a specific goal (Caughley 1977, Caughley and Sinclair 1994): conservation, sustained yield harvest, or control. Conservation involves treating small or declining populations to raise the density. If a population is to be harvested, the goal would be to maintain a sustained yield so the harvest could be conducted annually. Control involves treating a population that is too dense or has an unacceptably high rate of increase, to stabilize its density (Figure 8-1).

Increasing, stabilizing, or decreasing are the only three mutually exclusive things a population can do (Caughley 1977). In the past, terms such as "maintain," "preserve," and "perpetuate" have been used in stating population objectives. Caughley (1977) questions if a population is "maintained if it decreases by 10% or 20%?, increases by two times?, or if only one male or one female is left?" The point he was making is important and repeats Leopold's (1933) request for sound objectives. Terms like "maintain." "preserve," and "perpetuate" are not precise enough for sound management of wildlife populations.

Although wildlife managers are faced with these basic problems (i.e., to increase, decrease, simply observe, or stabilize populations), they are not the source of arguments over what should be done to populations, when action should occur, or how it should be administered. Problems in the management of wildlife develop when it does not suit someone's purpose and this is usually reduced to conflicting land use by humans and wildlife. For example, hunters usually want more animals to harvest, but ranchers want more forage allocated to livestock. It is only after a management decision is made that decisions can be made as to what should be done. In other words, the problem has to be identified and the objectives clearly stated. A lot of effort spent on the management of wildlife is devoted to finding out if a problem exists, what the problem is, and if there is a problem, how it can be solved. Until a problem is adequately defined, efforts to solve it are wasted (Caughley 1977).

Although problems of population manipulation have been reduced to only a few general areas, other concerns have created problems in being able to identify what should be done.

1. Objectives that change or objectives that are not clearly stated. Before any action is taken, managers should have a clear idea of what they are doing, why they are doing it, and some expectation of the management action.
2. Large study areas that are highly variable. Generally more variable areas require more sampling.
3. Projects that involve multi-objectives, with multi-species, and multi-habitat complexes (2 above) simultaneously require extensive manpower and coordinated activity.
4. Slow start-up times and low levels of funding for short period (1 – 2 years) are not adequate to investigate many natural systems.
5. Fluctuating and catastrophic alterations of habitats (e.g., fire, storms, floods) necessarily influence populations that are dependent on them.
6. Limited management options. If environmental factors account for 80 – 90% of animal density, the best management could do would be to operate on the residual.
7. When dealing with agriculture landscapes, habitat manipulation techniques are limited.
8. Until recently, there have been nonconsumptive users' attitudes and actions. Wildlife is a public commodity and the divergent views need to be considered. Setting objectives with a group of people that want different uses of the land and animals is a serious stumbling block.
9. Furthermore, increasing demands for human benefits increases habitat fragmentation, resulting again in population influences.

Regardless of these confounding obstacles, in order for a wildlife manager to be effective, written objectives make the job of population manipulation easier. Objectives are usually derived after baseline data have been collected, such as animal condition and weight, availability of forage, and reproductive condition. Or if data are not available, the object may simply be to collect baseline data so informed decisions can be made. Some examples of objectives follow:

1. To determine the factors that limit the red squirrel population on Mt. Graham
2. To harvest 300 elk in a specific area
3. To harvest 300 elk of either sex in a specific area
4. To determine the efficiency of the net gun in capturing white-tailed deer in oak forests
5. To minimize the reported wildlife-caused damage to soybeans in Lee County, Alabama
6. To increase the Sonoran pronghorn population in the United States to 500 individuals
7. To establish a harvest of at least 10 mature male bighorn sheep on the San Andreas National Wildlife Refuge, New Mexico

8. To determine those characteristics of urban housing areas with two homes per hectare that influence the nesting behavior of Cooper's hawks

Sound objectives also make it easier for others to determine what is being attempted. Numerous examples of management objectives (both clear and ambiguous) can be found in the wildlife literature (e.g., *Journal of Wildlife Management, Wildlife Society Bulletin*).

Being able to evaluate objectives is just as important as clearly stating them. For example, if an objective is to establish a herd of 100 pronghorn on a wildlife refuge, managers need to have some mechanism to evaluate their efforts. Without an evaluation it would be difficult to know if the objective has been realized. Leopold (1933) discussed the mechanisms of game management that have been the basis for sound wildlife management emphasized by Caughley (1977) and Caughley and Sinclair (1994). The remainder of this chapter is a summary of their ideas that have shaped how wildlife is managed in North America.

■ ALDO'S MODEL

Because of the numerous influences on a population, it is difficult to decide on what management can be done to reach a desired objective or maintain a certain population density. Many of the issues that biologists deal with are not technical but turn out to be problems of economics and sociology (Caughley 1977) that can easily be discovered. Getting people to take action seems to be the problem. Caughley (1977) discussed the decline of humpback whales from 1912 to 1962 as a direct result of excessive hunting. The excessive harvesting was a result of socioeconomic factors causing the governments of Japan, Norway, the then Soviet Union, and the United Kingdom to reject harvest restrictions, leading to a population collapse. Most conservation problems do not arise because of overharvesting, but because of other factors. Leopold (1933) developed a model for deer that is useful for biologists to determine how to manage and what to manage for, once a population objective has been established (i.e., Aldo's model).

Aldo's model is based on the biotic potential of the species, the desired population level, and the interaction between them. The model views the biotic potential of the species (i.e., unimpeded increase) as a flexible curved steel spring that bends towards the theoretical maximum but is pulled down at the same time by various factors (Figure 8–2). Of course, more than one of the factors operates at any one time, but the biologist's job is to determine which of the factors has the most influence on the population. Each one of the factors is constant in character and direction, but the value varies. Management attempts to control the values. And management is necessary because "A state of undisturbed nature is, of course, no longer found in countries facing the necessity of game management; civilization has upset every factor of productivity for better or for worse. Game management proposes to substitute a new and objective equilibrium for any natural one which civilization may have destroyed" (Leopold 1933:26).

Factors

The *factors* biologists seek to control are *decimating factors* and *welfare factors*. Decimating factors are those that kill directly and include hunting, predation, disease and parasites, accidents, and starvation. Welfare factors include food, water, cover, and special factors (e.g., mud wallows, dusting areas). They reduce the biotic potential indirectly by not being available in the necessary amount, which decreases reproduction and the population's defense against the decimating factors.

Decimating and welfare factors are not exclusive. For example, starvation could be a result of simple physiological breakdown in old animals or poor forage and water conditions for all age classes. Likewise, reduced cover could make species more susceptible to predation. Although there is interaction between the decimating factors and welfare factors, the distinction is useful in determining which factor should be manipulated to achieve a desired effect.

Because one factor exerts stronger influences on the population than the others do, it is referred to as the *limiting factor*. For example, if poor habitat conditions are preventing young from surviving because of a lack of cover that protects them from predators, cover would be considered the limiting factor. The lack of cover is responsible for the ability of predators to kill young in this case, and if habitat conditions improved, predation rates on young would decrease, if cover was the limiting factor. Another example to illustrate the limiting factor concept can be found in the San Andreas Mountains, New Mexico (Hoban 1990). In the late 1970s bighorn sheep became infected with psoroptic mites (Figure 8–1). The level of infection was so severe that animals were subjected to numerous decimating factors (i.e., mountain lion predation). However, the mite infection predisposed the animals to predation and is considered the limiting factor. The mites (and predation) have effectively reduced the population from more than 200 animals in 1978 to one female in 2000. Efforts are now underway to eliminate the mite from the mountain range and reestablish bighorn into the San Andreas Mountain Wildlife Refuge.

Influence

Decimating and welfare factors are of course affected by the environment. Environmental conditions usually operate on wildlife indirectly by influencing a factor. These environmental conditions include cuttings or clearings, fire, drainage, and grazing, and are called influences (Leopold 1933). Factors (i.e., welfare or decimating) are always unfavorable, but influences have positive or negative influences on the population. Grazing by livestock provides a good example. Domestic livestock grazing can interact with mule deer by altering plant succession to favor or reduce forage and cover (Peek and Krausman 1996). In some cases as forage and cover are reduced by domestic livestock, mule deer will alter their use of areas or habitats (Wallace and Krausman 1987) or experience a reduction in productivity, leading to altered population densities. On the other hand, livestock grazing can be used to improve vegetation conditions for mule deer. Foraging can stimulate production of twigs (Jameson 1963), and cattle have the potential to alter shrub form and productivity to promote subsequent use by mule deer (Gibbens and Schultz 1962). Understanding plant development, forage choices by mule deer and livestock, and habitat use patterns of both ungulates is key to the development of grazing management plans for any given area where mule deer and livestock share the same range (Peek and Krausman 1996).

COMPENSATORY MORTALITY

Studies of bobwhite quail (Errington and Hamerstrom 1935) led to the concept of compensatory mortality. Allen (1954) introduced the concept to the wildlife community. Errington and Hamerstrom (1935) reported that winter loss of bobwhite quail averaged 10% on areas that were hunted but 28% on areas that were unshot. They concluded than an autumn surplus of quail existed that would die primarily from predators, if not harvested. Their data suggested that if hunting could reduce the density to a certain level determined by forage and cover, there would be a compensatory reduction in predation that would allow all remaining birds to survive to spring. In other words, the habitat can support only so many animals. Animals in excess of that number will die. If they are not harvested, they will be killed by other limiting factors (e.g., predation). Bobwhite quail killed from limited hunting would be totally compensated as a result of the enhanced survival of the survivors.

If a limiting factor (e.g., hunting) is not compensatory, it may be additive. Additive mortality would suggest that hunting reduces the survival rate, but less so than predicted by adding the shooting mortality rate to the natural mortality rate. Partial compensation would lie between the extremes of totally compensatory mortality and additive mortality. Ellison (1991) provides an example that differentiates among total compensatory mortality, partial compensation, and additive mortality.

Consider a population that has a natural annual mortality rate of 40% and is harvested at 20%. Because some fraction of the population that was harvested would have died from natural causes without shooting (Roseberry 1979, Nichols et al. 1984), the definition of additive mortality is not the sum of the two mortality rates (i.e., 60%). The mortality from natural causes is similar to the product of the natural mortality rate and the shooting mortality rate (i.e., $0.20 \times 0.40 = 8\%$). Thus, if mortality were additive, the total annual mortality would be 52% (i.e, $60\% - 8\%$).

If mortality were totally compensatory, up to a certain threshold level of shooting, the annual mortality rate will remain constant at 40%. This would occur because natural mortality will decline sufficiently to compensate for all shooting mortality (i.e., natural mortality is perfectly density dependent) (Rosenberry 1979). If the natural mortality is only somewhat density dependent, then there will be only partial compensation for shooting and the total annual mortality would be between 40 and 52%.

The concept of compensatory mortality has been used in justifying and setting harvest levels. However, it is important to understand whether hunting contributes to compensatory or additive mortality. If the compensatory mortality hypothesis is correct, there should be no effect of shooting on annual mortality (i.e., survival) rate up to a certain threshold of shooting mortality (i.e., total compensation). If harvest exceeds this threshold, only partial compensation occurs. A second prediction of the compensatory mortality hypothesis is that there is a positive relationship between the nonhunting mortality rate and density at some time of year. The hypothesis also predicts density-dependent mortality in populations that are not hunted (Nichols et al. 1984).

There have been numerous studies that have tested these hypotheses. The most comprehensive have been with waterfowl, for which large landscapes and data sets are available. Most evidence supports the compensatory hypothesis rather than the additive hypothesis (Anderson and Burnham 1976, Rogers et al. 1979, Anderson et al. 1982, Nichols and Hines 1983, Burnham and Anderson 1984, Burnham et al. 1984, Nichols et al. 1984, Trost 1987). However, the population dynamics of wildlife are complex and their management requires more than the simple depictions of the completely additive and completely compensatory model (Smith and Reynolds 1992).

■ DIAGNOSIS OF PRODUCTIVITY

Aldo's model (Figure 8–2) depicts a flexible curve with an upward bend (toward the biotic potential) and decimating and welfare factors pulling the curve down. The diagnostic approach provides a method to determine how hard each factor pulls down, or which of the factors pulls the hardest (i.e., limiting factor), and why (Leopold 1933). Diagnosis in this manner is a basic component in understanding wildlife populations and requires knowledge and insight of the population. Leopold (1933) compared a doctor's ability to determine the cause of illness in his or her patient or the ability of a mechanic to fix a car. One has to understand the system, and it is no different in wildlife management. Because managers cannot manipulate all factors at once, they attempt to determine which is the limiting factor, essentially solve the problem, then move on to the next factor that is influencing the population until the desired effect is obtained. The basic question is which of the factors (i.e., disease, predation, starvation, hunting, accidents, food, water, cover, or special factors) has the strongest influence on the population at a point in time and why? What factors are keeping the population below the desired population level and what can be done to stabilize or increase the numbers of organisms in question? For example, what are the factors that prevent whooping cranes, red-cockaded woodpeckers, or desert races of mountain sheep from increasing to desired population levels? In the case of desert races of mountain sheep, population levels may be restricted by limited water. Thus, water would be the limiting factor and if added to the animals' habitat, the population should increase (Leslie and Douglas 1979). If, on the other hand, something else was limiting (e.g., adequate cover, predation, or disease), management could add all the water they wanted without the desired effect.

Caughley (1977) presents his views of solving management problems in two stages: an initial investigation and a stage 2 investigation. The initial investigation is

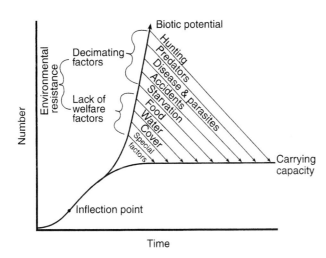

Figure 8–2 Aldo's population model (after Leopold 1933).

essentially the same as the water example for bighorn sheep. When managers notice a cause for a population response, they can take action to determine if they were correct. For example, if a manager suspected that domestic sheep were the cause of wild sheep mortality through disease, the concept could be tested by removing the domestic stock and noting how the wild sheep respond. Caughley (1977) argues that many conservation problems can be solved in this manner, but those that cannot require the stage 2 investigation. The stage 2 investigation is more complex than the initial investigation because the manager will know from the results of the initial investigation that the cause of the decline is not simple or obvious and more detailed study is required. This will result in a study of the reduced fecundity or increased mortality with the use of life tables and fecundity tables. Caughley's (1977) ideas are essentially a modification of Leopold's (1933:182) three essential steps in diagnosis of game productivity:

1. visualizing the population as it is, and as it should be,
2. making an intelligent guess as to what is wrong, then
3. testing whether the guess is correct without too heavy a risk of time, funds, or damage.

If the test verifies the population as it is and as the manager would like it to be, life tables, life equations, or computer models could be used. Life equations summarize the population dynamics and provide easily calculated models that can illustrate changes in the population. Leopold argues that wildlife managers are essentially practical ecologists (but not all ecologists have the observational skills for diagnosis) that use art and science. Much of the diagnostic aspect of the science of wildlife management is art. With life equations, managers can use empirical data and best guesses to develop their models.

The second step consists of making a biological estimate of the problem using all of the available information. The goal here is to find the limiting factor. This again depends on having sound objectives from which to work. Are you working with the entire range of a species or part of a species range? And what is the question to be answered (e.g., how animals are distributed and population density)? There are several ways solid biological estimates can be made. First, two similar areas can be compared where the suspected factor is only operative in one area. For example, if a manager suspected that bobcats were holding down fawn recruitment, two similar areas could be examined: one with bobcats and one without bobcats. If fawn recruitment to the population were higher in the area without bobcats, then predation might be the likely limiting factor. Furthermore, if water was suspected as the limiting factor for the distribution of a population, managers could examine how the population responds to added water sources. For example, in the Little Haruquahala Mountains in western Arizona, water was suspected as the limiting factor for bighorn sheep. The population was monitored for several years prior to the addition of water and then productivity and recruitment were monitored for five years after water was added to the mountain range. No measurable response was detected in productivity and recruitment, and Krausman et al. (1989) concluded that other factors were limiting. On the other hand, bighorn sheep populations have increased (Leslie and Douglas 1979) when water was added to their range.

When possible, habitat analysis of the area used by the population in question can address limiting factors. The analysis can indicate the juxtaposition of habitat components, availability and abundance of forage resources, and other habitat components. From this analysis, educated estimates can be made about the limiting factor.

The third way to estimate the limiting factor would be through the literature. More than likely, the situation managers are facing has been documented in the wildlife literature. There is no substitute for being current with the literature to fully understand what has happened in the past to assist with the present.

Leopold's third step in the diagnosis of populations is to test the estimate with a small-scale experiment. If managers rushed into a full-scale application to deal with their estimate, they could waste considerable time, funds, effort, and credibility if their estimate was wrong. By testing the estimate on a small scale, the managers have a better basis from which to make sound decisions. Of course, the tests should be conducted at the proper scale. Testing the influence of any factor on a small scale for a wide-ranging species would not yield useful information. At the same time, managers would have to consider the economics and time required for testing.

The development of water catchments in the Southwest provides a good example. Much of the Southwest is desert and water is not abundant. Intuitively it is a limiting factor. As a result, management agencies have developed water sources for wildlife throughout southern California, southern Utah, southern Nevada, New Mexico, Arizona, and southwest Texas. Unfortunately, they were established on a large scale without small-scale studies until recently. As a result, there is considerable controversy over the utility of water catchments for wildlife (Krausman and Etchberger 1995, Krausman and Czech 1998, Broyles 1995, Broyles and Cutler 1999). Even though catchments have been constructed since the 1930s (Glading 1943), they are only now being studied as to how they influence populations of desert mule deer (California Fish and Game Department), desert bighorn sheep (Arizona Game and Fish Department), and other species. Management of populations with developed water would be much farther advanced and less controversial today if preliminary studies had been conducted.

Monitoring the results of management practices is just as important as being able to correctly diagnose a problem. If monitoring is not included in the overall management plan, the mangers will not know the outcome of their efforts. If a management action is important enough to undertake, there should be a mechanism to follow up to make sure the objectives are reached. Unfortunately, and too often, management is initiated but there is no follow up, or the follow up occurs too long after the particular management has occurred to determine what really occurred. For example, if pronghorn were transplanted in an area to supplement a dwindling population, they should be monitored at least weekly to determine, at a minimum, distribution, habitat use, and survival. If not maintained until some time later (e.g., six months), the entire translocation could have been eliminated without any knowledge by biologists. Implementation of management events and subsequent monitoring are critical to better understand wildlife populations.

SUMMARY

Because of the complexity of animal populations, all of the influences they are exposed to cannot be examined at the same time. Aldo Leopold presented a simple model that incorporates biotic potential, environmental resistance, and carrying capacity to use in determining or estimating the factor that has the most influence on the population. His model has been used since first presented in 1933.

9

Decimating Factors: Predation

"Not far from camp, in the full heat of the day, five wild dogs were engaged in a wart-hog kill, yanking and tearing at the dying creature in an open grove of small yellowthorn acacia. In the dust and sun and yellow light, among skeletal small trees, the dogs in silhouette spun round and round the pig in a macabre dance in which the victim, although dead, seemed to take part. But within a few minutes, the wild pig had been rended for the dogs work fast perhaps to avoid sharing their prey with hyena or lion. The strange patched animals loped away into the woods, lugging . . . gobbets of fresh pig meat in the direction of their den, returning soon to fetch away the last wet scraps, gray now and breaded with dust in the hard, hot wind.

There's nature in the raw for you!"

P. Matthiessen (1981)

INTRODUCTION

Predation is a factor that has been difficult to obtain consensus about because of the very emotions it arouses in different people. Some are concerned about predators harming them, their livestock, or wildlife, and others are simply concerned about predators killing at all. This is not unexpected. After all, numerous children are raised on stories of Little Red Riding Hood, and the Three Little Pigs where the "big bad wolf" is the evil one. Even the father of modern wildlife management, Aldo Leopold, waged a war on predators in the Southwest before he realized the value of predators to biodiversity and ecosystem health (Flader 1974). Even the titles of some chapters of early books about wildlife management (Allen 1954) depicted predators as "varmints" in a condescending tone. In addition, the federal government established an entire branch to control predators across North America, and its roots were imbedded in the reduction of large predators. It is no wonder that studies of predator-prey relationships are controversial. The controversy occurs with laymen and in the professional ranks.

By examining the literature and selectively choosing articles, one can support most views on predation (Connolly 1978). As a result, the controversy continues from students of natural history who do not want any type of predator control to some ranchers and farmers who advocate elimination of some predators. The controversy will continue but one aspect of predation is clear: humans have created predator "problems" and their varied views on what should and should not exist. Wildlife biologists should not dive into the emotional issues as they will not be the deciding force on policies related to predation and are not referees to the controversy (Leopold 1933). The biologists' role is to understand a complex ecological relationship and present the facts to decision makers so informed decisions can be made.

In the past 25 years an abundance of knowledge has been generated about predators, yet the issue is as controversial today as in the past. Simply examine the range of views about the translocation of wolves to the Yellowstone ecosystem and to the Southwest. When predators are the topic of discussion, it is unlikely still that any consensus will be reached even though population studies in the past 25 years have clarified the role of predation. Unfortunately, some views will never be changed. Some people still do not believe in the advances of medicine, some farmers do not believe in the advantages of fertilizer, and there are those who will not change their views about predators, regardless of the research that has been conducted.

Nowhere in wildlife management is the practice of using scapegoats more prevalent than in the controversies involving predators. More money has been expended on efforts to control predators than to understand the basic biology and ecology of the animals. Unfortunately, the arena wildlife biologists often exist in is that of responding to the public attitudes. Fortunately, however, studies of wildlife led to studies of predators and several captured the attention of the American public. In particular, studies of large predators including wolves (Mech 1970), grizzly bears (Craighead et al. 1995), mountain lions (Hornocker 1970), and African predators (Schaller 1972) brought the issue of predation to the popular press, and the American public began an educational process that continues today. Although the data available on the interrelationships of animals is still imperfect, as it was in 1933 (Leopold 1933:230), the scientific basis to understanding them continues to develop. This chapter explores some of the basic issues relating to predation in wildlife management.

TYPES OF PREDATION

Predation occurs when animals kill and eat other live animals. Animals that eat dead material they did not kill are scavengers or detritivores. The early descriptions classified depredations based on how animals were killed; by accident, skill, or education (Leopold 1933:242). Based on this criteria, Leopold (1933) described five types of depredations: chance, habit, sucker list, starvation, and sanitary.

Chance predation occurs when a predator accidentally stumbles across prey and any repeat capture is simply random. The predator is not necessarily searching for the prey, but if it is present and the opportunity exists, a kill will be attempted. *Habit depredation* on the other hand is when a predator develops a habit of consuming a particular type of prey. *Sucker list depredations* occur when predators are able to take advantage of unsophisticated prey. Predators take advantage of prey at the lower end of the learning curve before they develop successful hiding strategies. The last two types are classified based on a weakness of the environment or prey. If the environment cannot provide the necessary cover or forage for animals in cover when they are attacked, education of the prey is possible. Leopold (1933:244) referred to this situation as *starvation depredation. Sanitary depredation* involves the culling of the weak, injured, or old animals often by predators that could not cope with healthy prey.

Leopold's (1933) classification was descriptive but difficult to quantify. Since then, more universal definitions to describe predation have been used: carnivory, herbivory, cannibalism, and parasitism.

Carnivory is the type of predation most people recognize. Carnivory is when a predator kills and eats animal prey (e.g., hawk kills and consumes a quail, or a mountain lion kills and eats a deer).

Herbivory occurs when animals consume living green plants or their seeds and fruit. The plants are not usually killed.

Cannibalism is a special case of carnivory where the prey and predator are the same species. *Parasitism* involves a parasite feeding on another animal (i.e., host) without killing it. The parasite is usually smaller than the host.

These definitions also help characterize types of predators, but to have a better understanding of predator-prey relationships, information about the animals and their habitat is needed. Errington (1956) also included compensatory and noncompensatory predation that considered habitat.

Compensatory predation is related to the availability of habitat that can protect an animal from predation. As the habitat is filled, those species without adequate cover are more vulnerable to predation and other forms of mortality.

Noncompensatory predation is linearly related to prey population density and depredation does not increase when a prey population exceeds the "security threshold" of its habitat.

■ UNDERSTANDING PREDATION

Clearly, more information is needed to understand how predators and prey interact. Leopold (1933:231) outlined five important variables: the density of the prey population, density of the predator population, the characteristics of the prey (e.g., ability to escape, health, habitat used), the characteristics of the predator (e.g., diet, mode of hunting), and the abundance of buffer species or alternative foods for the predator.

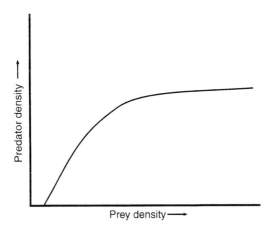

Figure 9–1 The numerical response is the trend of predator density as related to prey density.

Density

The density of the prey population and density of the predator population are needed to define the numerical response of predators to prey (Figure 9–1). When the prey density increases, more predators survive and reproduce, causing an increase in the predator population that eat more prey. Predator density increases to an asymptote determined by behavior such as territoriality that results in dispersal, causing resident numbers to stabilize. For example, 28% of 135 wolves dispersed from the area in which they were originally captured (Ballard et al. 1987). The dispersal contributed to the area used that was related to prey density.

Behavior

Biologists also need to understand the predator response to prey densities, predator densities, and distribution of prey. The response of predators to different prey densities depends on the functional response and numerical response (Solomon 1949). The *functional response* is how predators respond to different prey densities and varies with the feeding behavior of individual predators. The *numerical response* is how the predators respond to different prey densities through increases or decreases via reproduction or movement into or out of the area.

The functional response is determined by search patterns, appetite, and handling time (Holling 1959) of the predator. In general, there is a decreasing proportion of prey consumed as prey density increases (Figure 9–2). If the numbers of prey consumed per predator per time is slow at low prey densities and levels off at high densities, an S-shaped curve is produced (Figure 9–2). The S shape of the curve is caused by a behavioral characteristic of predators called *switching* or *prey switching*. Switching occurs when there are two prey types available and one is more abundant than the other. Predators will generally concentrate on the more abundant prey until their numbers are low, at which time they switch to the other prey until there is an upswing in the number of the first prey. For example, the main prey

Figure 9–2 Functional responses described by Holling (1959) for prey consumed/predator/time increasing with search time, search efficiency, and prey density to asymptote as prey density increases (---). When the numbers of prey increases, fast at intermediate densities and levels off at high densities, an S-shaped curve is produced (—).

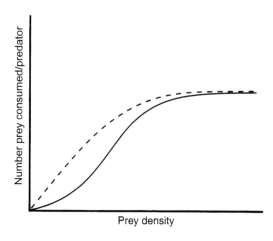

for mountain lions in Big Bend National Park, Texas, was mule deer and white-tailed deer. However, when deer numbers declined, mountain lions switched their diet to collared peccaries and small mammals (Leopold and Krausman 1986).

Predators also search for prey where prey concentrates in areas of high density. This has been documented for mountain lions hunting deer (Pierce et al. in press), wolves harvesting caribou, leopards hunting languars, or predators hunting snowshoe hares (Krebs at al. 1995).

The behavior of prey also affects predation rates in at least three ways: through migration and habitat use, herding and spacing, and birth synchrony. Prey can escape predator regulation if they can migrate beyond the range of their main predator (Fryxell and Sinclair 1988) (e.g., caribou herds following seasonal forage; Seip 1992). The theory is that predators have slow-growing nonprecocial young and are required to stay in a small area to breed and raise young. However, prey with precocial young can move within one hour after birth and can follow changing forage whereas predators cannot (Caughley and Sinclair 1994).

Predators may also cause prey to shift habitats to areas with poorer nutrition, resulting in a decline in productivity (Krebs et al. 1995). However, if adequate forage is associated with adequate cover from predators, prey will not have to make a trade-off between foraging benefit and predation risk (Pierce et al., in press). Habitat selection in herbivores likely influences reproductive fitness, and differences in age and sex can play an important role in habitat selection (Bowyer 1986, Bleich et al. 1997).

Animals should also be able to reduce their risk of predation by grouping (i.e., herds, flocks), so group size increases with increasing predator densities (Hamilton 1971, Alexander 1974). For example, predator density is likely the driving force for differences in group size in different herds of muskox (Heard 1992). Muskoxen can be found in groups of 30 or more animals. They live in open terrain where sociality likely evolved as a strategy to minimize predation (Klein 1992, 2000). At the opposite end of grouping is a behavior of many female ungulates when they give birth: isolation. Mountain sheep, mule deer, African antelopes, and other ungulates leave the herd to give birth. This behavior reduces the ability of predators to concentrate searches in areas of dense prey and provides some security for neonates.

Some species reduce predation of their young by synchronizing reproduction that synchronizes births (i.e., predators swamping) (Rutberg 1987). Birth synchrony is an important behavioral adaptation in prey and has been identified in primates (Boinski 1987), colonial nesting birds (Patterson 1965, Robertson 1976), wildebeest (Estes 1976), barren-ground caribou (Dauphiné and McClure 1974), bison (Berger 1992), and pronghorn (Byers 1997). Gregg et al. (in press) also found a relationship between birth date and pronghorn fawn survival. Fawns born during the period of peak fawning survived longer than fawns that were not born during the peak period. Advantages of birth synchrony are that the sheer number of young present at the same time enhances group defense and creates predator confusion (Estes 1976, Rutberg 1987). An additional advantage may be the wide spacing of females during parturition (Tinbergen et al. 1967, Etchberger and Krausman 1999, Gregg et al. in press). Wide spacing alleviates the concentration and encounter rates of predators. Breeding synchrony needs to be evaluated in terms of predation, condition of the females, and seasonality of the environment. If species rely on grouping behavior and birth synchrony, they could be vulnerable to excessive predation if human activities change any aspect that alters synchrony. This is especially important in broad-scale studies that do not examine small-scale, yet important, breeding sites that are used only for a short time (e.g., 1–2 weeks) (Etchberger and Krausman 1999).

▓ EFFECTS OF PREDATION

Understanding predator-prey relationships is complicated because the impact of predators on prey is complex. Furthermore, controlled, long-term experiments to answer specific questions are rarely conducted. When studies are made, they are usually short term, address a specific area, and are not universal. In addition, human and other influences complicate the results. Predator studies of ungulates in northern climates provide insight into predator-prey relationships because the distribution and habitats of predators and prey have not been impacted by humans as much as other native ungulates (Ballard and Van Ballenberghe 1997). Still, the interpretation of predation impact is fuzzy because of conflicting use of terminology. The terms *limiting* and *regulating* are used interchangeably (erroneously) throughout the scientific literature. Limiting factors include both density-dependent and density-independent factors that reduce the rate of population growth (Messier 1991). Regulating factors are only density-dependent factors and are a subset of limiting factors (Messier 1991). In other words, the point where an ungulate population is in approximate equilibrium with its long-term natality and mortality rates is regulation, and this equilibrium depends on density-dependent factors; the magnitude of impact depends on the number of animals per unit area. There is more of an impact at higher density and less at lower density (Ballard et al., in press). All mortality factors can be limiting but only those that are density dependent can be regulatory (Van Ballenberghe and Ballard 1994).

A review on the effects of predation on ungulates, as stated earlier, could reinforce almost any view on the role of predation (Connolly 1978). Based on this review, the answer to the question of whether carnivores cause ungulates to be more or less abundant than they would be without predation fell into three categories.

1. Predators control some ungulate populations.
2. Predators do not control some ungulate populations.
3. Predation increases the number of ungulates.

Although the answer that predators control some ungulate populations was a conclusion Connolly (1978) stressed, in one case (Mech and Karns 1977) a predator extirpated its prey from a large area. However, predators were rarely implicated as the sole direct cause of widespread decline of ungulates (except when humans were predators). Five other general points were raised (Connolly 1978).

1. Sometimes predators appear to be the agent by which ungulate numbers are adjusted to the carrying capacity, rather than a direct limiting factor.
2. Predators may increase prey mortality during severe winters and prevent or retard the recovery of ungulate herds that have been reduced by other factors.
3. When other factors are causing prey declines (e.g., habitat deteriorations), predation can accelerate the declines.
4. Low survival of neonates was the most commonly offered evidence of predation as a limiting factor.
5. Habitat was mentioned frequently as the ultimate limiting factor.

The issue of habitat was described as the cause of regulation when studies implied that predators do not control some ungulate populations. The review indicated that ungulate populations were often regulated by habitat conditions (e.g., forage, deep snow) and predators exerted minor influences to the population.

That predation increases the numbers of ungulates is largely theoretical and undocumented. The theory is that predators increase the vigor and reproductive success of prey sufficiently to more than replace the individuals killed (Howard 1974) by forcing a redistribution of ungulates on limited winter range, allowing for wiser use of forage (Hornocker 1970) and thus survival.

Over the past quarter century numerous advances have made it possible to understand predator-prey relationships, and there are four models that examine the role of predation in the population dynamics of ungulates (Ballard and Van Ballenberghe 1997, Ballard et al. in press). Each of these theories has been developed for moose in systems with large predators (e.g., wolves, grizzly bears, black bears) and in habitats where human impact is minimized.

Recurrent Fluctuation Hypothesis

In the recurrent fluctuation model (Figure 9–3) ungulate populations are characterized by fluctuating densities that are not at equilibrium. Changes in forage quality and quantity, weather, and human harvest influence ungulate densities, but predation is the strongest factor limiting ungulate density. At high densities predation

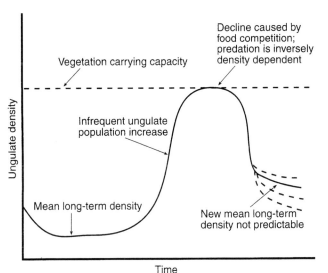

Figure 9–3 Conceptual model of ungulate density regulation under the recurrent fluctuations hypothesis (from Ballard and Van Ballenberghe 1997).

is inversely density dependent. Sometimes ungulates can escape the constraints of predation for high densities where forage competition causes population declines. These declines are accelerated or extended by inversely density-dependent predation. When a factor perturbs an ungulate population under this hypothesis, it does not return to a predictable density.

Low-density Equilibria

In the low density equilibria model (Figure 9–4) ungulate populations are regulated by density-dependent predation at low densities for extended periods and do not escape the constraints of predation for decades. Ungulates will remain at low densities until predation pressure is released (e.g., predator control, natural predator decline). Because ungulate density is below carrying capacity, forage competition is not important. After predators recover (e.g., after predator control), prey populations return to low densities (Figure 9–4). The low-density equilibria theory applies to multiple species of principal predators and principal prey (Ballard and Van Ballenberghe 1997).

Multiple Equilibria Hypothesis

Under the multiple equilibria hypothesis ungulate populations are regulated by density-dependent predation at low densities until predator pressure is released (e.g., predator control or natural predator reduction). When predator pressure is released, ungulate densities reach carrying capacity and prey are regulated by competition for forage (Figure 9–5). Competition for forage continues to regulate ungulate population growth at the higher equilibrium even if predators return to their

Figure 9–4 Conceptual model of ungulate density regulation under the low-density equilibrium hypothesis (from Ballard and Van Ballenberghe 1997).

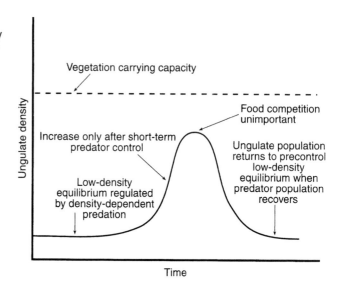

Figure 9–5 Conceptual model of regulation of ungulate densities under the multiple equilibrium hypothesis (from Ballard and Van Ballenberghe 1997).

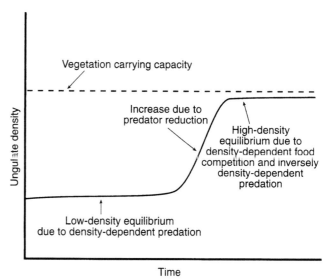

former levels (unless a natural catastrophe causes an ungulate decline). The area on the curve between the upper and lower equilibrium is referred to as a predator pit, where ungulates cannot increase because of density-dependent predation (Figure 9–5). Multiple equilibrium systems are characterized by multiple species of principal predators and prey.

Stable Limit Cycle Hypothesis

As the name implies, ungulate populations that operate under the stable limit cycle hypothesis experience regular 30 to 40 year cycles (Figure 9–6). Weather (e.g.,

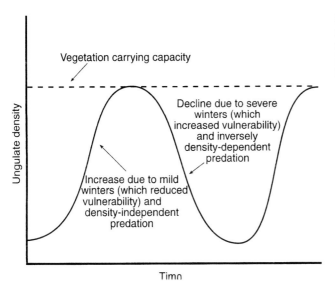

Figure 9–6 Conceptual model of regulation of ungulate densities under the stable limit cycle hypothesis (from Ballard and Van Ballenberghe 1997).

severe drought or winter) influences the viability and vulnerability of prey to predation from birth to adulthood. Predation is nonregulatory (e.g., density independent) as the ungulate population increases and inversely density dependent as ungulates decline (also nonregulatory). In the stable limit cycle hypothesis, regulation is caused by a combination of ungulate density, weather, and forage. These systems have one principal predator and one principal prey species (Ballard and Van Ballenberghe 1997).

Like most models, these four help to explain predator-prey relationships but do not fit all systems. Case history evidence supports the hypothesis that predation can and does limit many, but not all, ungulate populations. Recurrent fluctuations in simple predator-prey systems (i.e., one predator, one prey; wolf:moose) and single stable-state equilibrium in multiple predator-prey systems appear to be the most promising hypothesis for explaining ungulate-predator relationships (Ballard and van Ballenberghe 1997). However, each situation should be carefully examined. Inconsistencies and significant variations among studies (due to the relationships between ecological carrying capacity and the species, differing weather patterns, and short-term studies) decrease their usefulness in assessing the overall importance of predation. Furthermore, many studies simply document predation but offer no information about limiting or regulatory influences. For example, Ballard at al. (in press) reviewed studies that examined the effects of predation on mule deer and black-tailed deer but were not able to conclude with a generalized assessment of predation on deer. They did, however, document patterns from their review that summarize the state of knowledge of deer-predator relationships and illustrate the complexity of predator-prey systems.

1. Most fawn mortality occurs during mid-to-late winter.
2. In many studies where predation was responsible for high rates of mortality, coyote predation was a significant cause of mortality in fawns.

3. Mountain lions have been implicated as major predators of adult deer.
4. The impacts of predation are influenced by weather as it changes:
    a. forage and cover (Smith and Le Count 1979, Teer et al. 1991),
    b. alternate prey densities (Hamlin et al. 1984), and
    c. impacts on the physical condition of deer that influence vulnerability to predation (Unsworth et al. 1999).
5. Weather effects may be modified by predator densities, prey vulnerability, and the prey population's relationship to carrying capacity (Smith and Le Count 1979, Teer et al. 1991).
6. Wolves create a more pronounced effect on deer populations than do coyotes (Smith et al. 1987, Klein 1995).
7. In ecosystems where anthropogenic effects are minimized (i.e., arctic ecosystems), predation becomes an important mortality factor when severe weather or human overharvest initially causes population declines, and then mortality from predation retards or prevents population recovery (Gasaway et al. 1992, Ballard et al. 1997, Van Ballenberghe 1985, Bergerud and Ballard 1989).
8. In most systems predation does not cause ungulate population declines (Connolly 1978, 1981; Ballard et al. in press).
9. In altered ecosystems (e.g., contiguous United States, especially the Southwest) prey populations may not respond to predators as they would in less disturbed ecosystems. Because of disturbances (e.g., human influence on predators, prey, habitat including elimination of large predators, livestock grazing competition from livestock, fragmentation of habitats, human harvest), predation may or may not have limiting or regulatory influences. These anthropogenic factors may explain why in many deer populations predation is not an important mortality factor, or even when it is the most important limiting factor, population growth is not impacted.
10. Kill rates depend on whether predators are obligate or facultative carnivores. *Obligate* predators have consistent kill rates (Ballard and Van Ballenberghe 1997). Kill rates of *facultative* carnivores are assumed to be independent of prey densities except in certain situations (i.e. availability of neonate).
11. Criteria that can help determine when predation is an important (i.e. limiting or regulatory) mortality factor include at least two indicators.
    a. Predation is identified in seasonal or long-term changes in fawn:doe ratios. This can indicate when most losses are occurring but cannot be used to determine causes for change. Simple changes in fawn:doe ratios cannot be used to determine if predation is a limiting factor; that has to be determined through intensive studies.
    b. If predation is identified, the next step is to examine the population's relationship to ecological carrying capacity (e.g., examine natality rates, fawn:doe ratios, body condition, use of forage, prey densities).

ALTERNATE PREY

Alternate prey or *buffer species* (Leopold 1933) have not received much attention but may have an impact on the effects of predators. When moose, white-tailed deer, and desert mule deer declined, the major predators began consuming other prey to maintain their densities (Mech and Karns 1977, Leopold and Krausman 1986). In Arizona, livestock calves may have allowed mountain lions to maintain higher densities than would have been possible with native prey only (Shaw 1977). Mountain lions preyed on calves each year in late spring when deer numbers were low.

    Facilitative predators (e.g., coyotes) consume other prey that allows a higher predator:prey ratio than if ungulates were the sole or primary food. The reduction of the buffer species may cause higher predation rates on ungulates. On the other hand, an increase in buffer species could reduce or postpone coyote predation on deer. The role of alternative prey to predator-prey systems has not been thoroughly examined (Connolly 1978).

PREDATOR CONTROL

Predator control, the "unwanted stepchild" of wildlife management (Feldman 1996), is as controversial as other debates about predators and has as long a history as any other aspect in the field. Early humans took care of predator problems with whatever skills and tools were available. The federal government became involved in predator control in 1885, allocating funds to prevent and control wildlife damage (Hawthorne et al. 1999). This effort evolved to the development of the Division of Economic Ornithology and Mammalogy in 1886 to investigate bird and mammal damage. Over the years the name of the division changed (i.e., Division of Ornithology and Mammalogy to the Biological Survey to Bureau of Biological Survey) and their role increased from documenting damage caused by wildlife to developing solutions. The first direct control work began when funds were allocated to control plague-bearing rodents in national forests in California. The first congressional appropriations for federal predator operations occurred in 1915 to control wolves and coyotes. Appropriations continued and increased through World War I in response to the increased need for protein (i.e., beef and lamb ([Cain et al. 1972]).

    In 1916 acts authorized the killing of migratory birds that were injurious to agriculture (DiSilvestro 1985), and research laboratories were established throughout the United States to better understand ways to control predators. In 1931 the National Animal Damage Control Act was passed that authorized the federal government to conduct animal damage control activities and to enter into cooperative agreements with state governments. In 1939 President F. D. Roosevelt consolidated all federal activities dealing primarily with wildlife into the U.S. Fish and Wildlife Service in the Department of the Interior. Thus, the new organization was charged with controlling and enhancing different animal populations (DiSilvestro 1985).

The animal damage control program operated in relative obscurity through World War II and incorporated research, technical assistance, lethal control, non-lethal control, and other activities to various populations. New techniques (e.g., Human Coyote Getter, Compound 1080) were developed in the 1940s and became widespread in the West. Further cooperation with other federal and state programs enhanced the utility of the animal damage control program. However, public opposition against killing animals was growing, and in 1963 the animal damage control program was evaluated by the Advisory Board on Wildlife Management. The report (i.e., the Leopold report) was critical of the federal animal damage control program and charged it with indiscriminate, nonselective, and excessive predator control (Leopold et al. 1964). There were two basic premises of the report. "All native animals are resources of inherent interest and value to the people of the United States. Basic policy, therefore, should be (1) one of husbandry of all forms of wildlife; and (2) at the same time, local population control is an essential part of a management policy, where a species is causing significant damage to other resources or crops or where it endangers human health or safety. Control should be limited to the troublesome species, preferably to the troublesome individuals, and in any event to the localities where substantial damage or danger exists" (Leopold et al. 1964).

As a result of the report, the U.S. Department of Interior adopted their conclusions as guidelines for predator control and made numerous changes (e.g, name change to reflect philosophy such as "wildlife control" to "wildlife enhancement", development of a policy manual, hiring professionally trained wildlife personnel, reduction of predator control activities, and use of toxicants (Wagner 1988). Even with these changes, the new policy was not popular with the public. In the late 1960s and early 1970s there was increased awareness of the environment throughout the United States that generated numerous environmental organizations and political action groups. Predator control was not popular with the public, and numerous lawsuits (e.g., from Defenders of Wildlife, the Humane Society) demanded compliance with the National Environmental Protection Act. A second review of predator control was called for by the Secretary of the Interior. Although numerous changes were made from Leopold et al. (1964), the new report (Cain at al. 1972) was more critical of animal damage control and provided 15 major recommendations for program changes. The result was an executive order by President R. M. Nixon in 1972, banning the use of toxicants (e.g., 1080, strychnine, sodiumcyanide) for predator control by federal agencies or for use on federal lands.

Since then, animal damage control has been controversial within the federal government and with the public. The use of toxicants was de-emphasized over non-lethal methods of control, and there were attempts to eliminate the program and to remove it from the Department of the Interior and place it back in the Department of Agriculture. In 1986 the administration and personnel of the Animal Damage Control Program were returned to the Department of Agriculture (i.e., Animal Plant and Health Inspection Service (APHIS) and restructured. Major changes were to increase professionalism and training, improve relationships with other wildlife management agencies, improve data collection, monitor programs, develop new control methods, and protect public safety and threatened and endangered species (Acord et al. 1994, Berryman 1994, Hawthorne et al. 1999).

Many of the changes in the way predators were managed resulted from the general public. For example, leghold traps are especially unpopular, and recent ballot initiatives have banned or proposed banning and the use of snares in several states (Andelt et al. 1999). Clearly, certain methods for controlling predators are not accepted by many segments of society. However, under certain conditions predator control is publicly acceptable. Trapping in Illinois was acceptable for animal damage control, animal population control, or biological research (Responsive Management 1994). The general public in Utah was neutral on predator control (Krannich and Teel 1999) but more acceptable to specific rather than widespread programs (Messmer et al. 1999). More recently, there has been more public opposition to the control of predators. Wildlife managers are searching for additional information concerning public attitudes about all types of predator control. That predator control is controversial does not mean it has no place in wildlife management. Predator control is an important aspect of many wildlife management plans and programs. Managers should continue toward ecologically sound, sustainable programs that are in the best interests of wildlife and the majority of the people (Hodgdon 1991). "We should remember that wildlife damage management is likely to be an area of wildlife management that will always be controversial and complex — it is not a new problem or issue. It always has been, and probably always will be, a vital concern in the protection of human interests, needs, and desires; it rarely lends itself to simple and easy answers; it will not disappear or go away if we ignore it; and if not addressed by professionals, it is likely to force the landowner, manager, or community to take action that may result in chaos, environmental train wrecks, wasted resources, health hazards, or habitat elimination for many species" (Miller 1995).

When then is predator control warranted? The answer to this question will vary. A proportion of the populace will never accept predator control programs even if they are toward endangered species management. The wildlife biologist should approach the question by rephrasing the question to "when is predator control warranted in wildlife management?"

Because wildlife management consists of animals, habitats, and people, each has to be considered. Managers clearly need to better understand how the public views predator reduction programs and in what conditions predator control would be acceptable. Public controversy can stall or cancel the management of predators even when biologically justified (Stephenson et al. 1995).

Additional information is also needed about the long-term influence of predators on prey, when predation is limiting, when predator control should be implemented, and when it should terminate. Although additional information is needed, reviews of predator control in the past 25 years have illustrated important principles (Connolly 1978, Ballard et al., in press).

1. Predator control to increase prey populations can be successful when the prey population is below the ecological carrying capacity of the habitat.

2. Predator control can be effective if it has been correctly identified as the limiting factor. Prey declines due to habitat fragmentation, deterioration, or overuse cannot be reversed by predator control.
3. If predator control is implemented, the control efforts need to be severe enough to yield results (e.g., removal of $\geq$ 70% of coyote populations or 50% of 5–10-month-old wolves).
4. If predator control is implemented, it should occur just prior to predator or prey reproduction.
5. To be effective, control should be focused on a scale small enough to obtain results.
6. Predator control is expensive. There are limited data on benefit:cost ratios, and few studies have actually demonstrated that reductions in predators resulted in increased prey populations that can be harvested. With decreasing budgets and increased conservation issues managers need to have good information on the costs of effective predator control to solve problems. No agency has enough funds to squander on ineffective or inefficient management practices.

Connolly (1978) developed a decision matrix for the use of predator control in big game management that is a useful guide in determining what information is needed to decide on predator control and whether it is warranted. The management objectives need to be determined first and those listed by Connolly (1978) (Figure 9–7) are only a subset of what the objectives could be. However, the basic questions asked to arrive at a decision to control or not control predators would be the same (Figure 9–7). Upon initial examination the matrix may appear simple but upon further examination critical elements are addressed. What is the public's attitude towards predators: to manage, to hunt, or to maintain "natural" areas? How do predators influence prey? What is ecological carrying capacity? Are control efforts biologically and financially effective? These are questions managers still ask and for which the answers often come as biological guesstimates. In the 1950s there was an urgent need to determine, from controlled experiments, the impact of different levels of predator control (Cowan 1956). The same call was resounded in the 1970s (Connolly 1978) and in the new century (Ballard et al. in press). Until better data are available as to whether predator control should be implemented, at a minimum the following factors should be considered (Ballard et al. in press).

1. Prior to any predator control activity, a management plan needs to be in place with crystal clear objectives.
2. The status of the prey population resulting from predator control should be stated.
3. The estimated number of predators to be removed should be established.
4. The timing and methods of predator removal should be stated.
5. The size of the area to be treated should be determined.
6. Other factors that influence the prey population should be examined.
7. Monitoring the predator and prey populations is essential to determine when predator control should be eliminated or modified.

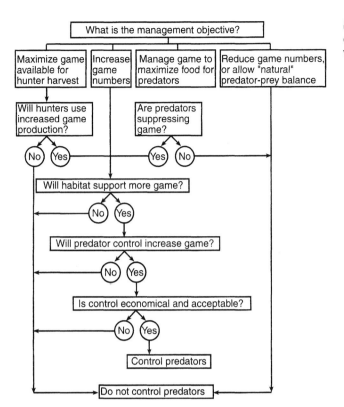

Figure 9–7 Decision matrix to determine when predator control is warranted (from Connolly 1978).

SUMMARY

Predation has been and will continue to be controversial, especially when it is influenced by wildlife management efforts. This chapter outlines the basic aspects of predation, discusses the effects of predation, and concludes with comments on predator control. When predator control is initiated, there should be specific objectives and a mechanism to monitor the results so managers can determine when the objectives have been met.

10

Decimating Factors: Hunting

"If you are a lousy hunter, the woods are always empty." E. O. Wilson (1994)

INTRODUCTION

Sport hunting has been the dominant tool used to manage game populations since modern wildlife management was advocated by Theodore Roosevelt and Gifford Pinchot in the "Doctrine of Wise Use" (Strickland et al. 1994). From 1900 to 1940 harvest was managed to reverse the trend of overharvest from market hunting (Leopold 1933). Since then, harvest management has been used to control increasing populations (Russo 1964) and to meet the public demands for recreation, animal damage control (Purdy and Decker 1989, Loker and Decker 1995), or commercial harvesting (Caughley and Sinclair 1994).

Hunting in the United States has been controversial as managers attempt to satisfy hunter demands, landowner concerns, other recreation, and the broader values of wildlife held by the general public (Carpenter 2000). Maintaining this balancing act is difficult because recreational hunting generates billions of dollars for the management and conservation of wildlife. For example, the 1996 expenditures for hunting in the United States exceeded $22 billion dollars (International Association of Fish and Wildlife Agencies no date) from which the majority of wildlife management programs (for game, nongame) are funded (Potter 1982). Because of the economic impact hunting has on local communities, there is now an economic importance to hunting that managers need to consider in harvest management strategies. Regardless of how or why populations are harvested, the basic component of successful harvest programs is a sustainable yield, a yield that can be removed annually without jeopardizing future yields (Caughley and Sinclair 1994).

This chapter is designed to discuss how hunting goals are set and how they are biologically and politically established. The excellent reviews by Carpenter (2000) and Caughley and Sinclair (1994) serve as the basis for much of the chapter.

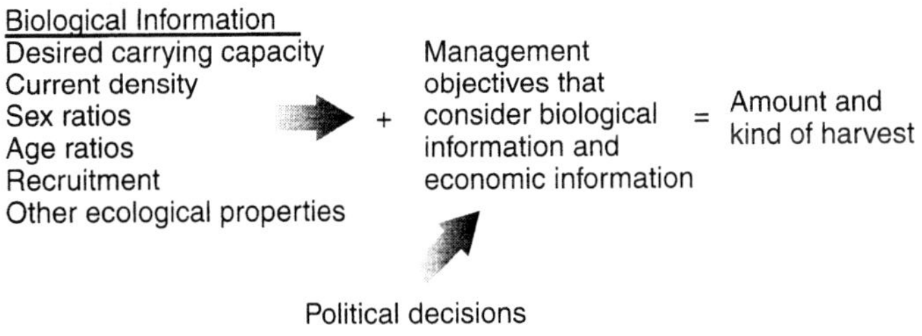

Biological Information
Desired carrying capacity
Current density
Sex ratios
Age ratios
Recruitment
Other ecological properties

+ Management
objectives that
consider biological
information and
economic information

= Amount and
kind of harvest

Political decisions

Figure 10–1 Biological information and management objectives are often influenced by political decisions in setting harvest levels.

SETTING HARVEST MANAGEMENT GOALS

According to the tenth edition of Webster's Collegiate Dictionary, hunting includes: to pursue for food or sport, to manage for the search of game, to pursue with an intent to capture, to search out, to drive or chase by harrying, to traverse in search of prey, to take part in a hunt, or to attempt to find something. Although aspects of this definition are used in the legal definition of hunting for law enforcement, this discussion relates to the pursuit of wildlife for food or sport. To do this and be able to maintain a sustained yield, managers have to determine when to hunt, where to hunt, what to hunt, and the methods of harvest to be able to preserve the brood stock, divide the huntable supply, provide for an orderly harvest, and control the size of the population. With the different views on how wildlife should be managed (and used) establishing objectives may be complicated but is critical for effective and scientific management (Figure 10–1). The planning process described by Strickland et al. (1994) follows the same logic as Leopold's (1933) diagnosis for game productivity: find out where you are, where you want to be, and test to see how well your objective was accomplished. The four basic steps of the planning process (Strickland et al. 1994) follow.

1. Determine the status of the resource.
2. Determine the objectives and goals.
3. Establish management strategies.
4. Determine how closely the management strategy achieved the population objectives.

Each of these steps is applied to the establishment of harvest management goals (Carpenter 2000).

Determine the Status of the Resource

The basis for determining the status of the population (i.e., step 1) is with a population census that will provide information on numbers and density and when possible on age and sex ratios. The geographical area for the species in question also has to be outlined. Most states have management units that are used to assist with population surveys (Figure 10–2). Game management units are also used to distribute hunters and for law enforcement. Population estimators are discussed in Chapter 6.

Figure 10–2 Game management units in Arizona. Game management units are used to help wildlife managers develop population estimates and distribute the harvest (from Arizona Game and Fish Department).

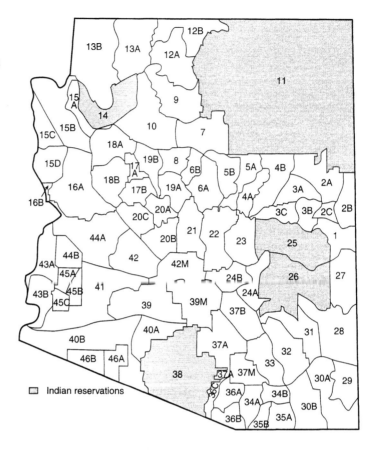

Determine the Objectives and Goals

For a harvesting program this step is determining the desired population level. Many populations are managed under the assumption that the population will continue to increase until it approaches the limits of the available resources to support it (i.e., K-carrying capacity) (Dasmann 1964, Caughley 1970, McCullough 1984), as discussed in Chapter 7. This is represented as the logistic curve, and the strategy for a maximum harvest would be to maintain the population near 50% of K-carrying capacity. To provide for maximum population growth and harvest, the population must remain below K-carrying capacity. Absolute values for K-carrying capacity are difficult to establish each year. As a result, managers often use long-term average levels of K-carrying capacity in environments without extreme weather variation. However, in environments with extreme weather changes from year to year, selecting for the average K-carrying capacity will result in too low a population level in mild winters (Strickland et al. 1994). The experience, past histories, and judgements of the wildlife manager are essential to the process of determining objectives.

Ecological carrying capacities often have to be replaced with cultural carrying capacities (Chapter 7) to account for tolerance levels of people. Ranchers with an objective of producing the maximum yield of livestock would certainly clash with an agency's desire to produce more elk. To compromise, the K-carrying capacity of elk would be reduced. The same type of relationship exists whenever the public has alternate uses of wildlife habitat (e.g., housing, recreation) or animals are considered a problem (e.g., deer and motorists, mountain lions or coyotes near urban housing).

Communication with and education of those that are influenced by and have an interest in wildlife are important as management agencies establish objectives. Understanding public attitudes toward the resource is often as important as the biological data in determining harvest management strategies (Brown and Decker 1979, Decker and Purdy 1988, Carpenter 2000).

After population objectives are agreed upon, they are compared with the current population to increase or decrease the harvest. When extreme weather or other natural mortality affects the population, the current population may be lower than the long-term goal. When this occurs, because of tradition and economic considerations, a minimum harvest occurs, but managers should carefully consider the implications of their harvest strategy decisions. Over the planning period (3–5 years) the strategy has to progress toward the established goal (Carpenter 2000).

Establish Management Strategies

Because it is so difficult to plan, communicate, and manage for different views, many state management agencies prefer to establish hunting seasons for several years at a time so annual changes are minimal (Carpenter 2000). If broad hunting guidelines can be established, alterations on an annual basis due to public comment will be minimal.

There are numerous considerations in establishing any hunting season.

1. Seasons should be set in accordance with the biological timetable of the hunted species (i.e., breeding period, parturition).
2. Impacts on grazing by livestock should be considered.
3. Impacts on timber management activities should be considered.
4. Recreational uses by nonhunters should be considered.
5. Hunter safety must be considered.
6. Predicted weather conditions should be considered.
7. Allocation of time to various hunting interests (e.g., archers, muzzle loaders, rifle hunters) should be considered.
8. Concurrent hunting seasons, or only one season for one species, should be decided.

These considerations (and others) are discussed by the game agency and public, and seasons are then established by a board or commission.

Other considerations in establishing the hunting season framework include the type of hunting license or permit and the effectiveness desired from the hunt. If the number of licenses is limited, they have to be allocated in a fair manner. Many states use a computer lottery that randomly selects licenses from the qualified applicants. Some states incorporate previous unsuccessful attempts at obtaining a license into the allocation process. For example, each year an applicant applies for an elk permit in Arizona but does not receive one, they are awarded points. In subsequent years the acquired points enhance their probability to obtain an elk permit. Licenses that are unlimited are distributed at wildlife agencies and authorized commercial vendors.

If the sex of an animal can be determined with a high degree of certainty, permits also list the authorized sex that can be harvested (e.g., cervids, bovids). If agencies can be more precise in setting hunting regulations with sound data, they can be more precise in population management. Obviously, species that cannot easily be sexed in the field by hunters would not have licenses that specify the sex.

Other license considerations include residence or nonresident permits and the associated restrictions that may be imposed on nonresidents (e.g., higher fees, hiring guides). Managers need to be aware of expected number of hunters and the different types of licenses issued to predict harvest levels; each license type results in different hunter success and time afield (Carpenter 2000).

The effectiveness and timing of the hunt (i.e., providing recreation or reaching a certain kill level) are other important considerations in establishing hunting seasons. These considerations involve the life cycle of the hunted species (e.g., independence of young, breeding) and hunter preferences (Lindzey 1981, Boyle et al. 1993, Squibb 1985). Managers would not want hunters afield during the breeding season if the hunting activity was detrimental to the population. Also, because the hunters want to have an opportunity to hunt more than two–three days, hunting seasons should be longer instead of shorter. When hunting seasons are shortened, more hunters are in a smaller area and the quality of the hunt is reduced. Hunting should be associated with a quality outdoor experience (Carpenter 2000).

The first days of the hunting season yield the highest hunting pressure and harvest (Murphy 1965), so when the season opens on or near the weekend, hunting pressure is more concentrated in the first few days of the season (because more people can hunt). If the season were to open early in the week, hunting pressure would be spread more evenly throughout the season. Managers are implementing midweek opening days in hunt units near urban areas to distribute the hunting pressure and reduce crowding. They also need to consider holidays and weather conditions, as both can increase hunter effort and harvest. If a holiday falls on Monday or Friday, the weekend is extended, resulting in additional hunting pressure, especially if the weather is good for hunting. If the weather is poor, the hunting effort and subsequent harvest will be effected. Keeping accurate records of hunters afield, weather conditions, and harvest records will allow better harvest predictions (Carpenter 2000).

■ MEASURING THE HARVEST

Measuring the number of animals harvested each year is essential to the scientific management of hunted populations. Leopold (1933) recognized the importance of measuring harvests and discussed simple systems that had been developed for some species. He was frustrated that a better accounting of harvested game was not available to consider all the necessary facts. "In short, the attempt to control hunting has suffered from ignoring economic and psychological facts and their varying relation to local conditions. It has especially suffered from the persistence of the concept that all hunting is the division of nature's bounty. We must replace this concept with a new one: that hunting is the harvesting of a man-made crop, which would cease to exist if somebody somewhere had not, intentionally or unintentionally, come to nature's aid in its production" (Leopold 1933:210).

As wildlife management matured, techniques were developed that provided random and adequate measures of the harvest. The success ratio discussed by Leopold (1933) is still used and others have evolved from earlier methods. Carpenter (2000) summarized the commonly used measures of annual harvests (e.g., hunter participation, number of harvested animals, composition of harvest by sexual age, hunter success rate).

Success Ratios

Success ratios were used for big game and measured the number of hunters afield and the number of animals killed. If these data are available, they can be used to establish harvest levels if the ratios are stable from year to year. If, for example, the long-term average indicated five hunters afield harvest one deer and the management plan calls for a harvest of 300 deer, the number of licenses issued should be limited to 1500 (300 × 5) licenses.

Mandatory Reporting

Mandatory reporting (e.g., with a report card attached to the license) from each hunter was a common method to obtain harvest data. At the end of each hunt the hunter would fill out the data sheet and send it to the managing agency (Figure 10–3).

Mandatory reporting can also be accomplished by requiring hunters to check in after their hunts at specific locations so managers can obtain the necessary information. This is effective when the hunt units are relatively small and have controlled access or when the number of hunters is small. For example, bighorn sheep hunters in Arizona are required to check in with the regional game and fish offices upon the successful completion of their hunt. This is effective because only 100 or so permits are allocated for bighorn sheep each year and compliance by hunters is functional. However, when the number of hunters is large and areas are large, mandatory checks can be used as an index at best unless managers can account for low compliance rates and response biases. Report cards alone do not indicate hunting pressure unless some index has been established. Mandatory reporting works best for small areas or limited hunters where some level of control can be obtained.

Check Stations

Check stations (Figure 10–4) are usually voluntary and were developed to obtain information on big game (Aney 1974, Rogers 1953, Sowls 1961). They have also been used on numerous other harvested game (e.g., small game, upland birds, waterfowl).

Figure 10–3
Report card that the California Department of Fish and Game requests hunters to return after their hunt.

<u>Hello Hunters</u>   **ZONE X-7B**

The Department of Fish and Game is conducting a survey of deer hunters and would like you to participate. Your name was chosen at random from a list of X-Zone deer hunters. Completing this short survey will help us manage deer. Please answer the questions below and return this postage paid card by October 15, 1999. Thank you for participating!

Did you hunt on **opening day**?   ☐ Yes   ☐ No

How many deer did you personally see on **opening day**? _____

Of the deer you saw on **opening day**, how many were bucks (including spikes)? _____

How many hours did you hunt on **opening day**? _____

Did you hunt on agricultural land on **opening day**?   ☐ Yes   ☐ No
If you saw any elk on **opening day**, how many did you see? _____

PLEASE RETURN THIS CARD EVEN IF YOU DID NOT HUNT

Figure 10–4 Check station for deer at Buenos Aires National Wildlife Refuge, Arizona (photo by Jim Hefflefinger).

Check stations are most effective when they are strategically located and they serve several purposes. First, biological data from harvested animals are collected (e.g., age, sex, body condition, harvest location) and any special body measurements or tissue samples can be obtained (e.g., samples of heart, liver, lung, muscle, blood, fecal pellets). In some situations hunters can bring in specific samples such as ovaries or stomach contents as long as instructions to the hunter are clear. The hunt is often the one time of the year that significant biological data can be collected for managing agencies.

Check stations also serve as communication, coordination, education, and public relations centers. Because animals are being weighed, measured, aged, and samples are being collected, it provides wildlife managers excellent opportunities to educate the hunting public, wildlife students, and anyone interested in why they are running the check station, how the data should be collected, the value of the data, overall goals, and the importance of solid public relations. Discussions usually move from the harvest to many different aspects of wildlife management.

Unfortunately, check stations are limited by the absence of a bias for a random sample, which prevents construction of confidence limits on the data (Carpenter 2000). Total checks can be obtained where access to hunting areas is controlled or the numbers of animals are limited. However, most check stations are time and labor intensive, precluding complete coverage for every hunt.

Telephone Surveys

Check stations have been used to compare data about harvest with telephone surveys. Information obtained from check stations and telephone surveys (i.e., total harvest, location of harvest, day of harvest) were similar for 60 to 80% of deer and elk hunters (Steinert et al. 1994). Compared to check stations, telephone surveys provided valid estimates.

Advantages of contemporary telephone surveys are that hunters can be contacted immediately after the season, and they are flexible so last-minute changes to questionnaires can be made. Being able to contact hunters immediately after the harvest minimizes biases due to memory loss. Disadvantages of telephone surveys include databases that have incorrect phone numbers, limited competence of some telephone surveyors, less accurate estimates of harvest location, and some hunters do not have phones, cannot be contacted, or consider phone surveys as "junk phone calls" (Carpenter 2000).

Mail Surveys

If managers have a random sample of the hunting population, successful mail surveys can be made (Kanuk and Bevenson 1975). Samples for mail surveys are obtained from license holders and can be stratified to answer specific questions and minimize variance (e.g., resident versus nonresident). There are two sources of bias with mail surveys: nonresponse (i.e., failure to respond) and response (i.e., failure to respond correctly) (MacDonald and Dillman 1968, Steinert et al. 1994). The latter is often caused by memory loss, so surveys should be mailed immediately after the hunt period (Carpenter 2000).

Questions on mail surveys have to be worded carefully and usually ask: species harvested, days spent hunting harvest, sex and age of harvest, location of harvest, and other specific questions. Surveys usually have two follow-up contacts to determine bias of the nonrespondent. A response rate of 70% is desired for adequate data. Mail surveys have been used extensively to estimate waterfowl harvest (Geissler 1990, Pendleton 1992).

Figure 10–5 Sustained rate of harvesting and sustained yield in relation to population size when no resource is limiting. Without harvesting, the population would increase 0.2/year. The dashed line is the harvesting rate (from Caughley and Sinclair 1994).

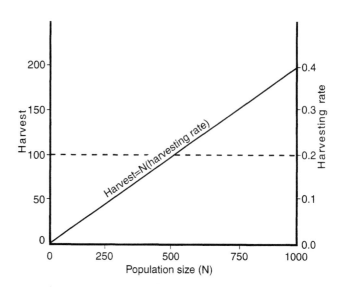

POPULATION MANAGEMENT WITH HUNTING

The basic strategy of harvesting is to harvest the population at the same rate of population increase. If a population increases at 20% per year, the harvest can approximate 20% if the management objective is to maintain a rate of increase of zero (Caughley and Sinclair 1994) (Figure 10–5). This model is useful when a species has been translocated into an area where it had been eliminated or is being introduced into a new area. In these cases it is likely the rate of increase will approach the intrinsic rate for several years after release and harvest can follow Caughley and Sinclair's model (Figure 10–5). However, how are hunting levels decided when over time the rate of increase is close to zero? In those situations the sustained yield would be zero also before harvesting could occur; for a sustained yield the population would have to be stimulated to increase.

This can be accomplished by correctly identifying the limiting factor and manipulating it so it is no longer limiting. For example, if burros are using the same forage and water resources as desert ungulates, their consumption may be reducing productivity and recruitment of wildlife. Removing the burros may be all that is needed to release forage and water so the amount available to desert ungulates increases. The direct consequence is that the fecundity of individuals is enhanced and mortality is reduced.

Most ungulate populations are managed based on density dependence. The concept of harvestable surplus was the driving force of management (McCullough 1979). This model measures the surplus then sets hunting regulations to harvest no more than that surplus. As discussed by Caughley and Sinclair (1994), the limitation of this harvest model is that as carrying capacity is approached, recruitment declines, which is the source of the harvestable surplus, and the surplus is minimized (McCullough 1979).

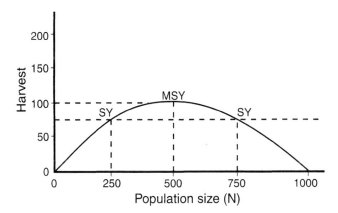

Figure 10–6 Sustained yield (SY) and maximum sustained yield (MSY). There are two levels of SY (for any level of SY other than MSY) from which the population can be harvested (from Caughley and Sinclair 1994).

McCullough (1979, 1984) provides an alternative to the above approach by increasing the harvest of populations near to carrying capacity to increase density-dependent response, to increase (or even maximize) the harvest. Maximizing the harvest is risky, and harvests should be adjusted below the maximum to maintain a safety valve against overexploitation (McCullough 1990, Caughley and Sinclair 1994).

Sustained Yield (SY)

Sustained yield is the number of animals that can be removed from a population year after year without jeopardizing future yields (Caughley and Sinclair 1994). The SY level is set by the forage supply and the number of animals that consume it. The trade-off between yield and density is an important aspect of SY harvesting. Overall, the more density is reduced, the higher SY becomes as a percentage of population size. The SY is highest at the population's intrinsic rate of increase. However, the intrinsic rate of increase is only obtained when the limiting factor (e.g., food) is at a maximum, which in turn usually occurs only when the population is at a minimum density (Caughley and Sinclair 1994). Caughley and Sinclair (1994) point out that density alone has no causal effect upon SY. The food supply influenced by the number of animals eating it sets the SY level.

There are essentially two levels of sustained yield (i.e., a harvest of a given size can be harvested from either of two densities) called a sustained-yield pair (Caughley and Sinclair 1994). The sustained-yield pair represent a lower and higher density from which the same harvest can take place (Figure 10–6). Harvesting at the lower density should be avoided because the harvest requires more effort than the same harvest at a higher density.

Maximum Sustained Yield (MSY)

Maximum sustained yield is the maximum number of animals that can be harvested without causing a population decline in theory. However, if harvest exceeds the MSY, the population will decline. Harvesting at the MSY should not be attempted because most managers do not have accurate measures of MSY and if the population is below MSY, harvest at that level would be an overharvest, further reducing the population's density. Further harvesting of the MSY will cause further declines. To maintain a margin of safety, populations should be harvested below estimated MSY (about 25% below, and more where weather variation fluctuates annually) (Caughley and Sinclair 1994).

SY and MSY in Theory

Harvest levels can be determined by the pattern of growth of a population, which is determined by the relationship between the population and resources. This relationship can be altered by changing the population's density, depending on population limitations (Caughley and Sinclair 1994).

1. When no resource is limiting. In this case the pattern of population growth is exponential and the harvesting rate or SY = hN, where h is the harvesting rate equal to the population's rate of increase when unharvested, and N is the population size (Figure 10–5).

2. When the population is limited by a consumable resource whose rate of renewal is not influenced by animals (i.e., mast crop), the rate of increase will follow the logistic curve. In this case MSY is taken at a harvesting rate of approximately $h_{msy} = r_m/2$ from a population size of $N_{msy} = K/2$, K being the mean size of the population when it is not harvested. In other words, MSY = $r_m K/4$ (Figure 10–7).

3. When the population is limited by a consumable resource whose rate of renewal is influenced by animals, growth will vary according to the interactions between the animal and the resource (Figure 10–8).

4. When the population is limited by a nonconsumable resource (e.g., space or nesting sites), growth will be a ramp in the form of an exponential curve truncated

Figure 10–7 Sustained rated of harvesting and sustained yield in relation to population size K/2 (from Caughley and Sinclair 1994).

by an asymptote (Figure 10–9). The yield represents the trend of SY on population size reaching a maximum when removed from a population size of $N_{msy} = K(r_m + 1)$. The MSY = $r_m K (r_m + 1)$. The N_{msy} will be less than K/2 only when the intrinsic rate of increase is greater than $r_m = 1$ on a yearly basis. Vertebrate populations seldom have that reproductive capacity, and the N_{msy} for wildlife populations will tend to be above K/2.

Overall, managing according to SY theory is based on population size, animal density, hunter harvest, habitat conditions, and overall K, to meet goals for harvest, nutrition, animal and habitat performance, or other factors of interest (McCullough 1984). The basic premise of SY theory is that as the population increases to ecological K, the nutritional plane of individual animals decreases, followed by a corresponding decrease in reproduction. Slow growth rates and relatively high mortality of juveniles, delayed puberty, reduced pregnancy rates, predominance of older age classes, and slow or zero population growth occur as populations reach ecological K

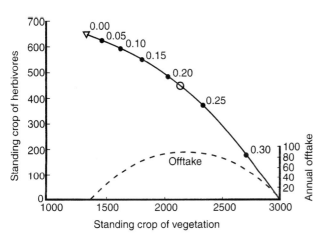

Figure 10–8 Relationship between plant and herbivore density and the rate of harvesting (i.e., offtake) needed to maintain equilibria. ∇ represents ecological carrying capacity and O represents economic carrying capacity (from Caughley 1976).

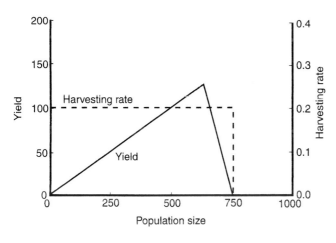

Figure 10–9 The relationship among population size, harvesting rate, and sustained yield for a population limited by a nonrenewable resource (e.g., nests) (from Caughley and Sinclair 1994).

(Wisdom and Cook 2000). Maintaining populations well below K should result in a more vigorous herd with a greater level of reproductive success.

Although SY theory is the basis for harvesting populations, in reality most harvesting of wildlife for recreational hunting has been by trial and error. The MSY is usually estimated from trend indices of the population, refined as often as possible, and the harvest is kept below MSY.

■ THE LEGAL BASIS OF HARVEST REGULATION

The regulation process and impacts of regulations have been thoroughly examined (Strickland et al. 1994). The interest relates to the numerous organizations that are involved in the management of wildlife (e.g., public, agencies, commissions and boards, state and federal government) (Carpenter 2000) and the concerns over how regulations can achieve harvest objectives (Denny 1978, Mohler and Toweill 1982) and how they can be improved (McCullough 1984).

Although harvest regulations are set differently from state to state, the general process is similar. State and federal statutes provide authority to establish regulations through administrative procedures, acts that include guidelines for due process and public review and comment. The general review process includes agency review, public review, and wildlife commission or board review as described by Carpenter (2000).

Agency Review

Regulations are cooperatively developed by law enforcement personnel, biologists, and public information specialists so they accomplish an objective, can be easily understood by the public, and are enforceable. Usually regulations are developed to address a management concern and forwarded through regional offices to agency headquarters. At each step the public and other state and federal agencies should be consulted to make sure objectives are met and lines of communication are kept open.

Public Review

Public review of regulations is critical to the process and many of the public concerns are social instead of biological (e.g., how to allocate the harvest among different groups). Wildlife managers need to stay in contact with all organizations interested in wildlife so they are well versed in what and why the departments regulating wildlife do what they do. The public has the opportunity to comment on wildlife in writing and orally at formal review sessions. Effective agencies are those that scientifically manage the resource and work toward being responsive to public desires that have been expressed. This is often a difficult task that can be made easier with solid and positive communication between the agencies and the public. This is often not the part of a manager's job he or she likes the most. However, it is one of the most important. An agency can have a brilliant and scientifically sound management plan to reach an objective, but if it is not accepted by the public, it is not likely to get very far.

Wildlife Commission or Board

Most states have a wildlife commission that represents the public and general organization of the state wildlife agency. They also establish policy and adopt hunting regulations after they have gone through the agency review and public review. Most commissioners are appointed by the governor and serve two- to four-year terms and serve as a buffer between the public and the managing biologists. The commissioners have the ear of the public, so it is again important that wildlife managers be able to express their views in a clear and logical manner. Commissioners can only make informed decisions with sound information. The importance of good communication skills cannot be overlooked.

SUMMARY

Sport hunting is an important management tool for many populations; it also generates funds for management. Setting management goals requires knowledge of the resource, clear objectives, strategies, and monitoring. Harvests are measured with success ratios, mandatory reporting, check stations, and telephone surveys. Hunting should always maintain a sustained yield and the maximum sustained yield should be avoided. Before hunting seasons are set, they have to be reviewed by the regulatory agency and the public before they are accepted by the wildlife commission.

11

Decimating Factors: Disease

"... diseases cannot be effectively controlled without timely and accurate
information about their occurrence (incidence and prevalence) with or without
the capability to measure the effects of control measures."

C. E. G. Smith

INTRODUCTION

Like most of the components that contribute to the science of wildlife management,
the study of disease has developed into a specialized discipline. Wildlife managers
are not pathologists but they are often the first to come in contact with dead or dying animals and can be the prime source of information from which an accurate diagnosis can be based. Understanding how decimating factors influence wildlife populations takes on additional importance when the decimating factor is disease and
can also influence those that handle the diseased animal and other wild and domestic animals, and can lead to sociological and political issues outside the realm of
biology. A prime example was the introduction of domestic stock to western rangelands during the latter half of the nineteenth century. As cattle, sheep, and goats arrived, so did a plethora of external and internal parasites, bacteria, viruses, and
other disease agents associated with them. Domestic stock were placed on rangelands of native ungulates (e.g., white-tailed and mule deer, pronghorn, bison, elk,
bighorn sheep, moose, reindeer, mountain goat, muskox) that had their own endemic parasites, bacteria, and viruses. Disease-causing agents of both livestock and
native wildlife were exposed to new susceptible hosts, resulting in some cases of die-
offs of wildlife and livestock (Jessup and Boyce 1996).

Wildlife diseases were further compounded in the latter half of the twentieth century as interest developed in ranching exotic wild ungulates (e.g., axis deer, red deer,
fallow deer, mouflon, oryx, barbary sheep). Many of the exotics were adjacent to habitat that supported native wildlife, escaped, and more disease problems arose. Managers recognized many of the potential disease-causing organisms shared by livestock,

native, and exotic wildlife, but mechanisms to control disease were limited, especially when the vastness of the area, limitations of treatment technology, complexity of the potential host and organism interactions, and legal and financial complications are considered. These complex issues lead to the sociological problems of disease.

Some livestock diseases (e.g., foot and mouth disease, scabies, and respiratory pathogens) shared by wild and domestic ovids have reduced or eliminated some wildlife populations (Krausman 1996). Also, some diseases of wildlife from North America (e.g., anaplasmosis, epizootic hemorrhagic disease) have caused financial losses to livestock owners. These conflicts have caused some conservation groups to blame livestock for most of the ills of wildlife and have called for the reduction or elimination of grazing on public land (Jessup and Boyce 1996). Everyone involved with disease of wildlife (e.g., livestock operators, private game ranchers, wildlife and land management agencies) would benefit by preventing the introduction and spread of disease among livestock, wildlife, and exotic wildlife. Few resources, however, have been allocated to research, preventive medicine, improved diagnosis, or cooperative approaches to management of disease problems (Jessup and Boyce 1996). Whether disease is studied in relation to the effects on man and domestic animals and other exotics or as a decimating factor to wildlife populations, further study is warranted to better understand the influences of disease on the entire community. This chapter discusses the cause of disease, classifies diseases, and reviews some important diseases of birds and mammals.

CAUSES OF DISEASE

The study of the structural and functional cause and effects produced by disease is *pathology*. Pathology involves the study of disease from the molecular level to the population. In most cases biologists are interested in the influence of diseases at the population level, although the individual organism is usually the focus of attention (Woolf 1976). *Etiology* is the study of the cause of disease (i.e., departure from health, or a state of imbalance) and the classification of the causes is varied (e.g., from infectious to noninfectious to intrinsic or extrinsic causes). Intrinsic causes include genetic influences, metabolic, and nervous disorders, and extrinsic diseases of animals are caused by viruses, rickettsiae, bacteria, fungi, protozoa, worms, and insects (Hopps 1964).

Woolfe (1976) characterized diseases by the nature of the environmental influence-producing disease as suggested by Smith et al. (1972) into six categories discussed below. These characterizations adequately define the nature of the disease.

Intrinsic Flaw

Intrinsic flaws include hereditary or congenital diseases. Hereditary diseases are those passed from parent to offspring genetically, and congenital defects are those abnormal conditions existing at birth. Hereditary defects that interfere with normal function usually are self-limiting and the individual has a reduced chance of survival.

Deficiency Diseases

Deficiency diseases are disorders of metabolism and include:

a. Inadequate nutrients in the diet
b. Poor quality diet
c. Interference with intake (e.g., anorexia, mechanical obstruction)
d. Interference with absorption of nutrients
e. Interference with storage or use of nutrients
f. Increased excretion
g. Increased dietary requirements associated with pregnancy or lactation
h. Inhibition of nutrients by inhibitors

The biggest deficiency disease is inadequate nutrients in the diet or poor quality diet. When this occurs, animals can easily be predisposed to other forms of mortality.

Exogenous poisons

Poisons act on organisms in different ways: producing local injury to tissues, destruction of epithelial cells in the kidney or liver after absorption, and upsetting metabolic and functional activities. If poisoning is suspected as a problem, attempts should be made to identify the chemicals of the suspected poison and clearly demonstrate its absorption into the body.

Plant poisons, under certain conditions, are a potential threat to wildlife. Some plants that are not normally poisonous can be, under the right conditions, and if consumed in large enough quantities (e.g., oak, wild cherry, palo verde).

Trauma

Trauma includes some accidents and predation that does not directly kill animals but because of the trauma animals were predisposed to disease. Bergerud (1971) reported that caribou calves died from secondary infections resulting from the bite of lynx. Also accidents with vehicles do not always result in death. If an organism survives the initial injury, there is a risk of secondary infection that could lead to death.

Tumors

Tumors are more common in older animals than younger animals. They influence individuals and not the population and are not considered important.

Living Organisms

Living organisms are the disease etiology biologists are most often concerned with. They are varied, and wildlife is susceptible to many in ways we are still learning about. Living organisms that cause disease include metazoan (i.e., multicellular) parasites, pathogenic protozoa, bacteria, fungi, and viruses.

The time these organisms involve the host determines if they are preacute, acute, or chronic. *Preacute* and *acute* diseases cause sickness for short periods of time and the illness is often not detected. *Chronic* diseases can last for months, but if the disease organism is highly virulent, death can occur in hours.

Because sick animals usually die unnoticed and in many environments decomposition of the carcass is rapid, obtaining accurate information about the role of the disease in population dynamics is difficult. Even in cold environments when the carcass does not decompose rapidly, repeated freezing and thawing hinder the ability of pathologists to obtain accurate diagnosis of the cause of death. The ability to examine sick and dying animals can provide valuable information to pathologists.

So, although managers are not pathologists, they should have the basic knowledge about diseases for at least six reasons (Woolf 1976).

1. The wildlife manager can provide information to the pathologist that will greatly assist in making an accurate diagnosis.
2. Because wildlife managers often represent the agency that is responsible for animal management, a knowledge of diseases of wildlife can be an important public relations tool. When the biologist can find the cause of death of a wildlife specimen, and then report the cause to the reporting individual, the stature of the agency and its professionalism are enhanced.
3. The health of the population can be used to some extent to serve as an indicator of the health of the habitat that supports the population in question. By maintaining healthy habitats, managers can usually minimize disease problems.
4. Wildlife can be an indicator for detecting and monitoring zoonotic diseases because wildlife are often abundant and widespread.
5. Diseases are important to understand because of the potential dangers to human health or economic losses involved (especially those associated with livestock).
6. Disease is a decimating factor, and managers should have some knowledge if and how it plays a role in the populations they manage.

Woolf (1976) categorized pathologic problems of wildlife with three broad categories.

1. Development of diagnostic procedures to be able to obtain information on the cause of death and the ability to determine high levels of potentially harmful substances
2. The development of studies that help understand disease problems that have the potential to negatively influence wildlife populations
3. Understand the ecology of diseases that are zoonotic, or transmittable to humans and domestic animals

A summary of selected diseases (Table 11–1) and discussion of these and other diseases follow. Other diseases that are related to nutritional deficiencies or accidents are mentioned in Chapters 12 and 13.

Table 11–1 Summary of selected diseases and parasites that are common in game animals (from Kristner and Bone 1993)

Animal	Disease or parasite	Body location	Causative agent	Human health importance
Antelope	Abscesses	Any body location (1)	Bacteria (2)	Possible infection of skin wounds (3)
	Lumpy jaw	Jawbone	Fungus, bacteria	Possible internal infection from pus in abscesses (3)
	Navel III	Abscesses at navel, joints	(2)	(3)
	Pinkeye	Eyes	Bacteria, virus	None (6)
	Adult tapeworms	Abdominal cavity from intestine (4)	Moniezia spp. of tapeworms	(6)
	Liver flukes	Bile ducts of liver	Fasciola hepatica	(6)
	Tapeworm cysts	Abdominal cavity, liver	Larval stage of Taenia hydatigena tapeworm	(6)
	External parasites	Skin	Ticks	Virtually none, unless tick attaches to human
Black bear	Abscesses	(1)	(2)	(3)
	Adult worms	(4)	Roundworms, tapeworms (5)	(6)
	Fleas	Skin	Pulex spp. of fleas	Primarily bite wounds from fleas.
	Trichinosis	Meat (7)	Trichinella spiralis (8)	Improperly cooked meat or sausage. Dangerous
Cougar	Abscesses	(1)	(2)	(3)
	Adult worms	(4)	(5)	(6)
	Trichinosis	(7)	(8)	Minimal from skinning
Deer and elk	Blood splotches	Under skin, in meat	Shocking power of high-velocity bullet	(6)
	Healed broken bones	Legs, jaw, ribs	Injury	(6)
	Arthritis	Joints	Injury, bacteria	(3)
	Abscesses	(1)	Bacteria (2)	(3)
	Adhesions	Chest, abdominal cavity	Bacteria, puncture wound	(3)
	Emaciation	Carcass very thin	Variety of causes	None, unsuitable for consumption

	Condition	Location	Cause	Ref
	Foot rot	Feet	(2)	(3)
	Navel III	Abscesses at navel, joints	(2)	(3)
	Necrotic stomatitis	Abscesses in mouth	(2)	(3)
	Pneumonia	Lungs	Bacteria, lungworms	(3), (6)
	Warts	Skin	Virus	(6)
	Abdominal worms	Abdominal or chest cavity	*Setaria* spp. of nematodes	(6)
	External parasites	Skin	Chiggers, fleas, lice, louse, flies, ticks	Flea bites and tick attachment
	Liver flukes	Liver, lungs	*Fascioloides magna*, North American liver fluke	(6)
	Lungworms	Lungs	*Dictyocaulus* spp. of lungworms	(6)
	Nasal bots	Nasal cavity, throat, lungs	Botfly larvae	(6)
	Adult tapeworms	(4) Bile ducts of liver	*Moniezia* spp. of tapeworms	(6)
			Fringed tapeworm	(6)
Deer and Elk (cont.)	Subcutaneous worm	Under skin, lower legs	Roundworm	(6)
	Immature tapeworms			
	Cysts small "rice grains" (deer only)	Abdominal cavity, liver	Larval stage of *T. hydatigena* tapeworm	(6)
		Lungs (7)	Larval stage of reindeer tapeworm	(6)
	Large white cysts 1	(1)	Hydatid cyst tapeworm	(6), Do not feed to dogs
Rabitts, hares, muskrats, and beaver	Abscesses	(1) Saliva, bite wounds Pinpoint white abscesses in lungs, liver, spleen, kidneys	(2)	(3)
	Rabies		Virus	Definite hazard
	Tularemia		*Franciscella tularense* bacteria	Potential human infection—dangerous

Continued

Table 11–1 Summary of selected diseases and parasites that are common in game animals (from Kristner and Bone 1993)
Continued

Animal	Disease or parasite	Body location	Causative agent	Human health importance
	External parasites	Pelage	Fleas, lice, ticks	Flea bites, tick attachment, tularemia
	Immature tapeworms	White cysts under skin; in liver and abdominal cavity	Several tapeworms of carnivores	(6)
Game birds	Emaciation	Generalized, keel prominent	Many causes	Unsuitable for food
	Imbedded shot	(1)	Lead pellets	Broken teeth
	Lead poisoning	Generalized (lead shot in gizzard)	Eating lead shot	Slight hazard
	Abscesses	(1)	(2)	(3)
	Aspergillosis	Body cavity yellow, gray, green	*Apergillus* spp. of fungus	Dangerous, discard carcass
	Avian pox	Wartlike growths on external body	Poxvirus of birds	(6)
	Salmonellosis	Abdominal cavity from gut	*Salmonella* spp. of bacteria	See text
	Tuberculosis	White, pearl-like nodules under skin; in internal organs	Tuberculosis bacteria	(6)
	Tularemia	Same as for rabbits		
	External parasites	Skin	Fleas, lice, ticks	Flea bites and tick attachment, tularemia
	Gizzard ulcers	Gizzard	Nematode (worm) parasites	(6)
	Sarcosporidiosis	Breast legs; white "rice grains"	*Sarcocystis* spp. of protozoa	Discard carcass; unsuitable for food

(1) Any body location
(2) Bacteria
(3) Possible infection of skin wounds
(4) Abdominal cavity from intestine
(5) Roundworms, tapeworms
(6) None
(7) Meat
(8) *Trichinella spiralis*

■ SELECTED DISEASES OF MIGRATORY BIRDS

Bacterial Diseases

Avian Cholera (Friend 1987a)

Avian cholera is also called fowl cholera or avian pasteurellosis and is a highly infectious disease caused by the bacterium *Pasteurella multocida*. Acute infections are common and can result in death 6–12 hours after exposure, killing more than 1000 birds per day. Other infections are more chronic and have longer incubation times with fewer losses. Transmission occurs by bird-to-bird contact, ingestion of contaminated food and water, and in aerosol form.

Avian cholera can occur in most birds and mammals, is more common in waterfowl, but annual losses constitute a small percentage of mortality. It occurs throughout the contiguous United States and losses can occur at any time of the year although predictable seasonal patterns exist where avian cholera is an established form of mortality for waterfowl.

Sick birds are lethargic or drowsy, can be approached closely before attempting to escape, and die quickly (seconds or minutes) when handled. Other field signs include convulsions; swimming in circles; erratic flight (i.e., flying upside down, attempting to land above the ground or water); mucous discharge from the mouth; soiling and matting of feathers around the vent, eyes, and bill; pasty, yellow droppings; and blood-stained droppings or nasal discharge.

Avian cholera should be suspected when numerous dead waterfowl are found in a short time, few sick birds are located, and the dead birds are in good flesh. Death can be so rapid the birds actually die in flight or while eating. None of these signs are unique to avian cholera.

Birds that have died of avian cholera usually have prominent lesions on the heart, liver, and gizzard, depending on how long they had the disease before death. Isolation of the causative agent is required for a definitive diagnosis. The best sample to provide the diagnostician is the entire carcass. If carcass collection is not possible, the most suitable tissues for culturing *P. multocida* are heart, blood, liver, and bone marrow. Samples should be frozen if they cannot be delivered within 24 hours.

The control of avian cholera is difficult because it is highly infectious and spreads rapidly, which is enhanced by gregarious birds. The prolonged environmental persistence (three weeks in water, three months in decaying carcasses, four months in soil) promotes new outbreaks. Early detection and control actions focused on migratory and scavenger birds can minimize costs and environmental contamination. Control includes collection and incineration of carcasses, preventing further exposure of the environment. For example, large concentrations of *P. multocida* are discharged from the mouth, so pick birds up by their heads to prevent further contamination of the soil, place them in double plastic bags, and burn them. In some cases the reduction of birds can be justified to control the disease if (1) the outbreak is discrete and not widespread, (2) techniques are available to allow for complete eradication, (3) methods only work on target species and do not pose a risk to nontarget species, (4) eradication is justified on the basis of risk to other populations if the outbreak is allowed to continue, and (5) the outbreak represents an extension of avian cholera into an important migratory bird population.

Habitat management is also used to control avian cholera by preventing use of specific wetlands that are focal points for infection by drainage, redistributing birds by impoundments, and adding water to problem areas to dilute concentrations of *P. multocida.* Other experimental techniques have been used and immunization has been used with control populations. However, there is no practical method of immunizing large numbers of free-ranging birds.

Avian cholera is not a high-risk disease for humans, but infections can be caused from dog bites or scratches. Therefore, the use of dogs to pick up carcasses during avian cholera outbreaks is not recommended. Humans should also avoid areas of high amounts of infected material with restricted air movement.

Avian Botulism (Locke and Friend 1987)

Avian botulism is also called limberneck, western duck sickness, duck disease, and alkali poisoning, and is a paralytic, often fatal disease of birds caused by ingestion of toxin produced by the bacterium *Clostridium botulinum.* There are seven types of *botulinum:* A-G. Type A affects domestic chickens; C, waterfowl; E, fish-eating birds; and the others are not known causes of avian botulism in North America.

C. botulism persists in wetlands in a spore form that is resistant to heat and drying and can remain viable for years. Optimum growth occurs at 25°C with a pH range of 5.7 to 6.2 in dead organic matter and a complete absence of oxygen. Especially potent toxins are produced in bird, mammal, and some invertebrate carcasses. Environmental factors contributing to botulism include water depth, water level fluctuations, water quality, the presence of carcasses, rotting vegetation, and high ambient temperatures. When outbreaks occur, they usually follow a well-documented bird-maggot cycle (Figure 11–1).

Waterfowl and shorebirds are most often affected by type C botulism in the wild. In captivity, waterfowl, pheasants, and mink are most often affected. Annual losses that exceed 50,000 in a single location are common; 4 to 5,000,000 birds died from botulism in the western United States in 1952. Avian botulism may be the most important disease problem of migratory birds.

Most outbreaks of type C botulism in the United States occur west of the Mississippi River but it has been reported nationwide. It has also been reported in Canada, Mexico, South America, Australia, New Zealand, Japan, Europe, and Africa within the past 20 years. Type E botulism outbreaks are less common and have been confined to the Great Lakes Region in the United States. July through September are the primary months for type C botulism outbreaks in the United States and Canada.

Lines of carcasses and receding water levels are typical appearances of major die-offs. Botulism-affected birds tend to congregate along vegetated peninsulas and islands. Sick and healthy birds are commonly seen together.

Avian botulism results in paralysis of voluntary muscles from affected peripheral nerves. Generally the power of flight is lost, followed by paralysis of leg muscles, then paralysis of the nictitating membrane and neck muscles. At this point death from

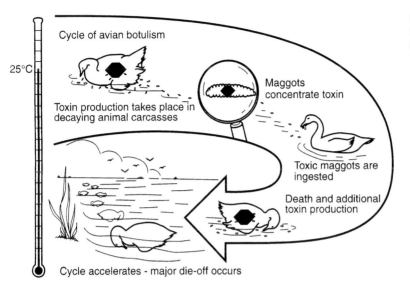

Figure 11–1 Avian botulism cycle (from Locke and Friend 1987).

drowning occurs before respiratory failure caused by the toxin causes death. Birds that have botulism can easily be captured by hand.

Because many birds die by drowning, lesions associated with drowning are observed during necropsy. There are no characteristic or diagnostic gross lesions in waterfowl dying from type C or E botulism. A presumptive diagnosis is often based on a combination of sick and dying birds with no obvious lesions upon examination. The most reliable test for avian botulism is the mouse protection test where one group of mice receives the serum fraction of blood from sick or freshly dead birds and another group receives the serum and a type-specific antitoxin. The latter group will survive but the former will become sick or die if botulism toxin is present.

Control efforts concentrate on those factors that contribute to avian botulism outbreaks that include fluctuating water levels during hot summers, an abundance of flies, and animal carcasses. The prompt removal of carcasses cannot be overlooked because thousands of toxic maggots can be produced from a single waterfowl carcass. Consumption of as few as two to four toxic maggots can cause intoxication.

Sick waterfowl can be saved (75–90%) if they are provided with fresh water and shade, or injected with antitoxin. American coots, shorebirds, gulls, and grebes do not respond to treatment.

Botulism (type C) is rarely a problem for humans, as thorough cooking destroys botulinum toxin in food. Botulism in humans is usually the result of improperly home-canned food (type A or B).

Avian Tuberculosis (Roffe 1987)

Avian tuberculosis is also called mycobacteriosis, tuberculosis, and TB and is usually caused by the bacterium *Mycobacterium avium*. There are numerous types of this bacterium but only three cause disease in birds. Transmission generally occurs by direct contact with infected birds, ingestion of contaminated feed, water, and the environment.

All avian species are susceptible to avian tuberculosis, as are humans, livestock, and other mammals. In free-ranging wild birds, tuberculosis occurs highest among those that live in close association with domestic stock and in scavengers. The prevalence of tuberculosis in wild waterfowl is less than 1% of birds examined at postmortem.

Avian tuberculosis is ubiquitous and occurs most commonly in the North Temperate Zone in small numbers of free-living migratory birds wherever there are major migratory concentrations. The disease is present year round and seasonal trends have not been documented.

There are no unique clinical signs to identify avian tuberculosis. Depending on the organs affected, birds may show no obvious clinical signs of being emaciated, weak, or lethargic. Other signs include diarrhea and lameness.

Upon necropsy, avian tuberculosis in migratory birds yields emaciated carcasses with solid to soft and yellow to gray nodules (less than 1mm to several cm) embedded deep in the infected organs and tissues (i.e., liver, spleen, lung, intestine).

Tuberculosis is often discovered during carcass examinations associated with die-offs due to other causes. Positive diagnosis is based on bacteriological isolation and identification of the organism. When possible, the entire carcass should be submitted for testing. The disease is transmissible to humans, so extra care should be taken when handling carcasses or tissues.

No methods have been developed to control avian tuberculosis in wild bird populations. The disease is rarely contagious from infected individuals.

Avian Salmonellosis (Stroud and Friend 1987)

Avian salmonellosis is also called salmonellosis, pullorum disease, and fowl typhoid and is a group of diseases caused by bacteria belonging to the genus *Salmonella*. The motile forms of salmonellae (i.e., parathyroid organisms) frequently cause illness and death in wild birds; they also infect all other warm- and cold-blooded vertebrates. Parathyroid infections are transmitted from fecal contamination of eggs, food, and water; mammals and exposure to contaminated environments; direct contact with infected animals; and ingestion of poultry and bird feeds that contain contaminated animal by-products.

Salmonella typhimurium is the most common parathyroid infection found in birds but results in only a small percent of mortality in wild birds. Larger die-offs have been reported for house sparrows, evening grosbeaks, northern cardinals, pine siskins, and American goldfinches at backyard bird feeding stations. Swans, ducks, American coots, and gulls have also experienced die-offs from *S. typhimurium*.

Infections of poultry exist wherever poultry are raised and in wild birds in the United States, Canada, Europe, New Zealand, Australia, and Antarctica. Infections in the United States are most common in the East and Midwest. Seasonal patterns to salmonellosis outbreaks have not been identified. Outbreaks are associated with environmental conditions that facilitate disease transmission.

Infected birds often appear to be in poor condition and have watery diarrhea, but other symptoms vary depending on the species of bird, age, and condition. The array of symptoms includes weakness, drowsiness, depression, drooping wings, gasping for air, pasted vents, swollen and stuck eyelids, and convulsions and other neurological signs.

Gross lesions also vary depending on species, age, and conditions. Livers are swollen, crumbly, and develop nodules. The intestinal lining becomes coated with fibrinous material, and impaction and enlargement of the rectum can be common. Diagnosis is possible by examination of entire carcasses and laboratory isolation and identification of *Salmonella* spp. from infected tissues.

Control of salmonellosis is generally not required in free-ranging migratory birds because the clinical forms of the disease show low prevalence in these species. Disease control at bird feeders should be encouraged by cleaning feeders and removing spilled and soiled feed. Because the disease infects humans, extra care with personal hygiene when handling birds or materials soiled by bird feces is warranted.

Chlamydiosis (Locke 1987b)

Chlamydiosis is also called parrot fever, parrot disease, and Louisiana pneumonitis and is caused by intracellular parasites that are a link between viruses and bacteria. The usual routes of transmission include direct contact with infected birds and inhalation of the causative organism in airborne particles. The disease influences birds and mammals but waterfowl, herons, and rock doves are the most commonly infected wild birds in North America.

Chlamydiosis is worldwide and can occur at any time. Infections range from acute to inapparent. The acute form is often fatal as birds become weak, stop eating, and develop pussy exudate around the eyes and nostrils. Birds become motionless and remain in a fixed position, huddled up with ruffled feathers. They also have respiratory distress and diarrhea. The most common anatomical change in infected birds is enlargement of the spleen by three to four times normal size.

Diagnosis has been conducted in the laboratory by isolating *Chlamydia* spp. from tissues of infected birds. Whole birds should be submitted for testing.

Control is related to human health considerations. When working in areas where there is a strong possibility of inhaling airborne avian fecal material, masks or respirators should be worn. Dry, dusty areas with bird droppings should be wetted down with bleach.

The correlation between parrots and chlamydiosis in humans resulted in importation bans on parrots and parakeets from 1930 to 1960 in the United States and Europe. Serious outbreaks have occurred among those working with poultry and among wildlife biologists. Biologists may have been infected from handling snow geese, great egrets, snowy egrets, white-winged doves, and ducks. Fatality rates have been rare since antibiotics have been available.

Viral Diseases

Duck Plague (Brand 1987)

Duck plague is also called duck virus enteritis or DVE, and is an acute, contagious, and often fatal disease caused by a herpesvirus. Duck plague virus can establish inapparent infections in birds surviving exposure (carrier birds). Outbreaks likely result from virus shedding by carrier birds to susceptible waterfowl. Only ducks, geese, and swans are susceptible to duck plague.

Duck plague is relatively new to North America. It appeared in New York in 1967 in the white Peking duck industry and has spread from coast to coast and from Canada to Texas. It has also been reported in Europe and Asia. Its influence on wild waterfowl is minimal and there has only been a single documented outbreak.

There is no prolonged illness associated with duck plague, but sick birds may exhibit hypersensitivity to light. Droopiness, thirst, and bloody discharge from the vent or bill may occur. Upon necropsy, hemorrhaging and free blood throughout the gastrointestinal tract are common because duck plague virus attacks the vascular system. A final diagnosis, however, can only be made by virus isolation and identification. Whole birds should be submitted for examination.

Because asymptomatic healthy duck plague carriers can shed the virus periodically for life but are not overtly identifiable, control is difficult. Obviously, waterfowl release programs should not use birds or eggs from flocks with a history of the disease. Furthermore, infected flocks and their eggs should be destroyed if they can be identified. There are no human health considerations.

Fungal Diseases

Aspergillosis (Locke 1987a)

Aspergillosis is also called asper or brooder pneumonia and is an infection of the respiratory tracts of birds and mammals caused by the fungi genus *Aspergellus*. This is one of the first bird diseases described in the literature. It is not contagious and usually results from inhalation of *Aspergellus* spores. All birds are susceptible to the disease and it is frequently seen in waterfowl and loons and captive raptors. It is worldwide except in Antarctica.

Most outbreaks in waterfowl occur in fall to early winter after hunting seasons. The disease is a complication of hunter-crippled waterfowl among birds with deficient diets or whose immune systems have been compromised by contaminants such as lead.

Aspergillosis-affected birds are typically emaciated and have difficulty breathing. If the infection reaches the brain, muscular coordination is lost. Gross lesions appear in the lungs and air sacs. Diagnosis can be confirmed by isolating fungus from tissues and microscopically examining material from fungal mats.

Aspergillosis can be minimized by maintaining a clean environment (i.e., avoid moldy feeds, straw, rotting agricultural waste). Aspergillosis is only of concern to humans with severe, preexisting lung damage or those few who are allergic to *A. fumigatus*.

Parasitic Diseases

Gizzard Worms (Tuggle 1987)

Gizzard worms are called stomach worms or ventricular nematodiasis, located beneath the surface lining and the grinding pads of the gizzard, and are most frequently found in waterfowl. Infestation occurs by ingestion of infective gizzard worm larvae.

Gizzard worms have worldwide distribution. Migratory birds are exposed to them on breeding grounds and can continue to be exposed throughout life.

Live birds do not exhibit obvious signs of the parasite, but in necropsy, sloughing, inflammation, hemorrhages, and ulceration of the gizzard lining can erode the grinding pads of the gizzard.

The condition can be diagnosed by identifying gizzard worm eggs in the feces. More thorough evaluations can be made by evaluation of carcasses.

There are no methods to control gizzard worms in free-ranging populations. Gizzard worms are not a threat to humans.

Nasal Leeches (Tuggle 1987)

Nasal leeches are also called duck leeches and are bloodsucking leeches (*Theromyzon*) that feed directly in the nasal passages, trachea, and under the nictitating membrane of the eyes of migratory birds.

Nasal leeches affect most aquatic birds north of the 30th parallel because these parasites are better adapted to cold water lakes. Peak parasitism occurs in spring and summer. Leeches can easily be seen protruding from the nares or mucous membranes of the eyes. They often resemble blood-engorged sacks of blood. Leeches cause birds to attempt to remove them (unsuccessfully) and cause labored breathing and gasping. Severe infections can result in blindness, eye damage, blocked nasal passages, throat blockage, or blocked trachea by engorged leeches.

Nasal leeches are 10–45 mm long when blood engorged, and have four pair of eyes. Diagnosis is usually made by seeing blood-engorged leeches protruding from the nares or attached to the eyes. There are no preventive measures developed for preventing leech infestation in wild birds.

Toxic Disease

Mortality from lead poisoning and oil is considered both accidents and disease. Although they were discussed as accidents, they will also be considered a disease.

Lead Poisoning (Friend 1987b)

Lead poisoning is also called plumbism and is an intoxication resulting from absorption of hazardous levels of lead into body tissues (e.g., ingested lead pellets from shot shells, lead fishing sinkers, mine wastes, paint pigments, bullets). Lead poisoning has affected every major species of waterfowl and a wide variety of other birds in

North America. Losses occur throughout the United States, Canada, the United Kingdom, Denmark, Sweden, Germany, Italy, and New Zealand at any time of the year. Most cases occur after the hunting season.

Birds that have lead poisoning are reluctant to fly and when they do, are weak and cannot sustain flight. They eventually become flightless and develop wing drop. They also become emaciated. Gross lesions include severe wasting of the breast muscle, reduced visceral fat, impacted esophagus or proventriculus, distended gallbladder, bile-stained gizzard contents, and the presence of lead pellets or parts of lead pellets. However, none of these signs and lesions is diagnostic by itself. A definite diagnosis is based on pathological and toxicological findings. When possible, entire carcasses should be submitted for diagnosis.

Lead poisoning can be reduced by denying bird use in problem areas and removal of dead and moribund birds. Other control measures include tillage programs, planting food crops other than corn or grains that aggravate the effects of lead ingestion, and use of nontoxic shot on hunting areas. The latter is the only long-term solution.

Lead poisoning related to affected birds has rarely been a problem in humans. However, there is always the possibility of breaking a tooth on lead pellets from biting a hunter-shot bird.

Oil Toxicosis (Gullett 1987)

Oil toxicosis results from animals exposed to spilled petroleum products. They suffer hypothermia and poisoning from ingestion, inhalation, or absorption of the product.

Human-caused oil spills on water and on land are the most damaging to birds. Responses to spills vary depending on species, species behavior, size of population, and location of the spill. Spills have been recorded in all 50 states.

Most spills occur in winter (i.e., January–March). This is also when sea birds and waterfowl congregate in wintering areas.

Birds exposed to oil look wet and chilled because the oil damages feather waterproofing and insulating properties. They are also thin because they stop feeding and use body fat and then muscle to produce heat in response to chilling.

Gross lesions are variable: oil may be in the trachea, lungs, digestive tract, and around the vent; the intestine may be reddened; salt glands may be swollen; and adrenal glands can be enlarged. Seeing oil on a bird or in the environment usually suffices as a diagnostic tool in large spills. Complete examinations of carcasses may be required for smaller spills or if birds die from ingesting oil in forage.

Control is expensive and complicated. Cleaning oiled birds is not justified unless the population is threatened or endangered. Proper cleaning techniques should be used while wearing protective clothing to prevent exposure of skin surfaces to oil.

■ SELECTED DISEASES OF MAMMALS

As with birds, there are numerous diseases of mammals. I discuss only a few and have selected those that likely interact with other animals (i.e., livestock) and have an impact on humans. Understanding the relationship between animals, disease, and their habitat has just begun. More attention is given to diseases of wildlife when they have a direct influence on humans (e.g., health or economics). The following is from a summary of important diseases shared by mammals and livestock (Jessup and Boyce 1996).

Bacterial Diseases
Tuberculosis

Tuberculosis or TB is a chronic disease caused by *Mycobacteria* spp. Tuberculosis can cause infections in captive members of the deer family and cattle. The disease process is usually slowly progressive and tubercles are formed as the body attempts to wall off the infection, thus the name.

Tuberculosis is found throughout the world and afflicts a wide variety of animals. In many underdeveloped countries tuberculosis in cattle and subsequent human infection with *M. bovis* are common. Bovine tuberculosis has nearly been eradicated from cattle in the United States by routine tests and slaughter. Many tuberculosis outbreaks have occurred in captive white-tailed deer, fallow, roe, axis, sika, red deer, and elk. Many large zoos observe strict testing procedures, but many small zoos, private game ranches, and auctions do not. Without testing there have been a number of tuberculosis outbreaks in bison, elk and other cervids, and exotic bovids on game ranches. Where range animals become infected, resident small carnivores (e.g., badgers in England and opossums in Australia) may also become infected and serve as reservoirs.

Tubercle bacteria are shed into exhaled air, in body fluids (e.g., saliva, urine, and milk) and in the draining exudates from tuberculosis abscesses (Davis et al. 1981a). These bacteria are hardy and can survive in soil for years. Bacteria usually enter a host by inhalation and ingestion and replicate locally and in regional lymph nodes. Pneumonia and tubercles that cause a chronic moist cough are other respiratory signs. The disease may result in emaciation, anorexia, and low grade fever but many infected animals appear outwardly normal (Miller et al. 1991).

Tubercles are often thick walled with a layered, onionskin appearance, but can look like abscesses caused by a variety of common bacteria. Postmortem diagnosis is based on gross and microscopic lesions that should be confirmed by isolation and characterization of the bacteria.

Diagnosis and screening of populations is done by dermal sensitivity testing. A thickening of skin at the site of an intradermal inoculation with tuberculin at 72 hours postinoculation is considered a positive test. A complete postmortem examination should be conducted on all animals that are culled due to dermal sensitivity to tuberculin. A battery of tests, including the ELISA antibody test, are used to screen red deer in New Zealand. The applicability of these screening procedures for North American wildlife is being explored (Thorne et al. 1992).

Tuberculosis is a difficult disease to eradicate and small pockets of infection remain in the United States. Elimination of positive herds, and in some cases, test and slaughter, remain the only management tools for tuberculosis control and eradication. Carcasses should be disposed of on-site, usually by incineration.

Tuberculosis is a persistent and infectious disease and can cause disease in humans. Exposure can cause permanent dermal sensitivity conversion. Millions of dollars have been spent to eradicate tuberculosis from livestock. It is only rarely seen in wildlife and keeping it from spreading to livestock or wildlife is important. Many state and federal agencies are tightening regulations on captive-owned wildlife and restricting their movement. If tuberculosis becomes more widespread, it will be difficult and expensive to control (Thorne et al. 1992).

Brucellosis

Brucellosis or Bang's disease causes a high fever that periodically reoccurs (undulant fever of humans), joint disease, abortion and damage to the reproductive system, and in some cases, death (Davis et al. 1981a). Many animals and humans are susceptible to infection from various *Brucella* spp. bacteria. Since 1934 state and federal agencies have attempted to eradicate *Brucella abortus* using serologic tests to detect infection, slaughtered infected cattle, and vaccinated calves (Thorne et al. 1992). The incidence of infection in cattle is less than 0.5% nationally. However, brucellosis in herds of elk and bison and a few herds of range cattle makes the eventual eradication of this disease problematic. Brucellosis also occurs in reindeer and wild pigs (Drew et al. 1992).

There are different types of brucellosis. *Brucella abortus* type 1 is primarily a disease of cattle, but there are foci of infection in bison and elk in northwestern Wyoming and Yellowstone National Park. Infection rates of approximately 30% have been reported from elk wintering on the National Elk Refuge and Grey's River Feed Ground (Thorne et al. 1982). The bison herds on the National Bison Range in Montana, in Wood Buffalo National Park, and in Elk Island National Park in Canada are infected also. Infections in moose in Minnesota, Montana, and Alberta, white-tailed deer in Ohio, and coyote in Texas have been reported (Thorne et al. 1982).

Other forms of brucellosis have been recovered from wild pigs, deer, reindeer, domestic sheep, and goats. Humans can become infected with any type of brucella by consumption of undercooked meat or unpasteurized milk, inoculation into tissue such as when cleaning a carcass, or through contact of the organism with mucous membranes (e.g., eyes, nose, and throat).

The oral route of infection (e.g., licking of the fetus, ingestion of fetal membranes and vaginal secretions) appears to be most important for cattle, elk, and bison. Unfortunately, elk in Wyoming often abort in winter when concentrated on winter feed grounds, exacerbating transmission.

The number of organisms and strains of bacteria to which an animal is exposed, its age, and innate resistance determine whether exposure results in infection. The bacteria penetrates the local mucous membranes, localizes in regional lymph nodes, and may then spread to other organs and joints, causing local inflammation. When brucellosis localizes in the uterus, the result is damage to the uterine lining and the placenta, and abortion. In male animals, the bacteria may localize in the testes (Thorne et al. 1982).

Abortion, premature birth, and neonatal mortality are the most important signs of brucella infection in elk and cattle. Chronically infected elk and reindeer may show lameness and signs of joint infections. Reproductive effects are reported to occur in deer, and moose infected with *B. abortus* become weak and emaciated and die of generalized infection. Wild sows infected with *B. suis* may abort or have nonviable piglets.

Diagnoses at necropsy yield different lesions for various species. In moose with generalized infections, lesions are prominent. In bison, swollen pus-filled scrotums and degenerated testicles are common. The testicular lesions in elk are less severe.

The lesions in elk, reindeer, and cattle are generally less striking than in bison and moose. A purulent exudate covering a thickened placenta can be visible following abortion. Microscopically, a necrotizing placentitis with mixed inflammatory cells and sometimes epithelioid and giant cells are visible. Lymph nodes of the head and pelvic region may be swollen and inflamed.

The lesions of brucellosis are characteristic but not definitive. Only culture of the organism from animals with typical lesions allows a firm diagnosis. Serosurveys of wild animals captured for a variety of reasons allow wildlife agencies to determine if infected populations exist, and are a relatively inexpensive way for wildlife and livestock interests to cooperate toward elimination of brucellosis.

Once free-ranging animals become infected, eradication or control of this disease becomes expensive and difficult. Infected cattle should not be allowed to use open range where wildlife may come into contact with them. Control of brucellosis in wildlife is only possible if eradication of brucellosis in cattle succeeds.

Management of infected herds involves culling, reduction or elimination of winter feeding, and vaccination. Brucellosis in free-ranging moose and deer appears to be rare and self-limiting.

Brucellosis can be a serious and life threatening disease in humans and is an economic disease in the livestock industry. Millions of dollars have been spent to eradicate brucellosis. Because the infection rate is low, and it is often not fatal in many species of wildlife, it is not a disease that is likely to destroy herds or populations. However, brucellosis must not be accidentally spread through translocation of infected wildlife or movement of infected privately owned livestock, or movement of wild animals. Routine testing prior to shipment and isolation of new stock prior to releasing them into a herd or the wild will help control this disease.

Paratuberculosis

Paratuberculosis, also called Johne's disease, is an infectious disease of dairy cattle and other livestock that may infect llamas, bighorn sheep, mountain goats, elk, reindeer, and other wild and domestic ruminants (Thorne et al. 1982). It is caused by the bacteria *Mycobacterium paratuberculosis* and is a chronic disease of the gastrointestinal tract that causes diarrhea and emaciation. It is not a highly infectious organism and has limited potential to spread widely or rapidly unless moved within infected animals.

In free-ranging wildlife, paratuberculosis has been diagnosed in a herd of bighorn sheep and mountain goats in Colorado and Tule elk in California (Jessup 1993). Captive moose and white-tailed deer have become infected. When wild animals are in captivity (e.g., zoos, wild animal parks) and once paratuberculosis has been established, it is extremely difficult to diagnose and eliminate.

Paratuberculosis is transmitted primarily from ingestion of bacteria shed in feces. More susceptible species and individuals (i.e., less than six months old) develop signs of disease up to a year or more following initial infection. Fluids and nutrients cannot be absorbed, causing chronic watery diarrhea, emaciation, and other signs related to malabsorption, and malnutrition. Diarrhea is not as prominent in bighorn sheep.

Gross lesions include thickening of the intestinal wall, blockage of lymph channels, and massive enlargement of regional lymph nodes due to chronic inflammation. Dissemination of the liver and other viral organs may eventually occur. The entire process from infection to death may take as long as two years. A positive culture of feces or tissue is the most sensitive way to reach a definitive diagnosis. Culture may take up to 12 weeks.

Once clinical manifestations are evident, paratuberculosis is fatal. There is no effective treatment. Mortality rates are not developed for wild animals, but may be 1–25% per year in infected herds.

Paratuberculosis is difficult to control in captive animals because animals in early stages of infection and subclinical carriers contaminate the environment. The lack of effective rapid diagnostic tests complicates management. Culling clinically infected adults, raising young animals away from infected adults, and good sanitation are only partially successful at reducing infection rates. There is no effective control of paratuberculosis in wild populations.

Prevention of infection is important with paratuberculosis. Infected wild animals should not be translocated. Feces and carcasses of infected animals should be treated as infectious waste. All applicable state and federal laws and policies for livestock should serve as a basis for wildlife and exotic animal management decisions.

Pneumophilic bacteria

Pneumophilic bacteria (e.g., *Pasteurella hemolytica* and *Pasturella multocida*) are major pathogens of wild and domestic ungulates. *Pasteurella* infections tend to take either a pneumonic or septicemic form. Pasteurella are common worldwide and most, if not all, species are apparently susceptible to disease. Pneumonias and septicemias and more chronic localized infections occur in cattle, domestic sheep, and goats. Pronghorn and bighorn sheep are susceptible to Pasteurella infections, particularly acute fatal pneumonias. Mule deer and elk are less commonly affected although Pasteurella septicemias are reported in elk, deer, bison, and other wild ungulate species.

Die-offs of bighorn sheep from acute bronchopneumonia following association with domestic sheep were first reported in 1982 (Foreyt and Jessup 1982). Nose-to-nose contact of bighorns with domestic sheep was followed by a die-off within weeks.

In numerous cases where bighorn sheep are in contact with domestic sheep, no sickness or mortality was reported in the domestic sheep (Onderka et al. 1988, Onderka and Wishart 1988, Callan et al. 1991, Krausman 1996), but bighorn sheep died (Foreyt and Jessup 1982).

Inherited differences in host resistance factors, including the pulmonary macrophages that guard the upper respiratory tract of bighorn sheep and are less capable of killing and processing bacteria than the macrophages of domestic sheep (Silflow et al. 1989), may account for some of the differences in susceptibility between domestic and bighorn sheep.

Pasteurella hemolytica is capable of causing progressive fatal pneumonia when established in the lung. *Pasteurella multocida* may require predisposing viral infections or stress immune suppression to cause fatal pneumonia. When an animal becomes infected, bacteria are shed in saliva and droplets form the upper respiratory tracts. Under moist and mild conditions Pasteurella can survive in the environment for months.

Animals with pneumonic infections often have labored breathing, coughing, and nasal discharge. Some animals may survive long enough to show signs of emaciation. Animals with acute septicemic pasteurellosis show few signs and die suddenly.

Lesions of septicemic infections include multiple hemorrhages on serous surfaces; red, wet lungs with froth in the trachea; and swollen, hemorrhagic lymph nodes of the head, neck, and chest (Davis et al. 1981a). Isolation of the organism on artificial media is needed for a definitive diagnosis.

Pasteurella pneumonias are difficult to treat. If given early, penicillins or tetracycline can be an effective treatment in livestock and in some bighorn sheep; however, some Pasteurella are resistant to these and other common antibiotics.

Pasteurella spp. has caused major die-offs of wild animals, and is an important disease that has caused conflict between wildlife and domestic sheep interest groups. Contact between domestic goats, sheep, and bighorn sheep should be minimized. The Desert Bighorn Council recommended a 15-km buffer zone between domestic and wild sheep to minimize the disease.

Caseous Lymphadenitis

Caseous lymphadenitis is caused by the bacteria *Corynebactrium pyogenes* and causes abscesses and suppurative infections in wild and domestic animals worldwide, especially wherever large numbers of animals are raised. Infections and abscesses due to *C. pyogenes* are common in livestock and have been reported in mule deer, white-tailed deer, pronghorn, elk, moose, reindeer, and bighorn sheep (Thorne et al. 1982).

C. pyogenes is a normal inhabitant of bacterial flora of grazing ungulates. Inoculation into tissues via puncture wounds, lacerations, ulceration, or bites may appear to be the most common routes of entry. Disease may result when the host cannot localize the initial infection. Joints can also be infected.

Animals that die from C. pyogenes infection are thin because the disease course is long and debilitating. Draining abscesses from the site of lymph nodes, swollen joints filled with thick pale green pus, and chronic infections of the jaw and mouth are common outward signs of C. pyogenes.

Abscesses due to C. pyogenes may be dry (caseated) and walled off or may contain large amounts of creamy pus. Joints are swollen and may be deformed. Infection should be confirmed by isolation of the organism.

This organism is ubiquitous and no form of management is effective. Isolation of infected animals to a "sick pen," removal of pus, and sanitation of contaminated pens and instruments will reduce spread.

Leptospirosis

Leptospirosis bacteria cause a disease that is transmissible between animals of the same and different species, from animals to man. Manifestations of infection range from subclinical to death.

Leptospirosis has a worldwide distribution and infects a wide range of species including rodents, raccoon, opossum, and skunk (Davis et al. 1981a). Dairy cattle are fairly commonly infected and range cattle and humans are only rarely infected. Illness and limited numbers of deaths of free-ranging pronghorn, black-tailed and mule deer, white-tailed deer, and red deer have been reported (Davidson et al. 1981). Cattle and wildlife may become infected by drinking from contaminated water sources.

Urine from infected animals is the most common source of infection. Leptospires can survive for weeks in stock ponds and slightly alkaline aquatic environments, but do not survive freezing weather well. They may penetrate abraded skin and mucous membranes of the eye, nose, or mouth. In the western United States, contaminated water holes in late summer may be important in transmission between wildlife and livestock.

After penetrating skin or mucous membranes, leptospires replicate in the bloodstream for a week. During this time a significant fever will be seen that may cause abortion. If the organism persists in the kidney or liver, damage to those organs will produce a variety of signs and usually death. Animals with leptospirosis can experience nervousness, uncoordination, aggression, dullness, inappetence, emaciation, diarrhea, and abortion.

Deer in terminal stages of infection have dark yellow to red wine-colored urine and generalized jaundice. The kidneys may be soft, pale, and swollen and have pale streaks. The liver can be swollen and soft and areas of coalescing central necrosis may be evident microscopically. Leptospires are difficult to isolate in artificial media and diagnosis is based on topical gross and microscopic lesions.

Clean water can be enhanced by supplying water in troughs or drinkers rather than from stock ponds and water holes. This will reduce the likelihood of transmission between wildlife and livestock. Water sources should be cleaned periodically. Avoid concentration of livestock and wildlife at point sources of water. Leptosirosis may be a density dependent disease and maintaining populations of wild animals at or below carrying capacity will reduce the chances of transmission or maintenance of leptosirosis. Vaccination of cattle will prevent clinical illness and abortion for 6 to 12 months, and is a good and inexpensive method to prevent spread to other animals. Infected cattle can be treated with antibiotics, but vaccination or treatment is not practical for free-ranging wildlife.

Leptosirosis can be transmitted to humans but is often self-limiting; it can be fatal. The most common sources of human infections are water, livestock, domestic pets, is wildlife. Leptosirosis is not a common disease of free-ranging wild ungulates.

Anthrax

Anthrax is a bacteria (*Bacillus anthracis*) that affects ruminants, and occasionally humans and other animals.

Anthrax occurs worldwide and has been reported from most areas of the United States where soils are neutral or alkaline and under appropriate weather conditions. Cattle are most often affected, but white-tailed deer, mule deer, and moose have died of anthrax (Thorne et al. 1982). In Africa, kudu and roan antelope are susceptible. A number of species of exotic wildlife on Texas game ranches have died of anthrax. Given sufficient contact with the infectious stage of the bacteria, almost any species of animal may succumb to anthrax.

Decomposing vegetation following flooding increases appropriate conditions for anthrax. A lack of good forage can force cattle or wildlife to graze close to the ground and increases the ability of grazers to ingest bacterial organisms. Infection can be from inhalation of spores or inoculation also.

When an outbreak occurs, biting insects help spread infection from animal to animal. Aggressiveness, agitation, depression, and death occur rapidly. Bloody fluid may leak from the rectum, nose, or mouth.

It is dangerous and inappropriate to do a postmortem examination on an animal that is suspected to have died of anthrax because it spreads spores and exposes humans to the disease. Aspiration from or removal of an eye, or nicking an ear to get a blood smear or culture is recommended. Rapid postmortem decomposition and gasification, thick tarry blood, edema, and hemorrhage in almost any or all organs are compatible lesions.

In endemic areas the signs and lesions described are sufficient for a presumptive diagnosis. Inoculation of blood of ocular fluid into enriched media will allow the typical medusa head or vegetative state bacteria to develop.

Cattle may be vaccinated but this is not practical for wildlife. However, thousands of roan, kudu, and other species have been vaccinated around outbreak areas in South Africa. Carcasses should be completely burned, but should not be moved or scattered.

The disease is rare but outbreaks of anthrax can kill large numbers of wild and domestic animals. Because it is a potential lethal human disease, hunters and people working around wildlife or cattle carcasses should be aware of the potential of exposure. Only very early aggressive treatment with antibiotics is successful in treating anthrax.

Yersinosis

Yersinosis is also called plague or bubonic plague. It is an infectious disease caused by the bacteria *Yersinia pestis.* If untreated, it is often fatal. Plague is primarily a disease of rodents in scattered epidemics in the western United States. It occasionally infects carnivores, and rarely infects humans and wild or domestic ruminants.

Plague is a relatively rare disease and is not of significance for wild ungulates or livestock. However, wildlife workers and ranchers in the western United States should recognize its presence in the areas where they work.

Rickettsial Diseases

Anaplasmosis

Anaplasmosis is a tick-transmitted disease of cattle, sheep, and a number of wild ruminants. It is usually a mild disease in wild animals; deer may be carriers. Anaplasmas is found throughout the world. In North America *Anaplasma marginale* and *A. ovis* infect cattle and sheep, respectively. Black-tailed deer, mule deer, white-tailed deer, elk, and pronghorn may be become infected with A. marginale and bighorn sheep may become infected with A. ovis.

Ticks (*Dermacentor* spp.) appear to be the most important transmitters of these rickettsia, which they ingest with a blood meal (Davis et al. 1981b). Affected animals may be weak, unwilling to eat, and show labored breathing. Recovered animals can become carriers of the organism.

The blood may be watery, the liver and spleen somewhat enlarged and pale, internal lymph nodes enlarged, and white tissues may appear yellow. A diagnosis is made by demonstration of marginal bodies in the cytoplasm of red blood cells. The disease is usually mild in wildlife and treatment is not necessary. Tick control in heavily infested areas can reduce the number of infected animals. Deer are a common carrier of anaplasmosis that infects cattle. However, if the tick vectors are present, it is possible that anaplasmosis can be transmitted to and from many species of wild and domestic ruminants.

Viral Diseases

Influenza Viruses

Influenza viruses are also called parainfluenza-III (PI-3) and respiratory syncytial virus (RSV). Both have a worldwide distribution and are common causes of upper-respiratory disease in cattle and sheep. Pneumonia from PI-3 and RSV has been reported in bighorn sheep.

Animals with viral respiratory infections appear depressed with nasal or ocular discharge, and may cough and sneeze. Lesions caused by influenza viruses are subtle and transient. Definitive diagnosis can be made by isolation of the virus. The only treatment for infected animals is to control bacterial infections that may follow the viral infection. Penicillins or tetracycline can be an effective treatment but is not practical. Keeping livestock and bighorn sheep separated is one potential preventative measure. Cattle and captive wildlife or small populations can be vaccinated for both of these viruses (Jessup et al. 1991). These are common viruses in livestock and their role in causing any disease process in wild animals is unclear. It is also unknown if the influenza viruses isolated from wildlife came from livestock or are indigenous to wildlife.

Influenza has been blamed for outbreaks of pneumonia in some herds of bighorn sheep, and some groups have tried to tie this to interactions with livestock. The evidence, however, is not consistent or convincing.

Hemorrhagic Disease

Hemorrhagic disease, also called bluetongue or BT, and epizootic hemorrhagic disease, called EHD, cause similar disease syndromes in a variety of wild and domestic ungulates. They can only be distinguished by isolating the causative virus or by serial blood sampling for antibodies, so both diseases are described together under a name more descriptive of the signs: hemorrhagic disease. There are four serotypes of bluetongue disease (i.e., 10, 11, 13 and 17) in the United States and two serotypes of epizootic hemorrhagic disease (i.e., 1 and 2). These diseases are transmitted by biting gnats and are not directly transmitted between hosts.

The viruses of hemorrhagic disease occur in livestock across most of the Untied States except the northeast and northwest coastal areas, and are most often seen in wildlife in the southeastern United States, California, and portions of the Rocky Mountain states. Bluetongue disease is primarily recognized as a disease of domestic sheep, but bighorn sheep, white-tailed deer, pronghorn, black-tailed or mule deer, and cattle can become infected (Thorne et al. 1982). Epizootic hemorrhagic disease is primarily recognized as a disease of white-tailed and black-tailed deer, but pronghorn, domestic sheep, bighorn sheep, elk, and cattle may be infected (Davidson et al. 1981). In zoos and on game farms an even wider variety of wild ungulates, including most deer species and many African antelope and wild bovids, have become infected by these viruses.

An outbreak of hemorrhagic disease requires a sufficient number of susceptible hosts, pathogenic strains of the virus, and populations of biting midges or gnats. Gnats become infected when they feed on a viremic host and in 10 to 14 days can transmit the virus to another host. Gnats prefer muddy, fecal-contaminated pond edges for breeding. Concentrating susceptible hosts around such ponds creates ideal conditions for an outbreak. In most areas of the United States late summer or early fall are the gnat season, but in the desert Southwest late winter and early spring may favor emergence.

Cattle appear to be potential reservoirs as they seldom show overt signs of disease and the virus persists longer in their blood than in wildlife. When inoculated into the host, the virus reproduces in the cells lining blood vessels. Most of the signs and lesions of hemorrhagic disease are caused by damage to small blood vessels.

Animals with hemorrhagic disease often show no overt signs of disease. However, animals with acute, severe disease may be depressed, appear to be blind, have swollen ears, eye lids, lips or tongue, may have froth exuding from the nostrils, and die within 24 hours. Large numbers of dead animals are found around water holes because they become very thirsty and febrile near death.

Animals with less severe signs of disease may have ulcers of the mouth, the tongue may be dark or ulcerated, or necrosis of the coronary band may lead to sloughing of the hooves and secondary foot infections. Some animals with less severe signs survive but tend to remain emaciated. Abortion and neonatal deaths can occur in deer, and reduced pronghorn fawn production has been reported in the year following an epidemic.

Hemorrhages occur on serosal and mucosal surfaces. Ruminal or intestinal ulcers may lead to bloody diarrhea and edema fluid may fill the lungs and pleural or peritoneal cavities. Microscopically, hemorrhage, vasculitis, and thrombosis in small arteries and associated necrosis are common. Diagnosis may have to be based only on a compatible case history and gross and microscopic lesions, because isolation of the virus is difficult. There is no good treatment for infected animals.

Gnat breeding habitat can be reduced by keeping pond sides steeply sloped and by reducing organic content of water. Troughs or spring boxes are less likely than stock ponds to serve as gnat breeding sites. During gnat seasons water sources should be kept clean, animals dispersed, and gnat breeding habitat minimized. Unfortunately, these precautions are not easy to implement or likely to be completely effective.

In some areas diseases have caused major die-offs of sheep, deer, antelope, and exotic wild animals. However, the full impact of an outbreak may not only be estimated by carcass count because birth rates and neonatal survival may be affected. In some locations these diseases may be population regulating, in others they may be common and benign.

Malignant Catarrhal Fever

Malignant catarrhal fever, also called MCF, is rare in the United States, not highly contagious, but often fatal. Malignant catarrhal fever has been reported from cattle in all regions of the United States, captive, exotic wild ungulates including Pere David's deer, red deer, axis deer, and several species of antelope, and in captive native wildlife including white-tailed deer, mule deer, bison, moose, and wildebeest in Africa (Castro and Heuschele 1992). Only two cases in free-ranging deer (i.e., black-tailed deer in California and white-tailed deer in the eastern United States) have been reported.

The virus that causes MCF is associated with white blood cells and it takes a large transfusion of infected blood to transmit the disease experimentally. The mode of natural transmission and pathogenesis are unknown.

High fever, inflammation of the nasal passages or ocular mucous membranes, photophobia, corneal opacity, diarrhea, dehydration, emaciation, and enlargement of lymph nodes are typical signs (Castro and Heuschele 1992).

The lesions of MCF include erosions of the lips, oral mucous membranes, tongue, and pharynx. Additionally tracheobronchitis, sinusitis, corneal opacity, generalized lymph node enlargement, hemorrhagic ulcers of the abomasum, bloody or watery diarrhea, and lesions of the heart, liver, or kidney may be present.

The lesion that differentiates MCF from other disease is fibronoid necrotizing vasculitis. The best organs to find these characteristic lesions in are the brain and abdominal organs (i.e., liver, kidney, and adrenal). Because so little is understood about this disease, there are few practical methods of control. Separation of susceptible species from wildebeests, especially during calving periods, and from domestic sheep is recommended. In some states the importation and private ownership of wildebeests is prohibited.

Wildlife are not commonly exposed to malignant catarrhal fever and it is not an important disease of free-ranging populations. However, its high case fatality rate has caused a number of states to discourage the ownership of wildebeest. It looks much like foot-and-mouth disease and other foreign animal diseases, so it is a cause for concern.

Foot-and-Mouth Disease

Foot-and-mouth disease influences many species of cloven-hoofed animals including cattle, sheep, goats, deer, antelope, and elk. This disease is caused by a virus, is considered to be a foreign animal disease, and an outbreak would likely cause the declaration of a national disease emergency. Infected animals are subject to quarantine and slaughter. The outbreak in the United Kingdom in 1967 – 1968 resulted in the slaughter of 450,000 livestock and the last outbreak in North America occurred in California in 1928 when more than 22,000 deer were slaughtered. The most recent outbreak of foot-and-mouth disease began in Asia in 2001 and has spread to the United Kingdom, Europe, and elsewhere. The influence of this outbreak is ongoing.

Domestic livestock and many species of North American wildlife are susceptible. African antelope held on game ranches are also susceptible. The only areas of the free world free of foot-and-mouth disease are North and Central America, Australia, New Zealand, Oceania, Japan, and portions of Europe.

The foot-and-mouth disease virus is shed in large amounts in the oral secretions and survives in manure, blood, and excreta for extended periods. Infected and recovered animals can become carriers. Foot-and-mouth disease is highly infectious and can be transmitted by aerosol over long distances.

The virus replicates in the superficial layers of the mucous membranes and coronary band causing the formation of blisters. These blisters rupture, releasing large numbers of virus particles followed by ulcers and secondary bacterial infections. The virus also causes systemic infection that often damages the heart.

Animals infected with foot-and-mouth disease are usually depressed, anorectic, emaciated, lame, and they salivate profusely due to painful blisters. Many infected animals will survive but seldom regain full health and can become carriers. Clinical signs and gross lesions of foot-and-mouth disease are not adequate for diagnosis but should trigger immediate investigation. A diagnosis of foot-and-mouth disease is based on the presence of virus-specific proteins, or antibodies in serum, detected by one of several tests or by isolation of the virus.

There is no treatment for foot-and-mouth disease. In North America any infection that occurs will be handled by quarantine, test, and slaughter. Any disease resembling foot-and-mouth disease should be reported immediately.

Contagious Ecthyma

Contagious ecthyma is also called sore mouth, scabby mouth, and orf (when it infects humans). It was first described in Europe over 200 years ago. Contagious ecthyma has a worldwide distribution and sheep and goats are commonly infected. Since 1954 periodic outbreaks in free-ranging bighorn sheep, mountain goats, musk-ox, and reindeer have been documented.

The infection is spread by direct contact between animals, but the virus may survive for greater than 20 years in scabs. Small breaks in the skin may allow entry of the virus. The mucous membranes of the nose, mouth, and rectum are most often affected, but also eyes, vagina, teats, and locations subject to trauma or inoculation. Salt blocks and alfalfa hay on which infected animals have previously fed have been the point sources of outbreaks in bighorn sheep.

Contagious ecthyma causes itching and pain, and infected sheep are nervous, lick their lips and nostrils, and rub affected areas. Grazing and suckling are difficult and lambs are often rejected by their mothers. If an outbreak occurs during lambing season, lamb losses can be high.

The early lesions of contagious ecthyma are rarely noticed, but when the tissues beneath the scabs become thick and cauliflower-like, they bleed profusely. Microscopically the epidermal cells in the basal layer can be seen undergoing ballooning degeneration with extensive rete-peg formation down into the dermis (Castro and Heuschele 1992). The species affected, time, and progression of lesions, their gross appearance, and histology are characteristic and, in combination, are sufficient to make a diagnosis.

A virulent live virus vaccine is used to inoculate lambs and kids on the scarified inner thigh to induce infection and subsequent immunity, but this is not applicable to free-ranging wildlife. Also, the vaccine virus can be spread into the environment, where it is persistent. Phenolic disinfectants and steam can be used to clean contaminated clothes and equipment. There is no effective treatment, but the disease is usually self-limiting and by the end of a month, most adult animals will recover spontaneously. Animals in active stages of infection should not be mixed with susceptible populations. Hay from fields that have been grazed by sheep and goats should not be fed to highly susceptible species.

This disease is transmissible to humans. Four of 65 people handling or working near infected bighorn sheep during a capture developed lesions. Contagious ecthyma is painful, unsightly, and can leave some scarring. Implements, clothes, and equipment should be disinfected thoroughly or discarded. This disease is probably not an important population limiting factor, although some mortality may occur. Infected populations tend to show recurrent signs when other stressing factors like malnutrition, overpopulation, or prolonged extreme weather are present.

Rabies

Rabies is also called hydrophobia, lyssa, or rage and is an acute contagious disease of mammals caused by a virus. Rabies occurs worldwide except in Australia. In the United States most cases of rabies occur in the eastern and southern states.

The disease is transmitted by the bite of rabid animals (usually dogs, skunks, foxes). Susceptible hosts include dogs, cats, horses, cows, sheep, pigs, wolf, fox, squirrel, skunk, raccoon, bats, rodents, and other mammals.

Infected dogs have symptoms of mental disturbance, nervous excitability, with subsequent paralysis and death. Symptoms are divided into three stages. The first stage indicates a change in disposition (e.g., troubled, distracted look, lack of desire to associate with others, restless). In the second stage the animal wanders, ceaselessly snaps or bites objects in its path, barks repeatedly without cause, has tremors or spasmodic muscle contractions, has a perverted appetite, and strays. In the third stage muscle paralysis, salivation, convulsions, and related nuerologic symptoms result in death. Duration is up to 10 days. This form of rabies is called furious rabies. Another form is called dumb rabies, where animals are lethargic but have tremors and die.

A history of being bitten by a rabid animal plus the symptoms of rabies in the animal that bit are usually sufficient for diagnosis. To be certain, diagnosis is confirmed by examining the brain of the rabid animal when it is in later stages of the disease (i.e., paralytic stage). Nergi bodies in the brain confirm the diagnosis in most species.

Rabies is a reportable disease in all states and dogs must be vaccinated in many states. There are fewer than 10 cases in humans per year in the United States. Most are from dog bites. Cleaning and cauterizing the wound and vaccination are the acceptable methods of treatment.

Internal and External Parasites

The few parasites covered in this section were selected because of their potential importance in the ecology and management of the interface involving wildlife and domestic animals. However, these selected parasites represent only a fraction of the array of parasites that occur in or on wildlife.

Most animals are infected with more than one different parasite including protozoa, nematodes, tapeworms, flukes, and arthropods. Some parasites occur in wildlife and domestic animals, whereas in other cases it is not clear whether or not the different hosts share identical parasites. Often, infected animals do not exhibit any outward signs of infection or disease. However, if an animal is weakened by some preexisting stress or condition such as malnutrition, then parasites may predispose the animal to clinical disease and death. In contrast, some parasites are so pathogenic that they can be primary causes of morbidity and mortality in the absence of any detectable preexisting condition.

Most infections in free-ranging wildlife cannot be treated because the animals are not handled individually. Therefore, managers should recognize that their efforts are most effective when they are aimed at preventing the establishment of new parasites and controlling the levels of existing infections.

Internal Parasites

Numerous parasites (e.g., gastrointestinal nematodes and tapeworm, lungworms, and liver flukes) are found in wild ungulates. Although some species of parasites in each of these groups are shared by domestic and wildlife hosts, most internal parasites are relatively host-specific and are not likely to be of major concern.

Gastrointestinal Nematodes

Several genera of nematodes (e.g., *Ostertagia, Haemonchus*) are in the gastrointestinal tract of domestic and wild ruminants. These parasites exist inside the host and on the pasture. Infected animals pass eggs in their feces and in warm, moist conditions infective larvae can accumulate on pasture. Infections are acquired by ingesting larvae with grass. Typical clinical signs include diarrhea, weight loss, and anemia. Clinical disease is most likely to be seen when large numbers of animals share a restricted grazing environment. The most effective way to manage these nematodes is through a preventive approach. Fecal examinations will reveal the approximate numbers and types of internal parasites, and new animals should be treated and examined before they are added to or mixed with other captive or free-ranging animals. Treatment of internal parasites in penned exotic and domestic animals is easy, but treatment is not effective if the pastures have already been contaminated with eggs and larvae. Treatment of free-ranging animals is impractical.

Lungworms

Lungworms (Protostrongylus stilesi) are one of the most important parasites of Rocky Mountain bighorn sheep (Davis et al. 1981b). Lambs are infected in utero by transplacental migration of larvae and heavy infections can cause high morbidity and mortality. Lungworm is a particular problem among high-density bighorn sheep populations that live in mesic environments supporting large populations of the snail intermediate host. The anthelmintic fenbendazole is an effective treatment to reduce morbidity and mortality in those situations where a substantial proportion of the population can be attracted to medicated feed.

Liver Flukes

Liver flukes (Fasciola hepatica and *Fascioloides magna)* are important internal parasites of domestic and wild ruminants in those areas of the Northwest, Midwest, and Southeast where the intermediate host and amphibious snail is found (Davis et al. 1981b). Both parasites are transmitted by accidental ingestion of infective stages present on vegetation. Both parasites are difficult to impossible to control in free-ranging animals. Domestic livestock are dead-end hosts for *F. magna* and do not pass eggs in their feces. Mule deer and white-tailed deer are the normal hosts for this parasite, and infections in domestic livestock only occur when infected deer are present in the same environment. Clinical signs include weight loss, diarrhea, and anemia. It is impractical to treat these infections in free-ranging wildlife.

External Parasites

Numerous arthropod ectoparasites, including ticks, mites, flies, lice, and fleas are in or on the skin of mammalian hosts. Many of these parasites can cause disease through their bloodsucking activities and serving as vectors of viruses, bacteria, and protozoa. Important tick and fly vectors (and their respective pathogens) include the ticks *Dermacentor* spp. (*Anaplasma* [anaplasmosis]), *Ixodes* spp. (*Borrelia* [Lyme disease, borreliosis]), *Boophilus* spp. (*Babesia* [Texas cattle fever]), gnats such as *Culicoides* spp. (bluetongue, epizootic hemorrhagic disease), and tabanid flies (*Elaeophora schneideri* [arterial worm]). Arthropod vectors and their associated pathogens can be found on domestic livestock and wildlife. These emerging potential problems include tick-transmitted disease of borreliosis, anaplasmosis, and babesiosis in bighorn sheep and deer.

Boophilus

The federal program for eradication of cattle fever ticks (*Boophilus annulatus* and *B. microplus*) provides an example of problems that can arise due to parasite interactions between domestic cattle and free-ranging wildlife. Boophilus annulatus and B. microplus are vectors of the blood protozoan *Babesia bigemina,* the causative agent of Texas cattle fever in North America. In 1906 the federal government instituted the United States Cattle Fever Eradication Program because economic losses (more than 40,000,000 per year) were being sustained by the cattle industry in the Southwest due to infections of *B. bigemina.* The program focused on cattle dipping to kill Boophilus ticks because these parasites are one-host ticks. However, B. micoplus and B. annulatus occur on cattle and white-tailed deer, and the only way to eradicate ticks on deer was through depopulation. Therefore, in a controversial program, thousands of deer were killed during the early 1900s to eliminate the wildlife host for ticks. Texas cattle fever no longer causes a significant problem in the United States.

Psoroptes spp.

Psoroptic scabies is an ectoparasitic disease of domestic and wild ungulates caused by mites of the genus *Psoroptes.* Scabies was a serious problem in the livestock industry that warranted a federally subsidized eradication program that was initiated during the turn of the twentieth century. The program involved isolating and treating infested domestic livestock and culminated in the 1973 federal acknowledgement that sheep scabies had been eradicated from the United States. Many observers are convinced that the introduction of domestic sheep and cattle onto bighorn sheep ranges in the late 1800s and early 1900s precipitated bighorn die-offs. However, it is of interest that bighorn sheep scabies is still relatively common in the western states, whereas domestic sheep and cattle scabies has decreased substantially (Boyce et al. 1991). These findings raise questions regarding the host specificity and species identification of Psoroptes spp. mites. During an outbreak of bighorn scabies in New

Mexico, several unsuccessful attempts were made to transmit Psoroptes spp. mites from bighorn sheep to cattle to domestic sheep.

Diagnosis of Psoroptes spp. infestations can be accomplished by microscopic examination of skin scrapings and ear swabs or by immunodiagnosis (Boyce et al. 1991). Ivermectin has played a major role in reducing the prevalence of scabies in domestic livestock. The potential of these mites to infest different host species may preclude any control strategy that focuses on treatment as a management tool for wildlife.

SUMMARY

Although diseases play an important role in the ecology of wildlife, biologists need more information on their role. The study of disease is growing and, as more is learned, the information will help in the management of populations. This chapter briefly discusses selected diseases of migratory birds and mammals.

12

Decimating Factors: Accidents and Starvation

"Accidental mortality is of greater concern if the affected wildlife population is small, so that a few accidentally killed animals constitute a fairly large proportion of the population."

J. A. Bailey

INTRODUCTION

Accidents and starvation are sources of mortality that are often undetected but need to be accounted for. This chapter discusses the more common types of accidents and starvation in wildlife.

ACCIDENTS

Leopold (1933) defined *accidental mortality* as all losses from physical causes other than diseases, parasites, hunting, predators, or lack of welfare factors. Accidents are not rare in nature and are a constant source of mortality. Wildlife populations reproduce at a higher rate than necessary to replace mortality, and most individuals die young, before reaching breeding age (Caughley 1966, Ricklefs 1979:499). Accidents contribute to mortality and may include, or are the result of, human development. Common types of accidents include collisions with human-made objects (e.g., birds colliding with television towers, picture windows, fences), mortality from oil spills, wildlife-car accidents, drowning, electrocution, falls, entanglement in fences, and a host of other unusual situations (Figure 12–1). Accidents generally are not the major form of mortality for a population unless the population is small. For example, since 1973, 62% of mortalities of the Florida panther in the wild have been associated with human interaction (i.e.,

(a)

(b)

(c)

(d)

(e)

(f)

Figure 12–1 Mortality of wildlife caused by accidents: a. locked antlers; b. fall; c. road kill; d. fence entanglement; e. hoof in fence (photos by P.R. Krausman); f. elephant killed by train (photo by A. C. Williams).

road kills, illegal hunting, or injuries) (Roelke et al. 1993), and vehicle collisions are a significant cause of mortality for the endangered ocelot in southern Texas (Tuovila 1999). The main source of mortality for all threatened and endangered vertebrates in Florida is road kills (L.D. Harris, University of Florida, personal communication). Road kill is also the limiting factor in the recovery of the endangered American crocodile in southern Florida (Kushlan 1988) and in the cooperatively breeding Florida scrub Jay (Mumme et al. 2000). Road kill also contributes to mortality of the prairie garter snake in Ohio (Dalrymple and Reichenbach 1984) and peninsular bighorn sheep in California (J. De Forge, Bighorn Institute, personal communication). Road kills can influence larger populations also. Vehicles are the primary cause of death for moose in the Kenai National Wildlife Refuge, Alaska (Bangs et al. 1989), for barn owls in the United Kingdom (Newton et al. 1991), and contribute to the mortality of Iberian lynx in Spain (second highest form of mortality after hunting) (Ferreras et al. 1992), white-tailed deer in New York (third highest form of mortality) (Sarbello and Jackson 1985), and wolves in Minnesota (Fuller 1989).

A major cause of accidents worldwide is the increase of humans moving into habitats occupied by wildlife. In 1950 only 750,000,000 people lived in urban settlements. That figure reached 1,350,000,000 in 1970; 2,450,000,000, in 1990; and 3,200,000,000 in 2000 (Rathore 1995). This human expansion rapidly depletes the natural capital. Forest lands are cleared to accommodate growth, causing destruction or fragmentation of habitats, which in turn endangers the survival of flora and fauna. The cities are high consumers of natural capital (i.e., water, energy, raw materials) and high producers of pollution (i.e., contaminated water, air, soil), causing the loss of nearly 1,000,000 hectares each year to urbanization (Rathore 1995). With this type of increase directly influencing wildlife habitats, accidents will continue to increase and take a larger toll on wildlife.

Accidents have been considered mortality factors independent of disease. However, Woolf (1976) classified accidents as trauma, a major cause of disease. He used 1976 data from Pennsylvania, when at least 24,183 white-tailed deer were killed by vehicles, more than the number of deer harvested in 35 other states that year. High deer mortalities have been reported in other states (e.g., 25,722 in 1983–84 in Wisconsin; Chizek 1985) and have steadily increased over the past 25 years (Romin and Bissonette 1996). Over 530,000 deer were killed by vehicles in the U. S. in 1991. This is a conservative estimate because mortality data from all states were not reported and many animals are hit, wander from roadways, and are undetected (Lehnert and Bissonette 1997). Conover et al. (1995) estimated that more than 1,000,000 deer-vehicle collisions occur in the United States each year. Woolf (1976) classified many accidents as trauma because bodily reaction to initially sublethal traumatic insult commonly is inflammatory. If an animal survives the initial injury, a risk of secondary infection is high until tissue repair is complete. Biologists should not discount trauma as the etiologic agent of disease when a secondary infection is apparent (Woolf 1976).

Major Accidents that Kill Birds

Nearly 2% of all wild birds in the continental United States that die each year are killed by humans. Hunting accounted for 61% of human-related bird deaths, but collision with man-made objects was the greatest form of human-caused accidents leading to avian deaths (32%). Pollution and poisoning caused the death of about 2% of the total (Banks 1979). Because only a few species account for most of this mortality and continue to maintain large, harvestable populations, the numbers of most bird species are essentially unaffected by human activities (e.g., accidents). However, accurate estimates of accidental mortality have not been made. It is important to understand how accidents influence the population, whether increases in accidents influence the population, whether increases in accidents are caused from human-induced increases in the population, and how accidents replace or add to other mortality (Murton 1972, Banks 1979). Because accidents are essentially a by-product of some other action, figures relating to them are estimates with less reliability than direct cause of death such as hunting.

Mortality on Roads

The number of lanes, nature of the road surface, traffic density, traffic speed, and vegetation along roads and through which roads go all influence birds killed on roads (McClure 1951, Finnis 1960, Hodson 1962, Murton 1972). Other factors include weather, time of day (Scott 1938), and season of year. Seasons when bird populations are high or they tend to congregate along roads because of available resources are particularly important. Agricultural activities (e.g., combining) can force pheasants and other birds to seek alternate cover during the harvest. In these cases birds are attracted to roads where grit and spilled grain are abundant (Buss and Swanson 1950). Birds most frequently found in studies of road kills include house sparrows, ring-necked pheasants, American robins, song sparrows, song thrushes, and blackbirds (Scott 1938, Zimmerman 1954, McClure 1951, Channing 1958, Hodson 1962, Dunforth and Errington 1964, Hodson and Snow 1965, Evenden 1971, Sargent and Forbes 1973). Over 780 individuals of more than 30 avian species were likely killed by vehicles along a 32-kilometer section of road in Texas over two years (Tuovila 1999). Banks (1979) estimated the annual mortality of birds on U.S. roads was over 57,000,000.

Collisions with Towers, and Other Obstructions

Thousands of birds die when they strike structures in flight, often at the same time. Johnston and Haines (1957) reported 25 instances of mass avian mortality between 5–8 October 1954, resulting in 106,804 dead birds in the eastern and southern United States. These collisions were associated with a weather pattern that affected the entire area in a short time. Other reports have documented birds killed by striking the Washington Monument, television transmission lines, navigational aid towers, and wind generators (Banks 1979). Birds from the families Vireonidae, Parulidae, and Fringillidae are most frequently affected. Banks (1979) estimated that at least 1,250,000 died each year from collisions with structures. Birds have been killed from collisions with fences (McCarthy 1973, Edeburn 1973), electrical transmission and utility wires (Stout and Cornwell 1976, O'Connell 2000) aircraft (Solman 1974, 1978, Richardson 1994; Thorpe 1996), and anything else man puts in the air, including tennis balls and golf balls (Washington Star-news, 26 July 1973; Lincoln

1931). Annual mortality to birds from these factors is low, but costs to human lives and property can be extensive. Since 1986, bird collisions with aircraft have caused 33 human fatalities and $500,000 in damage to U. S. Air Force aircraft (Lovell and Dolbeer 1999). Another study documented 104 human fatalities resulting from 19 bird strikes with nonmilitary aircraft from 1960 to 1988 (Conover et al. 1995).

Picture Windows

Most picture windows have the imprint of a bird on them at one time or another from birds flying into them. Birds that are killed in these collisions are likely removed by pets or scavengers so the number of actual deaths has not been estimated. However, Banks (1979) assumed a low estimate of one bird killed per 3.2 kilometer^2 for a minimum estimate of 3,500,000 bird kills per year from striking windows.

Lead Poisoning

Tons of lead shot and sinkers have been distributed over wetlands each year by hunters and fishermen (Banks 1979, DeStefano et al. 1995, DeStefano 1999). Lead poisoning from ingested shotgun pellets has been a source of accidental mortality since the 1950s (Bellrose, 1959). More recently, lead poisoning from ingesting fishing sinkers has been identified as a source of mortality for mute swans and common loons (Simpson et al. 1979, Sears 1988, Kirby et al. 1994, Pokras and Chafel 1992).

Other species are likely killed from the effects of ingesting lead as they are weakened and predisposed to other forms of mortality. Mallards, for example, that have ingested lead have a reduced ability to migrate and are more likely to be shot than healthy birds (Bellrose 1959). Furthermore, mourning doves ingest lead on areas managed for public hunting (Lewis and Legler 1968). These authors also reported nearly 100,000 lead shots per half hectare in the top three centimeters of soil. This amount of lead in the soil may pose a hazard to numerous vertebrates. Banks (1979) estimated that more than 2,000,000 birds died each year from lead poisoning.

Oil-Related Mortality

As humans utilize natural capital at a higher rate, the incidence of oil spills increases. Oil spills occur in oceans, lakes, and ruptured pipelines, and are becoming a common occurrence. Large spills that kill large numbers of birds can have strong influences on local populations. For example, at least two-fifths of the world's African penguin population was endangered by an oil spill in the waters off Cape Town, South Africa, in June 2000. The oil soiled or threatened to soil more than 40% of the estimated 150,000 to 180,000 African penguins in the world after a tanker sank 19 kilometers off Cape Town and two of its fuel tanks ruptured (Arizona Daily Star, 30 June 2000, Tucson, Arizona). Most spills cause few mortalities spread over many species. Banks (1979) estimated that at least 15,000 birds die from oil spills each year, but with the increase in oil spills in the past 20 years, this figure is conservative. A more serious problem is caused by oil sumps, (i.e., a pit or excavation in which fluids produced from gas field operations are collected or stored). These sumps vary from a square meter to several hectares in size and are filled with wastewater covered with a layer of oil up to several centimeters deep. Sumps occur in oil fields throughout the country, look like ponds, and attract wildlife. Once attracted to the sumps, birds can become fatally oiled or entrapped. This problem was first identified shortly after World War II by the California Department of Game and Fish. Since then, corrective action has been taken to reduce their attractiveness to wildlife but they still are responsible for at least 1,500,000 deaths per year (Banks 1979).

Other Accidents

The accidents above represent only the most commonly documented accidents of birds. Other accidents that have been described include reproductive failure due to pesticide and herbicide ingestion (Hickey 1966), banding casualties that kill approximately 10,000 birds per year (Banks 1979), electrocution by power transmission lines (Laycock 1973), choking on food (Holte and Houck 2000), electric fences (Stewart 1973), entanglement in commercial fishing nets and sport fishing lines (Tarshis 1971, Tull et al. 1972), entrapment in buildings, haying and logging operations, and traps set for mammals (Laycock 1973), among other factors. If all of these factors contribute to one bird death per 3.2 kilometers square for all of these miscellaneous factors combined, the estimate would be 3,500,000 deaths per year in the United States (Banks 1979). No systematic study of accidental mortality has been conducted on a landscape level, so this estimate is likely conservative. Domestic animals, especially cats, probably consume many ground-nesting birds. George (1974) documented that domestic cats may remove 2,500,000,000 non mammalian vertebrates from U.S. populations each year. The estimates of annual avian mortality caused by humans are summarized by Banks 1979 (Table 12–1).

Major Accidents that Kill Mammals, Reptiles, and Amphibians

Roads

There is a 6,200,000-km system of public roads in the United States used by 200,000,000 vehicles, so it is not surprising that roads contribute to accidental mortality of animals (National Research Council 1997, Forman 2000). Approximately 20% of the U.S. land area is directly affected ecologically by the system of public roads (Forman 2000), and their influence on wildlife is being examined in detail (Hourdequin 2000). Roads influence populations in at least seven ways: mortality from road construction, mortality from collisions with vehicles, modification of animal behavior, alteration of the physical environment, alteration of the chemical environment, spread of exotics, and increased use of areas by humans (Trombulak and

Table 12–1   Summary of annual avian mortality in the United States related to human activity (From Banks 1979)

Mortality Factor	Number of birds	% of total deaths	Number of total deaths	% of total deaths[a]
Hunting				
Anseriformes	16,854,730	8.60		
Galliformes	58,416,000	29.81		
Gruiformes	1,327,580	0.67		
Charadriiformes	1,818,190	0.92		
Columbiformes	42,123,000	21.49		
Subtotal			120,539,500	61.50
Depradation control			2,000,000	1.02
Research and propagation				
Depredation control research	87,000	0.44		
Propagation	1,820	0.00		
Scientific research and other permit purposes	21,190	0.01		
Subtotal			894,010	0.46
Other direct mortality			3,500,000	1.79
Pollution and poisoning				
Lead poisoning	2,000,000	1.02		
Oil spills	15,000	0.00		
Oil sumps	1,500,000	0.77		
Subtotal			3,515,000	1.79
Collision				
Roads	57,179,300	29.18		
TV towers, etc.	1,250,000	0.64		
Picture windows	3,500,000	1.79		
Subtotal			61,929,300	31.60
Banding casualties			10,000	0.00
Other indirect mortality			3,500,000	1.79
			196,887,810	100.00
Total				

[a]Percentages do not total exactly because of rounding.

Frissell 2000). Each of these could contribute directly or indirectly to accidental mortality of animals, but I will only mention mortality from road construction and collisions with vehicles.

Road construction has destroyed nearly 5,000,000 hectares that formally supported flora and fauna in the conterminous United States (U.S. Department of Transportation 1996). This construction obviously killed numerous organisms but direct mortality has not been estimated (Trombulak and Frissell 2000). However, the toll is considerable at, and removed from, construction sites. The transfer of sediment and other material to aquatic systems at road crossings causes high concentration of suspended sediment that may directly kill aquatic organisms (Newcombe and Jensen 1996).

Mortality from collisions with vehicles is well known and the accidental kills include most terrestrial animals (Trombulak and Frissell 2000). Reptiles (Rosen and Lowe 1994) and anurans are especially vulnerable to vehicles because they are slow moving and need to cross roads to migrate between wetlands and uplands (Fahrig et al. 1995, Trombulak and Frissell 2000). Recent increases in traffic volumes worldwide are likely contributing to declines in wildlife populations, particularly in populated areas (Fahrig et al. 1995, Trombulak and Frissell 2000). In one 32-km section of road in Texas, 2500 mammals and reptiles were likely killed by vehicles over two years (Tuovila 1999).

Other Accidental Mortality

Types of accidental mortality are as varied as the species that die from accidents. Leopold (1933) summarized documented accidents, which included death by fire, steel traps, oil, drowning, goring by horns or antlers, locked antlers, miring in bogs, falling off cliffs, becoming wedged between trees, electrocution by lightning, feet caught in wire fences or brush, falling into aqueducts, anus being blocked by vegetation, or accidents involving agricultural machinery. Since then, other types of accidental mortality have been documented and further examples mentioned by Leopold (1933) have been reported.

Moose calves have been held in place by scrub trees with numerous trunk branchings at heights that entangle their long legs (LeResche 1968). J. J. Hervert (Arizona Game and Fish Department, personal communication) observed pronghorn fawns that were prevented from nursing because of chain-fruit cholla burrs on their snouts. Neonates have been stepped on, kicked in the head by the female (LeResche 1968, Etchberger and Krausman 1999), and deserted (LeResche 1968). Trains have collided with moose, elephants, pronghorns, and other wildlife (Atwell 1962, Schwartz and Bartley 1991, Yoakum and O'Gara 2000, Williams et al. in press). Mammals have died after being trapped in pipes and water catchments (Mensch 1969), entangled in fences (Halls 1978, Child 1998), drowned in eel nets (Lode 1993), involved in boat collisions (Shackly 1992), and electrocution (Shaw 1973). Miring in mud, being stranded on ice, hail, fighting, falls, and encounters with other large ungulates have also caused deaths (Yoakum and O'Gara 2000). Avalanches, ungulate-aircraft collisions, and forest fires (Child 1998), falls into mine shafts, and entrapment in water canals (Rautenstrauch and Krausman 1989, Krausman and Etchberger 1995), and being fatally damaged by porcupines (Pulling 1945) are other sources of accidental mortality. Even though some of these accidents (i.e., falls or train accidents) can kill 100 or more animals at a time (i.e., pronghorn), the cumulative effect on the population is usually minimal (Yoakum and O'Gara 2000).

▒ STARVATION

Starvation, or the lack of food due to the inability to consume enough (e.g., old age) or the simple lack of food, is another decimating factor. The public is well aware of starvation as it is sensationalized by the press, as illustrated below.

> "YELLOWSTONE NATIONAL PARK, Wyo. (AP) — Extreme winter weather may decimate the buffalo in Yellowstone National Park, a herd already being killed by man in record numbers once the beasts leave the safety of the park.
>
> Deep snows that came early to the park froze into a thick crust that the buffalo can't get through to graze. They are surviving on bark and pine needles.
>
> 'That's starvation food'..." (20 January 1997, Arizona Daily Star, Tucson, Arizona).
>
> "ANADYR, Russia — About 700 reindeer have died of hunger in Russia's remote Chukotka Peninsula and another 150,000 are facing starvation, officials said yesterday.
>
> The region's civil defense chief, Yuri Zhulin, said the animals are going hungry because rain and snow during the past two weeks have turned up to 70 percent of their pastures into ice sheets. Four helicopters and 25 crosscountry vehicles were combing Chukotka in search of green pastures, Zhulin said, according to news agency reports.
>
> Zhulin had earlier estimated the number of reindeer facing starvation at 30,000. The new estimate brought the number to about one-half of the local population.
>
> The Chukotka Penisula is directly across the Bering Strait from Alaska. The native Chukchi people are nomadic, reindeer-herding group dependent on the herds for their livelihood" (29 February 1996, Arizona Daily Star, Tucson, Arizona).

Starvation is not uncommon and can have serious impacts on populations. From 1.4 to 27.3% of radio-collared wolves starved in different areas throughout North America from 1969 to 1994 (Peterson et al. 1984; Messier 1985; Ballard et al. 1987, 1997). Malnutrition and winter starvation directly related to forage limitations have killed muskox (Klein 2000), pronghorn (Yoakum and O'Gara 2000), and other ungulates (Sauer and Boyce 1979, Houston 1982) and birds.

When animals fail to obtain enough energy to maintain the metabolic processes of the body, they are forced to use their bodily reserves for energy. At some point the body structures become weakened so normal activities cannot continue and the animal dies. Usually, before actual starvation occurs, the weakened animal will be killed by another decimating factor. Starvation is the most important contributing cause of death in these cases but the proximate cause is usually different (Dasmann 1981).

Robbins (1993) summarized the three phases of weight loss leading to starvation in birds and mammals that accumulate large fat reserves. The first phase is characterized by rapid weight loss, emptying of the gastrointestinal tract, little fat loss, and use of glycogen and protein to meet energy requirements. The first phase is relatively short. The second phase is characterized by fat utilization, reduced protein loss (i.e., protein sparing), and energy conservation by reducing basal metabolism and activity. The result is reduced rate of weight loss. The second phase is very long with more weight loss than the first phase. The duration of the second phase is related to the initial body fat content because fat represents more than 80% of the energy metabolized

Figure 12–2 Relative contribution of fat-to-fat and protein losses during starvation as a function of total body concentration (from Robbins 1993).

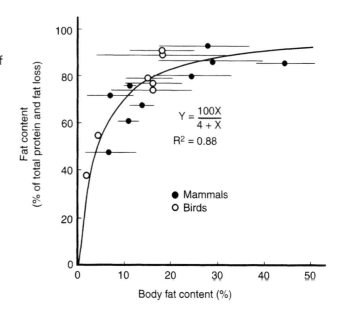

Figure 12–3 Energy content of weight loss related to the contribution of fat-to-fat loss (from Robbins 1993).

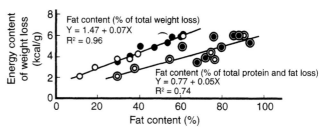

(Figure 12–2). The second phase ends when most of the reserve fat stores have been mobilized. Thus, the last phase is characterized by the increased use of protein to meet energy requirements, which leads to rapid weight loss. The energy content of the weight loss can range from more than 6 kcal/g when fat is the main energy source to 2 kcal/g when protein or glycogen is used (Figure 12–3); thus the composition of the weight loss is a major factor of the rate of weight loss. Animals recover from the first, second, and early stages of the third stage, but later in the third stage, starvation is irreversible.

Many small birds and mammals do not store much fat, so the three phases of starvation do not occur, or they occur over hours instead of days. Most nonhibernating small mammals and birds can survive only one to three days of fasting. As a result, winter nighttime survival depends on feeding just prior to roosting to increase the body's total reserves so fasting endurance can be prolonged. If these animals are deprived of energy during ice storms, crusted snow, or other climatic conditions that remove forage supplies, sudden and extensive mortality can occur (Robbins 1993). Some grouse and hares can survive fasting for three to seven days (Whittaker and Thomas 1983, Hissa et al. 1990), but larger animals generally store more fat, have a lower metabolic rate per mass, and have lower critical temperatures; thus survival times generally increase with body mass (Lindstedt and Boyce 1985).

SUMMARY

Accidents are responsible for numerous deaths of wildlife and include road kills, collisions, lead poisoning, and oil-related mortality, among others. However, accidents have little influence on the population unless the population is small.

Starvation can account for high (more than 25%) mortality rates. There are three general phases of weight loss leading to starvation: rapid weight loss from emptying the gastrointestinal tract, fat utilization, and increased use of protein to meet energy requirements.

13

Welfare Factors: Forage

"Food limitation is the ultimate control if the others fail."

D. R. McCullough

INTRODUCTION

Welfare factors (i.e., food, water, cover, special factors) are included in Leopold's population model (Leopold 1933) as decimating when they are not present in the proper amount. Welfare factors are generally considered the basis for the habitats animals use and often considered as a gain. Leopold (1933) recognized the importance of welfare factors but only recently have they been seriously examined on a large scale in their relation to populations. For example, the number of diet studies that have been published every two years in the 1990s is more than the total number published from 1935 to 1960 (Robbins 1993). Also, numerous studies have documented that wildlife used free-standing water (Halloran 1949, Monson 1958, Krausman and Czech 1998) but only recently have researchers and managers examined seriously the role of different sources of water to wildlife as a management practice (Krausman and Etchberger 1995, Broyles 1995, Rosenstock et al. 1999). The role of cover is also being studied with more intensity than in the past (Cook et al. 1999). Still, obtaining information on the role of welfare factors on free-ranging populations has been, and will continue to be, challenging. Managers often have little control over free-ranging animals and their environment. For example, wildlife management agencies in the South and East indicated that at least 90% of the total northern bobwhite production and harvest occurred without deliberate management (Frye 1961). In many cases nutrition is not considered as an important limiting factor for wildlife because of the apparent abundance of forage (Hairston et al. 1960) and the allocation of decimating factors to other causes (i.e., predation and disease), when the animal in fact was predisposed to them by poor nutrition (Thorne et al. 1976, Clutton-Brock et al. 1987, Adams et al. 1996). Clearly, an abundance of vegetation does not represent an abundance of forage for wildlife. Plants present significant restrictions on animals' ability to harvest food, their ability to process sufficient quantities of forage or sufficient nutrients, and their ability to metabolize or convert digested plant nutrients into animal products (Spalinger 2000). Understanding the requirements of the animal and restrictions of forage demands, the wildlife managers utilize other disciplines: environmental physiology, range and forest management, animal science, and nutrition. Wildlife nutrition has developed into a science because the "nutritional interactions between the animal and its environment are not random events but highly predictable interactions forming the basis for the science of wildlife nutrition" (Robbins 1993:5). Understanding wildlife nutrition is critical to effective management because many of the problems in wildlife management are related to forage, water, and special factors that relate to nutrition: starvation, competition, winter feeding of wildlife, diet formulation, habitat manipulation (e.g., food plots, clearcuts, fertilization, reseeding), predator-prey interactions, and nutritional carrying capacity (Robbins 1993). In short, the main areas of wildlife management (i.e., increase, decrease, maintain populations) require a basic knowledge of the role nutrition plays in wild populations. As indicated, the field of wildlife nutrition has expanded to include numerous aspects of nutrition and wildlife. However, the basics needed in management include an understanding of the nutritional requirements of the animals managed and the availability of those resources in the habitats used (Caughley and Sinclair 1994). That forms the basis of this chapter.

FOOD, FORAGE, OR NUTRITION

Nutrition is a process an animal uses to procure and process portions of its external chemical environment for the continued functioning of internal metabolism. All animals are located somewhere on a tissue metabolism gradient (Robbins 1993) determined by energy, protein or amino acids, minerals, vitamins, essential fatty acids, and water.

Energy is most commonly measured as heat in nutrition represented by the calorie (i.e., amount of heat needed to raise the temperature of 1 gram of water from 14.5°C to 15.5°C) (1000 cal = 1 kcal) and the joule (1 cal = 4.184 joules). Measurements of energy in plant or animal tissue are made with a bomb calorimeter, which measures the amount of heat released when a sample of plant or animal tissue is completely oxidized.

The differences in the amount of energy of flora or fauna are related to the energy contents of the specific chemical compounds they contain (Table 13–1). Fats have the highest energy content (9.45 kcal/g), followed by protein (5.65 kcal/g), and carbohydrates (i.e., starch) (3.96 kcal/g).

Although plants contain numerous compounds, the gross energy plant tissues are uniform (range from 3.90 kcal/g in tropical rainforest to 5.07 kcal/g in seeds; Table 13–1). Oils, fats, waxes, and resins contribute to the higher energy contents.

The energy requirements of an animal for basal metabolism, activity, thermoregulation, growth, reproduction, lactation or brooding, pelage or plumage

Table 13–1 Average energy content of different food components (From Brody 1945, Maynard and Loosli 1969, and Robbins 1993)

Element, plant part, or community	Average energy content (kcal/g)
Starch and glycogen	3.96
Cellulose	4.18
Fats	9.45
Protein	5.65
Urea	2.53
Leaves	4.23
Stems and branches	4.27
Roots	4.72
Seeds	5.07
Perennial grass	3.91
Conifer forest	4.79
Alpine meadow	4.71
Tropical rain forest	3.9

growth, support of parasites, and all other energy-demanding processes include thousands of energy transformations. Estimation of the animals' energy requirement for each of these would be tedious and difficult, if not impossible, if each biochemical reaction were measured. Biologists generally refer to the energy requirements of the entire animal because the measurements of the parameters in the living animal indicate the energy required for all ongoing metabolic transformations and net energy retention as growth or other productive processes equals the gross energies of the tissues or products (Robbins 1993)

Protein

Proteins are high-molecular-weight compounds that are the major components of an animal's body (e.g., cell walls, enzymes, hormones, and lipoproteins). Proteins are made up of mixtures of up to 25 different amino acids that are linked with peptide bonds to form plant and animal proteins. During digestion animal and plant proteins are usually degraded to their constituent amino acids so the specific kind of protein is not as important as the amino acid content and availability relative to an animal's requirement (Robbins 1993).

There are two types of amino acids: *essential* and *nonessential*. Those that must come from the diet because they cannot be produced in adequate amounts by the animal are essential amino acids. Ruminants can synthesize all amino acid requirements (Perry 1984). However, monogastric animals (i.e., animals with simple stomachs) require 10 amino acids that cannot be synthesized in sufficient quantities: arginine, histidine, isoleucine, leucine, lysine, methionine, phenylanine, threonine, tryptophan, and valine. Nonessential amino acids are those that can be produced in the body.

Deficiencies of most essential amino acids will hinder growth and reproduction. In some animals (i.e., cats) the absence of arginine in food (e.g., a single meal) can lead to convulsions and death within one hour (Morris 1985).

The nitrogen content of forage is used to provide an index of the protein content. Proteins generally have an average of 16% nitrogen (range = 15.7% for milk protein to 18.9% for nuts), so the proportion composed of nitrogen is multiplied by 6.25 (100/16) and called crude protein. This is a rough measure of protein because not all plant or animal nitrogen is in the form of protein, the nitrogen-to-protein ratio may not be known (Sedinger 1984), and amino acids can be deaminated and used as an energy source when there is not enough dietary energy (Robbins 1993). As more is learned about amino acids, more sophisticated measures of protein may be available to management.

Minerals

Minerals are a diverse group of nutrients essential to proper maintenance of organisms but usually make up less than 5% of an animal's body. Those minerals that are required in relatively large amounts (i.e., mg/g) are called *macroelements* (e.g., calcium, phosphorous, potassium, sodium, magnesium, chlorine, sulfur). Those minerals that are required in small amounts (i.e., ug/g) are called *trace elements* (e.g., iron, zinc, manganese, copper, molybdenum, iodine, selenium, cobalt, fluoride, chromium). An animal's requirement for minerals varies with age, sex, species, season, maturity, reproductive, and productive condition (Robbins 1993). Unfortunately, biologists have not obtained the data necessary to understand mineral requirements in wildlife. Much of the data in wildlife studies have been derived from domestic or laboratory animals, assuming those animals have similar requirements to wildlife (Robbins 1993). That assumption is likely false (Robbins et al. 1985, Samson et al. 1989, Robbins 1993). However, until more specific data are available from wildlife, biologists will continue to rely on data that have been collected from similar species.

Many aspects of mineral requirements, metabolism, and deficiencies (Table 13–2) have been studied in wildlife when acute deficiencies have been observed. There is still a lot be learned and deficiencies may be more common as humans interrupt mineral cycles (e.g., cropping, removing domestic livestock, wildlife, trees, forage) (Fleuck 1989). The field of wildlife-related mineral studies is an area of future development to better understand wildlife nutrition and management (Robbins 1993).

Vitamins

Vitamins are organic compounds, occur in food in minute amounts, cannot be synthesized by animals, are distinct from carbohydrates, fat, and protein, and are essential for normal life and functioning. When vitamins are not available in the proper amounts in the diet of free-ranging wildlife, animals exhibit characteristic

Table 13–2 Functions, requirements, deficiencies, and dietary sources of macroelements and trace elements

Macroelements	Function	Requirements	Deficiencies	Dietary sources
Calcium (Ca)	98% contained in bone. Blood clotting, nerve and muscle function, Ph balance, eggshell formation, enzyme activation	0.4–1.2% in dry diet for growing mammals and birds (at least 4000 ppm/diet)	Retarded growth, decreased food consumption, high basal metabolic rate, reduced activity, rickets, and osteoporosis, abnormal posture and gait, eggshell thinning, retarded feather growth, retarded antler growth, death	Bone meal, calcium, phosphate, limestone, or oyster shells, snail shells, bones, eggshell fragments, calciferous grit, insects, fungi
Phosphorous (P) Ca:P in bone approx. 2:1	80% contained in bone. Most aspects of metabolism (e.g., energy metabolism, muscle contraction, carbohydrate fat, and amino acid metabolism)	0.3–0.6% in dry diet for growing mammals and birds (at least 2000 ppm/diet)	Loss of appetite, abnormal appetite (pica), reduced antler growth and strength, rickets, reduced body growth, weakness and death	See dietary sources for Ca
Sodium (Na)	Regulation of body fluid volume and osmolarity, acid-base balance and tissue Ph, muscle contraction, nerve impulse transmission, growth, and reproduction	0.5–0.4% of dry diet (at least 500–1800 ppm/diet)	Reduced growth, softening of bones, corneal keratinization, gonadal inactivity, loss of appetite, weakness and incoordination, adrenal hypertrophy, impaired energy utilization, decreased plasma and fluid volumes, producing shock and death	Animal tissues, some soils (coastal and desert), mineral licks, urine, aquatic plants
Potassium (K)	Nerve and muscle excitability, carbohydrate metabolism,	At least 6000 ppm/diet	Muscle weakness, intestinal distension, cardiac and respiratory	Growing plants and animals

	Function	Requirements	Deficiencies	Dietary sources
	enzyme activation, tissue Ph, osmotic regulation		weakness and failure, retarded growth, tubular degeneration of the kidneys	Plants and animals
Magnesium (Mg)	70% in bone, enzyme activation	At least 2000 ppm/diet	Hyperirritability, convulsions, loss of equilibrium, tetany, increased heat production, reduced appetite, weight loss, impaired blood clotting, liver damage, defective bones and teeth, death	
Chloride (Cl)	Acid-base relations, gastric acidity, digestion	At least 500–3000 ppm/diet	Reduced growth and feed intake, hemoconcentration, dehydration, nervous disorders, reduced blood chloride levels	
Sulfur (S)	Constituent of sulfur-containing amino acids, and the hormone insulin	At least 2000–6000 ppm/diet	Synonymous with a deficiency of sulfur-containing metabolites (e.g., amino acids) and not a simple deficiency of inorganic sulfur	
Trace elements	**Function**	**Requirements**	**Deficiencies**	**Dietary sources**
Iron (Fe)	Metalchelate of hemoglobin and myoglobin, and is in many enzymes	At least 30–60 ppm/diet	Anemia, listlessness, weight loss, impaired brain and immune function	Milk, red meat, forages, soil
Zinc (Zn)	Synthesis of DNA, RNA, and proteins; components of enzymes	10–70 ppm/diet	Enlarged hocks of birds, poor feathering, retarded growth, rough hair coat and hair loss, weight loss, poor reproduction, loss of appetite, increased susceptibility to infections	Animal tissue, forages

Continued

Table 13–2 Functions, requirements, deficiencies, and dietary sources of macroelements and trace elements. *Continued*

Trace elements	Function	Requirements	Deficiencies	Dietary sources
Manganese (Mn)	Bone formation, energy metabolism enzyme activity	At least 10 ppm/diet	Slipped tendon, weight loss or reduced growth, impaired reproduction, weakness, nervous disorders, loss of equilibrium, bone malformation	Seeds, forages, animal tissues
Copper (Cu)	Hemoglobin and melanin formation, component of several blood proteins and enzyme systems	At least 5–10 ppm/diet	Anemia, diarrhea, loss of appetite, nervous disorders, loss of hair color, reduced hair growth, defective keratinization of hair and hooves, fragile bones, bone deformities, impaired reproduction, cardiovascular disorders, sudden death	Liver, heart, brains, kidney, milk, forage
Molybdenum (Mo)	Component of enzymes	< 0.1 ppm/diet	Renal calculi, reduced growth	Plants, soil

Mineral	Function	Amount	Deficiency symptoms	Source
Iodine (I)	Hormone function	At least 0.08 ppm/diet	Reduced growth, goiter, reduced reproduction and energy metabolism, mental deficiency, dwarfism	70–80% of the iodine in the body is in the thyroid gland; soil; vegetation
Selenium (Se)	Interacts with vitamin E to maintain tissue integrity	At least 0.01/ppm/diet	Nutritional muscular dystrophy, labored breathing, difficulty feeding, diarrhea, liver necrosis, reduced fertility, lung edema, reduced disease resistance	Soil, plants
Cobalt (Co)	Vitamin B_{12}	At least 0.07–0.11 ppm/diet	Anemia, wasting away, listlessness, loss of appetite, weakness, fatty degeneration of the liver, reduced hair growth	Plants
Fluoride (F)	Specific biochemical roles uncertain		Reduced rate of growth, infertility, dental problems	Bones, teeth, antlers
Chromium (Cr)	Glucose uptake and metabolism		Reduced growth and longevity, hyperglycemia	

Sources: Church (1988), Perry (1984), Robbins (1993), Caughley and Sinclair (1994), Van Soest (1994).

diseases; thus they continually adapt their diets to avoid vitamin deficiencies. The physiological function of vitamins is poorly understood so their functions are defined as preventives for certain dysfunctions (e.g., vitamin C is the antiscurvy vitamin).

Vitamins are classified as fat-soluble (i.e., vitamins A, D, E, K) and water-soluble (i.e., vitamin B complex, C, and others). Fat-soluble vitamins are absorbed in fat and can be used when not in the diet. Most water-soluble vitamins (B_{12} and riboflavin can be stored in the liver) are not stored in the body and are needed in the daily diet. Excesses of water-soluble vitamins are excreted in the urine so toxic overdoses are rare but excesses of fat-soluble vitamins (especially A and D) can be toxic.

Vitamin A is a major constituent of visual pigments, vitamin D transports calcium and prevents rickets, vitamin E is needed for metabolism, and vitamin K is required to make proteins for blood clotting (Table 13–3). These fat-soluble vitamins are generally abundant in the diet of wildlife. Very little research has been conducted on the relationships of water-soluble vitamins and free-ranging wildlife; most deficiencies have been described for captive animals (Robbins 1993).

Seasonality and Secondary Plant Compounds

Forage used by wildlife varies with the seasons. Herbivores have more forage when plants grow (i.e., summer in temperate polar regions, rainy season in tropics and subtropics) and more energy. Young grass contains up to 20% protein but has only 3% and 2% protein in mature flowering grass and dry senescent grass, respectively (Caughley and Sinclair 1994).

Animals generally adjust breeding patterns so their highest physiologic demands for energy and protein occur during the growing season. For example, most birds time breeding to occur when the forage supply is highest (Perrins 1970) and ungulates give birth at times when lactation can be enhanced. Northern ungulates give birth in spring (i.e., lactation occurs during the growing season) but tropical ungulates give birth after rains so the mother can build up fat supplies to support lactation (Spinage 1973).

Plants contain energy for many wildlife species but they also attempt to deter wildlife from foraging on them through evolution at the chemical level. Lignin, cutin, suberin, and biogenic silica offer structural agents to plants to prevent degradation. Lignin provides rigidity to the plant cell wall (Harkin 1973) and is largely indigestible. Cutin is the structural component of the plant cuticle and extracellular, and suberin occurs between the cell wall and cytoplasm. Cutin is a major component of bark (Martin and Juniper 1970). Cutin and suberin physically block some digestibility. Silica in plants also reduces cell wall digestibility and increases tooth wear. There are also over 33,000 compounds in plants that create mixtures of molecules that can interfere with growth, neurological and tissue functioning, reproduction, and digestion of organisms that consume them (Robbins 1993). These are called *secondary compounds* because most do not have primary metabolic functions in plants. However, their role is important in the evolutionary struggle between plants and their consumers (Levin 1976). Some even have beneficial roles in animal nutrition (e.g., tannins can reduce viral action) but most act to deter consumption. The three common groups of secondary compounds include soluble phenolics, alkaloids, and terpenoids.

Soluble phenolics include over 8000 flavonoids (i.e., contribute to flower, fruit, and leaf color), isoflavonoids (i.e., pytoestrogens), and tannins (i.e., generalized defensive plant compound). Excessive consumption of flavonoids and isoflavonoids can produce abortions, sterility, or liver damage. Tannins (derived for their use to tan animal hides to leather) occur in 17% of nonwoody annuals, 14% of herbaceous perennials, 79% of deciduous wood perennials, and 87% of evergreen wood perennials (Rhoades and Cates 1976). They can produce internal physiological damage (McLead 1974) to consumers.

Alkaloids occur in approximately 20% of flowering plants and 10,000 different alkaloids have been identified (Robbins 1993). Well known alkaloids include nicotine, morphine, delphinine of larkspur, conine of poison hemlock, tomatine of tomato plants, atropine of deadly nightshade, and lupinine of lupine (Robbins 1993). Alkaloids have some physiologic effects but act more as toxicants or poisons than as digestion inhibitors.

Terpenoids have low molecular weight and are soluble in organic solvents. Common terpens include essential oils from sagebrush, evergreens, citrus fruits, eucalyptol of eucalyptus, carotene, papyriferic acid in paper birch, camphor of white spruce, and gossypol of cottonseed. Terpenoids are bitter tasting, inhibit activity of rumen bacteria (Schwartz et al. 1980), and are effective feeding deterrents (Schwartz et al. 1980, Reichardt et al. 1984).

Other Nutritional Considerations

Feeding Rate

The amount of time it takes for animals to consume forage and for forage to pass through their system is a significant factor affecting nutrient assimilation. Time is important because the large animals need to satisfy their large daily nutritional requirements and small animals need to satisfy their energetic lifestyles (Spalinger 2000). The feeding rate is generally related to plant density, or the availability of the plant to the animal (e.g., bite size) (Hudson and Watkins 1986, Spalinger et al. 1988, Gross et al. 1993, Rominger and Robbins 1996). The relationship between plant density and bite size is complex and in most cases feeding rates are determined by the bite sizes animals can obtain from plants. Thus, bite size is dictated by animal morphology (e.g., incisor dimensions, molar surface area) (Illius 1989, Shipley et al. 1994). When plant morphology (e.g., thorns) restricts bite size, large and small herbivores have similar feeding rates and large animals are at a disadvantage relative to smaller ones (Spalanger 2000). This partially explains why the largest mammals in North America are not found where high-quality plants are limited (e.g., deserts, alpine).

Time is critical also. When plants are growing, feeding rates are sufficient for adequate diets with minimal foraging (e.g., 3 – 6 hours per day). However, foraging time increases to 10 hours per day or longer in winters or hot, dry periods when forage is not as abundant and intake rates still cannot meet energy demands (Hansen

Table 13–3 Function, requirements, and deficiency symptoms of fat-soluble and water-soluble viamins (from Robbins 1983)

Fat-soluble vitamin	Requirements[a] Birds	Requirements[a] Mammals	Functions	Deficiency signs
A (Retinol, retinal, and retinoic acid).	4000–5000	500–23,000	Major constituent of visual pigment, maintenance, differentiation and proliferation of epithelial tissue, glycoprotein synthesis	Nervous disorders, reduced fertility or sterility, birth defects, reduced egg hatch ability and chick survival, reduced growth or loss of weight, oral and nasal pustules, weakness, night blindness, impaired eyesight because of copious lacrimination or none at all, corneal degeneration, eye infections and eyelid adhesion, bone and teeth abnormalities, lack of alertness, visceral gout, unsteady gait and incoordination, ruffled droopy appearance
D (D_2—ergoalciferol; D_3—cholecalciferol)	900–1200	150–2484	Necessary for active calcium absorption from the gut, calcium metabolism and resorption from bone	Rickets, osteomalacia, nervous disorders
E (Tocopherol)	10–25	3–50	Antioxidant	Yellow fat disease (steatitis–orange or brownish yellow discoloration of body fat or organs), sudden death when subjected to stress, dystrophic lesions of intercostal and myocardial muscles, hypersensitivity, decreased activity and depression, anorexia, fever, lumpiness of subcutaneous fat, severe edema, exudative diathesis or fluid accumulation in the plueral and abdominal cavity and body tissues, nutritional muscular dystrophy, severe hemolytic anemia with blood in urine, weight loss, abnormal pelage molt, unsteady gait, fur discoloration, reproductive failure, ataxia, electrolyte imbalances
K (Phylloquinone and menaquinone)	0.4–1.0	0.05–5.0	Necessary for blood clotting	Stillbirth or death of neonates soon after birth, hemorrhaging

	Requirements[a]			
Water-soluble vitamin	Birds	Mammals	Functions	Deficiency signs
B₁ (Thiamine)	2.0	1.0–20.0	Necessary coenzyme in carbohydrate metabolism	Anorexia, weight loss, weakness, lethargy, unsteady gait, ruffled feathers, impaired digestion, diarrhea, seizures, and other neurological disorders, liver degeneration, death
B₂ (Riboflavin)	2.5–4.0	1.6–15.0	Functions in two enzymes—flavin adenine dinucleotide (FAD) and flavin mononucleotide	Curled toe paralysis in birds, anorexia, weight loss or reduced growth, perosis, rough hair coat, corneal vascularization, atrophy of hair follicles, diarrhea, leg paralysis, reduced fertility, death
Niacin (nicotinic acid and nicotinamide)	20.0–70.0	9.6–90.0	Functions in two coenzymes—nicotinamide adenine dinucleotide (NAD) and nicotinamide adenine dinucleotide phosphate (NADP)	Reduced growth, enlarged hocks, poor feathering, anorexia, diarrhea, dermatitis, drooling and tongue discoloration, death
B₆ (pyridoxine, pyridoxamine, pyridoxal)	2.6–4.5	1.6–6.0	Functions in enzyme systems of protein metabolism	Testicular atrophy, sterility, anorexia, retarded growth, roughness and thinning of the hair coat, muscular incoordination, convulsions and neurological disorders, death
Pantothenic acid	9.0–16.0	7.4–40.0	A component of coenzyme A, necessary for fat, carbohydrate, and amino acid metabolism	Skin lesions, crusty scabs about the beak and eyes, emaciation, weakness, reduced growth, leg disorders, poor feathering, intestinal hemorrhages, enlarged fatty degeneration of the liver, enlarged and congested kidneys, reproductive failure, death
Biotin	0.1–0.3	0.12–0.6	Functions as a coenzyme in carbon dioxide fixation and carboxylation.	Fur discoloration, hair loss, degenerative changes in the hair follicle, thickened and scaly skin, conjunctivitis, fatty infiltration of the liver, eye infections.
Folicin (folic Acid)	0.8–1.0	0.2–4.0	Transfer of single carbon units in molecular transformations	Anorexia, retarded growth, decreased activity, weakness, diarrhea, profuse salivation, convulsions, adrenal hemorrhages, fatty infiltration of the

Continued

235

Table 13–3 Function, requirements, and deficiency symptoms of fat-soluble and water-soluble vitamins (From Robbins 1983) *Continued*

Water-soluble vitamin	Requirements[a]		Functions	Deficiency signs
	Birds	Mammals		
B$_{12}$ (cyanocobalamin)	1000–2000	600–2000	Functions as a coenzyme in single carbon metabolism and in carbohydrate metabolism	liver, reduced hemoglobin, hematocrit, and leukocytes, death Anorexia, weight loss, fatty degeneration of the liver, neurological and locomotion disorders
Choline	1000–2000	600–2000	Nerve functioning	Diffuse fatty infiltration of the liver, rupture and hemorrhaging of the liver, enlarged hocks, slipped tendon, reduced growth of the leg bones, awkard gait, growth retardation, muscular weakness, lowered hematocrit, pale kidneys
C (absorbic acid)	—	200	May reduce infections, necessary for phagocytic activity, bone and collagen formation, functions in hydroxylation reactions, antioxidant	Scurvy, severe necrotic stomatitis, anorexia, weight loss, gingivitis, glossitis, pharyngitis, hemorrhages throughout the body, weakness, stiffened hind legs, lowered body temperature, diarrhea, bone fractures, enteritis, retarded growth, death

[a]Requirements for fat-soluble and water-soluble vitamins are in IU and mg/kg of dry diet and mg/kg of dry diet respectively (except K, which is in mg/kg dry diet, and B$_{12}$, which is in µg/kg of dry diet) (National Research Council 1978a, 1978b, 1982, 1984).

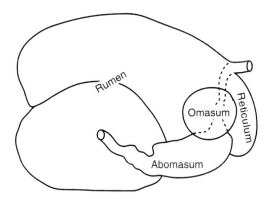

Figure 13–1 The four-chambered stomach of a ruminant.

1996, Rominger and Robbins 1996). Snows can easily cover forage supplies (Hansen 1996) and hot, arid environments can force animals to search over large areas for adequate forage. Both situations require more time and energy to search for and acquire forage resources necessary for survival.

Gut Capacity and Passage Rate

Gut capacity (i.e., the physical restriction of the digestive tract), passage rate (i.e., the rate at which the gut empties), and the digestive kinetics and toxicity of plants also influence forage processing in herbivores. However, biologists have not studied these constraints as much as others (Spalinger 2000).

One can learn a lot about the life history of an animal by simply examining the stomach to understand how they make a living. In vertebrate herbivores, for example, there are two basic digestive systems depending on the placement of their fermentation chamber: *hindgut fermenters* and *forgut fermenters*. Hindgut fermenters have their fermentation chamber in a postgastric position (i.e., calcum or large intestine), have few barriers to food passage, and have the capacity for high food intake rates (Illius and Gordon 1990). Hindgut fermenters include the introduced wild pigs, horses, and burros.

Foregut fermenters have their fermentation chamber anterior to the gastric stomach (i.e., reticulum, rumen). Most North American foregut fermenters are ruminants (except collared peccaries), regurgitate their food, and process it. The fermenters have bacterial digestors that help them break down and use complex carbohydrates of the cell wall as energy (Pritchard and Robbins 1990). Animals with simple stomachs (single-chambered) cannot ferment plant foods as do ruminants with their four-chambered stomachs (i.e., rumen, reticulum, omasum, abomasum) (Figure 13–1).

Ruminants can digest the cell wall components of plants (i.e., fiber) because of their complex stomach. The rumen and reticulum (i.e., reticulorumen) make up 60 – 77% of the capacity of the entire digestive tract (Staaland and White 1991, Jenks et al. 1994) (6 – 20% of body mass; Van Soest 1994) and most fermentation occurs in the reticulorumen (Parra 1978). The omasum is small, between the reticulorumen and abomasum, and acts as a filter that sorts particles and liquids that enter from the

Figure 13–2 The amount of energy from the gross energy available to white-tailed deer on two diets (from Mautz et al. 1976).

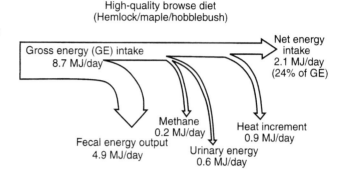

reticulorumen and passes smaller ones to the abomasum. Larger particles are passed back via contractions for further digestion (Stevens and Hume 1995).

The abomasum is the gastric stomach of the ruminant and initiates bacterial protein digestion produced in the rumen. Each of these steps restricts passage of forage through the system by size of forage particles and time. The passage of forage from the reticulorumen to the lower tract cannot take place until particles are processed to 0.5 to 1.0 mm (McLeod et al. 1990). The ability to break down forage depends on the fiber content.

Even though the ruminant animal has a complex mechanism to metabolize nutrients, the amount of nutrient consumed does not equal the nutrients that can be used for maintenance and productivity. Nutrients are lost from the urine, belched from the gut as methane, lost during fermentation, and general reductions of energy from transformation (Figure 13–2). Making a living from plant consumption is a time consuming and complex process; understanding the mechanisms of energy requirements is even more complex.

■ ENERGY REQUIREMENTS

Energy requirements for wildlife have been studied for years but accurate measurements are difficult to obtain. To obtain accurate values, biologists have to rely on captive animals. Unfortunately, the maintenance energy requirements of captive animals likely do not represent the requirements of free-ranging animals. Captive animals do not have to search for food, move between forage patches, or watch for predators. Activity and thermoregulatory requirements are minimized in captivity (Robbins 1994). Energy expenditure of free-ranging ducks, for example, was 15% higher than captive birds (Owen 1969, Wooley and Owen 1978) and free-ranging ungulates can expend 25 to 100% more energy than confined animals (Holleman et al. 1979). These data confound the nutrition picture for wildlife but there are basic components that have been established.

Maintenance

Maintenance requirements are the intake at which the animal's weight remains constant and the animal remains healthy (Robbins 1994, Spalinger 2000). Energy and protein are the most likely limiting nutrients for wildlife populations, except at certain times of the year when sodium may be required for lactation or phosphorus and calcium are required for lactation and antler growth (Spalinger 2000). Basal metabolism requirements for many mammals have been summarized as:

Basal metabolic rate (kilocalories/day)=70 $M^{0.75}$ where M is body mass (kg)

In other words, for mammals the basic metabolic rate varies directly with body size and larger animals have relatively lower energy requirements (per unit of body mass) than small animals, making them more efficient than small animals relative to basal metabolic energy needs (Spalinger 2000). Although this is the general equation, the values change as we learn more about the requirements and diet of animals. For example, the equation can change from 91.9 $M^{0.813}$ for vertebrate eaters to 23.3 $M^{0.451}$ for nectarivores (McNab 1986). Basic metabolic rates are not constant for birds or mammals and obviously change with activity (e.g., standing, locomotion, burrowing, flying, swimming, foraging) and thermoregulation (Robbins 1994).

Reproduction and Growth

Reproduction demands are some of the most expensive aspects of an animal's lifetime. Males can expend so much energy during breeding that they can exhaust all energy reserves (Gavin et al. 1984, Van Ballenberghe and Ballard 1997) and die during or shortly after the breeding season. Females have to cope with breeding, gestation, and lactation or incubation. Lactation is two to three times more expensive than gestation (Robbins and Moen 1975) in ungulates and depends on the number of neonates nourished. In general, milk output rises rapidly after parturition, peaks several weeks later, and drops off as the neonate begins to forage for itself.

Birds have minimal requirements for gamete synthesis but high demands for egg production (e.g., 220% in waterfowl, gulls, and terns; Robbins 1994). These costs are met by reducing nonessential costs, accumulating fat prior to laying, and increasing food intake (Robbins 1994). The costs of incubation in wild birds have not been determined accurately but one of the likely costs is the reduced time available for foraging. For example, arctic nesting female Canada geese lose up to 46%

Figure 13–3 Condition classes of bighorn sheep based on rump characteristics. A and B = good condition; C and D = medium condition; E and F = poor condition (from McCutchen 1990).

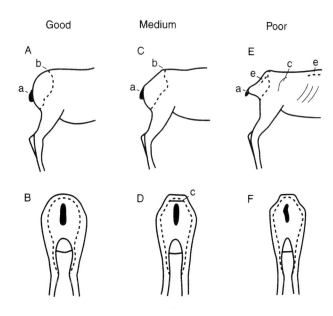

of their peak spring weight by the time their eggs are hatched (Raveling 1979). As a result, females are emaciated and feed continuously after egg laying.

■ ANIMAL CONDITION INDICES

For years biologists have attempted to measure the nutritional health or condition of animals with an array of techniques (Kirkpatrick 1980). Unfortunately, animals are too variable for the indices to be universal. Many indices are applicable only for young or old animals in a localized area, for entire carcasses, or only for healthy animals. Furthermore, the value of indices between studies is difficult to determine because of small sample sizes; effects of, age, sex, and season; different laboratory techniques; use of drugs to capture animals; and different techniques and procedures used when handling blood, tissue, or digesta samples. Although there are numerous limitations to the various techniques to assess animal condition, several general classification schemes have been used: visual appraisals, physical measurements, fecal analysis, and blood characteristics (Brown 1984).

Visual Scores

Most biologists feel confident in judging the condition of animals, but visual scores are subjective. Piloerection (i.e., erect hair) during cold weather can make large animals appear heavier and healthier than they are, or a rough coat associated with pelage change can make an animal appear to be in poor condition. McCuthchen (1990) has provided guidelines to the visual condition of bighorn sheep by examining the roundness of the rump (Figure 13–3).

Physical Measurements

Body Mass. Body mass is one of the oldest methods of quantifying animal condition (Severinghaus 1955) because it is related to survival and reproduction in mammals (Hanks 1981, Dark et al. 1986) and birds (Johnson et al. 1985). Total body mass can be easily measured for small mammals and birds but becomes cumbersome with large mammals in remote areas.

Numerous factors influence body mass besides animal condition: age, species, sex, pregnancy, and genotype (Anderson et al. 1974). As such, managers often take out the effects of body mass by using a ratio of weight to some body measure that is a function of size. For example, in cottontail rabbits, there is a relationship between predicted body mass in grams (PBW) and total length in centimeters (L) (Bailey 1968):

$$BPM = 16+5.48 \ (L^3)$$

Also, fat mass (F) is related to body mass (BM) and wing length (WL) in female mallards by the formula derived by Ringleman and Szymezak (1985):

$$F=(0.571 \ BM) - (1.695 \ WL)+59.0$$

When measuring fat in ducks from the northern hemisphere, managers need to be aware of the four general strategies for storing fat before laying: fat is deposited before migration, then supplemented with local forage on the breeding grounds; reserves are formed entirely before migrating to the breeding grounds; and reserves are built up on the breeding area and supplemented by local food (Thomas 1988). In addition, ducks and game birds can alter the length of their digestive system as food supplies change (Whyte and Bolen 1985).

Large mammals' fat depots begin in the gut mesentery, around the kidneys and heart, in the narrow long bones, then on the rump. Fat depots are also used up in that order (Mech and DelGiudice 1985). Thus, no single fat index is a perfect indicator of total body fat. However, some fat indices such as kidney fat and bone marrow fat can be used to obtain specific indications of reproduction and starvation, respectively (Caughley and Sinclair 1994, Mech and DelGuidice 1985).

Although information from body mass is limited it is often the only consistent record big game mangers have available to them. In those cases weights should be collected in a consistent manner (i.e., whole body, bled, eviscerated) in the same season and corrected for age and sex. These data can then be used for trend indicators for animal and range condition.

Physical Measurements. Body Fat Indices. Whole body fat, carcass density, back fat depth, kidney fat index, femur marrow fat, and mandibular marrow fat have all been used to assess the condition of animals. Whole body fat (Robbins et al. 1974) and carcass density (Anderson et al. 1972) are excellent measures of animal condition but require extensive laboratory equipment, exact measurements, and time. Both are impractical for field biologists but better suited for researchers.

Back fat depth measurements are too variable to be a useful indicator of condition. Back fat is the last to be deposited and the first to be used, so even variable measures are of little value. Furthermore, standardizing the point of measurement has been problematic (Brown 1984).

The Kidney Fat Index (KFI) is perhaps the most used and controversial of the fat indices. Because ungulates accumulate fat around the kidney in anticipation of the demands of reproduction, the KFI has been used as an index of ungulate condition (Smith 1970, Finger et al. 1981). The percentage of fat in the body of white-tailed deer is related to KFI by

$$\% \text{ fat} = 6.24 + 0.30 \text{ KFI (Finger et al. 1981)}$$

Kidney fat is measured in several ways but the most useful is to pull the kidney from the body wall by hand. The surrounding connective tissue and fat easily tear away also. The KFI is the ratio of connective tissue and fat to kidney mass for both kidneys.

Femur marrow fat declines after kidney fat has been used, so a decline in bone marrow is a severe depletion of energy and provides an index of nutritional stress. The use of femur marrow fat was first suggested by Cheatum (1949). He and others (Riney 1955) estimated the condition of the femur marrow fat from its color and consistency. If the femur marrow fat is solid, white, and waxy, and the marrow can stand on its own, it contains 85 – 98% fat and animals are not suffering from malnutrition. If the fat is white or pink, opaque, gelatinous, and the marrow cannot stand on its own, it likely contains 15 – 85% of fat and suggests depleted fat reserves. Yellow, translucent, and gelatinous femur marrow fat indicates less than 15% fat and is an indication of starvation. Visual estimates can be subjective but are practical when other parts of the animal are destroyed or have been removed. Femur marrow fat can remain useful for up to 10 days after death (i.e., roadkills, predated carcasses) (Kie 1978). However, caution should be used when making decisions based on femur marrow fat because the femur fat is not a gauge reflecting general body fat conditions; it is a gauge of the lost fat store (Mech and DelGiudice 1985).

Mandibular marrow fat can also be used as an index of condition in deer (Nichols and Pelton 1972) and is often more available because mandibles can be collected in the fall at check stations. If mandibular marrow fat is expressed on a dry weight basis, it may be an early indicator of declining dietary energy (Warren and Kirkpatrick 1982).

Physical Measurements

Organ Weights. The thymus (Welch 1962), adrenal (Hughes and Mall 1958), and thyroid glands (Hoffman and Robinson 1966) have been cited as organs that can indicate animal condition. Unfortunately, the results are varied by organ, species, sex, and age. Also, studies of organs as an indicator of animal condition have been done on growing animals under laboratory conditions, or on shot or road killed animals with an unknown nutritional history. Until better data are available, the use of organ weights is not warranted as a condition index (Brown 1984).

Fecal Analysis

Feces deteriorate rapidly when exposed to the environment, so fecal analysis has not been used as a good method to estimate dietary quality until recently. Fecal nitrogen content is a common method to estimate digestibility of forage (Holechek et al. 1986), when forage has been clipped and measured and can be contrasted with fecal nitrogen. The indigestible nitrogen in feces is separated, allowing an estimate of the total metabolic nitrogen fraction by difference (Van Soest 1982). However, the estimates are crude and error is up to 15%. Bacterial matter contributes more than 80% of the total fecal nitrogen and causes variation.

Diaminopimelic acid (DAPA) is an amino acid unique to bacteria and has been used as a marker to estimate the microbial output from fermentation and the proportion of microbial matter in feces (Nelson et al. 1982, Van Soest 1982). Because DAPA content in individual strains of bacteria can vary, DAPA can only provide general estimates. More studies with wildlife are needed (Brown 1984).

Blood Indices

Blood indices have been useful indicators of animal condition when blood characteristics for a species are available. Early studies of blood indices were largely descriptive (Brown 1984), but in the 1970s researchers found that animals on better diets had higher levels of blood urea nitrogen (BUN), hemoglobin, total serum protein, albumin, fibrinogen, cholesterol, packed cell volume, red blood cells, potassium, and phosphorus (Seal et al. 1972). Subsequent research supported BUN as the best predictor of dietary protein (Seal et al. 1978). However, BUN is confounded by dietary energy and differs seasonally (Bahnak et al. 1979).

Blood urea nitrogen is the main blood constituent affected by dietary protein levels. However, BUN should not be used as an index of dietary protein unless additional indices of protein intake, vegetation analysis, or energy content of the diet are available (Brown 1984).

Recently, other blood and urine indicators have been used as indicators of malnutrition and protein loss. Protein loss from the body was correlated with body weight loss in white-tailed deer. The best blood and urine indicators of malnutrition and protein loss were serum urea nitrogen levels and the ratio of urinary urea nitrogen to creatinine (DelGiudice and Seal 1988, DelGiudice et al. 1990).

Combining Measures

Any of these techniques to assess animal condition used alone will be limited. Kidney fat is appropriate to estimate the upper range of body fat values, bone marrow fat represents lower fat levels, and other indices need to be associated with diet or some other aspect of the animal's life history. Combining different values (e.g., body mass, body fat) has been used for complete carcasses (Kistner et al. 1980) but is of no value for animals that have been predated, scavenged, or decomposed. Determining the nutritional state of a population is difficult especially when sampling from the live population (Caughley and Sinclair 1994). When a population is sampled, the collection is biased toward healthy animals because those in poor condition or dying are not available, and the age groups most sensitive to density-dependent restrictions in forage (i.e., young and old) form a small part of the population and will likely be underrepresented in a sample. Biologists need to continue to refine the methods used to assess animal condition and habitat health.

SUMMARY

Forage is not usually a limiting factor but can be, and poor nutrition can predispose animals to other forms of mortality. Biologists know more about the nutritional requirements of domestic animals and often assume that nutritional requirements of wildlife can be related to domestic animals. There is a wide-open field for those interested in studying the nutritional requirements of various wildlife species.

14

Welfare Factors: Water, Cover, and Special Factors

"*Wildlife management has been concerned with water requirements because of the opportunities to increase wildlife populations when water is limited by improving natural watering facilities or providing artificial watering facilities. . .. Although building water facilities in arid areas may be worthwhile public relations efforts for conservation agencies, the rewards on a strictly biological basis may not be as apparent in all cases.*"

C. T. Robbins

INTRODUCTION

Water, cover, and special factors make up the rest of the welfare factors required for wildlife. This chapter introduces each one and provides samples of how they are important in management.

■ WATER

Water makes up 99% of all molecules within an animal's body (MacFarlane and Howard 1972) and is used for hydrolytic reactions, temperature control, transportation of metabolic products, excretion, lubrication of skeletal joints, and sound and light transport within ears and eyes (Robinson 1957). Birds and mammals often have water concentration between 71 and 88% of their body weight. As they grow, mature, and accumulate water, concentrations are reduced to 50 to 65% in healthy animals (Moulton 1923). The physiological requirements of animals for water do not reveal the entire water requirements an animal may have. How much water and how often water is required by animals is dependent on ambient air temperature, solar and thermal radiation, vapor pressure deficits, metabolic rates, forage intake, productive processes, amount and distribution of activity, and physiological, behavioral, and anatomical water conservation adaptations (Robbins 1993).

Obviously, water is an essential nutrient. Water for all the requirements comes from three sources: *free water,* (e.g., streams, lakes, puddles, rain, snow, dew), *preformed water* in food, and *metabolic water* that is produced as a product of the oxidation of organic compounds that contain hydrogen (Bartholomew and Cade 1963). Free water is most abundant. Preformed water ranges from 2–3% of the weight of air-dried seeds to 70% or more of the fresh weight animal tissue or succulent plant parts (Robbins 1993). When completely oxidized, anhydrous carbohydrates, proteins, and fats yield 56, 40, and 107%, respectively, per gram metabolized (Bartholomew and Cade 1963).

When determining the water requirements of animals, measurements of all three water sources need to be considered. Measurements of only free water, for example, would underestimate the total water requirements because of the omission of preformed and metabolic water (Robbins 1993). Estimates of preformed and metabolic water have been made from detailed studies of forage intake (usually with captive animals) and isotopes of water injected into animals. By injecting animals with a known amount of isotopic water, researchers can obtain an estimate of total *water flux* (i.e., water intake and excretion). The injected isotopic water mixes with the animals' total water pool (one hour for small mammals and birds to seven to eight hours in ruminants; MacFarlane et al. 1969, Mullen 1971, Hughes et al. 1987). Once complete mixing has occurred, a blood or urine sample is taken to determine the isotopic water concentration. The loss of isotopic water after its injection (i.e., from urine, feces, evaporation) provides an estimate of total water flux (Robbins 1993). Many studies on water requirements have been conducted with the animals in laboratory conditions (Robbins 1993). However, diet and adaptations to stressful environments may not be duplicated under captive conditions, reducing the utility of these studies to free-ranging animals (Alkon et al. 1982).

Free-ranging birds (Bartholomew and Cade 1963, Calder 1981, Alkon et al. 1982, Tidemann et al. 1989), predators (Golightly and Ohmart 1984), marsupials (Kennedy and Heinsohn 1974, Morton 1980), rodents (Yousef et al. 1974), and ungulates (Taylor 1969, Beale and Smith 1970, Taylor 1972, Zervanos and Day 1977, Krausman et al. 1985, Fox and Krausman, in press) can meet their water requirements, at least part of the year, from preformed and metabolic water and do not need to drink free water. However, determining when these periods occur has been difficult.

The water requirements between birds and mammals are different because of physiological differences outlined by Robbins (1993). Small passerines generally dehydrate during flight when air temperatures exceed 7°C (Torre-Bueno 1978) because of their more intense metabolism and high rates of water loss. This water loss during prolonged migrations by birds may limit their flight range or require them to fly at moderate temperatures or high altitudes (Hart and Berger 1972). The high evaporative water loss by birds is partially offset by their reduced urinary water excretion relative to mammals. Urea excretion in mammals requires 20 to 40 times more water than that required to excrete a similar amount of uric acid in birds (Bartholomew and Cade 1963). Also, birds evaporate more water than they produce because of their diurnal behavior where they live at or above their thermoneutral zone (Bartholomew 1972, Dawson et al. 1979). As a result, few seed-eating birds can meet their water requirements from only metabolic and preformed water (Williams and Koenig 1980), unlike many fossorial and nocturnal mammals.

Physiological adaptations are only one way animals can conserve water. Besides the production of dry feces and concentrated urine, the storage of heat by passively raising body temperature that restricts evaporative heat loss (Nagy 1987), and reducing basal energy utilization and forage intake when water is limited (Bartholomew 1972, Reese and Haines 1978, Lautier et al. 1988), animals also conserve water by storing and selecting forage and habitats and restricting activity. Desert rodents store dry seeds in humid burrows to increase preformed water content and select seeds based on water content. Collared peccaries, mountain sheep, and other animals select succulent plants and plant parts, (Zevanos and Day 1977, Warrick and Krausman 1989), and restrict activity to cooler parts of the day and cooler microclimates that reduce thermal loading (Davies 1982). When preformed water in the diet decreases, desert granivores and herbivores increase consumption of succulent insects (56–82% water) (Morton 1980, Karasov 1983, Goldstein and Nagy 1985, Degen et al. 1986).

Management of Water for Wildlife

Water is a fundamental requirement of life, and wildlife managers have added water to habitats (where water was believed to be limiting) to increase populations. The addition of water for the management of wildlife has been especially common for species in arid environments, and since the 1940s, managers have concentrated on providing water for mule deer (Elder 1954, Krausman and Etchberger 1995), chukars (Degen et al. 1984), lagomorphs (Cooke 1982), Dorcas gazelle (Ghobrial 1970), bighorn sheep (Blong and Pollard 1968, Krausman and Etchberger 1996), pronghorn (Sundstrom 1968), and other species. Early water developments were designed for upland game birds and ungulates (Glading 1947, Wright 1959), but in the 1980s and 1990s developments have been used to mitigate for the loss of naturally occurring waters (Krausman and Etchberger 1995, Rosenstock et al. 1999), and supply water for other species (Sanchez and Haderlie 1990).

Because water is a fundamental requirement, the addition of water in arid habitats was relatively unquestioned. However, as early as 1967, biologists began questioning the addition of water for wildlife without considering reviews of water requirements relative to existing supplies for all sources (Snyder 1967). Unfortunately, that has not been accomplished and water developments continue with little information available as to how they influence wildlife populations. The assumption has been that water has been the limiting factor for many populations in arid environments and by adding water wildlife populations would benefit by expanding animal distribution, increasing productivity, reducing mortality, and increasing fitness (Rosenstock et al. 1999).

Recently, the use of water catchments has been further challenged as a valid wildlife management tool (Broyles 1995, Brown 1998, Broyles and Cutler 1999). As a result, Rosenstock et al. (1999) examined the effects of water developments on wildlife based on an extensive literature review and discussion with resource managers throughout the United States. They concluded that water developments have benefited some wildlife populations in arid habitats by increasing the distribution and abundance of game and nongame species. The results of their review in part follow (Rosenstock et al. 1999) with other examples from Krausman and Etchberger (1996) and Krausman and Czech (1998). Payne and Bryant (1998) summarize the importance of water developments to some ungulates and game birds in the western United States and Canada (Table 14–1).

Birds
Upland Game Birds

The impacts of water catchments for quail may vary because Gambel's quail and scaled quail meet most of their water needs with succulent food (Hungerford 1960, Schemnitz 1994). However, catchments were beneficial to quail when they were placed in areas characterized by drought during the breeding season (Campbell 1961). Chukar partridge populations also have been increased, and the establishment of new (i.e, introduced) populations were facilitated with the addition of water (Benolkin 1990).

Because mourning doves and white-winged doves require surface water, their populations may be enhanced with developed water catchments. Populations in Idaho may be increased by establishing permanent water sources approximately 6 kilometers apart (Howe and Flake 1988). Merriam's turkey populations also have increased with the development of permanent water sources within each 2.6 square kilometers of turkey habitat (Shaw and Mollohan 1992, Hoffman et al. 1993).

Waterfowl

Earthen tanks, primarily developed for livestock, are also used by wildlife including waterfowl (Cutler 1996). However, the use by waterfowl could be enhanced if additional vegetation cover were available (Menasco 1986, Scott 1998). The nesting habitat of human-developed tanks is influenced by water surface area and emergent and bank vegetation (Rumble and Flake 1983).

Nongame birds

Water developments are used by passerines, shorebirds, waterfowl, and raptors in arid areas. However, how water influences these populations has barely been studied and the few studies that have been conducted do not yield consistent results. In Arizona, bird abundance was negatively correlated with distance from water at one of two sites studied (Cutler 1996). Other studies in the Southwest did not find a difference in bird abundance between watered and unwatered control sites (Smith and Henry 1985, Burkett and Thompson 1994).

Raptors in arid environments use catchments and surrounding vegetation for drinking, bathing, perching, nesting, and foraging. Some developments have enhanced population expansion of hawks into previously unoccupied habitats (Dawson and Mannan 1991).

Mammals

Desert Bighorn Sheep

Providing human-made water sources for desert bighorn sheep is a major part of their management in the Southwest (Tsukamoto and Stiver 1990). It is well documented that desert bighorn sheep usually consume freestanding water when it is available (Graves 1961, Welles and Welles 1961, Crow 1964, Olech 1979). However, Halloran (1949) found most sheep studied were close to water, but others (male and female without lambs) inhabited areas without water. Halloran (1949) speculated that sheep obtained some water from succulents. Krausman et al. (1985) support these findings, documenting that two females did not consume water for 10 days or more during the hottest and driest part of the year. Warrick and Krausman (1989) report that bighorn sheep use freestanding water when available but are able to survive hot, dry periods by consuming succulents. Some populations of bighorn sheep are independent of freestanding water (Watts 1979, Krausman et al. 1985, Alderman et al. 1989).

Although some populations can survive with less available freestanding water than others, bighorns will drink during hot, dry periods (Wilson 1971), and waterholes serve as focal points for drinking and other activities (Olech 1979). Use of waterholes may be dependent on the moisture obtained from succulent plants (Wilson 1971, Leslie 1978, Krausman et al. 1985), but when water is available to sheep, they generally use it. Usually, as temperatures increase without rains, the use of freestanding water sources increases.

The mean time spent at water sources by bighorn sheep was recorded between 3.3 minutes (Campbell and Remington 1979) and less than 5 minutes (Knudsen 1963). Hailey (1967) reported that sheep prefer natural water sources to human-created water sources and that they consumed 2.6 – 3.1 liters per animal per day. Knudsen (1963) reported that sheep consumed 3.8 liters per visit. There is also variability in the frequency of watering by sheep during warm months. Bradley (1963) reported that desert bighorn sheep drank every two to three days. Knudsen (1963) also reported that sheep visit water once every three days. To meet these demands, water should be available at a distance of 3.2 (Blong and Pollard 1968, Leslie and Douglas 1979) to 8.0 km (Halloran and Deming 1958) in their habitat.

When humans create disturbance around watering sources, bighorn sheep change their watering patterns to periods when the disturbance is not present (Campbell and Remington 1981). Leslie and Douglas (1980) also reported that construction activity changed watering patterns for desert bighorn sheep but did

Table 14–1 Importance of water to selected gamebirds and ungulates in the Western United States and Canada (from Payne and Bryant 1998)

Species	Need for open water	Movements relative to known water source	Optimum spacing of water developments	References
Native				
Scaled quail	Uncertain; succulent vegetation might supply needs	Range 0.4 to 2.4 kilometer from water	1.6 kilometer	Schemnitz (1961) Campbell et al. (1973)
Gambel's quail	Might congregate at water sources; might require open water if succulent vegetation is unavailable	Range 1.6 to 3.2 kilometer from water	1.6 to 3.2 kilometer	MacGregor (1950) Hungerford (1962) Hungerford (1960)
California quail	Might need open water if succulent vegetation is unavailable	Broods: 0.4 kilometer from water Adults: 0.8 to 1.2 kilometer from water	1.6 kilometer or less	Leopold (1977)
Mountain quail	Might need open water if succulent vegetation is unavailable	Range 3.2 kilometer from water	1.6 to 3.2 kilometer	MacGregor (1950)
Northern bobwhite	Use free water in hot, dry periods; succulent vegetation usually is adequate; need 3.47% BM[a] per day	Range 0.4 to 1.2 kilometer from water	1.6 kilometer or less	Prasad and Guthery (1986) Koerth and Guthery (1990) Goodwin and Hungerford (1977) McNabb (1969)
Blue grouse	May need open water during nesting and brooding in drier areas of the cruising range	None reported	Might be beneficial on dry ranges	
Sage grouse	Open water used when and where available.	None reported	3.2 to 8.0 kilometer	Dalke et al. (1963)
Lesser prairie chicken	Open water used when and where available.	None reported	1.6 to 3.2 kilometer	
Greater prairie chicken	Open water used when and where available	None reported	1.6 to 3.2 kilometer	

Continued

Species	Water needs	Distribution	Distance	References
Wild turkey	Open water when and where available; 1900 liters per flock per summer	Most hens nest within 0.8 to 1.6 kilometer of open water	1.6 to 3.2 kilometer in arid regions	Payne and Copes (1988)
Introduced				
Chukar partridge	Open water influences distribution; 2800 liters per covey per year	None reported	0.8 to 1.6 kilometer	Christensen (1954) Mackie and Buechner (1963)
Gray partridge	Might need open water if succulent vegetation is unavailable	No major distribution pattern associated with water; usually found within 0.4 to 1.6 kilometer of water	0.8 to 1.6 kilometer	Porter (1950) Oliver (1969)
Ring-necked pheasant	Usually food supplies water needs	Might nest near water sources	0.8 to 1.6 kilometer	Payne and Copes (1988)
Ungulates				
Pronghorn	25.33 milliliter per kilogram BM[a], dry conditions; 95–190 milliliter per kilogram BM[a], dry conditions, dependable water needed during periods of dry forage; can meet requirements from preformed or metabolic water	95% of 12,000 Wyoming pronghorn were within 4.8 to 6.4 kilometer of water	1.6 to 8 kilometer	Boyd et al. (1986) Beale and Smith (1970) Robbins (1983) Sundstrom (1968) Yoakum (1978)
Desert mule deer	100–200 milliliter per kilogram.	3–5 km; pregnant females usually are within 0.8 kilometer of open water	4 to 4.8 kilometer gentle terrain; <3.2 kilometer rough terrain	Hanson and McCulloch (1955) Roberts (1977) Severson and Medina (1983)
Mule deer	42–63 milliliters BM[a], normal conditions; 100–200 milliliters per kilogram BM[a], dry conditions; move to open water if supply diminishes	Migratory	0.8 to 3.2 kilometers	Elder (1954) Mackie (1970) Roberts (1977)

Table 14–1 Importance of water to selected gamebirds and ungulates in the Western United States and Canada (from Payne and Bryant 1998) *Continued*

Species	Need for open water	Movements relative to known water source	Optimum spacing of water developments	References
White-tailed deer	16 milliliters per kilogram BM[a], dry vegetation 31 milliliters per kilogram BM[a], succulent vegetation	1.6 to 3 kilometers; associated with riparian habitats and areas with water uniformly distributed	1.6 to 3.2 kilometers	Boyd et al. (1986)
Coues white-tailed deer	Might survive on moisture in succulent vegetation			Severson and Medina (1983)
Desert bighorn	Dependable water required	In dry seasons, most bighorns were within 1.6 to 2.4 kilometer of water	Minimal 8 kilometers; optimal, 2 kilometers	Graff (1980) Boyd et al. (1986) Halloran and Deming (1958)
Elk	42–63 milliliters per kilogram BM[a]	Migratory	1.6 to 3.2 kilometers	Mackie (1970) Roberts (1977) Boyd et al. (1986)
Collared peccary	Do not depend entirely on free water; will use if available			
Bison	Need water daily	Areas inhabited by bison have readily available water sources	3.2 to 8 kilometers	Meagher (1978) Boyd et al. (1986)
Caribou	Free water is not an important limiting factor	Found near riparian habitats at least most of the year; migratory	Water developments rarely needed	Bergerud (1978)
Moose	Riparian zones important though not indispensable for habitat use; use aquatic plants in summer	Along and around riparian zones	Water developments rarely needed	Boyd et al. (1986)
Muskoxen	Drinking free water is rare among adults; snow is adequate	Unrelated to free water	Water developments rarely needed	Lent (1978)
Mountain goat	Free water is not an important limiting factor; snowbanks are adequate	If introduced, water might restrict goat movements in southern ranges		Rideout (1978)

[a]BM = body mass

252

not influence productivity. Likewise, Jorgensen (1974) documented a 50% reduction in watering activity because of vehicle activity adjacent to water sources. Other activities can be detrimental to watering patterns by desert bighorn sheep. Weaver (1959) found that burros destroy some water sources and discourage sheep from drinking at others (Weaver 1959).

The amount of water desert bighorn sheep can obtain through vegetation is not clear. Sheep need 4% of their body weight in water per day (Turner 1970, 1973). Individual sheep have relatively constant drinking rates (Reffalt 1963) and may be able to concentrate urine as a water conservation measure (Horst 1971, Horst and Langworthy 1971). However, few physiological adaptations to water deprivation have been documented.

When sheep have been deprived of freestanding water that was in their habitat, mortality has occurred. Mensch (1969) documented the death of 34 sheep in and around a dry water source. Allen (1980) also documented mortality related to the elimination of a water source.

Although sheep can fulfill water requirements in cooler months by consuming forage with 1.5 – 3.0 mililiters of preformed water per gram dry weight (Turner 1970, 1973), during warmer months sheep need to obtain approximately 4% of their body weight in water per day. However, the amount of freestanding water used is dependent on the activity of individuals and the amount of water available (Leslie and Douglas 1979). Watts (1979) suggested that some populations could be independent of freestanding water, and Krausman et al. (1985) and Warrick and Krausman (1989) documented that sheep can obtain moisture from plants during the hottest and driest periods, until the monsoons provide freestanding water. This idea had been proposed by Gross (1960). In various studies and reviews of bighorn sheep, authors discuss the importance of water as a habitat component (Irvine 1969, Simmons 1969, Ferrier and Bradley 1970, Wilson 1971, Merritt 1974, Leslie and Douglas 1979, McCarthy and Bailey 1994).

Most studies examining the relationships between desert bighorn sheep and water have been descriptive, anecdotal, and usually conducted in conjunction with some other aspect of the ecology of desert bighorn sheep. As a result, very few studies have examined the important influence the availability of freestanding water has on productivity and recruitment. Gross (1960) suggested that lamb crops may be dependent on freestanding water, and Mahon (1971) indicated that water was the limiting factor for desert bighorn sheep. Douglas and Leslie (1986) reported that 87% of lamb survival could be accounted for by autumn precipitation. Krausman and Etchberger (1993) were not able to document situations where productivity and/or recruitment were enhanced by the addition of water to the environment. However, agencies responsible for bighorn sheep and their habitat spend thousands of dollars (Mouton and Lee 1992) and resources on development of water, including natural tanks, springs, natural sites, dams, wells, horizontal wells, tinajas, rain catchments, earthen dams, seeps, perforated pipes, aprons, caves, and other mechanisms to supply water to desert bighorn sheep (Halloran and Deming 1958, Kennedy 1958, Schadle 1958, Weaver et al. 1959, Mahon 1971, Parry 1972, Gray 1974, Bleich et al. 1982, Brown and Johnson 1983, Werner 1984, 1985, Tsukamoto and Stiver 1990). Broyles

(1995) questioned the continued development of water sources for bighorn sheep and concluded that, "Water development too long has been an uncontrolled experiment: unproven, undocumented, and unreplicated. It is time to scientifically assess water development for desert wildlife so we can either energize it with more and better water holes or so we can redirect the enormous expense and energy to management projects which in fact help wildlife."

Clearly, water developments have been beneficial to some bighorn sheep populations. Additional research is needed to determine how and if catchments are important to different populations of bighorn sheep.

Elk

In the western United States elk have expanded in arid shrubsteppes, woodlands, and forests. Because female elk have high water requirements during lactation, the addition of water for livestock may have influenced the expansion.

Mule Deer

Leopold (1933) and Andersen (1949) proposed that desert mule deer do not require free water, but deer movements in response to the loss or establishment of a water source have been documented (Clark 1952, Hanson and McCulloch 1955, Johnson 1962, Evans 1969, Hervert and Krausman 1986, Bellantoni and Krausman 1991). Rautenstrauch and Krausman (1989a) documented that a population of desert mule deer in King Valley, Arizona, migrated during the dry season to areas with permanent water sources. They provided evidence that deer are capable of detecting rainfall at long distances and moving to those areas. Hervert and Krausman (1986;674) proposed that "desert mule deer appear to be at least behaviorally dependent on freestanding water." Bickle (1969) asserted that water was the factor most influential on density and movements of deer on the Fort Stanton Range in New Mexico, and Woods et al. (1970) suspected the same. Brownlee (1979) thought that water was a limiting factor in much of the Trans-Pecos region of Texas. Deer require freestanding water, but their diurnal home ranges may not include freestanding water (Maghini and Smith 1990).

Deer tend to be closer to water during the summer dry season (Ordway 1985, Ordway and Krausman 1986, Rautenstrauch and Krausman 1989b). Krausman and Etchberger (1995) noted the same with female deer west of Phoenix, Arizona, although they found males farther from the nearest water catchment during summer than during other seasons. In the latter study, however, deer were always within 5 kilometers of water. Krausman and Etchberger (1995) found that deer were closer to water at all seasons than they would be to random locations. McNab's (1963) hypothesis that animals in arid regions have larger home ranges than conspecifics in mesic regions appears to apply to mule deer (Rautenstrauch and Krausman 1989a).

Deer tend to visit water sources once each day, usually around sunset or at night during the summer (Hazam and Krausman 1988). During winter, they most commonly drink just before sunrise. In the Belmont and Picacho Mountains, Arizona, about 38% of female deer watered during the daylight hours (Hervert and Krausman 1986).

Elder (1954) found that female deer in the Tucson Mountains, Arizona, consumed 6.6 liters per catchment visit during summer. Hervert and Krausman (1986) detected an average consumption of 5.1 liters per visit from study areas in the Picacho and Belmont Mountains, Arizona, during summer. Females in late summer consumed the most. Hazam and Krausman (1988) found that deer in the Picacho Mountains consumed from 1.5 to 6 liters water per catchment visit during summer, and the average consumption was 3.7 liters. Females drank more (3.3 liters in early summer, 4.16 liters in late summer) than males (2.7 liters in early summer, 3.55 liters in late summer).

Water requirements of mammals are inversely proportional to body size (Richmond et al. 1962), and Hervert and Krausman (1986) pointed out that males should require less water per body weight than females, and fawns should require the most. Hervert and Krausman (1986) reported the males were found farther from water than females during all months except July, and that females were more predictable in their watering habits than males, which visited catchments once every one to four days. Deer may water more than once per day when under heat stress (Clark 1952, Swank 1958, Hervert and Krausman 1986). Hervert and Krausman (1986) thought the lower frequency of male visits to water in late summer compared to other seasons was due to the greater water requirement of females, which usually lactate in late summer (Short 1981). Fox and Krausman (1994) found that females during fawning were closer to water than during the rest of the year in Maricopa County, Arizona, and fawning coincides with the monsoon season (mid to late summer), when there are many ephemeral pools of water and plants contain more moisture. On the other hand, the monsoon season during the study was unusually dry, and only one of seven fawns monitored survived (Fox and Krausman 1994).

Many animals have a thermoneutral zone in which a change in ambient temperature causes no change in water demand (Taylor and Lyman 1967, Schmidt-Nielsen 1981). Hervert and Krausman (1986) estimated the thermoneutral zone of desert mule deer at 25 to 30° C. When temperatures remained in that zone, deer in the Belmont Mountains of Arizona watered no more than once every four days. However, Hervert and Krausman (1986) listed a number of factors that complicate that estimation, including efficiency of evaporative heat loss, size of animal, nutrition, relative humidity, and wind velocity. They also identified 38° C as the maximum body temperature, above which evaporative cooling must occur for desert mule deer to survive.

Mule deer remain inactive during hot days to conserve water (Hervert and Krausman 1986). They incorporate succulent vegetation (e.g., cacti fruits) in their diets to reduce their need for free water, especially during the dry season, but perhaps in general (Short 1977, Krausman 1978, Sowell et al. 1985, Krausman et al. 1997).

Desert mule deer appear readily adaptable to the use of artificial water developments (Elder 1954). However, when water is plentiful, they tend to use natural sources and obtain more of their water requirements through foraging (Hervert and Krausman 1986). In the Picacho Mountains, Arizona, younger

Figure 14–1 Desert mule deer that drowned in the Central Arizona Project canal (photo by P.R. Krausman).

males used artificial catchments more frequently than older males, which used earthen stock ponds (Hazam and Krausman 1988), but that may have been a function of other factors influencing distribution. Some artificial sources of water have negative and positive effects on deer populations. Thousands of deer have drowned in water canals in the West (Busch et al. 1984; Rautenstrauch and Krausman 1986, 1989*b*) (Figure 14–1).

White-tailed deer

Freestanding water has been identified as an essential habitat component for white-tailed deer in arid areas (Krausman and Ables 1981, Maghini and Smith 1990). Although comprehensive studies on white-tailed deer and their use of freestanding water have not been conducted, water developments may have benefited their populations in areas. Coues white-tailed deer selected areas less than 0.4 kilometers from water sources and avoided areas more than 1.2 kilometers from water sources. When forage moisture is low, supplemental water may enhance fawn survival and recruitment (Ockenfels et al. 1991).

Pronghorn

Pronghorn obtain water by drinking, foraging, and from oxidative metabolism (Yoakum 1994, Fox 1997). Deblinger and Alldredge (1991) found that pronghorn distributions in Wyoming did not change with the availability of free water, but Boyle and Alldredge (1984) noted that water content of pronghorn forage in Wyoming was inversely related to pronghorn distance from water and that pronghorn were found closest to free water during the driest seasons. Beale and Smith (1970) found an inverse relationship between the use of free water and forage moisture content in Utah. When moisture content of forage was more than 75%, pronghorn did not drink water even when it was available.

Forage selection by Sonoran pronghorn appears to be influenced by water content (Hughes 1991), but the only research that focused on the water requirements of Sonoran pronghorn was conducted by Fox (1997). She was concerned primarily with determining if pronghorn at the Cabeza Prieta National Wildlife Refuge, Arizona, could meet their water requirements through consumption of forage. Site, season, and species affected the preformed water content of forage. However, pronghorn could not increase their consumption of preformed water by feeding at a specific time of day, as was often assumed, because preformed water content did not vary with time of day.

The water requirement model of Richmond et al. (1962) indicates that pronghorn require from 1.8 liters per 29 kilogram animal per day to 3.4 liters per 64 kilogram animal per day, while Robbins' (1993) model indicates that pronghorn require from 2.6 liters per 29 kilogram animal per day to 5.0 liters per 64 kilogram animal per day. (Smaller individuals require less water, but more water per body mass). Fox (1997) estimated the amount of preformed water intake from 1.4 to 6.9 liters per animal per day, and metabolic water production at from 0.2 to 0.5 liters per animal per day. However, she derived the low preformed water intake estimate (i.e., 1.4 liters animal per day) under the assumption that pronghorn do not select plants in proportions different from those available. She rejected that assumption and surmised that free water was not a limiting factor for pronghorn at one of her study sites, but that the forage at the other sites did not contain enough preformed water to support pronghorn at any season.

Also at Cabeza Prieta, Cutler (1996) monitored the use of two water developments constructed specifically for pronghorn. She detected no use by pronghorn. Furthermore, the catchments were used by coyotes and therefore may have comprised a net detriment to pronghorn survival. Water developments may improve pronghorn distribution in some situations (Beale and Smith 1970, Heady and Bartolome 1977), but they are not the only factor influencing pronghorn density or distribution (Deblinger and Allredge 1991).

Predators

Because most predators obtain their needed moisture from their prey, they are not likely to be dependent on freestanding water. However, predators like other animals often use catchments when available (Cutler 1996).

Small Mammals

Small terrestrial mammals in arid environments are relatively unaffected by the development of water catchments (Smith and Henry 1985, Cutler 1996). They have physiological and behavioral adaptations that reduce their requirements for free water (Mares 1983). Small vertebrates were consumed by ground squirrels in part for moisture when succulent vegetation was not available (Hudson 1962). Texas antelope ground squirrels also consumed small vertebrates (e.g., young desert cottontails) in part for water demands (Friggens, in press).

Bats in arid areas, however, use water catchments to drink and as focal areas for foraging. Bat activity has also been higher at water developments then at control areas without water (Schmidt and DeStefano 1996). It is likely that water catchments have expanded the distribution of bats in areas that also had suitable roosts (Geluso 1978, Schmidt and Dalton 1995).

Reptiles and Amphibians

Most reptiles do not require free water (Mayhew 1968) but will drink it when available. Water developments have not enhanced reptiles (Smith and Henry 1985, Cutler 1996) other than creating debris, left after construction, that they can use for cover (Burkett and Thomson 1994).

Amphibians may benefit from water developments, as studies have reported a higher use of stock tanks compared to control plots (Smith and Henry 1985, Burkett and Thompson 1994, Rosen et al. 1995, Sredl and Howland 1995, Collin 1998). Water developments are also used by turtles (e.g., Sonoran mud turtles , yellow mud turtles) (Menasco 1986, van Loben Sels et al. 1995).

The Benefits of Water Developments

We still have a lot to learn about how water developments influence wildlife populations. The assumption that all water sources benefit all wildlife does not hold up, but there is little evidence that the development of water is detrimental to wildlife (e.g., predation, competition, direct mortality, disease). However, if the goal of the development of water sources is to increase populations, the effort will only be successful if water is the limiting factor and other welfare factors are present in the proper amount. Actual increases in long-term density as a result of added water are rare (Nish 1964, Gullion 1966, Bradford 1975, Christian 1979, Adbellatif et al. 1982). Furthermore, if water is already adequately present in the area, the addition of more water will not likely influence populations (Krausman and Etchberger 1995).

Because the public can see water catchments (and assist with their development), these waters are viewed as positive wildlife management efforts. As a result, the development of water has been a valuable public relations tool for wildlife and managing agencies. However, the benefit to wildlife on a biological basis may not always be apparent (Robbins 1993). Because water developments are expensive, time consuming, and of questionable value, the development of water catchments should have specific management objectives, some basis to indicate that water is the limiting factor for the population(s) in question, a formal economic benefit: cost analysis, and a program to monitor the success of the project (Rosenstock et al. 1999).

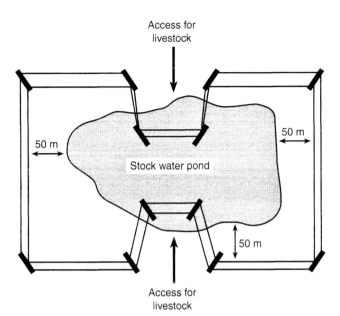

Access for
livestock

50 m

50 m

Stock water pond

50 m

Access for
livestock

Figure 14–2 Fencing design around a pond to allow livestock access but prevent trampling of shoreline vegetation (from Payne and Bryant 1998).

Types of Water Catchments

Water harvesting techniques are not new; they were used as early as 4500 B.C. in the Middle East (Hardan 1975) and water harvesting systems used for runoff farming in Israel's Negev Desert were used 4000 years ago (Evenari et al. 1961). Indians in the southwestern United States used similar systems 700–900 years ago (Myers 1975). Water development for wildlife began much later and the first efforts to provide ranges with water primarily for wildlife began with modifications in rain storage devices devised by early prospectors (Halloran and Deming 1958). In the early 1930s catchments became a widely used and effective tool for game-bird managers, referred to as "gallinaceous guzzlers" (True 1933, Rahm 1938, Glading 1947). Modifications and improvements in watering devices have provided water to many arid lands for wildlife. However, some of the first artificial water sources used by wildlife were provided by the livestock industry (Krausman and Shaw 1984).

Livestock Watering Developments

The efforts of providing water on arid ranges in western North America have been facilitated by artificial development of watering places for livestock. As ranching spread into desert valleys and mountains, reservoirs were excavated on flood plains and dams were constructed to catch and store the floodwater of thunderstorms (Halloran and Deming 1958). Water has been developed for livestock on millions of hectares of western land; most of that water is available and used by wildlife. Many livestock watering developments in use now have been or can be modified for use by wildlife (Wilson and Hannans 1977) (Figure 14–2).

Figure 14–3   (Top) Rainwater catchments with collection tank and drinker. (Bottom) Rainwater lands on the asphalt or corrugated steel apron and is stored in a tank (under the corrugated steel apron) then fed to the drinker (photos by P.R. Krausman).

Wildlife Watering Developments

Water developments were one of the first techniques game managers used to improve habitats; numerous types of water harvesting systems have been developed for wild animals. Standardization in construction is not practical because no water harvesting system is suitable for all applications (Frasier 1980). Factors that need to be considered in selecting the appropriate systems include the need for available water, suitability of the habitat for forage and cover, construction and costs, and accessibility of the land for development, other land use, depth of water table, topography for drainage, yearly and seasonal rainfall, evaporation rates, soil characteristics of the site, and construction materials used (Frasier, 1980). Each of these should be carefully considered before deciding if water developments are desirable and if they can be constructed properly. Of all the types of water harvesting devices for wildlife, rainwater catchments are the most popular.

Rainwater catchments

There are two basic components to most water harvesting systems: a catchment apron for collecting precipitation and water storage facility (Figure 14–3). Rainwater catchments are structures consisting of a storage reservoir with an apron designed so that falling rain is drained into the reservoir. The reservoir is sunk to ground level and covered with boards or concrete slabs and then covered with earth. Other models collect water in an aboveground tank. Water is then piped to a watering trough. Watering ramps are used to guide wildlife to the water and often extend across the mouth of the drinking pool. When the catchments are designed primarily for small game, they are often covered to prevent exposure to the sun and reduce evaporation. If covered, the top of the cover should be high enough to avoid behavioral changes in the animal, prevent crowding, and provide enough clearance for drinking. Generally, small rainwater catchments hold 3800 liters and should be placed every 3 kilometers depending on objectives.

Larger catchments with open ramps can be used for both big game and small game. The water storage capacity varies but should be at least 7600 liters. The type of animal they are designed for is important in spacing. In general, if for big game, one catchment every 8 km is adequate.

The type of apron and storage facility for water are important considerations. Popular catchment treatments used in the southwestern United States were asphalt-fiberglass membranes and paraffin wax-soil treatments (Fink et al. 1973, Myers and Frasier 1974). Theoretically, rain storage from 0.13 cm of precipitation should produce 117 liters of stored water per 93 square meters of runoff area on an impermeable surface. In these calculations no allowance is made for losses due to evaporation and adhesion to the collecting surface. Small amounts of precipitation will produce collectable water.

Another type of semipermeable catchment surface used successfully is the paraffin wax-soil treatment. This is used on catchment areas up to 1115 square meters and consists of spraying molten paraffin wax onto a smoothed catchment surface at a rate of 1.6 kilogram per square meter . Eventually, the wax will penetrate into the soil to a depth of 1.3 centimeters, coating each soil particle and creating a water-repellent layer that resists infiltration.

This treatment is particularly applicable with lighter, coarser textured soils in hot areas where soil surface temperatures exceed 54°C for short periods during the day. If the site is not selected properly, soil erosion may result, but if applied properly, this treatment will yield 80–95% runoff with rainfalls of 0.25 centimeters (Frasier et al. 1978).

Other materials have been used successfully as aprons including galvanized corrugated sheet metal, which has a runoff efficiency of 90–100%. Frasier (1980) summarizes other catchment treatments that have been used including long clearing (runoff efficiency, RE = 20–30%), soil smoothing (RE = 25–35%), silicone water repellents (RE = 50–80%), concrete (RE = 60–80%), and artificial rubber (RE = 90–100%).

An adequate storage facility for collected water is as important as an adequate apron and often accounts for over 50% of the system's cost. Typical storage containers include butyl bags, steel tanks, and excavated pits with waterproof linings (Frasier 1980).

Steel tanks come in a variety of shapes and sizes and can be covered in numerous ways to prevent water loss. These are expensive but often can be constructed on-site, can hold as much water as required and last for many years with a minimal amount of maintenance.

Steel rim tanks with plastic sheet liners can be used to cut costs but these have to be covered and there have been problems with liner deterioration. Also, earthen pits

Figure 14–4 Water is pumped from the ground by a windmill to a storage tank (e.g., half of a 55-gallon barrel) (photo by P.R. Krausman).

have been lined to hold water but these systems are not as efficient as the other type of tanks. All storage tanks should be covered in some manner to prevent evaporation.

All types of rain catchments can transform ranges that can be used throughout the year when rainfall varies from 20 to 50 centimeters. With less than 20 centimeters a year it is usually necessary to haul water to maintain a permanent supply of water in catchments. By collecting runoff at 100% efficiency and improving storage capacities, this problem can be reduced.

Wells and windmills

Wells and windmills have been used successfully to provide water to wildlife along old river beds or in other areas where the water table is less than 15 meters deep. Windmills are used to pump water to a sunken reservoir and a drinking ramp similar to those used with rainwater catchments. Return pipes are often installed so that when the water trough is full the overflow water returns to the reservoir or to the well (Wright 1959).

Windmills are currently being used to provide desert mule deer with water in southeastern California to divert them from using water in the Coachella Canal System (Figure 14–4). Problems have occurred in the past as deer drink from the canal, fall in, and are unable to escape. Windmills and wells have advantages over catchments in areas where rainfall is too low to fill catchments, providing water is available to be pumped.

Retention dams

Retention dams can be used to form artificial springs. The reservoir basin is often formed in a wash and dammed with sand, gravel, and other wash material. Water is drained off into a sump or through a pipe inserted in the base of the dam. Advantages of retention dams are that water does not become stagnant, water flow can be controlled mechanically, the storage basin does not have to be cleaned, and evaporation is minimized.

Water-cut canyons can offer suitable sites for concrete dams. Dams should be firmly keyed to bedrock on the sides and bottom and rock sealing may be required. Concrete dams should be under 12 meters long and not over 3 meters high (Halloran and Deming 1958, Wright 1959).

Spring development

Water can be increased in springs simply by eliminating phreatophytes (Biswell and Schultz 1958). The loss of cover is offset by the availability of permanent water. Other simple spring developments may involve only the protection from livestock and feral animals and shading to reduce evaporation (Halloran and Deming 1958). Most spring development is tedious work requiring the use of hand tools and extensive manual labor. Also, it is not possible to control water flow, and spring water is exposed and readily susceptible to contamination (Bleich et al. 1982).

Types of spring development range from developing ramps to make water accessible in mine shafts and abandoned wells to the construction of basins or pools to conserve water and make it available to big game. Basins are usually made of rock, cement, or masonry. When small seeps are found coming from rock strata, pools may be constructed in the rock to catch and store water.

When natural tanks or tinajas are used as water sources, deep and well shaded tanks retain water the longest. If exposed tinajas are used, covering them may reduce evaporation, providing more water for wildlife.

Whatever type of spring development is employed, maintenance requirements should be minimal. Periodic rechecking will be required to ensure debris has not accumulated in basins or pipes and developments are not buried, destroyed by cloud bursts, or eliminated by water-dissipating plants. Periodic checking also allows for evaluation of the development.

Horizontal wells

The use of horizontal wells has recently been applied to the management of wildlife habitat. Horizontal wells are an effective method of developing water supplies in arid regions at sites that historically produced water but no longer generate surface flows (Bleich et al. 1982). With horizontal wells, water is brought to the surface by gravity flow and the only required moving parts of the well are a float valve, vacuum relief valve, and a shut-off valve. The greatest advantage of this well, however, is that substantial water yields can be developed where little or no surface water previously was available. Site selection is important and geologic formations creating impervious barriers, such as clay or rock that forms natural barriers to aquifers, are likely suitable. To develop the spring, the impervious barrier must be penetrated below the seep to tap the stored water. Site selection, equipment, and operations of the well are detailed by Bleich et al. (1982). There are several advantages of this system that make it an attractive form of water development. It can increase the success rate in spring development, the water flow can be controlled, the spring area is protected and not readily subject to contamination, they are relatively inexpensive to develop unless equipment has to be transported by helicopter, maintenance costs are low, and there are few operational costs.

▓ COVER

Cover is a shelter for wildlife and consists of vegetation and topographic features that provide places to feed, hide, sleep, play, and raise young (Leopold 1933). Cover has been classified as winter, refuge, loafing, nesting, roasting, thermal, escape, bedding, and other types that are important in the life histories of different species. Winter cover, as the name implies, consists of vegetation or other features that provide an animal protection during snow and cold. Leopold (1933) defined refuge cover as vegetation from which game cannot be driven by hunters. Loafing cover offers protection from the elements where animals can rest. Nesting, roosting, and bedding cover are described by their names. Thermal cover has been used recently and is defined in different ways but generally implies a vegetation association that protects animals from the cold (e.g., synonymous with winter cover) or heat. Escape cover is generally an area where an animal can out-maneuver a predator (Gionfriddo and Krausman 1986). These generalized categories are sometimes useful but in reality gloss over the complexity of the needs of cover (specific and general) for wildlife. Management of cover becomes even more complex because the management actions for one resource may be detrimental to another .

For example, the structural diversity of forests provides cover for *edge, interior*, and *ubiquitous* species. Species that have higher densities along edges (ecotones) would benefit from some degree of fragmentation because more edge would be created. However, that increase would come at the expense of interior species that have more homogenous cover requirements. Plethodontid salamander density is lowest in newly regenerated forests and highest in forests more than 120 years old (Herbek and Larsen 1999). A reduction of old growth likely reduces microhabitat availability. By cutting the forests, habitat, which includes cover, would be enhanced for edge species, reduced for interior species, and likely have only minimal influences on ubiquitous species because as generalists they can exploit either the edge or interior (Whitcomb et al. 1981).

How humans use the landscape directly influences cover for wildlife. For example, forest-breeding birds are more successful when forest cover exceeds 10×10 kilometer blocks. Thus, conservationists should focus on preventing a decrease in forest cover and not be concerned with the spatial pattern of remaining forest (Trzcinski et al. 1999) in Canada. Likewise, in Arizona grass cover for birds is more than four times taller, has a basal area two and one-half times larger, and canopy cover more than twice as large, and reproductive canopy cover more than 10 times larger on areas not generally grazed by livestock. Furthermore, areas not grazed by livestock supported more birds than areas that had been grazed (Bock and Bock 1999). Small mammals also are influenced by agricultural practices that alter cover.

Pasitschniak-Arts and Messier (1998) reported that rodents were more abundant in dense nesting cover, intermediate in delayed hay and along rights-of-way, and lowest in idle pastures.

In some situations considerable expenditures of time, effort, and money are allocated to the management of specific types of cover. Over the past 50 years wildlife biologists have managed vegetation, especially dense coniferous forests, for ungulates as a mechanism to enhance survival in cold climates. "Thermal cover has been credited widely with moderating the effects of harsh weather, and therefore, may improve overall performance of populations (i.e., survival and reproduction) by reducing energy expenditures required for thermostasis" (Cook et al. 1998:6). This led to the widespread belief that thermal cover was a key component of ungulate habitat in the western United States. As a result, thermal cover was incorporated in the development of large scale national forest plans, and other management agencies made numerous site-specific, case-by-case decisions as to how to harvest forests or prescribe fire based on the perceived view that thermal cover was required by ungulates (Edge et al. 1990, Cook et al. 1998). Recently, however, Cook at al. (1998) documented that thermal cover had little influence on the herd productivity and demographics for elk in a series of controlled experiments. This example illustrates the importance of understanding how cover is used by wildlife. Prior to the studies by Cook et al. (1998), thermal cover was an important component of elk habitat management on millions of hectares of national forests, at a high cost. More effort should be placed on forage quantity and quality and the ability to evaluate forage conditions across landscapes (Cook et al. 1996, Cook et al. 1998) instead of managing for thermal cover for elk.

Management for cover is actually the management of succession. For example, farming essentially establishes plant communities in early successional stages and range management attempts to establish rangelands in grassland stages. If succession is not interrupted, shade-intolerant plant communities will be replaced by more tolerant plant communities until climax is reached (i.e., the last stage of plant succession in a biome). Climax flora include specific dominant plants in each biome. The vegetation and related cover at the climax stage are different from the edges or transition zones (i.e., ecotores). Specific animal communities associate with specific succsessional stages (i.e., sere) and all related ecotones (Figure 14–5). Seres are often replaced by silvicultural treatments or brush control that subsequently replaces one animal community with another (Payne and Bryant 1998).

Succession varies with soil type, soil moisture, microclimate, climate, topography, slope, aspect, elevation, temperature, and types and degrees of interference (e.g., *autogenic* forces caused by flora and fauna or *allogenic* forces caused by outside influences such as fire or wind). In grasslands successional stages are:

1. bare soil,
2. annual grasses and forbs,
3. perennial forbs,
4. short-lived perennial grasses,
5. until sod-forming grasses or bunch grasses

Figure 14–5 Successional stages of a riparian landscape. As the pond is silted in conditions become favorable for different vegetation, which contributes to different habitat components for wildlife species.

Figure 14–5 *Continued*

establish, which maintain a relatively stable equilibrium (Coupland 1992). Shrub-land and woodland vegetation advances from:

1. bare soil,
2. grass and forbs, to
3. dominant shrubs or woody plants.

For example, secondary succession after fire in a pinyon-juniper ecosystem follows:

1. bare soil,
2. annuals,
3. annual and perennial forbs,
4. perennial forb/grass half shrubs
5. shrub stage or perennial grass stage, and
6. pinyon-juniper woodland (Payne and Bryant 1998).

Forest plant communities have four seral stages (Patton 1992):

1. bare soil,
2. grass/forb,
3. shrub, and
4. tree.

The climax stage (i.e., tree) progresses through four stages of stand development (Oliver and Larson 1990) that are further divided into six structural stages (i.e., stand development, Thomas et al. 1979, Hall et al. 1985):

1. Stand initiation stage, where new species and individuals appear for several years after a disturbance
2. Stem exclusion stage, where no new individuals appear, some die, and survivors grow to a brushy stage, shading the forest floor
3. Understory reinitiation stage, where shade-tolerant herbs, shrubs, and tree seedlings again appear from 60 to 150 years later, depending on site and forest type, but have limited growth
4. Old growth stage, where some overstory trees die and some understory trees grow to the overstory

The structural stages or stand conditions are: 1) grass-forb, 2) shrub, 3) open sapling/pale sawtimber, 4) large sawtimber, and 5) old growth. Each of these can be further refined as related to objectives for stand condition or wildlife distribution. Wildlife will be influenced by the cover provided and vegetation in each stand. Species richness is higher in early and late successional stages with fewer species in the intermediate seres (Hall et al. 1985).

All successional stages from each type in each biome should be represented in various sizes, shapes, and distribution to enhance species richness and biodiversity. These landscape conditions can be advanced or retarded by natural or human means (e.g., mechanical, chemical, fire, wind, flooding, insects, disease, grazing, irrigating, planting, fertilizing; Thomas et al. 1979). Leopold (1933) referred to planting, fencing against livestock, and protecting areas from fire as tools used to speed up succession, whereas plowing, burning, grazing, and cutting were tools used to set back succession. These "tools" are not unique to wildlife management as they are also incorporated in agriculture, forestry, and range management. It is as important to coordinate activities that influence habitats with various agencies and organizations today as it was in 1933 when Leopold first sounded the alarm about cooperative management.

SPECIAL FACTORS

Special factors are welfare factors that are important enough to specific species, usually in small quantities and for short periods of time that they have been provided with their own category (Leopold 1933). Most special factors could be placed with forage, water, or cover but are singled out because of their unique importance.

Likely, there are numerous special factors that have not been identified and those that have may be more important than has been documented. Well known special factors include "gravel for gallinaceous birds and waterfowl, salt licks for herbivores and some birds, mineral springs for pigeons, dust baths for various birds, mud baths and hibernation places for bear, caves or dense shade for sheep and quail to reduce water loss during the heat of the day in arid climates, open wind-swept parks or deep water for the relief of moose and deer in fly season, and sandy knolls for 'booming grounds' of prairie chickens" (Leopold 1933:27). Other special factors include lambing areas with adequate protection for desert bighorn sheep, breeding knolls for swamp deer, minerals for antler development, and vitamins. Areas of the United States (e.g., the Piedmont, the Ozarks, Michigan, the Adirondacks) have soils of low fertility (i.e., low mineral content) that do not generally produce large antlered deer even though the genetic potential exists for large antlers. Some have attributed small antler size in these areas to a lack of minerals in the diet (Jones and Hanson 1985).

Minerals and vitamins are required in minute amounts but are critical to life. Their presence or absence likely influences the geographic distribution of species (Leopold 1933) and have been called special factors.

SUMMARY

Wildlife depend on water, cover, and special factors for survival. Water is available to wildlife as free water, preformed water, and metabolic water. In arid areas wildlife managers enhance water availability with the development of water catchments.

Cover provides shelter for wildlife and special factors are unique requirements of species (e.g., dust baths for some birds).

15

Censusing Wildlife Populations

"Despite the variety of techniques used to estimate elk (and other cervid) population sizes, validation of any single method is currently impossible in free-ranging populations."

L. C. Bender and R. D. Spencer

INTRODUCTION

An estimate of the number of animals in a population provides the foundation upon which effective wildlife management plans are based. Remember, the basic goals of wildlife management are to increase populations, decrease populations, or maintain a sustainable yield. Evaluating these goals would not be possible without some measure of the number of animals in the population. If you cannot measure it, you cannot manage it. Knowing the number of wildlife species present and their density is as important to the wildlife manager as knowing the kinds and numbers of livestock on the range for a range manager. If the management objective is to allocate a percentage of the natural forage produced to wild and domestic herbivores, then the numbers of each must be determined. If the management objective is to produce only huntable surpluses of wild animals, then seasons and bag limits must be set and these are based on population parameters. Even if the area under consideration is not managed to produce a harvestable surplus of any kind (e.g., national park), the resource manager should know something about the populations in his care. It is important to know when a rare or endangered species is becoming less numerous or when a common one might be getting out of control. Few, if any, ecosystems are so free from anthropogenic influences that they can be left unmanaged completely. Population size is the currency used to evaluate many wildlife management programs (Lancia et al. 1994). Leopold's (1933:170) statement in the 1930s that "continuous census is the yardstick of success or failure of conservation" is as true today is it was 70 years ago.

The term *census* means an enumeration of numbers in a population but implies more than just a count. Counts of only the number of individuals present in an area provide a minimal amount of useful information. Of equal importance is a knowledge of mortality, reproduction, emigration, and immigration. Of course, managers need to have information about all four fundamental demographic variables to determine mechanistic explanations of population response to some types of manipulation (Lancia et al. 1994). If the management objective is to have a harvestable surplus, information would also be needed on the age structure, sex ratios, and physical condition of the population. Productivity depends on these population parameters and the harvestable surplus is directly related to annual production. In instances where only a single census is possible, sex and age ratios may provide clues to the health and vigor of the population and may indicate whether the population is expanding, stable, or declining. Repeated censuses can provide proof of such changes and are much more reliable than a single census at some point in time. The amount of information obtained during a census will depend on numerous variables including constraints of time and money. However, it is usually possible to obtain data other than just numbers in a census such as habitat use, health or vigor, sex and age ratios, and other useful information. Every effort should be made to obtain the maximum amount of information possible about the population when conducting a population census.

Although the term census is defined as a count and implies total enumeration of absolute numbers present, activities commonly referred to as censuses are actually samples from which estimates are made. Few new census methods have appeared in recent years. However, there have been numerous refinements of accepted techniques. Census models have been redesigned to allow for more rigorous definition of estimates. It is now possible to place confidence intervals on estimates obtained from all of the common sampling techniques and refined statistical treatments have enhanced their usefulness. This chapter explores some commonly used census techniques and defines census terminology.

CENSUS TERMINOLOGY

A terminology has developed around census and sampling techniques and an understanding of the meanings of certain terms is imperative. The following terms have been commonly used and standardized in censusing literature (Davis 1963, Overton and Davis 1969, Caughley 1977, Verner 1985) and have been summarized by Lancia et al. (1994).

1. *Population.* A population is a group of animals of the same species that occupies a specified area at a certain time as defined by those interested in the group (e.g., the Sonoran pronghorn in Mexico, the continental mallard population, or kudu on a Texas ranch). This definition is specific to censusing.

2. *Abundance or population size.* Abundance or population size refers to the number of individual animals (e.g., 14 pigmy owls). Abundance is expressed as relative

abundance (i.e., populations are ranked according to size) or absolute abundance (i.e., the number of individuals in the population is known or estimated).

3. *Population density.*   Population density is the number of individuals per unit area (e.g., one elephant/10 km^2). Because animal abundance and area are difficult to measure, density is often difficult to measure. Abundance is often easier to estimate than density and is adequate for many management decisions.

4. *Relative density.*   Relative density is the ranking of populations by density (e.g., area A has 40% more mice/ha than area B).

5. *Census.*   A census is a complete count of all animals in a population (e.g., all bighorn sheep in a mountain range or all rattlesnakes in a neighborhood).

6. *Population estimate.*   A population estimate is an approximation of the true population size based on some sampling method. Most population estimates are characterized by some degree of bias. True population size is represented by N. An estimate of population size calculated from sample data is represented by $\hat{N}$ (i.e., "N hat").

7. *Population closure.*   Population closure consists of demographic (i.e., births or deaths do not occur between sample periods) and geographic (i.e., there is no immigration or emigration of the population between sample periods) closure. Population closure is often a basic assumption for population estimations based on repeated observations over time.

8. *Open population.*   An open population may have births, deaths, immigration, or emigration.

9. *Population index.*   The population index is a statistic related to population size. They are used as comparisons between populations on the same area over time or between different areas at the same time. The exact relationship between the index and the true population frequently is unknown.

10. *Frequency of occurrence.*   The frequency of occurrence is a count of something that has a particular attribute (e.g., the number of traps that caught a certain number of animals). Frequency of occurrence is an either/or situation and is commonly expressed as a proportion of the total possible count and thus, ranges from 0 to 1.

11. *Accuracy.*   Accuracy is a measure of how close a population estimate is to the true population size (Figure 15–1). Accuracy is measured by mean square error (i.e., average of the squared deviations between the true population size and the population estimate, repeated many times).

12. *Precision.*   Precision is a measure of how close a population estimate can be repeated (Figure 15–1). Precision is measured by variance (i.e., average of the squared deviations between a population estimate, repeated many times, and its expected value) or standard error (i.e., square root of the variance).

13. *Bias.*   Bias is the difference between the expected value of a population estimate and the actual population size; a systematic distortion away from true density. If the expected value of the population estimator equals the actual population, the estimator is unbiased. Most census techniques result in biased estimates. If a constant proportion of animals is missed by aerial surveys, then the estimate is biased. If one sex or age class is less observable than another, then the estimate is biased. No

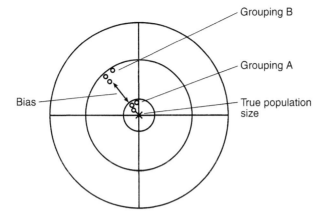

Figure 15–1 An analogy between firing a rifle at a target and estimating population size. If the center of the target is the true population size, Grouping A would be accurate. Grouping B would be precise but not accurate. Bias is the difference between hitting the center of the target the hitting the target elsewhere (after Overton and Davis 1969).

amount of mathematical manipulation of the census data will correct for bias unless the magnitude of the bias is measured and incorporated into the census technique. The relationship among mean squared error, variance, and bias is:

$$\text{mean squared error} = \text{variance} + \text{bias}^2$$

14. *Confidence interval.*   A confidence interval indicates the reliability of a population estimate. The confidence interval implies that if an estimate were repeated many times, then a certain percentage (e.g., 95%) of the confidence intervals corresponding to each estimate would include the true value of the parameter.

15. *Parameter.*   A parameter is a constant but usually unknown quantity characterizing a population (e.g., capture probability).

16. *Model.*   A model is an abstraction and simplification of reality that includes important features necessary to develop a population estimate. Models are constructed so that unknown quantities (e.g., population size) can be expressed in terms of known quantities (e.g., counts or captures of animals).

17. *Sampling error.*   Sampling error refers to random variations that reflect the inherent variation in the population being sampled.

18. *Statistic.*   A statistic is a value obtained from observed values.

The analogy of a shooter firing a rifle at a target (Overton and Davis 1969) has been helpful to demonstrate the usefulness of some of these statistical terms (Figure 15–1). The bull's-eye is analogous to the parameter being estimated (i.e., number of animals). The actual aiming and firing the rifle under a particular set of conditions is analogous to the process of collecting data and calculating an estimate under a specific set of conditions. Where the bullet hits the target is analogous to the value of an estimate.

If numerous shots are fired at the target, the average point of impact is analogous to the expected value of the estimator (i.e., where the rifle is firing on average). The precision of the estimator is analogous to the spread of the group about the mean point of impact. A greater spread results in less precision. Accuracy is analogous to the spread of the group about the bull's-eye. The closer the shots are to the center, the better the accuracy of the estimator. And, the distance from the mean point of impact to the bull's-eye is analogous to the bias of the estimator. The concept of hitting the bull's-eye is analogous to the desired census. Population estimates are more useful when they are precise and accurate.

To be able to obtain accurate estimates of any population, the biologist has to overcome the common problems of observability and sampling. Observability is the ability to estimate the proportion of the population seen or counted in some manner because censusing techniques do not provide for the capture of all animals in the population. Also, entire areas cannot usually be sampled because of limitations on time and money. As a result, sample areas need to be selected that represent a part of the area of interest (Lancia et al. 1994). Because only a part of the entire area is sampled, there are numerous types of sampling designs (i.e., simple random sampling to double sampling).

The methods used to census wild animals can be arranged into several categories and sequences. I have arranged them in this chapter according to the type of information obtained and by the techniques used to outline some of the more common techniques used, and discuss their assumptions, advantages, and disadvantages. For more detailed information about censusing techniques, the reader is referred to Caughley (1977), Burnham et al. (1980), Seber (1982), Verner (1985), Pollock et al. (1990) and Lancia et al. (1994). Davis (1983) also presents census estimators for amphibians, reptiles, birds, and mammals.

Before commencing any wildlife census, the pertinent literature should be reviewed and a competent statistician should be consulted. The manager should also have a specific objective in mind when selecting census methods. If the time and effort are to be expended to estimate population size, there should be some usefulness in obtaining the data. Lancia et al. (1994) suggest that managers carefully consider the need for population estimates and recommend that managers ask what they will do with the estimate once obtained. For example, if a population estimate of 40,000 elk versus one of 60,000 elk is not likely to lead to different management responses, then devoting extensive resources and effort to obtain a very precise and accurate population estimate may not be necessary (Lancia et al. 1994).

ESTIMATES OF ANIMAL ABUNDANCE

Complete Counts (i.e., All Individuals Observed)

Complete counts involve the actual counting of animals observed, or in some cases, of the images on electronic devices or photographic film. If such counts are complete, that is if every individual present is seen, a measurement of absolute density is obtained. Such procedures have almost always been limited in their applicability to one of three conditions: 1) species that are conspicuous and occupy open range such as caribou on the tundra, 2) species that congregate in restricted habitat such as waterfowl, or 3) small study areas that can be intensively searched or observed. At present complete counts are rarely possible in wildlife investigations. Some animals are likely missed even in small study areas (e.g., < 3.2 km^2 enclosures; McCullough 1979). However, new technology (i.e., remote sensing) offers the possibility of detecting all individuals over large areas. One advantage of complete counts is that statistical treatment of the data is not needed because the entire population is counted. Some of the commonly used complete counts include drives, territorial mapping, aerial surveys, aerial photography, remote scanning with thermal scanners, and population reconstruction. These and other types of direct counts are discussed by Overton and Davis (1969), Eberhardt et al. (1979), Seber (1982), and Miller (1984).

Drives

Drives are perhaps the oldest method of counting animals. However, drives require so much manpower that they are infrequently used unless numerous volunteers are available. The assumptions are that all individuals can be seen and counted, either by the line of drivers or by observers stationed at vantage points. In practice a line of beaters moves through an area and is spaced close enough so that any animal between them can be seen. Usually each member counts only the animals that pass to its right and back through the line. Other observers count animals that leave the area to the sides and ahead of the line of beaters. Spacing of the beaters will depend upon the density of the vegetation and the conspicuousness of the animal. One of the most common uses of drive counts is to substantiate the accuracy of the sample-census methods (Gross et al. 1974). Unfortunately, at low population levels drive counts underestimate the true population, and at high population levels they may overestimate the true population by as much as 20–30% (McCullough 1979). As a result, drive counts are viewed as an index of population size.

Territorial Mapping

Plotting territories of breeding birds (Odum and Kuenzler 1955, Verner 1985) is a common form of complete census. The method works well on birds or mammals that defend territories by being conspicuous, either through calls or displays. Many common songbirds can be censused in this manner. Patterson (1952) used a variation of the techniques to count sage grouse on booming grounds and felt that the results were accurate. Prairie chickens, ruffed grouse, and some species of quail lend themselves to this technique. Caution should be exercised when interpreting the results because only the conspicuous sex holding territories, usually the males, are being counted. In order to extrapolate to the total population, the sex ratios must be known. It must also not be assumed that all males are being counted; there is frequently a floating reserve of nonterritorial males is not being detected.

Spot-mapping or territorial mapping is a modification of total mapping and involves plotting locations of individual birds on a gridded map over repeated visits to the study area. A pattern emerges of individual territories that is used to estimate the total number of birds. The number of territories is multiplied by the number of birds per territory to obtain the estimate of population.

Assumptions are that populations are constant and birds remain in their territories during surveys; birds on territories are conspicuous enough (e.g., song or visual display) to be detected; all territories or portions of territories along study area boundaries can be detected; the estimated number of birds in each territory is accurate; and birds are correctly identified (Verner 1985).

Disadvantages are met in trying to meet the assumptions. Determining the spatial arrangement of territories varies among observers and those plotting territories (Verner and Milne 1990). At best, this technique is an index (Lancia et al. 1994).

Aerial Surveys

Aerial surveys may be expensive alternatives but are often indispensable. Aerial surveys may be the only mechanism available to census animals such as elk, deer, or mountain sheep on winter ranges that inhabitat inaccessible terrain. Surveys by fixed-wing aircraft have found wide acceptance by both federal and state wildlife agencies. Helicopters make much better observation platforms because of their maneuverability and slow flight speed, but they become prohibitively expensive if used for long periods of time and they can disrupt animal behavior.

Assumptions behind aerial counts are deceptively simple. If the animals are present, they can be seen and counted. In practice the assumptions are very difficult to meet. LeResche and Rausch (1974) conducted controlled experiments on known numbers of moose. Only 68% were seen by experienced observers under the most ideal conditions (i.e., no leaf cover and fresh snow). Inexperienced observers saw only 43% and their counts varied widely. Observer bias is the major problem with aerial surveys. The human eye is incapable of picking out detail at speeds normally encountered in fixed-wing aircraft. Humans are also not accustomed to looking at objects from above but from a more or less horizontal angle. Most surveys are conducted by personnel who fly only a few times per year and are conditioned to looking at a printed page. Another problem seldom mentioned is that of air sickness. Many biologists who conduct or assist with aerial surveys are not comfortable in a light aircraft. Such distractions erode further the reliability of aerial counts. The use of aircraft in wildlife census work is discussed by Swank et al. (1968).

Aerial Photography

Aerial photography is especially useful on concentrations of animals and has been used for years to count flocks of waterfowl on wintering grounds (Haramis et al. 1985). The equipment need not be expensive or complex though bigger negatives provide better detail if enlargements are to be made. Aerial photography, with 35-mm cameras, has been used to photograph an array of animals including concentrations of wildebeest in East Africa (Talbot and Talbot 1963), sandhill cranes on the Muleshoe National Wildlife Refuge in Texas (Leonard and Fish 1974), domestic sheep (Dudzinski and Arnold 1967), and canvasbacks (Haramis et al. 1985). Advances in lens design, filters, and film in recent years have made aerial photography a very useful tool for census work.

Remote Sensing with Thermal Scanners

Photography is a form of remote sensing, though it is passive in that reflected light is converted into an image on photographic film. Remote sensing with thermal scanners involves detection of some form of radiation emitted by an animal. The radiation is sensed by some kind of electronic device, processed into electrical energy, and usually recorded on photographic film. Ninety-eight of 101 deer were detected on the George Reserve in Michigan with infrared scanning (Croom et al. 1968) but in other areas the system exhibited large errors (Wyatt et al. 1980). Other researchers have had similar problems surveying other species with infrared scanners (Strong et al. 1991). The method should not be abandoned, however, because of a few failures. If a reliable device can be made available to the wildlife community, some major disadvantages of aerial censusing would be eliminated (e.g., observer bias and effects of animal activity on observability).

Population Reconstruction

When all the dead animals in a population can be identified by year and age at death, the population can be "reconstructed" based on the time each animal was in the population. From these data the population size can be determined for a given year in the past only after all the individuals alive in the year have died (Lancia et al. 1994). Population reconstruction is accurate when all individuals are accounted for. Unfortunately, this is rare in most field applications. Population reconstruction has been used to estimate population size in deer (McCullough 1979) and elk (Bender and Spencer 1999).

Complete Counts Without Counting All Individuals

When a representative portion of the entire area under consideration is censused instead of the entire area, the method is commonly referred to as a *sample-area census*. Sample censuses are the application of a total census to a sample area. All of the assumptions for total counts apply. An area representative of the total geographic range occupied by the population is selected, usually by a random or stratified random sampling process. The number of animals counted in the known sample area is expanded to estimate the entire population. Reliability depends upon all of the individuals in each sample unit being counted and the representative area of the sample being known exactly.

The estimation methods used for complete counts from samples are derived from statistical sampling theory and have been summarized by Lancia et al. (1994). The estimated population and density are determined from simple relationships.

A = total area occupied by the population.
N = total population size.
s = number of randomly selected sample plots where counts are made.

a = area of each sample plot.

S = A/a = total number of potential sample plots in A from which the s plots are selected.

x^i = number of animals counted on plot i.

$$\bar{x} = \sum_{i=1}^{s} \frac{x^i}{s} = \text{mean number of animals counted per sample plot.}$$

$$\hat{var}(x^i) = \sum_{i=1}^{s} \frac{(x^i - x)^2}{(S - 1)} = \text{the estimated sampling variance of the } x^i$$

The total population size (N) can be estimated as:

$$\frac{N = \sum_{i=1}^{s} x^i = \bar{x}S}{\dfrac{s}{S}}$$

The variance of $\hat{N}$ can be estimated as:

$$\text{var}(\hat{N}) = S^2 \left(\frac{\text{Var } x^i}{s} \right) \left(1 - \frac{s}{S} \right)$$

Density can be estimated because the area of each plot (a) and total area (A) are known, so

$$\hat{D} = \frac{\hat{N}}{A} \text{ and } \text{var}(\hat{D}) = \frac{1}{A^2} \text{var}(\hat{N})$$

Proper sampling design and valid statistical models are important for reliable estimates to be obtained. Though mathematical models are necessary, they should be used with caution because their validity depends on certain assumptions that are difficult to investigate and verify. Some models are sensitive to departures from assumptions made (Seber 1982).

Ground Transects

Line or strip transects are useful when animals can be seen or flushed into the open. In variable strip transects, the observer travels along a predetermined line and records the angle and radial distance of each animal sighted, or the right angle distance from the point where the animal is observed to the path of the observer. The average of all sighting or flushing distances provides an estimate of the width of the strip censused. King (1937) first developed this method and it is commonly referred to as the King Strip Method. King calculated the number of ruffed grouse per square mile according to the following formula:

$$N = \frac{AZ}{2 \text{ FL}}$$

where N is the estimated number of grouse per square mile, A is the total area of the unit being sampled (43,560 ft^2 x 640), Z is the number of grouse flushed, F is the average flushing distance in feet (doubled because the observer disturbs animals on both sides of the center line), and L is the length in feet of the line traveled.

Perpendicular distance data or sighting distance and sighting angle data are necessary to estimate sighting probabilities. Information needed includes the perpendicular distance (xi) from the line to the detected animal (i), the distance from the observer to the animal at the moment of detection (r^i), and the angle ($\emptyset$) between the line of travel and the line of sight to detected animal i at the moment of detection (Figure 15–2). Data needed for any line transect survey are the total number of animals detected (n), the perpendicular distances (x') or the sighting distances (r^i) and angles ($\emptyset$). Estimation of the estimated density ($\hat{N}$) is then made with these data (Gates et al. 1968, Lancia et al. 1994):

$$\hat{N} = \frac{n-1}{2L(\overline{xx_i})},$$ where L is the length of the transect and $\overline{\chi}\chi_i$ is the mean perpendicular distance from the transect line to the object being counted.

A number of a variations and refinements of the original King method have been applied to most wildlife species that are commonly counted (Burnham et al. 1980, Lancia et al. 1994). Regardless of the application or variation, there are certain assumptions that must be met: all animals are located on the line and detected, the distribution of animals being counted is random, sightings are independent of each other, animals are counted only once, behavior of the population does not change when the transect is being run, response behavior is homogeneous, the probability of an animal being seen is a simple function of distance, and distances and angles are measured correctly (Burnham et al. 1980). It would be very rare for all of these assumptions to be met on a single census. Even the assumption that the area being sampled is known was not met by the original King method. The average flushing distance was really an underestimate of the area being counted, and some birds were flushing outside of the calculated strip width. The resultant population estimate was too high.

Robinette et al. (1974) tested ten strip census methods including the King method. Although the latter appeared subject to biases, these biases were less critical than in most of the other methods. These authors warned that if vegetation changes and sighting conditions become too variable, then problems with the method increase. Gates (1969, 1980) used computer simulation to examine and compare a number of transect methods and the variables that influence them. Gates (1969) proposed a new estimator based solely on radial distances.

Fixed-width strips are frequently used on large animals. Such strips are predetermined and are traversed on foot, horseback, or by vehicle. They are usually conducted at a specific time of day and during a specified season. The width of the strip may be determined by a variety of methods. Hahn (1949) was one of the first to describe such a method in which the effective radius of observability was calculated by having a person carry a handkerchief in his back pocket and walk at right angles to the line of travel. When the handkerchief disappeared, the assumption was made that white-tailed deer would have also disappeared from sight. The method suffers

Figure 15–2
Measurements for
sighting distances (r$_i$),
sighting angle (Ø$_i$) and
perpendicular distance (x$_i$)
from the transect line (– –)
to the sighted animal
(after Burnham et al.
1980).

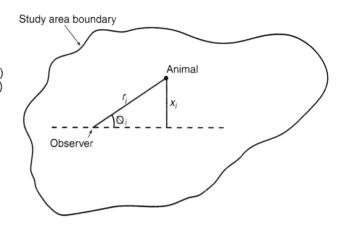

from several biases but seems to work reasonably well because biases tend to cancel each other.

Lamprey (1964) used ground transects in East Africa with measured visibility profiles, a variation of the Hahn method. Lamprey (1964) was able to validate the method by aerial counting and total counts on the ground. Hirst (1969) in South Africa compared the Hahn method and Kelker's belt transect, in which perpendicular distances were measured with a range finder. He found the modified Hahn method most accurate. Hahn (1949), Lamprey (1964), and Hirst (1969) worked in vegetation that was an interspersion of low woody forms and grasslands. The method was developed for use in such vegetation and probably should not be applied to either dense woodlands or open grassland.

Flinders and Hansen (1973, 1975) used an interesting variation of the fixed-width transect. They trained themselves to estimate a 100-meter distance and counted cottontails and jackrabbits out to 100 meters at night by the use of a hand-held spotlight. The 100-meter distance could be estimated accurately and the accuracy was checked frequently to obtain corrections if needed.

Aerial Strips

Aerial strip censuses are among the most widely used methods where overstory cover is minimal and animals tend to congregate (e.g., caribou; Bergerud 1971). A fixed-wing aircraft was flown at 128–160 kilometers per hour at a constant height of 156 meters in strips 0.80 kilometers apart across inhabited range. Observers scanned 0.40-kilometer wide strips on each side of the aircraft and an assistant recorded the observations. The outer boundaries of the transect were predetermined by marks on each wing strut, through which the observer projected an imaginary line. The inner boundary was determined in a similar manner using a landing wheel or ski. Intensity of the sampling varied with animal density, with denser units receiving more intensive sampling (Bergerud 1971). Strips up to 1.6 kilometers in width (Bayless 1969) may be used on species such as pronghorn that inhabit open range. Helicopter strip censuses (Sheffield et al. 1971) are more practical where vegetation density varies. Strip width can be adjusted by varying the speed and altitude of the aircraft.

The major disadvantage of aerial strip censuses is observer bias. Evans et al. (1966) counted moose within 3.2 square kilometers quadrants within strips and discovered that only one-fourth of the animals had been seen in a continuous linear strip. Aerial strip censuses are inaccurate when observers miss a significant number of animals (Caughley 1974). Accuracy deteriorates progressively with increasing width of transect, speed of aircraft, altitude, and fatigue. There is no technical solution to these errors and bias must be measured and the estimates corrected (Caughley 1974). Some animals attempt to hide from the aircraft, further complicating the bias problem. White-tailed deer on the Welder Wildlife Refuge in Texas circled around clumps of brush in attempts to avoid a helicopter. Bighorn sheep in western Arizona also grouped up under small trees in attempts to avoid helicopters during surveys.

Variations in position of the aircraft can also influence bias (Watson and Graham 1969, Watson et al. 1969). Inaccuracies can occur due to changes in altitude of the aircraft and other variables. There are mathematical models to correct for some of these variables (Watson and Graham 1969, Watson et al. 1969).

Plots are not used widely in aerial censuses because of the difficulty in locating boundaries and the cost of searching for plot locations. Siniff and Skoog (1964) estimated caribou numbers by searching blocks that were located by landmarks. They stratified the study area into density classes and allocated more blocks to areas having denser populations. Bergerud and Manuel (1969) used a randomized block technique and 3.6 square kilometer blocks to census large animals under an open forest canopy. The method works best immediately after a fresh snow. Other kinds of aerial plot censuses, such as randomized blocks, demand considerable flying time. Numerous estimation models have been developed for use with line transects (Gates 1979, Burnham et al. 1980) and computer programs. TRANSECT (Laake 1979) and LINETRAN (Gates 1980) compute estimates under different methods. Biologists designing and conducting line-transect surveys should review Anderson et al. (1979) and Burnham et al. (1980).

Complete Counts but Incomplete
Relative Density

Measurements of relative density are easier to obtain than measurements of absolute density. Rather than expressing numbers per unit area, they are expressed relative to time elapsed, distance traveled, or other units such as mice per trap night, deer killed per hunter, or animals seen or heard per area or time. Results obtained are useful for showing trends over time or for comparing one area with another. If estimates of absolute density are available, then conversion factors can be calculated and used to convert relative estimates into absolute estimates. Accuracy is not as important as consistency of the technique. It does not matter greatly if 40% or 80% of the animals present are seen, but rather that the same percentage is seen from one time to another or in different areas being compared. To insure consistency and more valid comparisons, such counts are made according to specific rules concerning dates, starting time, and other factors that affect the count. Measurements of relative density are used in large animal species but are much more widespread on small game species or on songbirds and rodents. Glover (1969) pointed out that an accurate total census of breeding birds requires a census repeated five times on a 183-hectare study area. A sampling method that yields relative density is much more practical when large geographic areas are concerned.

Roadside Counts

Because of their ease and convenience, roadside counts are a common census method. Roadside counts are widely used for rabbits, game birds such as bobwhite quail or pheasants, and for songbirds. The advantages are that large geographic areas are quickly and easily traversed. Problems arise primarily from variations in observability caused by animal activity, vegetative cover, season, time of day, and weather factors.

For birds the technique works best on species that are conspicuous (Howell 1951). Frances (1973) found that roadside counts provided acceptable accuracy for estimating absolute densities of red-winged blackbirds, a species easily seen and heard. Davis et al. (1974) surveyed a complex of birds along roadsides by using visual and auditory methods. The mourning dove call-count survey (Dolton 1993), the woodcock singing-ground survey, and breeding bird survey (Robbins et al. 1986) are common roadside counts. Campbell et al. (1973) counted scaled quail along horse transects rather than roadsides and compiled the data as quail seen per 1.6 kilometers.

Several authors (Newman 1959, Kline 1965) have investigated the factors that influence roadside counts of rabbits. Their results are pertinent to many species of small mammals other than rabbits. In general, rabbit activity increases at the onset of darkness and declines at sunrise. Season of the year causes some change in activity, with more rabbits active during early morning in summer. However, spotlights produced higher numbers than early morning counts, even in summer (Lord 1959). Increases in wind velocity and degree of cloudiness caused the largest decline in rabbit activity. Weather factors in general seem most influential in producing variations in roadside counts among many species.

Baskett et al. (1978) examined the influence of dove pair status, position in the nesting cycle, time of day, population density, and weather on mourning dove surveys from road counts. None were found to cause serious problems but they cautioned that more research was needed.

A variation of the roadside count is the rural mail carrier index. These surveys have been conducted for many years. The mail carrier is simply asked to record the number of animals of a specified species that he or she sights during the course of driving their route. Allen and Sargeant (1975) felt that an index based on sightings of red foxes by rural mail carriers was adequate to measure changes in fox densities from one year to another.

Roadside counts can be used by anyone who repeatedly drives the same route over a period of months or years. Animals hit by cars provide a useful index but some sexes may be overrepresented, such as male deer during the breeding season.

Some western states use a conversion factor to convert deer seen per kilometer of survey to deer per square kilometer. Brown (1966), for example, estimated 720 desert mule deer in the Tortolita Mountains of Arizona as follows:

$$\text{Deer/mile}^2 = (\text{Deer/mile of survey})(\text{square miles of habitat})(\text{conversion factor})$$

where:

Deer/mile of survey equaled 1.28,
square miles of habitat equaled 96, and the
conversion factor equaled 5.86.

Such estimates have limited accuracy and are probably more reliable as methods of sampling population structure. However, there is reason for caution in accepting sex or age ratios based on roadside counts. In the Serengeti Plains of East Africa, sex ratios based on roadside counts of Thomson's gazelle are biased in favor of males. Males are bolder than females and tend to remain closer to roads.

Roadside counts have been used to estimate counts of birds seen or heard (e.g., raptors, crowing male pheasants, drumming ruffed grouse, and whistling bobwhites) (Bull 1981), and to spotlight counts of deer and alligators (Woodward and Marion 1978). Statistics based on the counts may provide reasonable indices to abundance sometimes. However, the question of consistency of the proportion of the population seen or heard has not been adequately addressed (Lancia et al. 1994).

Trap Records

Small mammals are traditionally trapped, either in snap traps or live traps, and the results used to express relative densities. The unit is commonly the number of captures per trap per night or trap/night. A wealth of information has been accumulated on methods of studying small mammals. Blair (1941) discusses techniques for studying small mammal populations and presents information on trapping methods. Stickel (1954) describes trap layouts that are commonly used in home-range studies. Calhoun (1950) summarizes the results of a nationwide small mammal survey, and most of the trap layout designs are described briefly in various sections of his paper.

Trapping results are influenced by most of the variables that affect roadside surveys plus variation in capture probability over time (Nichols and Pollock 1983), species captured (Nichols 1986), the type of trap, and the effect of trapping. Snap traps remove individuals, allow immigration, and thus do not provide accurate density figures, especially if conducted for several nights in a row (McCulloch 1962).

Live traps work well on some species, such as pocket gophers (Howard and Childs 1959), but not on others. Some individuals become "trap happy" and are captured repeatedly, thus preventing the capture of other members of the population. Wiener and Smith (1972) pointed out that the type of trap is a significant variable in determining results of small mammal population studies. Pit traps are more effective on shrews than are snap traps. Briese and Smith (1974) caught a larger number of total species and more individuals per species in pit traps than in live traps. The social hierarchy can also influence trap results. Joule and Cameron (1974) caught larger rats in greater numbers on the first night and smaller ones the second. Social status influences the age class being captured and variation in the age structure can thus lead to errors in estimating the population.

Some capture-recapture studies of fur bearers demonstrate year-to-year variation in capture probability (Smith et al. 1984). When that is the case, the number of animals caught generally will not represent a constant proportion of the population. Thus, the statistics will generally be a poor index of abundance (Lancia et al. 1994).

Incomplete Counts (Estimates of Relative or Absolute Density)

Methods based on counts of something other than the animals themselves are commonly used to census species difficult to observe (e.g., big game on inaccessible ranges or inhabiting dense cover) and for obtaining estimates over large geographic areas where direct censuses are not practical. Presence is inferred from some sign, sound, or other clue made by the animal. The relationship of the sign being observed or heard is assumed to be proportional to density, but the numerical nature of this relationship is usually not known. In some instances the relation between the sign being observed and the population density is quantified to provide a means of estimating absolute density. This has been done for the two most widely used index methods, dove call counts and pellet group counts.

Call Counts

Call counts provide an auditory index to obtain population trends and are widely used on game and nongame birds. A predetermined route that has been selected by a random procedure is established along country roads. The route is run according to standardized rules relating to season, time of day, number of stops, and listening period. More recently, the observer has been recorded to be used as a covariant in analysis of index data (Geissler and Sauer 1990). Mourning doves have been censused with nationwide mourning dove call counts (USDI 1952, Foote et al. 1958, Blankenship et al. 1971) since 1953 throughout the United States. Each call-count route has 20 3-minute listening stations spaced at 1.6-kilometer intervals, usually on lightly traveled secondary roads. The number of individual doves heard during each listening period is totaled for each route and used to construct the index. A stratified random process is used, with major strata being physiographic divisions of the United States. A similar procedure is used by the U.S. Fish and Wildlife Service to provide an annual index to breeding birds (North American Breeding Bird Survey) for the entire United States and southern Canada (Peterson 1975, Robbins et al. 1986).

Since the initial system was designed, there have been several reexaminations and refinements. Blankenship et al. (1971) analyzed five years of call-count data and demonstrated that strata based upon natural vegetation reduced error variance by approximately 30%. Calls can be influenced by a wide variety of factors such as season of year, yearly variations in chronology of the reproductive cycle, daily variations in weather, time after sunrise, and abilities of listeners. Gates (1969) and Gates and Smith (1972) analyzed the influence of many of these variables and constructed statistical models to correct for some influences such as time after sunrise. They also presented a model for calculating absolute densities from call-count data.

Spontaneous calls and calls elicited by the investigator have been used on a variety of species. Drumming by ruffed grouse has been used to estimate the population (Gullion 1977) and courtship displays and calls by woodcock have been used to estimate their density (Duke 1966). Cottam and Trefethen (1968) described a method for use on large concentrations of white-winged doves. Volume of noise made by nesting doves was measured and calibrated by counting nests. Smith and Galliziolli (1965) demonstrated a linear relationship between call counts of Gambel quail in the spring and the fall hunting success. Scott and Boeker (1972) established a correlation between Merriam's turkey gobbler calls and the sex ratio. Thus, if the sex ratio is known, the population density can be estimated. Braun et al. (1971) played tape-recorded male challenge calls to locate white-tailed ptarmigan. The technique was very effective and censused all males in some trials. The technique is not restricted to birds. Pimlott and Joslin (1968) played recordings of howls or imitated the howls to census red wolves in the southeast United States.

Even though attempts have been made to account for some of the variables with standardization of some techniques, detection probability varies over time and space. Ideally, the variation in the detection probability is small enough so index values are adequate to detect substantial changes in populations over large landscapes (Lancia et al. 1994).

Pellet-group Counts

The need to have some index to the numbers of deer, elk, and other herbivores using a habitat too dense for aerial census or other direct approaches to census has led to wide acceptance of pellet-group counts. The method has also been used for estimating absolute density. Most investigators have converted pellet-group data into estimates of animal density. Basic assumptions are: 1) the defecation rate (number of pellet groups per day per animal) is known, 2) all pellet groups can be located and correctly identified, 3) all groups are present that were deposited over the time period under consideration, and 4) animal use of an area is reflected by the number of pellet groups in the area.

A major advantage of the method is that pellet groups can be sampled by standard field techniques commonly used in studies of vegetation. The major problems are: 1) defecation rates vary with age of animal, changes in diet, moisture content of forage, and amount of feed intake; 2) observers miss pellet groups and have difficulty deciding the limits to a group; 3) pellets of some species are difficult to distinguish from each other; and 4) losses of pellets are great enough to preclude use of the technique in some areas (Neff 1968).

The procedure is as follows: 1) establish a sample design and mark sample plots, 2) remove or mark (i.e., paint) old pellets from the plots, 3) revisit plots and count pellet groups when vegetation is at a minimum, and 4) use a defecation rate that has been accepted for the area under consideration. Deer per square kilometer are calculated as follows:

$$\text{Pellet group/kilometer}^2 = \frac{\text{Mean number of groups per plot} \times 100}{\text{Area of plots in hectares}}$$

$$\text{Number of deer/kilometer}^2 = \frac{\text{Pellet groups per kilometer}^2}{\text{Interval in days} \times \text{pellet groups per animal per day}}$$

The time interval is the number of days since the plots were last cleared of pellets.

Assumptions underlying the pellet-group technique have been investigated by several authors. Eberhardt and Van Etten (1956) evaluated the pellet-group method in an area where deer numbers were known and determined that the method underestimated the population. Rodgers (1987) documented variations in numbers of pellet groups deposited per day by mule deer. Kind of foods consumed and age structure of the herd may influence the number of pellet groups produced. Van Etten and Bennett (1965) found that people differ significantly in their abilities to find pellets and to assess their age. Pellets can last up to five years.

Neff (1968) reviewed the pellet-group technique and concluded that the pellet-group method is valid and can be made to yield reliable data under field conditions. He pointed out that sampling design and analysis of data need careful study by a biometrician. Smith (1964) advised a rectangular plot and Van Etten and Bennett (1965) proposed a long rectangular plot and based their recommendation on several theoretical and practical considerations. All investigators should examine closely the assumptions and make corrections for their particular studies. Biologists should not assume that all deer deposit 13 pellet groups per day under all conditions on all ranges, but this seems to have been done repeatedly. The pellet-group technique is used to provide index data. However, Fuller (1991) did not find a relationship between pellet-group counts and aerial surveys that were corrected for observability bias.

Tracks and Other Signs

Track counts have been used extensively as indices to abundance. Sampling design for track counts is similar to that used in other methods. Lord et al. (1970) used a tracking board coated with a mixture of newspaper ink and mineral spirits to estimate rodent numbers. Linhart and Knowlton (1975) used scent stations to attract coyotes and counted tracks around the stations. The method provided an index to relative abundance useful for comparing population densities between geographic areas and between years. Tyson (1959) and Connolly (1981) established a ratio between deer tracks and the number of deer seen on total drives. Tyson (1959) then constructed a model to convert tracks into an estimate of absolute density. However, he could not demonstrate that the relationship was constant throughout the year. Downing et al. (1965) also compared track counts with known numbers of deer but found considerable variation.

Sometimes an object other than tracks can provide clues to abundance. Some common examples are nests of tree squirrels (Uhlig 1956) and alligators (Taylor and Neal 1984), houses of muskrats or beavers (Dozier et al. 1948, Hay 1958), slides of otters, trees cut by beavers, mounds of pocket gophers, and nests of various bird species (Fuller and Mosher 1981). When these types of structures or activity are used to index animal abundance, the assumption is that the structures are all detected. This is rarely the case and formal estimation methods should be used to properly estimate animal numbers from tracks, nests, or other signs (Nichols et al. 1986, Lancia et al. 1994).

Ratio Methods

Capture-Recapture

The application of this method is very simple but the theory underlying the technique is detailed (Pollock et al. 1990). A sample of a population (n_1) is marked in some way, usually by being captured and banded or collared, the sample is allowed to intermingle with the free-ranging population, a second sample (n_2) is taken that includes both marked (m_2) and unmarked animals, and the total population is calculated according to the following formula:

$$\frac{m_2}{n_2} = \frac{n_1}{N},$$

where N is the total population size.

$$\text{Thus, } \hat{N} = \frac{n_1 \times n_2}{m_2}$$

Suppose that 50 rabbits are captured and marked with ear tattoos. During a hunt two weeks later, 150 rabbits are shot, 20 of which are marked. Substituting into the above formula, we calculate the total rabbit population as follows:

$$\hat{N} = \frac{(50)(150)}{20} = 375 \text{ rabbits}$$

This simple relationship was modified with less bias (Chapman 1951) as:

$$\hat{N}_c = \left[\frac{(n_1 + 1)(n_2 + 1)}{(m_2 + 1)} \right] - 1,$$

with the variance of $\hat{N}_c$ (Seber 1982:60) as:

$$\text{var}(\hat{N}_c) = \frac{(n_1 + 1)(n_2 + 1)(n_1 - m_2)(n_2 - m_2)}{(m_2 + 1)^2(m_2 + 2)}$$

An approximate 95% confidence interval is obtained with the following:

$$N_c \pm 1.965 \sqrt{\hat{var}(\hat{N_c})}$$

The above method is known as the Lincoln-Petersen model. It was first used on fish by Petersen in 1896 and later used by Lincoln (1930) to estimate the continental duck population from band returns.

There are several assumptions of this model: 1) the population is closed (i.e., there is no recruitment during the experiment), 2) all animals have the same chances of capture and recapture, 3) marking does not affect the chances of survival or recapture, 4) the second sample is random, 5) marks are retained and are not overlooked, and 6) the sample mixes randomly with the unmarked segment of the population. Some assumptions are more likely to be true than others. Over short periods of time, marks will likely be retained and the population is not likely to increase due to recruitment, either by immigration or births. Thus, the second sample should be taken as soon as possible after the initial marking. Capturing and marking an animal is very likely to affect its chances of being recaptured, especially if trapping is used in both instances. The assumption that the probability of capturing an animal remains constant is considered to be the greatest source of error in the Lincoln-Petersen model. The second sample should never be taken by the same method as the first.

Violation of some of the assumptions and the consequences are discussed by Seber (1982). Random natural mortality does not affect the Lincoln-Petersen index because the ratio of marked to unmarked animals does not change. If more marked animals than unmarked die or if marks are lost, the index will overestimate the population. Incomplete tag returns will produce the same effect, as will recruitment. Thus, the Lincoln-Petersen index is very likely to overestimate the population under conditions normally encountered in the field. Specific problems associated with the Lincoln-Petersen index are discussed by Eberhardt et al. (1963).

Because the probability of capturing an animal is a property of individual animals with unique capture probabilities (i.e., heterogeneity), the variation in capture probabilities among individuals can occur due to sex, age, social status, spatial distribution of animals, and capture efforts. Each of these can alter census results for closed and open populations. As a result, a series of models has been developed to account for different sources of variation in capture probabilities (Lancia et al. 1994). Computer programs (e.g., CAPTURE) have been developed to assist in selecting and using the best model that fits specific data sets.

Change in Ratio

Kelker (1952) began what developed into another series of ratio methods for big game that depend neither upon capture nor on the marking of individuals. To understand the principle of Kelker's ratios, think through the following example: if before a hunt there are only 50 males for every 100 females in a herd, and after the hunt we find only 30 males per 100 females, then 20 males were killed by the male-only hunting season for every 100 females in the herd, or one male was killed for every five females. If 150 males were killed from the herd, then $150 \times 5 = 750$ females in the herd before the hunt, plus 375 males (one-half as many females), or a prehunt total of 1125 adult deer. This basic idea applies to all of the ratio methods using this principle, but they may get more complicated if either-sex hunts are held.

For example:

$$\text{Males AHS} = \frac{D_b K_b - K_d}{D_a - D_b} \text{ where}$$

AHS is after hunting season,
D_b is number of females per male before the hunting season,
K_b is kill of males,
K_d is kill of females,
D_a is number of females per male after the hunting season.

The following data are from a hunt on the Kaibab, Arizona (Russo 1964).

Survey	Males	Females	Fawn	Female:male ratio	Fawn:female ratio
Prehunt	60	103	84	1.72:1	0.815:1
Posthunt	93	308	214	3.31:1	0.695:1
Kill	1748	1117	374		

$$\text{Males AHS} = \frac{1.72(1748) - 1117}{3.31 - 1.72} = \frac{3007 - 1117}{1.59} = 1189$$

On the posthunt survey, there were 3.31 females per male, so females = 1189×3.31 = 3936 females. There were 0.695 fawns per female, so fawns equal $3936 \times 0.695 =$ 2736 fawns. Thus, the herd total after the hunting season was 7861 deer. Variance estimates can also be applied to these data (Seber 1982). Classification surveys must be complete, accurate, and precise before this kind of population determination is useful (Russo 1964). Although the mathematical approach is accurate, if the field data are incorrect, misleading results can be obtained.

The methods are developed further and discussed by Paulik and Robson (1969), and Conner et al. (1986). Change-in-ratio methods can be applied when the population can be classified into two categories and when a change occurs in the relative abundance of the two kinds. Normally these data are obtained before and after a hunt in which there is a differential kill of either a sex or age class. The major advantage is that enumeration of animals in the field is not necessary. Paulik and (Robson 1969) constructed confidence intervals for the estimators, presented techniques for calculating biases, showed how to plan field experiments, and presented several examples of applications. They speculated that a reason why the method has not been used widely was related to sex or age ratios being expressed in terms of a base. The assumptions of change-in-ratio methods (Lancia et al. 1994) are:

1. The observed proportions of animals are unbiased estimates of the true proportions in the population.
2. The population is closed (except for the removals).

3. The number and type of removals are known.
4. The proportion of the animals removed is different from that in the population. If they are removed in the same proportion in which they occur in the population, the method fails (Lancia et al. 1994).

Distance Methods

Distance methods have been used for years by plant ecologists but only recently have been applied to animals (because of their mobility). However, many animals, although mobile, stay within circumscribed areas such as territories. Ellis et al. (1969) used the nearest-neighbor method of Clark and Evans (1954) to calculate the mean distance between coveys of quail. The mean distance was then squared to give a representative space occupied by an average covey. Dividing the area in hectares occupied by the average covey into 100 gave the number of coveys per square kilometer. If the size of the covey is known, then quail per square kilometer can be calculated. Koenen and Krausman (in press) applied distance sampling (Buckland et al. 1993) to desert mule deer. It is likely the distance methods will be tried on other populations and may prove to be superior to techniques commonly used.

Miscellaneous Methods

Questionnaires are used widely by state game and fish agencies to gain knowledge of trends. These surveys are subject to a number of biases such as memory, nonrespondents, guessing, and a higher rate of return by successful hunters. However, general trends in populations can be detected. Harvests can also be estimated by questionnaire responses. The number of licenses sold is known, the number of animals killed by a sample of hunters is obtained, and by use of a ratio, the total kill is estimated.

Several methods have been devised based on animal removal. Sarrazin and Bider (1973) estimated small mammal populations by calculating the number that would have to be removed to reduce activity (measured by tracks) to zero. They listed the advantages as: 1) changes in animal activity due to climate, season, food, population fluctuations, and other factors do not influence the method and 2) the problem of calculating size of trap plot is eliminated. For such methods to be accurate, a major portion of the population must be removed. This portion is sufficiently close to the total population that no further calculations are necessary. The portion removed becomes a good estimate of the population size.

Pelt records (Keith 1963) and other hunting or trapping records can be used to estimate historical changes in populations. Cameras that take photos at measured intervals have been used to monitor animal use of certain habitats. Such devices are practical around water holes (Cutler and Swann 1999). Toggle switches or pressure switches hooked to chart drums are useful when placed at fence crawls or burrow entrances. Because every animal possesses a distinctive scent, a "sniffer" such as developed by the military has potential. The only real limitation to censusing wild animals is the imagination of the investigator. However, the design must have some statistical validity if inferences are to be made from the results.

SUMMARY

The selection of an inventory procedure is an important decision that must be made by today's resource manager. Management is now confronted with the problem of wildlife census that considers the total ecosystem and the population changes of many species. Continuous census, the yardstick of success or failure of conservation (Leopold 1933), must be done with the best available method for each species. The method selected will often be a compromise between the need for accurate data and the constraints of cost and time. Hopefully, information presented in this chapter will be useful in developing a framework of understanding the complexities of estimating animal populations.

16

The Basics Of Habitat

"Yes, the sparrow has found a house, and the swallow a nest for herself, where she may be by her young."

Psalms 84

INTRODUCTION

The evolution of wildlife management has progressed through five stages: laws and regulations, predator control, the reservation of land and refuges, artificial replenishment, and environmental controls (Leopold 1933). The last stage could be expanded or a sixth step added: habitat management. In the simplest form, the habitat of an organism is the place where it lives (Odum 1971:234). This simple definition is informative but one needs to go further when discussing habitat in relation to wildlife management. Remember that the wildlife management triad (Figure 1–1; Giles 1978) includes animals, people, and the habitat of animals as the cornerstones of effective management. Thinking about any species is difficult without considering its habitat. As an example, try to visualize a polar bear, elk, and bullfrog without also visualizing some component of their respective habitats. The animal and its habitat go hand in hand. However, understanding habitats and managing them is not as simple even though there is abundant literature that addresses habitat (Verner et al. 1986, Hall et al. 1997). Leopold (1933:20) stated that "science had accumulated more knowledge of how to distinguish one species from another than of the habits, requirements, and interrelationships of living populations."

One of the earliest works examining the habitat of a species was Stoddard's (1931) study of bobwhite quail. Since then, the field has advanced significantly and habitat management is the cornerstone of contemporary wildlife management. Textbooks on the subject are in use (Morrison et al. 1998, Payne and Bryant 1998) and the latest edition of the "techniques" manual published by the Wildlife Society (Bookhout 1994) emphasizes habitat. Furthermore, most universities and colleges that offer degrees in wildlife management offer courses directly related to wildlife habitat and some offer degrees specifically in wildlife habitat and its management. Clearly, an understanding of the habitats of wildlife is important and gaining momentum. Habitat use and selection by animals, however, is complex.

Consider humans as a common example. Humans have the greatest ability to select our "habitats" of all the organisms. Habitat selection might involve picking out an ideal site somewhere on the globe and moving there. However, that is rarely done. If it were, those areas with cool summers and mild winters would likely have even more people than they do now and more of us would likely be more familiar with the French and Italian Riviera. We do not move wherever we want because of an array of limitations: food, money, jobs, culture, family obligations, language barriers. . . . People are where they are because of their ability to fit or successfully or unsuccessfully use an area. Some individuals may be marketable worldwide and could select any habitat they wanted, but most humans face limitations given specific circumstances and desires. For example, students interested in wildlife made a set of decisions that dictate where they study: the area of wildlife selected, particular universities and faculty, opportunities, desire to leave home, and too many other variables to list here. How, then, if we cannot articulate the conditions under which humans select habitats, can we determine the mechanisms that are used by wildlife to use and select areas in which to live? It is a complex arena but one that needs serious study because habitat is essential for healthy populations.

Most people (professional and laypeople) have an understanding of habitat and recognize that it includes food, water, cover, and all necessary special factors required for life but similarities end there. As biologists' and land managers' understanding of habitat increases, it becomes apparent that the use of concepts and terms is not consistent. This distorts the communication among scientists in our disciplines and the layperson, and confuses the public because we give ambiguous, indefinite, and nonstandardized responses to ecological inquiries in legal and public situations. All one has to do is quickly glance at the literature to see the different uses of terminology in relation to habitat (Hall et al. 1997).

▓ HABITAT TERMINOLOGY

Hall et al. (1997) examined how recent (i.e., 1980–1994) authors used habitat-related terms by reviewing 50 papers from peer-reviewed journals and books in the wildlife and ecology fields that discussed wildlife-habitat relationships. In their review of each paper, Hall et al. (1997) noted whether habitat terms were defined and evaluated the definition(s) against standard definitions derived from Grinnel (1917), Leopold (1933), Hutchinson (1957), Daubenmire (1968), and Odum (1971). Of the 50 articles reviewed, only nine (18%) correctly defined and used terms related to habitat. Hall et al. (1997) proposed the following as standard terminology.

Habitat

Habitat is the resources and conditions present in an area that produce occupancy, including survival and reproduction, by a given organism; in other words, habitat is species specific. Habitat implies more than vegetation or vegetation structure. It is the sum of the specific resources that are needed by organisms (Thomas 1979). These resources include food, cover, water, and special factors needed by a species for survival and reproductive success (Leopold 1933). Wherever an organism is provided with resources that allow it to survive, that is habitat. Thus, migration and dispersal corridors and the land that animals occupy during breeding and nonbreeding seasons are habitat. For example, until recently, desert bighorn sheep in southeastern California were managed by mountain range and little to no consideration was allocated to the desert between mountains. Schwartz et al. (1986) argued that the deserts between the mountain ranges were important for intermountain movement and the series of ranges and connecting desert should be managed as a metapopulation. Similar management strategies should have been carried out around bighorn sheep ranges around Tucson, Arizona, that may have precluded the elimination of sheep in much of the Tucson Basin (Krausman 1997).

Habitat Use

Habitat use is the way an animal uses the physical and biological resources in a habitat. Habitat may be used for foraging, cover, nesting, escape, denning, or other life history traits. These categories (e.g., foraging, escape) divide habitat but overlap occurs in some areas. One or more categories may exist within the same area, but not necessarily. An area used for foraging may be comprised of the same physical characteristics used for cover, denning, or both (Litvaitis et al. 1994).

The various activities of an animal require specific environmental components that may vary on a seasonal or yearly basis. A species may use one area in summer and a different one in winter. These same areas may be used by another species in reverse order (Hutto 1985, Morrison et al. 1985).

Habitat Selection

Habitat selection is a hierarchical process involving a series of innate and learned behavioral decisions made by an animal about what habitat it would use at different scales of the environment (Hutto 1985). Wecker's (1964) classical studies of habitat selection by deer mice revealed that heredity and experience play a role in determining selection. Small organisms with low mobility have evolved means for using water and air currents for dispersal. Spores, seeds, ballooning spiders, and some insects drift in the upper portions of the atmosphere. Some planktonic forms ride the waves. Even terrestrial organisms accidentally cross long stretches of water, and live fish have been dispersed by hurricanes. Some of these species transplanted in these random events eventually reach areas where they can survive and reproduce (Wecker 1964). Most die in this dispersal type of *inactive* "habitat selection."

Most animals *actively* select habitats by reacting to specific cues in their environment (at least in part). For example, single-celled protozoans up to beetles and salamanders select habitats based on physicochemical gradients (e.g., temperature, moisture, light) in the environment. Some beetles, moths, and butterflies select egg-laying sites that are appropriate for the larvae but not the adult form. Birds often select habitat based on height, spacing, vegetation form, and availability of song perches or nest sites. These attributes have been called "psychological factors" in habitat selection and they may be as important to higher forms of life as physicochemical gradients are to lower forms of life (Wecker 1964).

Rosenzweig (1981) asserted that habitat selection was generated by optimal foraging theory. However, foraging is only one behavior a species may select for. Habitat may be selected for cover availability, forage quality and quantity, and resting or denning sites. Each of these may vary seasonally. If an individual or species demonstrates disproportional use for any factor, then selection is inferred for those criteria (Block and Brennan 1993). Reproductive success and survival of the species are the ultimate reasons that influence a species to select a habitat (Hilden 1965). The ability to persist is governed by ultimate factors such as forage availability, shelter, and avoiding predators (Litvaitis et al. 1994).

Many interacting factors have an influence on habitat selection for an individual (e.g., competition, cover, and predation). Competition is involved because each individual is involved in intraspecific and interspecific relationships that partition the available resources within an environment. Competition may result in a species failing to select a habitat suitable in all other resources (Block and Brennan 1993) or may determine spatial distribution within the habitat (Keen 1982).

Predation also complicates selection of habitat (Block and Brennan 1993). The existence of predators may prevent an individual from occupying an area. Survival of the species and its future reproductive success are the driving forces that presumably cause an individual to evaluate these biotic factors. With a high occurrence of competition and predators, an individual may choose a different site with less optimal resources. Once predators are removed, areas with necessary resources can then be inhabited (Rosenzweig 1981).

Habitat selection is therefore an active behavioral process by an animal. Each species searches for features within an environment that are directly or indirectly associated with the needed resources that animal would need to reproduce, survive, and persist. Habitat selection is a compilation of innate and learned behaviors that lie on a continuum of closed to open (i.e., learning) genetic programs (Wecker 1964). Innate behaviors give an individual preadaptation to behave in a certain manner. Therefore, preadaptation to certain environmental cues plays an important role in habitat selection, but the potential for learning exists in many species (Morrison et al. 1992).

Habitat Preference

Habitat preference is the consequence of habitat selection, resulting in the disproportional use of some resources over others. For example, if four different vegetation associations were available in an animal's habitat and the animal selected one (or more), more than could be expected by chance alone, those associations would be preferred over others.

Habitat Availability

Habitat availability is the accessibility and procurability of physical and biological components of a habitat by animals. Availability is in contrast to the abundance of resources, which refers only to their quantity in the habitat, irrespective of the organisms present (Wiens 1984:402). Theoretically, one should be able to measure the amounts and kinds of resources available to animals, but in practice it is not always possible to assess resources availability from an animal's point of view (Litvaitis et al. 1994). As an example, one could measure the abundance of a prey species for a particular predator but could not say that all of the prey present in the habitat are available to the predator because there may be factors (e.g., ample cover) that restrict their accessibility. Similarly, Morrison et al. (1992:139) suggested that vegetation beyond the reach of an animal is not available as forage, even though the vegetation may be preferred. Measuring actual resource availability is important to understand wildlife habitat, but in practice it is seldom measured accurately because of the difficulty of determining what is and what is not available (Wiens 1984:406). Consequently, quantification of availability usually consists of *a priori* or *a posteriori* measures of the abundance of resources in an area used by an animal, rather than true availability.

Habitat Quality

Habitat quality refers to the ability of the environment to provide conditions appropriate for individual and population persistence. Hall et al. (1997) suggest that habitat quality is a continuous variable, ranging from low (i.e., based on resources available only for survival), to median (i.e., based on resources available for reproduction), to high (i.e., based on resources available for population persistence). Habitat quality should be linked with demographics, not just vegetative features, if it is to be a useful measure. The terms *unused* or *unoccupied* habitat are useful when biologists and managers are discussing threatened, endangered, or rare species that are reduced in number to the point they cannot use some areas of what appears to be adequate habitat. If their numbers were greater, they would likely use the "unused" habitat.

Critical Habitat

Critical habitat is primarily used as a legal term describing the physical or biological features essential to the conservation of a species, which may require special management consideration or protection (U.S. Fish and Wildlife Service 1988). Because critical habitat can occur in areas within or outside the geographic range of a species (Schreiner 1976, U.S. Fish and Wildlife Service 1988), the definition is not specific enough ecologically to allow for easy and rapid delineation of critical areas for threatened and endangered organisms. Also, it is not definitive enough to satisfy many public interest groups concerned with U.S. Fish and Wildlife listing decisions. Critical habitat should be specifically linked with the concept of high quality habitat (i.e., ability of an area to provide resources for population persistence. This definition would make it an operational and ecological term, not a political term; [Murphy and Noon 1991]).

As exemplified by Hall et al. (1997), habitat terminology has been used in the literature vaguely and imprecisely. However, to be able to communicate effectively and obtain accurate information about habitats, land managers and biologists should be able to accurately measure all aspects of habitat.

GENERAL CONCEPTS RELATED TO HABITAT USE

Definitions only help us understand how organisms interact with their habitat. To be even more meaningful, there are basic concepts that have evolved with the importance of habitat: habitat has a specific meaning, is species specific, is scale dependent, and measurements matter. Some of these concepts are implied in the definitions provided but additional emphasis is warranted.

Habitat Has A Specific Meaning

That biologists use the term habitat numerous ways is not useful, and is confusing to the public. Of course, habitats are variable but they all include the specific resources and conditions in an area that produce occupancy. This includes survival and reproduction. Habitat is frequently used to describe an area that supports a particular type of vegetation (Morrison et al. 1992). Vegetation is important but is only part of habitat that includes food, cover, water, temperature, precipitation, topography, other species (e.g., presence or absence of predators, prey, competitors), special factors (e.g., mineral licks, dusting areas), soils, and other components in an area important to species that managers may not have identified.

Habitat is Species Specific

When I hear someone state, "This is great wildlife habitat," it is like walking into a brick wall and I can only guess what they mean. All the components necessary for reproduction and survival are not the same for all species, and "great wildlife habitat" for one species may not even come close to serving as appropriate habitat for others. This has been, and will continue to be, a problem because manipulations of the landscape will favor the habitats of some species but be detrimental to the habitats of others. A lot of effort has been placed on ecosystem management (Czech and Krausman 1997) in the 1990s, but when considering specific organisms one has to consider their unique array of requirements for survival. With a knowledge of habitat requirements for the species under a manager's care, that manager can make informed decisions as to how landscape alterations will influence plant and animal communities.

As an example, consider the relationship between an animal and the habitat it uses (Figure 16–1). Each animal requires an arrangement of food, water, cover, and special factors and has adapted to that juxtaposition (i.e., *habitat niche*). The animal also plays a specific role within the environment that contains the necessary components for persistence (i.e., *ecological niche*).

Plant communities are unique combinations of vegetation that exist in specific places due to the soil characteristics, temperature, elevation, solar radiation, slope, aspect, and precipitation that influence vegetation (Daubenmire 1976). Plant communities progress to climax through succession; each stage is a *sere* or *successional stage* (Figure 14–5).

Various combinations of plant communities create part of the habitat for specific wildlife, and the wildlife supported by the vegetation (and related habitat components) is the *animal community*. The animals in turn influence the vegetation community.

Other terms (used primarily in forestry) that help identify the relationship between an animal and its habitat include site types, stands, stand condition, edges, ecotones, edge effect, richness, diversity, and stability (Thomas 1979). *Site types* are roughly analogous to plant communities as they are areas (i.e., *sites*) classified by climate, soil, and vegetation. Each site type contains one or more *stands* (i.e., plant communities [e.g., trees] with sufficient uniformity to distinguish them from adjacent plant communities). The *stand condition* is a description or measure of the size, density, age, spatial arrangement, and condition of stands.

The area where two communities or successional stages overlap or produce a distinct combination of plants or structure is an *ecotone*. *Edges* are the boundaries of plant communities, successional stages, or stands. Both provide a higher array of vegetation for wildlife because they include the attributes of the adjoining communities or succession stages and that of the edge or ecotone (Leopold 1933). The intermixing of different vegetation from specific arrangements is an *edge effect*.

By increasing the edge, additional wildlife can inhabit an area (i.e., *richness*) because of the added vegetation types and structure. Mixing communities or successional stages is often referred to as *interspersion*. More interspersion may increase edge that may enhance the variety of plant and animal communities (i.e., *diversity;* Patton 1975). If an increase in diversity allows a community to withstand catastrophe (i.e., *stability;* Margalef 1969) or to return to its original state after severe alteration, conservationists would have an important index to the practice of conservation (Odum 1971:256). Unfortunately, the relationship between an animal and its habitat is complex and dependent on other species. Modification of the plant community may benefit some species but not others. As a result, managers need to carefully evaluate the habitat modifications they call for to ensure that the alterations for the species in question are not creating unacceptable conditions for other species. Managers need to think beyond simple definitions in the complex management of wildlife. For example, the edge concept is not necessarily a positive feature of management. If edge is increased beyond certain levels, it can lead to fragmented habitats, which could cause increased predation (Morrison et al. 1998). Furthermore, edge does not generally apply to mobile species and is beneficial to "edge-obligate species" (Leopold 1933, Guthery and Bingham 1992). Animals are not able to increase their abundance above a maximum density, so the relationship between edge and density has limits (Morrison et al. 1998).

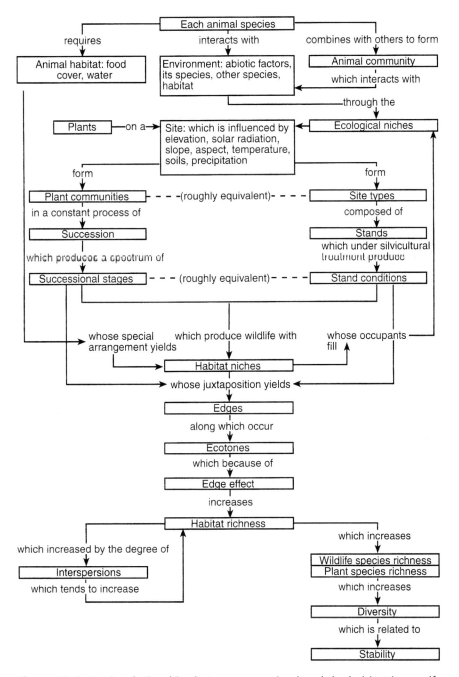

Figure 16–1 Basic relationships between an animal and the habitat it uses (from Thomas 1979).

Figure 16–2   Present distribution of desert races of mountain sheep in the Southwest. This distribution would be first-order habitat selection. The home range of an individual or social group within this area would be second-order habitat selection.

Habitat Is Scale Dependent

Macrohabitat and microhabitat are common terms but actually relate more to the landscape scale at which a study is being conducted for a specific animal than to a type of habitat. Generally, macrohabitat is used to refer to landscape-scale features such as seral stages or zones of specific vegetation associations (Block and Brennan 1993). Microhabitat usually refers to finer-scaled habitat features. Johnson (1980) recognized this hierarchical nature of habitat use where a selection process will be of higher order than another if it is conditional upon the latter. He summarized four natural ordering habitat selection processes (Johnson 1980).

First-order selection

The first-order selection is essentially the selection of the physical or geographical range of a species. For example, the geographical range of desert bighorn sheep in the Southwest would be first-order selection (Figure 16–2).

Second-order selection

The second-order selection is the home range of an individual or social group within their geographical range.

Third-order selection

The third-order selection relates to how the habitat components within the home range are used (i.e., areas used for foraging, bedding, breeding, parturition).

Fourth-order selection

The fourth-order of habitat selection relates to how third-order selection is carried out. If third-order selection determines a foraging site, the fourth-order selection would be the actual procurement of food items from those available at that site.

Based on these criteria, macrohabitat is first-order habitat selection and microhabitat is similar to the second, third, and fourth levels in Johnson's (1980) hierarchy. Understanding these levels can have profound influences on the management of a species. For example, Etchberger and Krausman (1999) found that desert bighorn sheep used most portions of the Little Harquahala Mountains in western Arizona (second-order selection) throughout the year, but individual females used specific individual sites for lambing (third-order selection). In addition, site fidelity was strong for each site used by each female. Understanding the importance of these smaller areas at specific times to the population would influence the way the population is managed. This example also demonstrates that habitat use is temporal.

Measurements Matter

As indicated, habitat is not ambiguous and to understand how it interacts with a species one has to ask the correct question: what component is being measured, when is it being measured, and how many samples are necessary for meaningful results? Obviously, to even pose these questions, one has to have knowledge of an animal's total life history strategy. Without it, measurements of habitat could be meaningless or erroneous. This is not always easy, even with well studied species such as elk. For example, for years many biologists accepted the concept that weather-sheltering effects of dense forest cover or thermal cover reduced energy expenditure and enhanced survival and reproduction. As a result, providing thermal cover for elk was a key habitat objective on elk ranges in the West. Cook et al. (1998), however, demonstrated that energetic status and reproductive success were not enhanced with thermal cover, and suggested that habitat management based on the perceived value of thermal cover should be reevaluated. The majority of the empirical support for the thermal cover hypothesis was derived from observational studies of habitat use. Cook et al. (1998) demonstrated the difficulty associated with determining habitat requirements from empirical observations of habitat use. They also demonstrated the need for scientific studies within a clear conceptual framework with adequate sampling rigor.

Determining the important components to measure when describing an animal's habitat is related to the question asked and scale of habitat being examined. For example, there is little use for describing microhabitat variables if the goal of a study is to describe the distribution of an animal across general vegetation associations (Morrison et al. 1998). Likewise, the measurement of the distribution of an animal across general vegetation association would not reveal important microhabitat variables (e.g., nesting or breeding sites).

One hierarchic method to examine wildlife-habitat relationships involves a series of specific and general questions (Van Horne 1983). Level one calls for the direct evaluation of habitat for a single species at a specific site. The second level uses more general variables but the application of the data collected can be used in other locations. The third level is a very broad approach that examines relationships for an array of wildlife (i.e., community) (Van Horne 1983). Each of these levels could be subdivided but this simple model indicates how habitats can be studied (Morrison et al. 1998). For example, black bears require five basic habitat components (i.e., escape cover, fall sources of meat, spring and summer feeding areas, movement corridors, and winter denning habitat) (Pelton 2000). Describing these variables for a black bear population would fit Van Horne's (1983) second level, but an in-depth evaluation of one or more would be similar to level one. A level three would have to consider the habitat of bears with the other species in the area.

MANAGING HABITAT

A thorough discussion of managing habitat is not possible within the context of this manuscript. The reader should consult Bookhout (1994), Morrison et al. (1992, 1998), and Payne and Bryant (1998) for treatments of contemporary management. However, much of what is addressed in this book is important to adequately manage habitat. Leopold (1933) developed the basic tenets of habitat management: that organisms require the essentials of food, water, cover, and special factors for survival. Giles (1978) and others built on this concept and developed the wildlife, habitat, and human triad that is so critical to management today. The triad forces one to examine wildlife in the context of its evolutionary origins and see how it changes in relation to human disturbances. There are numerous models and techniques biologists can use to manage habitats (that are readily available in the literature), but for them to be realistic they should consider the animal and the relationship it has within the environment (Thomas 1979, Morrison et al. 1998).

Reducing animals' habitat to a numerical value without relating it to the niche an animal occupies does little to advance the understanding of habitats. As demonstrated by Cook et al. (1998), biologists need more carefully designed studies and greater attention to the assumptions they make (i.e., realistic assumptions) (Noon 1986). Without those basic considerations, research findings will fail to be incorporated into species management.

Warner and Brady (1994) outlined habitat management programs for farmlands that are also applicable to other habitat management projects.

1. The needs of the landholder should be accommodated.
2. Management agencies should deliver the technical and material assistance (where applicable) to make habitat programs work.

3. Habitat improvement projects should have clear goals that can be evaluated.
4. Habitat development and maintenance should be viewed from a long-term perspective.
5. The sustainability of the habitat also should be long term.

These five considerations appear to be obvious but one or more are often ignored, causing the program to fade away without any benefit to society (or wildlife) (Warner and Brady 1994).

Successful habitat programs that produce measurable responses usually include eight components (Warner and Brady 1994:650–651).

1. Compatibility with the objectives of the landholder and with the primary use (e.g., agricultural production, forests, rangeland) of the land.
2. A determined resolve by management agencies to succeed. This attitude has to be supported by careful planning with realistic goals, costs, and strategies for accomplishing tasks.
3. Cooperation among the principal agencies and groups that influence the decision-making environments of landholders.
4. A program of information and education that includes promotion to capture the interest of landholders and positive reinforcement of desirable land practices essential to the program that are also rewarding to the landholder. The education should lead to a land ethic.
5. A thorough understanding of the ecology and behavioral biology or target species and the impacts of management programs on plant and animal communities.
6. A habitat development and maintenance program that has been successful for targeted wildlife species, including the establishment of suitable plant forms and configurations of cover and their maintenance over appropriate physiographic regions.
7. Scales of time and space large enough to allow for buffering of the changeable physical environment including weather, chemical and biotic factors, policies and programs, and the land use practices of individual landholders.
8. An ongoing evaluation and refinement process that permits the program to respond to changing circumstances.

Basic Habitat Management Techniques

The basic objectives of most management programs are to maintain habitat for species as it exists in undisturbed ecosystems or provide habitat where it has been depleted or where a specific component is missing. Most habitat alteration resulting in deteriorating habitats for specific wildlife was caused by the axe, plow, cow, fire, or gun (Leopold 1993). These basic tools are also used to restore habitats. For example, the axe and fire have restored the prairie habitat for sharp-tailed grouse and waterfowl, the plow is used on waterfowl refuges to produce food patches, cattle grazing on some western rangelands have had some benefits to elk, and the gun has been instrumental in legal hunting to raise funds for conservation efforts.

Before any habitat management project is attempted, the manager should consider several basic principles.

1. The project should have a specific objective: to maintain, improve, or completely alter the habitat for a particular species and a particular purpose. Projects with an objective of improving habitat for wildlife in general are too broad and ambiguous, and cannot be monitored to determine if the objective is ever met.
2. The project should be biologically justified. To add a habitat component that has been lost or altered by anthropogenic forces would be reasonable, compared to completely developing habitat for a species that was not indigenous to the area.
3. The habitat manipulation or alteration should be evaluated for its influence on other natural resources, species, and land uses.
4. The project should be economically practical.
5. The project should simulate natural conditions as much as possible.
6. All habitat alteration projects should be evaluated to determine how the objectives have been accomplished. This is an extremely important step but often ignored due to the budgets or limited personnel. However, if the project is not monitored, money may be wasted, there is no measure of whether the objectives have been met, and the management activity could be a waste of time and even detrimental to the resources the project was intended to benefit.

Habitat Evaluation

There are numerous models used to describe and examine the habitats of species (Mannan et al. 1994) (Table 16–1) but every one should have basic information. At the most general level a site description is important so one can obtain a general overview of the habitat being described. Habitat is more than vegetation, so brief descriptions of the geology and topography, soils, weather and climate, cultural features, history of the land, and general indices of site quality are useful in describing habitats. Survey maps should include surface features of the landscape important to wildlife. Soils influence vegetation so are obviously of interest and the weather and climate have direct effects on all organisms. Cultural features also need to be considered. How close are towns, villages or other urban centers, power lines, fences, roads, and any other human-created structure? Also, what is the history of the area: has it been burned, subjected to drought, cutting, flooding, or grazing? Finally, what are the general indicators of the usefulness of the habitat to different species? For example, specific plants are indicative of differ-

Table 16–1 Summary of models of forest habitat-wildlife relationships (Mannan et al. 1994)

Model name	Description	Reference
	Stand growth and yield models	
CLIMACS	Projects long-term effects of disturbances	Dale and Hemstrom 1984
Douglas fir Similator (DFSIM)	Even-age Douglas fir yield	Curtis et al. 1981, Fight et al. 1984
FORCYTE		Kimmins 1987
FOREST		Ek and Monserud 1974
FREP		U.S. Forest Service 1979
Prognosis	Managed stand yield model	Wykoff et al. 1982
Stand Projection System (SPS)	Even-aged managed stand yield model	Arney 1985
STEMS		Belcher et al. 1982
VARP	Assesses stand cruise data on HP–41C	Tappeiner et al. 1985
WOODPLAN		Williamson 1983
	Forest succession models	
DYNAST	Projects multiple stand growth	Boyce 1980
FORPLAN	Linear programming of stand growth	U.S. Forest Service 1979
	Statistical models	
CORRELATION MODELS	Associates habitat parameters with species parameters	Many studies
MULTIVARIATE MODELS	Associates multiple habitat parameters with single or multiple species parameters	e.g., Capen 1981
	Species-habitat models	
HABITAT SUITABILITY (HSI) MODELS	Relates three habitat variables to species parameters	Schamberger et al. 1982
HABITAT CAPABILITY MODELS	Relates habitat variables to capability of habitat to support a species; similar to HSI models	
HABITAT EVALUATION PROCEDURES (HEP)	Assesses habitat condition	U.S. Fish and Wildlife Service 1980
PATTERN RECOGNITION (PATREC)	Predicts probabilities of wildlife effects from habitat conditions	Willams et al. 1977
SPECIES-HABITAT MATRICES	Lists wildlife species by habitat types and conditions	e.g., Thomas 1979, Verner and Boss 1980
GUILD AND LIFE	Denotes response of guilds to habitat conditions	e.g., Severinghaus 1981, Short 1983, DeGraaf et al. 1985
COMMUNITY STRUCTURE	Denotes wildlife species distribution, abundance, and diversity as a function of vegetation structure	Raphael and Barrett 1981, Schroeder 1987

Continued

Table 16–1   Summary of models of forest habitat-wildlife relationships (Mannan et al. 1994).
Continued

	Forest landscape models	
HABITAT DISTURBANCE	Simulates effects of disturbance on habitat composition and structure	Shugart 1984, Pickett and White 1985, Shugart and Seagle 1985
FOREST FRAGMENTATION	Displays effects of forest stand fragmentation on species' distribution and abundance	Dueser and Porter 1986, Askins et al. 1987, Bock 1987, Stamps, et al. 1987, many others (see text)
	Cumulative effects models	
DYNAST	Integrates species habitat relationships	Benson and Laudenslayer 1986, Holthausen 1986, Sweeney 1986
ECOSYM	Tracks multiple stand development	Davis 1980, Davis and DeLain 1986
FORHAB	Projects stand with species data	Smith et al. 1981, Smith 1986
FORPLAN	Integrates species' needs as constraints	Davis and DeLain 1986, Holthausen 1986
FSSIM	Projects multiple stand growth and species' requirements	
HABSIM	Habitat capability model	Raedeke and Lehmkuhl 1986
STEMS		Belcher et al. 1982
TWIGS		Belcher 1982, Brand et al. 1986
Grizzly Bear	Assesses spatial effects from management activities on occurrence and viability of grizzly bears	Weaver et al. 1985
	Decision-aiding models	
DECISION SUPPORT	Helps advise, weigh, and prioritize management decisions; includes expert systems	Marcot 1986, 1988
	Monitoring models	
ADAPTIVE MANAGEMENT	Allows amending management direction from monitoring results	Walters and Hilborn 1978, Walters 1986

ent conditions that may be useful in describing different sites (Table 16–2). When selecting areas for habitat analysis, two basic steps are followed: the reconnaissance and the primary survey.

During the reconnaissance, the area is generally surveyed via characterizing the landscape and vegetation. This process should be done on the ground with the aid of aerial photographs, topography maps, orthophotoquads, and geographic information system techniques. This stage provides a general overview of the area with enough detail to make it meaningful. The amount of detail will depend on your objectives. For example, what do you want to know: species composition, density, forage available, forage used, ground cover. . . ?

Table 16–2 Plants indicative of human disturbance, high moisture, and high temperature in four vegetation associations in Arizona

Vegetation association	*Plants indicative of:*		
	Disturbance	**High moisture**	**High temperature**
Saguaro-paloverde	Snakeweed	Seep willow	Ironwood
	Burroweed	Catclaw acacia	Saguaro
	Cockleburr	Salt grass	Jojoba
Desert grassland	Burroweed	Hackberry	Ocotillo
	Mesquite	Sycamore	
	Three-awn		
Oak woodland	Lecheguilla	Willows	Prickly pear
	Bear grass	Sycamore	Cholla
Pine woodland	Lankspur	Strawberry	Shrubs
	Mullen		Lupine

The primary survey will take more staff and time as specific habitat data are collected for the specific area (e.g., an entire region or state down to a small area). Some of the data collected generally includes vegetation; vegetation cover; density of cover; browse intensity; penetrating light; abundance and use of cliffs, caves, snags, cavities; available food; food used; water; disturbance; and any array of these and other variables depending on the question and objective. For example, Rich (1986:549) studied nest site selection of burrowing owls and described occupied burrows based on "presence of rock outcrops with a 1-km radius, type of burrow, compass orientation of burrow opening, slope and aspect of the ground within a 50-m radius, height of outcrop (if present) above the burrow, maximum right-angled dimensions of the outcrop, estimated percent ground cover within 50 m of the burrow, and number of alternate burrows." Other classification schemes could be as complex as the hierarchic classification of key environmental correlates for wildlife species presented by Morrison et al. (1998) (Table 16–3).

Regardless of the variables used to describe the habitat or part of a habitat for a species, it is critical that an efficient sampling design is selected so objectives can be measured and determined whether and when they have been met. To obtain information of sampling design, see Morrison et al. (1998).

Evaluation of an animal's habitat is complex even when a simple description of the habitat is desired. How animals use habitats is even more complex because that requires information about animal behavior (e.g., dispersal, ability to detect new habitats) and how behavior changes with human-altered habitats. An important consideration in habitat studies is how animal populations respond to the resources in the habitats used (Hobbs and Hanley 1990).

Table 16–3   A hierarchic classification of key environmental correlates for wildlife species (Morrison et al. 1998)

1. Vegetation elements
    1.1 cover types
    1.2 structural stages
    1.3 forest or woodland vegetation substrates
        1.3.1 down wood
        1.3.2 snags (entire tree dead)
            1.3.2.1 bark piles at base of snag
        1.3.3 mistletoe brooms
        1.3.4 litter
        1.3.5 duff
        1.3.6 shrubs
        1.3.7 fruits, seeds, mast
        1.3.8 dead parts of live trees
        1.3.9 moss
        1.3.10 live trees
            1.3.10.1 exfoliating bark
        1.3.11 flowers
        1.3.12 lichens
        1.3.13 bark
        1.3.14 forbs (including grass)
        1.3.15 cactus
        1.3.16 fungi
        1.3.17 roots, tubers, underground plant parts
        1.3.18 peatlands
    1.4 herbaceous vegetation elements or substrates
        1.4.1 herbaceous vegetation cover
            1.4.1.1 aquatic submergent vegetation
        1.4.2 fruits, seeds
        1.4.3 moss
        1.4.4 cactus
        1.4.5 flowers
        1.4.6 shrubs
        1.4.7 fungi
        1.4.8 forbs
        1.4.9 bulbs, tubers
        1.4.10 cryptogamic crusts
    1.5 diversity of vegetation cover types
    1.6 edges
        1.6.1 openings
        1.6.2 meadows
    1.7 mycorrhizal associations

2. Biological (nonvegetation) elements
    2.1 presence of prey species
        2.1.1 carrion
    2.2 presence of predators
        2.2.1 absence of predator
    2.3 presence of exotic species
        2.3.1 exotic plants
        2.3.2 exotic animals
    2.4 insect irruption areas
        2.4.1 mountain pine beetle
        2.4.2 spruce budworm
        2.4.3 gypsy moth
    2.5 presence of burrows or presence of burrowing mammals
    2.6 grazing
        2.6.1 direct effects (trampling, consumption)
        2.6.2 indirect effects (habitat degradation)
        2.6.3 seasonality of grazing
    2.7 presence of beaver or muskrat ponds or lodges
    2.8 presence of nesting structures
        2.8.1 cavities
        2.8.2 platforms
    2.9 presence of other species (specify)
    2.10 forest pathogens
    2.11 colonial nester

3. Nonvegetation terrestrial substrates
    3.1 rocks
        3.1.1 gravel
    3.2 soils
        3.2.1 soil class
        3.2.2 soil depth
        3.2.3 soil texture
            3.2.3.1 sand, dunes
            3.2.3.2 soil suitable for burrowing vertebrates
            3.2.3.3 soil suitable for burrowing invertebrates
        3.2.4 soil pH
        3.2.5 soil temperature
        3.2.6 soil moisture

Table 16–3 *Continued*

3.2.7 soil chemistry

3.2.8 soil organic matter

3.3 lithic (rock) substrates

    3.3.1 lithic series or types (including lithic formations)

    3.3.2 avalanche chutes

    3.3.3 cliffs

    3.3.4 talus

    3.3.5 boulders, large rocks

    3.3.6 caves

    3.3.7 rock outcrops, crevices

    3.3.8 lava flows

    3.3.9 lava tubes

    3.3.10 canyons

    3.3.11 barren ground

    3.3.12 rugged terrain

    3.3.13 rocky ridges

    3.3.14 ravines

    3.3.15 cirques or basins (also see entry 5.7 below)

3.4 snow

    3.4.1 snow depth (winter)

    3.4.2 glaciers, snow fields

3.5 water characteristics

    3.5.1 dissolved oxygen

    3.5.2 water depth

    3.5.3 dissolved solids

    3.5.4 water pH

    3.5.5 water temperature

    3.5.6 water velocity

    3.5.7 water turbidity

3.6 environment space (typically for foraging) above tree canopy

4. Riparian and aquatic bodies

4.1 rivers

    4.1.1 riverine wetlands

    4.1.2 oxbows

4.2 streams (permanent or seasonal)

    4.2.1 intermittent

    4.2.2 rocks in streams

4.3 seeps or springs (including warm seeps or springs)

4.4 exposed mudflats, sandbars

4.5 sandbars, unconsolidated shore

4.6 gravel bars

4.7 shallow water

4.8 lakes or reservoirs (lacustrine)

    4.8.1 lakes with submergent vegetation

    4.8.2 lakes with floating mats

    4.8.3 lakes with silt or mud bottom

    4.8.4 lakes with emergent vegetation

    4.8.5 alkaline lake beds

4.9 ponds (permanent or seasonal)

    4.9.1 ponds with submergent vegetation

    4.9.2 ponds with floating mats

    4.9.3 ponds with silt or mud bottoms

    4.9.4 ponds with emergent vegetation

4.10 wetlands, marshes, or wet meadows (palustrine)

    4.10.1 bulbs or tubers in wetlands marshes, or wet meadows

    4.10.2 Phragmites

4.11 bogs of fens

4.12 swamps

4.13 islands

4.14 waterfalls

4.15 hyporheic zone

4.16 irrigation ditches

4.17 ephemeral pools

4.18 deciduous riparian areas, including willow and cottonwood

4.19 vernal or seasonal flooding or flood plains

4.20 bottomlands

4.21 water table

5. Topographic or physiographic elements

5.1 elevation

5.2 slopes

5.3 aspect

5.4 slope position

5.5 ridgetops

5.6 plateaus

5.7 convex or concave basins (also see entry 3.3.15 above)

Continued

Table 16–3 A hierarchic classification of key environmental correlates for wildlife species (Morrison et al. 1998). *Continued*

5.8 flats
5.9 mima mounds
6. Climate
   6.1 precipitation (amount, pattern, seasonality)
   6.2 Mediterranean influence (dry summers)
   6.3 maritime influence (higher humidity and more moisture)
   6.4 temperature
   6.5 humidity
   6.6 wind
7. Fire
   7.1 fire recency
      7.1.1 recent fire
      7.1.2 old fire
   7.2 effects of fire-suppression activities
   7.3 fire frequency
   7.4 fire intensity
      7.4.1 overstory lethal
      7.4.2 overstory nonlethal
   7.5 prescribed fire
      7.5.1 spring prescribed fire
      7.5.2 late summer or fall prescribed fire
   7.6 historic fire suppression
8. Human disturbance elements (positive or negative effects)
   8.1 recreation areas and activities (including dispersed camping areas)
   8.2 roads or trails
   8.3 residential developments
   8.4 buildings
   8.5 bridges
   8.6 tunnels

8.7 agriculture and croplands
8.8 livestock (disease)
8.9 mines and mining activities
8.10 harvest (including legal hunting, legal trapping, and illegal poaching of animals)
8.11 fences
8.12 bird feeders
8.13 winter recreation
8.14 garbage
8.15 logging
8.16 nest boxes
8.17 perch structures
8.18 platforms
8.19 guzzlers
8.20 pesticide use
8.21 exotic plant effects
   8.21.1 direct displacement
   8.21.2 indirect competition
   8.21.3 inhibited recruitment
   8.21.4 habitat structure change
8.22 livestock grazing strategies
   8.22.1 season long
   8.22.2 spring grazing
   8.22.3 summer grazing
   8.22.4 fall grazing
9. Barriers to movement
   9.1 forest management (clearcuts)
   9.2 canopy closure
   9.3 agriculture
10. Other natural disturbances—floods, scouring, openings in forests

SUMMARY

Understanding the mechanisms linking the performance of animal populations to resources in the habitats they use will make objectives clearer and perhaps easier to evaluate and allow for predictions to be made. ". . . Habitat evaluation systems capable of predicting the effects of human impacts or other forms of disturbance require understanding the mechanisms by which the environment influences the distribution and abundance of animals. Forecasting the consequences of habitat change depends on understanding the processes that control animal response to changes" (Hobbs and Hanley 1990). Prediction is the key. Understanding habitats of wildlife is still developing, and managers need to be able to predict the responses of wildlife to human activities. We may never understand all the cues animals use in selecting habitats, but in knowing how animals use habitats, managers can have a head start in making sure that management plans for landscapes are at least predictable and the multiple use is beneficial for all resources being managed.

17

Economics In Wildlife Management

"Every head of wildlife still alive in this country is already artificialized, in that its existence is conditioned by economic forces."

A. Leopold

INTRODUCTION

Economics is the "study of the way mankind organizes itself to tackle the basic problem of scarcity. All societies have more wants than resources . . . so that a system must be devised to allocate these resources between competing ends" (Pearce 1997:121). The system that has been derived in the United States and much of the world is a form of capitalist democracy in which people, markets, and government interact to allocate resources. These resources include virtually all things physical (including wildlife and their habitats), and money, a medium of exchange. Because our definition of wildlife management includes the habitat, the animal, and humans and their interactions, it may be informative to review some of the ways funds have been generated for the management of wildlife resources and the economic impact wildlife has on society above and beyond the funds generated from the sale of hunting and fishing licenses (Tables 3–2, 3–3). However, a simple examination of the economics in wildlife management that is universally accepted has not yet been developed. Environmental economists have produced numerous textbooks (e.g., Tietenberg 1988, Bromley 1998), hundreds of monographs, and more than 250 articles published annually in approximately 24 scientific journals (de Steiguer 1989). Furthermore, economics is an independent field that biologists need to embrace to effectively allocate the resources generated from and used for the management of wildlife (Czech 2000*a*).

The management of wildlife is complex. Although wildlife in North America belongs to the public, even this view is challenged. Some have argued that wildlife is part of the ecosystem of which humans are part, and free-ranging wildlife is a product of the soil and belongs to nobody (Bubenik 1991).

This concept is rooted in Roman law as "Res nullius" (Anderson 1985). In the United States wildlife is considered a public commodity, but the habitat features it depends on are often controlled by private landowners, complicating the effective management of wildlife. Even in the West, where much of the land is federally regulated (e.g., U.S. Forest Service, National Park Service, Bureau of Land Management, U.S. Fish and Wildlife Service), the state game and fish agencies have the responsibility of managing most species.

Since the beginning of the twentieth century, wildlife has been treated as a public commodity in North America. At the end of that century there have been numerous discussions about the commercialization of wildlife in North America. This in part generated from successes in Zimbabwe and South Africa for elephant conservation via commercialization (Rasker et al. 1992). Both countries increased their declining elephant populations (from 47,000 to 49,000 in Zimbabwe and from 500 [in the nineteenth century] to over 8000 in South Africa) by reducing poaching (poachers are shot on sight). Excess animals are culled and their ivory and hides sold. These markets have provided private landowners an incentive to manage their land for the benefit of wildlife. The price of an average trophy hunt in Zimbabwe is about $25,000 (Simmons and Kreuter 1989). Local tribes are also provided with incentives to participate in the conservation efforts by selling hunting permits and harvesting elephants. In other parts of Africa (e.g., Kenya) there is no market, elephants compete with poor rural farmers for resources and land and elephants are viewed as a menace. The elephant population has decreased over 10 years (from 1979 to 1989) from 65,000 to 19,000. Some blame poaching and the international ivory market for the decline and others find Kenya at fault for not creating monetary incentives for elephant conservation (Simmons and Kreuter 1989). Game-ranching elephants in Zimbabwe and South Africa appears to have been a successful strategy to reduce the decline of the elephant population, perhaps the only option (Rasker et al. 1992).

Similar problems are facing the conservation of bighorn sheep in Mexico. Poaching has been a serious concern, and the government has allowed private landowners the opportunity to capture bighorns, place them behind fences, and at a later date translocate sheep to specified areas. Offspring from the "farm-raised" bighorn can be harvested and sold as an incentive to protect the animals. How this program enhances the bighorn sheep resources has yet to be determined.

In the United States a unique interaction of people, markets, and governments occurs on large Native American reservations. For example, on the San Carlos Apache Reservation in Arizona, the sale of hunting permits is more profitable than the tribal timber and cattle operations (Czech and Tarango 1998). Elk permits at San Carlos sold for $43,000 in 1992; all money was earmarked for habitat improvement. Tribal wildlife economics has burgeoned in recent decades for a variety of political and legal reasons (Czech 1995, 1999).

Figure 17-1 Musk deer, India (photo by P.R. Krausman).

One thing is clear. "The apparent success some countries have had in wildlife conservation via market incentives does not imply, however, that wildlife commercialization is the best solution for North America. In some instances markets may indeed work. In others, public ownership may work best. There is no single mechanism that will work in all instances" (Rasker et al. 1992:346).

Regardless of the philosophies about how wildlife should be managed, the reality is that wildlife does have a commercial value that cannot be ignored. A few examples can illustrate the commercial value of wild animals.

■ COMMERCIAL VALUES

The most valuable product from a wildlife species is musk. The musk deer (Fig. 17–1) produces a thick, oily secretion that has sold for three times the price of gold (i.e., musk). Musk is used in oriental medicine and in perfumes but is only produced by males (Vietmeyer 1991). Unfortunately, due to its value, musk deer have been overharvested to the brink of extinction.

Numerous other species are also "collected" for their commercial value (e.g., lizards, parrots, and cats; Figure 17–2). Thousands of ungulates have been introduced and translocated to North America (Table 17–1) to be used for a game species, hobby, or as a food source. Animals "raised" on private lands for hunting demand high fees (Table 17–2, 17–3), and many state agencies auction off wild ungulates to pay for conservation programs (e.g., auction of bighorn sheep tags in the western states for hundreds of thousands of dollars).

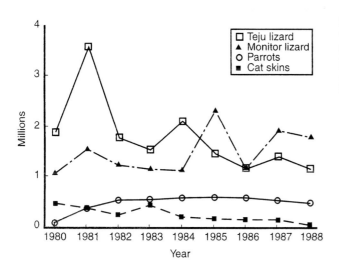

Figure 17–2 Net trade in selected wildlife products from 1980 to 1988: skins of all cat species (*Felidae*), live parrots (*Psittacidae*), skins of monitor and teju lizards. Data from 1988 are incomplete (from Luxmoore 1991).

Fourteen million hunters in America spend over $22,000,000,000 each year in pursuit of their sport. If hunting were ranked as a corporation, it would be in thirty-fifth place on the Fortune 500 list of American businesses, between J. C. Penney and United Parcel Service (International Association of Fish and Wildlife Agencies 1997:1). Unfortunately, all of the expenditures do not contribute to wildlife conservation and are part of the "bloating" economy (Czech 2000b) that expands at the competitive exclusion of wildlife in the aggregate (Czech 2000a, Czech et al. 2000; Figure 17–3). Furthermore, not all financial returns from fee hunting can be allocated to conservation (Benson 1991). For example, less than 25% of the landowners in Utah who charged a fee for hunting improved wildlife habitat (Jordan and Workman 1989). Also, in Texas, one of the largest fee-hunting regions in North America, where over half of the ranches offer fee hunting, few landowners invested funds in habitat management (Freese and Trauger 2000).

de Steiguer (1995) summarized the three schools of economic theory in relation to the environment (including wildlife). Although each theory is unique, together they can provide a comprehensive economic guide towards solving environmental problems (de Steiguer 1995).

ECONOMIC PHILOSOPHIES

Classical Economists

The classical economists attempted to view economics based on natural law. In other words, they described the order of human life and community based on reason of a naturally endowed set of principles. Thomas R. Malthus was a leader of the classical school and predicted that, although society can produce agriculture crops at an

Table 17–1 Results of a mail survey of exotic large herbivores in North America and Canada (from Teer 1991)

Province or State	Species and date of introduction	Present numbers	Free ranging (yes or no)	Purpose of introduction	Introduced by	Regulations for game ranching	Policy for introductions
Alabama	Fallow deer—1912	200–300	Y	Hunting	Private	Game breeders' license required	Legislation pending
	Wapiti—1916			Hunting	Private		
Alaska	Reindeer—1891	9 herds-25,000		Food source	U.S. Government	Wapiti ranching statute	None, but generally opposed
	Wapiti—1926	5 herds-1340		Hunting	State	AS 16.05.331	
	Bison—1928	4 herds-850		Hunting	State		
	European boar			Hunting	Private		
Arizona	Bison—1900's	210	N	Hunting	Private	Private game-farm license R12-4-413	Generally opposed
Arkansas	Wapiti	2 herds-100	Y	Hunting	State	Game breeders' license required	Generally opposed
California	Fallow deer	300	Y	Hunting	Private	Game breeders' license required	By permit
	Sambar deer	Uncommon	Y	Hunting	Private		
	Axis deer	Several hundred	Y	Hunting	Private		
	Barbary sheep	1200	Y	Hunting	Private		
	Himalayan thar	Few hundred	Y	Hunting	Private		
	Rocky Mt. wapiti						
Colorado	Barbary sheep	120	Y	Hunting	Private	Ranching of native game by permit under review	Opposed—laws and regulations
	Russian boar	25	Y	Hunting	Private		
	Mouflon sheep	30	Y	Hunting	Private		
Connecticut	0	0				Game breeders' license required	Generally opposed, importations for release never approved
Delaware						Department of Agriculture regulates introductions	Generally opposed

State	Species		Import	Purpose	Ownership	Holding regulations	Release policy
Florida	Axis deer		Y	Hunting	Private	By permit after careful study	Generally opposed
Georgia	Sambar deer	0	Y	Hunting	Private	Native wildlife can be held for scientific or educational purposes only	Can release wildlife on shooting preserves by permit
Idaho	0	0				Game breeders' license required	Permit required to import any species
Illinois	Fallow deer		Y	Farm hunting	Private	Exotic game hunting area permit required	Generally opposed
Indiana	- -	-				Game farms for meat production are permitted	Wildlife code does not specifically address non-native wildlife
Iowa	Fallow deer	-	N	For meat	Private	Game breeders' license required	Must have permit for re-release of any leases; no laws covering escapes
Kentucky	Red deer—1950's	0	Y	Hunting	State	Illegal except for game birds	Exotic animals not permitted
Louisiana	Fallow deer	0		Hunting	Private except for sika deer stocked by state	Game breeders' license required	Written permission from state required to release exotics
	Sika deer	0		Hunting			
	Blackbuck antelope	0		Hunting			
Maine	0	0				Only game birds may be produced in captivity. Law requires importation permit and veterinary certification	Importation not permitted for any species that poses a threat to native wildlife

Continued

Table 17–1 Results of a mail survey of exotic large herbivores in North America and Canada (from Teer 1991). *Continued*

Province or State	Species and date of introduction	Present numbers	Free ranging (yes or no)	Purpose of introduction	Introduced by	Regulations for game ranching	Policy for introductions
Maryland	Sika deer					Only game birds may be raised in captivity	Importation of any species for which there is no USDA-approved vaccine against rabies is prohibited
Massachusetts	Fallow deer—1940's–1960's Sika deer			Hunting	Private	Game breeders' permit required	Generally opposed
Michigan	Rocky Mt. wapiti—1918	1,100	Y	To fill vacant niche	State	Game breeders' license required	Importation is regulated by state and federal laws
Minnesota	0	0				Game breeders' license required	Regulations and policies are under review
Mississippi	Axis deer Barasinga deer Fallow deer Chinese water deer Munjac Scimitar horned oryx Eland Llama Blackbuck antelope Red deer	300 16 26 9 12 6 5 2 60 15	All of these are on a private ranch in penned conditions	Meat and hunting	Private	State has exclusive control of the propagation and distribution of wildlife; no specific laws relating to exotics	No policies on nonnative wildlife
Missouri	Moose Wapiti		N N		Private Private	Wildlife breeders' permit required	Releases for nonnative wildlife would not be approved
Montana	Fallow deer Russian boar	10 30	Y Y	Hunting	Private	Game breeders' license required	Generally opposed Can be introduced

State	Species	Number	Hunted	Use	Ownership	Regulations	Attitude
Nebraska	Fallow deer	2 herds—35	N	Unknown	Private	No regulations	Generally opposed to release of exotics
Nevada	Rocky Mt. goats—1960's	100–300	Y	Hunting	State	No specific laws	Approval for any releases must be obtained from state
New Hampshire	0					Need permits to import, hold, and propagate	Generally opposed
New Jersey	0					Permit required to import and possess	Generally opposed
New Mexico	Gemsbok	500	Y	Hunting	State	Permit required to possess native game; exotics no longer permitted	Opposed to releases of exotics
	Persian wild goat	600	Y	Hunting	State		
	Aoudad	4000	Y	Hunting	State		
	Siberian ibex	75	Y	Hunting	State		
New York	0					Farming of white-tailed deer by permit	Exotic mammals may be imported, but not released without a license
North Carolina	Sika deer		Y	Hunting	Private	Exotics may be possessed but not hunted or slaughtered for meat	Illegal to release any exotic to the wild
	Fallow deer		Y	Hunting	Private		
North Dakota	Sika deer	Rare	Y		Private	Game farming permitted	Under review at present
	Fallow deer	Rare	Y		Private		
	Barbary sheep	Rare	Y		Private		
	Bighorn sheep from California	200–250	Y	Hunting	State		

Continued

319

Table 17–1 Results of a mail survey of exotic large herbivores in North America and Canada (from Teer 1991). *Continued*

Province or State	Species and date of introduction	Present numbers	Free ranging (yes or no)	Purpose of introduction	Introduced by	Regulations for game ranching	Policy for introductions
Ohio	Wild boar	Small group	Y	Hunting	Private	Regulate possession of native game animals	Do not regulate possession of exotics; generally opposed
Oklahoma	A variety of exotics		Y	Hunting	Private	Game breeders' permit required	Exotic may be released by written permission of the state
Oregon	Mouflon sheep Barbary sheep Fallow deer Llamas	200–400 18,000	Y Y Y	Hunting Hunting Hunting Hobby	Private Private Private Private	Permit required to hold and propagate wildlife	Generally opposed
Rhode Island	Fallow deer	18	N	Hobby	Private	Permit required	Generally opposed; laws currently under review
South Carolina	European boar		Y	Hobby	Private	Farming of game species not addressed; sale of game meat is prohibited	Agency has no policy addressing exotic animals
South Dakota						If over two generations from the wild, all cervids considered an agricultural pursuit and may be raised under captive conditions	No laws or policies for exotics
Tennessee	European boar— 1960–1988 (over 800 released) Black-tailed deer—1966–1967 (75 released)		Y Y	Hunting Hunting	State State	Permit required to raise exotic and native wildlife	Generally State approval is needed to release exotics into wild

State	Species	Number	Purpose		Ownership	Regulation	Policy
Texas	Axis deer	39,040	Hunting	Y	Private	Native species cannot be farmed	No official policy
	Nilgai antelope	36,756	Hunting	Y	Private	Exotic species are not regulated and are considered domestic livestock	
	Blackbuck antelope	21,232	Hunting	Y	Private		
	Barbary sheep	20,402	Hunting	Y	Private		
	Fallow deer	14,163	Hunting	Y	Private		
	Sika deer	11,879	Hunting	Y	Private		
	(A total of 123 species on 486 ranches totaling 164,257 in 1989, see Table 17-3)						
Utah	Sika deer	3		N	Private	Must have certificate of clean health and must be properly confined	Control imports of mammals into state
	Mouflon sheep	25		N	Private		
	Waterbuck	2		N	Private		
	Zebra	1		N	Private		
	Himalayan tahr	3		N	Private		
Vermont	0	0					
Virginia	Sika deer—1921	700	Hunting	N	Private	Only fallow deer may be kept in captivity and then only for commercial purposes	No importation permits given for release of any species into the wild
	Fallow deer	500	Meat	N	Private		
Washington	Sika deer				Private	No authority to regulate exotics unless they are considered harmful	No policy
	Fallow deer				Private		
West Virginia	European boar—1969–73, 43 released		Hunting	Y	State	Annual license required to hold exotics	No importation permits given for release of any species into the wild
Wisconsin	0	0				Game farms for deer permitted under license	Policy for captive wild-life now being developed

Continued

321

Table 17–1 Results of a mail survey of exotic large herbivores in North America and Canada (from Teer 1991). *Continued*

Province or State	Species and date of introduction	Present numbers	Free ranging (yes or no)	Purpose of introduction	Introduced by	Regulations for game ranching	Policy for introductions
Wyoming	Barbary sheep Ibex Red deer Himalayan tahr Guar		Y Y Y Y		Escaped from another state	Ownership prohibited except by permit. Animals must be suitably confined	Opposed to any introductions. Remove any free-ranging exotics
Alberta						A person may possess and hold exotics by complying with regulations for captive wildlife. May traffic in parts but cannot sell meat	Covered by license and rules
British Columbia	Fallow deer—1908 Red deer		Y N	Hunting Hunting	Private Private	Bison, reindeer, and fallow deer can be held captive by permit	Under review for interest in game farming
Manitoba	None on crown land				Private	Individuals may import exotics and stock private land Manitoba Department of Agriculture regulates health requirements	Currently under review
Newfoundland	Moose Plains bison Reindeer			Hunting Hunting Hunting	Province Province Province	Permit required to hold wildlife in captivity; Minister may issue permit	No exotic may be released until careful studies are made as to their potential impact
Nova Scotia	0					Animals may be held captive under permit	Import permit required

Province/Territory	Species	Number		Use	Ownership		
Ontario	Wild boar			Hunting	Private	No legislation to regulate imports	Under review Deliberate releases are prohibited
Prince Edward Island	0					None at present	Permission is required to import any species
Quebec	Muskox	290 in 1986—probably 500 in 1989	Y		Province	Under review Regulations now being prepared for game farming	Under review Dangerous and undesirable species will be banned under new laws
Saskatchewan	Fallow deer	600		Meat	Private	"The Game Farming and Products Merchandising Regulations" allows game farming for meat	Opposed, does not allow any imports for releases into wild
Northwest Territories	Reindeer	12,000	Y	Meat	Government of Canada	Supports development of game farms and ranches for native species only	Opposed to nonnative species for release
Yukon Territory						Considered an agriculture activity	Opposed to any introductions of exotics

Figure 17–3 General relationship between human economic growth and nonhuman species conservation, given principles of competitive exclusion and niche breadth. K may be conceptualized as carrying capacity for human economy or as carrying capacity for species in the aggregate. The latter concept would entail converting the Y-axis into a scale of the economy of nature (e.g., biomass) (from Czech 2000).

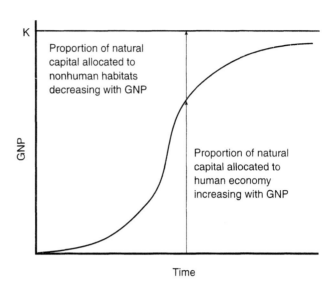

Table 17–2 Selected kill fees charged by private ranches in Montana, 1991. Generally, daily rates are charged if an animal is not taken (from Furniss 1991)

Common name	Kill fee (U.S. $)
Trophy deer	
White-tailed deer	2750–3000
Mule deer	2500–3000
Sheep	
Dall's	8000
Stone's	8000
Bighorn	10,000
Bison (Bull)	2000–3500
Wapiti	
Super trophy (high quality)	8000 and up
Normal trophy (5x5 or larger)	6000–6500
Nontrophy bull	3500–5500

arithmetic rate, the human population was increasing at a geometric rate. If true, food supplies would not be able to keep up with the human population and at some point human populations would crash. Obtaining food would be the dominant force in our lives until populations were reduced by wars, plagues, famine, and other catastrophes and would periodically reduce populations until "the cycle diminishing marginal returns to human effort would then recommence" (de Steiguer 1995).

Table 17–3 Selected South African animal kill fees charged by professional safari companies for game-ranched animals, 1991 (a per diem of U.S. $300–600/day/hunter is also charged) (from Furniss 1991)

Species	Kill fee (U.S.$)
Ostrich	350–400
Impala	150–180
Red hartebeest	600–700
Rhino (10 day minimum)	25,000–30,000
Fallow deer	300
White Springback	350–400
Common springbuck	150–250
Black wildebeest	550–800
Blue wildebeest	375–700
Black Springbuck	350–400
Bontebok	900–1000
Blesbok	275–325
Zebra	550–700
Giraffe	1000–2500
Sable	3500
Red lechwe	21,200
Waterbuck	800–1400
Gemsbuck	650–700
Eland	900–1900
Nyala	950–1100
Kudu	700–950

Steady-state Economy

John Stewart Mill shared the views of Malthus but was more hopeful that a "stationary" state would be reached where the human population and consumption were stabilized but at a low level of human satisfaction (de Steiguer 1995). Mill argued that to reach the desired steady state, humans would have to practice a less consumptive existence in developed countries so there could be a more equitable worldwide distribution of the earth's resources.

Whereas Malthusian doctrine portended a drastic decline of the human population, Mill believed disaster could be avoided with a less consumptive existence. His views were later shared by Thoreau, who advocated that human lives needed "simplicity, simplicity, simplicity" (Thoreau 1854), and Leopold (1949), who called for a land ethic where humanity would use less resource exploitation to maintain the integrity of the environment.

Neoclassical Economists

Alfred Marshall was the leader of neoclassical economics that employed analytical geometry and differential calculus to develop abstract models of market economies. Neoclassical economy, using supply and demand curves, prices and quantities, and market equilibria, is the economic theory used today. These economists use empirical data of prices, costs, and other market-related phenomena (de Steiguer 1995) for their theories of supply and demand. "According to the theory, as producers and consumers meet in the marketplace, voluntary exchange prices are established to set production and consumption at optimal equilibrium levels. Furthermore, increasing prices spur discovery to new technologies and new materials, and they encourage efficiency of production and consumption. Any departure from the market system, according to the neoclassicists, results in a less than optimal allocation of resources, and likewise, lower human satisfaction" (de Steiguer 1995:553–554).

What is the Best Economic Theory?

de Steiguer (1995) addressed the question of what each of the economic theories (i.e., Malthusain, Mill, or the neoclassicists) contributes towards solving environmental problems. Malthus had a hypothesis that yielded a state of emergency to environmental problems: one that forecasts doom to society. Mill, on the other hand, presented a philosophy, whereas the neoclassicists have emphasized market-related issues (e.g., cost and demand). Their models have been useful in environmental issues (e.g., determine the economic importance of environmental damage, examine trade-offs required to control losses, and suggest specific policy) but have not been able to account for other important aspects of economics involving wildlife (e.g., existence value). This leads us back to the ideas of Mill (and Leopold) that humans are led by the heart and soul to make ethical and intelligent choices that supersede material goods (de Steiguer 1995).

However, when considering economics, most people are exposed to and take the neoclassical approach. Neoclassical theories are not applicable when wildlife is put into the equation, for at least eight reasons (Erickson 2000, Gowdy 2000, Hall et al. 2000).

1. Economic development often exploits resources, normally in a nonrenewable manner.

2. Natural scientists use models as tentative as ideas develop, data are collected, and new models are developed and old ones discarded. On the other hand, economic growth theory lacks hypothesis formulation and comprehensive testing. Natural scientists do not universally apply laws unless they have been tested. "Unfortunately, major decisions that affect millions of people and vast areas of wildlife habitat are often based on economic growth models that, although elegant and widely accepted, are not validated" (Hall et al. 2000:20).

3. The basic model of neoclassical economics views the economy as if it has no limits and does not consider biology, or the limits of the biological world, in the production of the raw materials required for growth. Should one resource decline, neo-

classical economists argue that alternate technologies will develop others (via higher prices).

4. The gross domestic product (GDP) (i.e., all wealth produced within a nation's boundaries) is used by economists as an indicator of the quality of life for humans. The GDP, however, is not a complete measure because:

a. It does not measure all conditions that relate to human well-being.
b. It does not indicate which consumer needs are being satisfied (e.g., required versus luxuries).
c. The GDP does not accurately measure production (e.g., a new dam will generate electricity that will be included in market evaluation but will ignore homes, societies, wildlife, and habitat for wildlife that is destroyed).
d. Nonmarket transactions are not evaluated (e.g., properly functioning ecosystems).
e. The GDP does not measure actual wealth enjoyed by citizens, only the flow of new wealth.

Furthermore, the GDP does not include costs for wildlife conservation and ecosystem functions (e.g., positive aspects of earthworms, birds, pollination, soil). Consequently, the GDP underestimates the influence of human activities on wildlife populations and vice versa. The result, especially in developing countries, is to encourage higher GDP at the expense of natural resources.

5. Investment policies based on neoclassical economics encourage developing countries to borrow from developed countries. However, this causes indebtedness that the developing countries reduce by spending their natural capital below its true value. Forest products in some tropical countries are sold for 50 to less than 10% of their market value (Repetto 1988). When this occurs, neoclassical economics can lead to environmental degradation and not even meet the primary objective of enhancing human life.

6. Discount rates (i.e., weighing the time value of money) value the near future more than the far future and are not advantageous to ecosystems and wildlife. Discount rates can yield increased money but nothing to invest in, and the destruction of natural systems for short-term, large profit investments. If systems are destroyed, money may be made in the short term, but not comparable to the long-term gains of a healthy ecosystem.

7. Geist (1988) argues that the success of wildlife management in North America was, in part, the result of the removal of markets for wildlife. As discussed earlier, market decisions have been detrimental to wildlife (e.g., passenger pigeon, buffalo, whales) and the market is not the correct way to make large-scale decisions (especially for wildlife).

8. Nonmarket values are important but are usually ignored or dealt with in neoclassical models in abstract ways. "A more holistic approach is required for the purposes of wildlife conservation" (Hall et al. 2000:23).

Hall et al. (2000) classified the alternatives to neoclassical economics as sustainable development (i.e., development without alteration or destruction of wildlife and its habitat), biophysical economics (i.e., understanding economics from a biophysical perspective that focuses on the land and its resources rather than a social perspective), and ecological economics.

Czech (2000a, 2000b) posited that ecological economics is the paradigm most conducive to wildlife conservation. Ecological economics embraces a broad interdisciplinary array of professionals that includes the natural and social sciences. Ecological economics intentionally connects with ecologists by using ecological terms and concepts (Czech 2000). Those that practice ecological economics consider it the science and management of sustainability (Costanza 1991). Ecological economists advocate a steady-state economy with a sustainable gross national product, "whereby stocks of natural capital (e.g., soil, trees, fish populations) remain constant in the long run" (Czech 2000).

Maintaining sustainable resources on private land is a serious challenge. The landowner needs to be considered and provided with some motivation for maintaining wildlife habitat. If not, a profit-motivated landowner may be better off financially reducing or eliminating habitat for wildlife (Rasker et al. 1992). This is not a new idea. The landowner has the habitat wildlife depend on and the landowner can most efficiently conserve it (Leopold 1933:395). Problems arise for landowners (and wildlife) when people trespass on private land because liability can become a problem (Hyde 1986, Wright and Fesenmaier 1988), when wildlife consume crops and cause other damage (e.g., broken fences) (Nielsen and Lytle 1985, Wade 1987), or when the returns from ranching and farming are more than the benefits received from preserving wildlife and wildlife habitats (Rasker et al. 1992).

Financial gain is not the only incentive for ranchers and farmers. They may want wildlife produced on their lands for an array of reasons (e.g., aesthetic appreciation, desire to hunt, pride and public recognition) (Rasker 1989). Kellert (1980) described nine general attitudes Americans had towards animals: naturalistic, ecologistic, humanistic, moralistic, scientific, aesthetic, utilitarian, dominionistic, and negativistic.

Naturalistic attitudes were held by those interested in wildlife and the outdoors, whereas ecologistic attitudes were held by those who had a primary concern for the environment as a system and were interested in the interrelationships between wildlife and their habitats. Those that have a strong affection for individual animals (principally pets) are humanistic, and those interested in the proper treatment of animals, with strong opposition to exploitation or cruelty to animals, are moralistic. Scientific attitudes were held by those with a primary interest in the physical aspects and biology of species. Those with aesthetic views are primarily interested in the aesthetic views and symbolic characteristics of animals, whereas utilitarian attitudes view animals for their practical and material values. Dominionistic views are held by those who want to control animals, especially in sporting events. People who have fear, dislike, or are indifferent to animals and actively avoid them have a negativistic attitude.

Regardless of the views landowners have about wildlife, society needs to be practical and not expect landowners to produce wildlife or protect its habitat without some type of motivation, especially if habitat fragmentation is the alternative. Rasker et al. (1992) outlined several contemporary concepts used to provide incentives to manage wildlife or private lands other than fees.

MANAGEMENT INCENTIVES

Land Acquisition

State, federal, and private departments and organizations have purchased private land to be held in the public trust. This activity requires sizable expenditures but can also ensure long-term productivity of the land with proper management versus short-term gains that landowners often require to make a living. As with other habitat management programs, cooperation among surrounding landowners may be essential to meet all the needs of wildlife. Cooperation and land acquisition incentives may also be induced by incentive arrangements or by government regulation. The Nature Conservancy has maintained numerous conservation easements that have banned mining and timber harvesting, and protected riparian areas and important elk, deer, and other wildlife habitats. Easements have been especially useful as an alternative to ranch subdivisions.

Farm Programs

Since the 1930s, the United States has provided incentives to increase agricultural productivity, often at the expense of wildlife. For example, the incentives often resulted in reduction or elimination of marshes, swamps, woodlands, or prairie grasslands and the establishment of agriculture monocultures.

By the1980s, there were agricultural surpluses, declining commodity prices, and increased international competition. These conditions have led to new policies designed to reduce agricultural surpluses, promote agricultural exports, and more importantly for wildlife, encourage conservation practices (Martin et al. 1988). The latest programs under the recent farm bills are beneficial to wildlife and provide opportunities for cooperation between farmers and wildlife interests.

Under the Conservation Reserve Program, for example, farmers can receive payment for retiring eligible cropland on a 10-year basis and plant the area with cover crops for wildlife (e.g., upland game, ground-nesting birds, grassland species, waterfowl). Millions of erodible hectares have been targeted for the Conservation Reserve Program and most have been enrolled into the program.

Taxation

Tax programs can also encourage conservation practices if they are well designed.

> "To illustrate how taxes can be implemented to alleviate market failures, imagine a situation where the actions of an individual result in a cost to others, such as a farmer draining a wetland valuable for waterfowl production. In order to equate social and private costs, a tax could be levied on the farmer equal to the amount of damage done to the public good, the waterfowl. Theoretically, the landowner will respond to avoid or minimize this additional cost. If the tax is large enough, the price of draining the wetland may be too high and it may be more profitable to leave it intact" (Rasker et al. 1992:345).

Another, more positive, way to encourage the maintenance and enhancement of wildlife is through positive tax incentives (e.g., conservation easements). Conservation easements allow the landowner to donate an easement for conservation to an organization in return for a tax reduction on the land. Limitations are put on land use that benefit wildlife (Small 1990) while the landowner enjoys income-tax savings.

Macroeconomic Approaches

Farm programs, tax incentives, and land acquisition are microeconomic tools. Czech (2000a, 2000b), however, suggested that a macroeconomic approach is ultimately required for wildlife conservation. Macroeconomic tools tend to comprise social and political movements and sweeping legislative reforms. For example, Czech and Krausman (2001) identified the Endangered Species Act of 1973 as a macroeconomic tool for wildlife conservation because, strictly enforced, the act would tend to replace an expanding economy with a steady-state economy.

There are numerous potential solutions to deal with the economics of wildlife besides commercialization or privatization of wildlife. It is important to recognize, however, that various programs may work well in some areas but not others. As with other aspects of wildlife management, "There is no single mechanism that will work in all instances" (Rasker et al. 1992:346). Keep an open mind!

It is encouraging that Americans have a desire to maintain wildlife resources. In a nationwide survey of public opinion toward species conservation and socioeconomic institutions that affect conservation, the availability of resources for future generations was rated highest among all categories (Czech and Krausman 1999) (including the maintenance of economic growth). That opinion is important because, "For biodiversity conservation in North America to succeed, the nonmonetary values placed on wildlife and natural ecosystems must form the backbone of our conservation ethic, policies, and practices" (Freese and Tranger 2000).

SUMMARY

The classical forms of economics do not apply to natural resources, including wildlife. A new brand of economics, ecological economics, is developing that has promise to treat natural resources in a manner so they can be sustained and not depleted.

18

Contemporary and Classical Issues in Wildlife Management

"Agriculture will not be able to support the majority of the people and innovative solutions will have to be found to enhance rural production without further environmental damage. The challenge is both economic and political, national and international."

G. Child

INTRODUCTION

Changes in the field of wildlife management have been well documented, but managers still continue to deal with some problems on a regular basis. Other issues are relatively new and theory and techniques are still being developed to better understand new conflicts. At the center of most conflicts are humans and their different desires to increase or decrease populations. Once that hurdle is cleared, the technical aspects of wildlife management can proceed. Unfortunately, it is not that way because humans are not unified in their desires. As a result, livestock grazing on public lands, native ungulates grazing on public lands, estimating biodiversity, maintaining biodiversity, ecosystem management, endangered species, exotics, international wildlife, fires, winter feeding, urban wildlife, nuisance birds and mammals, predators, wilderness, and just about any aspect of wildlife management has been, or will be, controversial. In this chapter I will briefly explore several issues in wildlife management that cut across views and values about wildlife that often make the issues controversial and management a challenge.

▦ EXOTICS

All species living outside their normal area of distribution are *exotics*: brown trout in British Columbia, chukar and ring-necked pheasant in North America, or Lehman's lovegrass in Arizona. Technically, domestic livestock, most humans, and animals in

331

zoos are exotics but the word is not usually applied to them. Humans, however, are the ones responsible for the distribution of animals outside of their normal ranges, whether intended or unintended (Mungall 2000). Some exotics have filled vacant areas and become important game species (e.g., ring-necked pheasant, chukar, ibex) in parts of the United States, whereas others have become pests and burdens to the system they were introduced into (e.g., English sparrows, starlings, Lehman's lovegrass).

There are numerous reasons people would want to associate with exotics but they boil down to four main topics: enjoyment or curiosity, revenue from fee hunting, preservation or introduction, and a willingness to provide homes for surplus animals (Mungall 2000).

Enjoyment or Curiosity

People are interested in animals from around the world and have been motivated to import animals for viewing, hunting, or simply enjoying different species. From the time of the ancient Egyptians to the development of zoos in the Americas, people have had an interest in learning about different species (Mann 1934). Those that had land (and money), and enjoyed seeing animals in areas that were not otherwise being used, imported species for varied viewing. Because native wildlife was controlled, exotics are often easier to obtain and work with than native game that is owned by the public and controlled by the government. There are broad areas in North America that do not have any or have few native species and some would like to have exotic cervids and/or bovids introduced (Decker 1978). In the United States numerous birds have been successfully introduced, including the ring-necked pheasant, Hungarian or gray partridge, and chukar partridge (Long 1981). Each of these has been classified as gamebirds. There are numerous private lands, especially in the Southwest, where exotics have been established.

Desire to Make Money

Exotics can be profitable as trophies (e.g., more than $1000 for axis deer or blackbuck antelope) or breeding stock (Mungall and Sheffield 1994) for those that have the infrastructure already established from livestock operations. The exotic operation is then able to supplement the livestock industry, especially when the profits from the latter are low. Exotics have become extremely popular in Texas for game ranching (of exotics), gourmet food in upscale restaurants, antler harvesting, and game farming. Their management is more along the lines of animal husbandry than wildlife management. And because the landowner owns the animals, they have flexibility in terms of what can be done and when, without government intervention.

Preserving Rare and Endangered Species

Some maintain exotics simply to be involved in conservation efforts to prevent animals from vanishing. Private breeding efforts on Texas ranches have been instrumental in enhancing captive populations of blackbuck antelope, Arabian oryx,

Père David's deer, and barasingha (Mungall and Sheffield 1994). The intent is to maintain these species until translocation is possible on their native ranges.

Other landowners have entered into cooperative agreements with zoos that have limited space but want to work with declining species. The American Zoo Association has a "Species Survival Program" that uses private ranches to perpetuate species that are hardy enough to exist on ranches. The Scimitar-horned oryx and Grévy's zebra were the first species released to ranches under the Species Survival Program. Animal brokers interested in enhancing rare species have seen how intensive management has benefitted some that are headed towards extinction (e.g., Sömmering's gazelle). However, they are frustrated that their efforts are often shunned by government regulations that preclude involvement from the private sector.

Surplus Animals

When zoos have more animals than they have space for, the excess may be euthanized if other options are not available. In some cases ranchers have opened their lands to these species. Many of these excess species were responsible for fueling the establishment of the exotic trade in North America (Mungall and Sheffield 1994).

Although these may be valid reasons to maintain exotics, and thousands of exotics exist in North America (Table 18–1), there are more reasons why exotics

Table 18–1 Origins, locations, and population estimates for major exotics (mammals) in North America[a] (from Mungall 2000)

Animal	Origin	American locations	Major North Estimated number
Deer			
Axis deer	India and Sri Lanka	California	350
		Florida	6500–9500
		Hawaii	4535–5235
		Hawaii	> 500
		Louisiana	700–800
		Mississippi	300
		Texas	55,424
Barasingha	India	Florida	50
		Louisiana	50
		Mississippi	16
Brocket deer	S. America	Texas	b
Chevrotain	Africa?	Texas	b
Chinese water deer	Asia	Florida	15–20
		Mississippi	9
Elk's deer	Asia	Texas	130
Elk, Rocky Mountain (wapita)	N. America	Hawaii	85

Continued

Table 18–1 Origins, locations, and population estimates for major exotics (mammals) in North America[a] (from Mungall 2000). *Continued*

Animal	Origin	Major North American locations	Estimated number
Fallow deer (Euro-pean and possibly a few Persian hybrids)	Europe and Asia Minor	Saskatchewan	600
		(Canada total)	(20,000)
		Alabama	200–300
		California	350
		Florida	2000
		Georgia	500
		Kentucky	200–300
		Louisiana	700–800
		Mississippi	26
		New York	> 2500
		Oregon	200–400
		Texas	27,177
		Utah	75
		Virginia	500
Hog deer	Asia	Texas	69
Muntjac	Asia	Florida	40–50
		Mississippi	12
		Texas	120
Père David's deer	China	Florida	70
		Texas	291
Red deer	Europe	Canada	5000
		Florida	4000
		Louisiana	100
		Mississippi	15
Red deer (other)	Europe?	Texas	4802
Red deer-elk hybrids	Hybrid	Florida	53
		Texas	3975
Rusa deer	Asia	Texas	23
Sambar	Asia	Florida	75–100
		Texas	82
Sika deer	Asia	Florida	1500
		Louisiana	75
		Maryland	4300
		Texas	11,966
		Utah	3
		Virginia	700
Other deer		Texas	949
Antelopes			
Addax	North	Florida	5
	Africa	Texas	1824

Table 18–1 *Continued*

Animal	Origin	Major North American locations	Estimated number
Indian blackbuck antelope	India	Florida	200
		Louisiana	300
		Mississippi	60
		Texas	35,328
Blesbok	Africa	Texas	57
Bongo	Africa	Florida	21
		Texas	1
Bontebok	Africa	Florida	16
		Texas	36
Bushbuck	Africa	Texas	b
Dik-dik	Africa	Florida	15–20
		Texas	37
Duiker	Africa	Texas	b
Eland, common	Atrica	Florida	50
		Mississippi	5
		Texas	919
Gazelle, dama	Africa	Florida	26
		Texas	91
Gazelle, dorcas	Africa	Texas	b
Gazelle, Grant's	Africa	Texas	104
Gazelle, Persian goitered	Asia	Texas	b
Gazelle, slender-horned	Africa	Florida	8
		Texas	b
Gazelle, Thomson's	Africa	Texas	143
Gazelles, other		Texas	99
Gerenuk	Africa	Florida	13
Hartebeest	Africa	Texas	b
Impala	Africa	Florida	13
		Texas	341
Kudu	Africa	Florida	13
		Texas	341
Lechwe	Africa	Florida	13
		Texas	537
Nilgai	India	Florida	40
		Texas	8153
Nyala	Africa	Florida	24
		Texas	49

Continued

Table 18–1 Origins, locations, and population estimates for major exotics (mammals) in North America[a] (from Mungall 2000). *Continued*

Animal	Origin	Major North American locations	Estimated number
Oryx, Arabian	Arabia	Texas	288
Oryx, East African (both beisa and fringe-eared)	East Africa	Texas	26
Oryx, scimitar-horned	Africa	Florida	40 – 50
		Mississippi	6
		Texas	2145
Oryx, South African (gemsbok)	Africa	New Mexico	2000
		Texas	456
Roan antelope	Africa	Florida	12
		Texas	[b]
Sable antelope	Africa	Florida	4
		Texas	215
Sitatunga	Africa	Texas	93
Springbok	South Africa	Florida	3
		Texas	87
Suni	Africa	Texas	[b]
Waterbuck	Africa	Florida	14
		Texas	432
		Utah	2
Wildebeest, black	Africa	Texas	[b]
Wildebeest, blue	Africa	Texas	[b]
Wildebeest, white-bearded	Africa	Texas	32
Wildebeest, other or unspecified	Africa	Louisiana	1
		Texas	166
Other antelopes		Texas	134
Cattle			
African Cape buffalo	Africa	Texas	6
Banteng and gaur	Asia	Florida	43
		Texas	22
Forest buffalo (dwarf buffalo)	Africa	Florida	21
Water buffalo	Asia	Florida	750
		Texas	30
Watusi	Africa	Florida	9
		Texas	370
Yak	Asia	Florida	2
		Texas	11
Zebu	India	Texas	38
Other buffalo		Texas	86

Table 18–1 *Continued*

Animal	Origin	Major North American locations	Estimated number
Goats and relatives			
Catalina	Domestic	Florida	170
		Texas	1714
Chamois	Europe	Texas	b
Ibex, Iranian (Persian ibex, wild goat)	Asia	New Mexico	600
Ibex, Siberian	Asia	New Mexico	75
Ibex, (unspecified)	Asia, some Africa?	Florida	10
		Texas	1042
Ibex hybrids	Texas	Texas	1697
Markhor	Asia	Texas	221
Tahr, Himalayan	Asia	California	> 200
		Texas	165
		Utah	3
Tur, Caucasian	Asia	Texas	b
Other goats		Texas	51
Sheep and relatives			
Aoudad (Barbary sheep)	Africa	California	1200
		California	Eliminated
		Colorado	120
		Florida	26
		New Mexico	4000
		Texas	12,292
Argali	Asia	Texas	b
Barbados	Africa	Texas	3917
Corsican sheep (mouflon hybrids)	Texas	Florida	500
		Texas	4953
Texas Dall	Texas	Florida	45
		Texas	b
Four-horned sheep	Asia and Europe	Florida	3
		Texas	491
Mouflon	Europe	Colorado	30
		Florida	40
		Texas	7574
		Utah	25
Red sheep	Asia	Texas	924
Urial	Asia	Texas	b
Other sheep		Texas	277

Continued

Table 18–1 Origins, locations, and population estimates for major exotics (mammals) in North America[a] (from Mungall 2000). *Continued*

Animal	Origin	Major North American locations	Estimated number
Pigs			
Wild boar, Eurasian[c]	Europe and Asia	Canada	2000
		Colorado	25
		Kentucky	300
		New Hampshire	> 1000
		North Carolina	700–900
		Tennessee	2300
		Texas	1417
Giraffes and relatives			
Giraffe	Africa	Florida	14
		Texas	98
Okapi	Africa	Florida	3
Camels			
Bactrian	Asia	Texas	16
Dromedary	Middle East	Texas	43
Camel (unspecified)		Florida	2
Llamas and relatives			
Alpaca	S. America	Texas	48
Guanaco	S. America	Texas	49
Llama	S. America	Florida	5
		Mississippi	?
		Oregon	18,000
		Texas	12
Vicuna	S. America	Texas	b
Other	S. America	Texas	1327
Zebra			
Zebra, Chapman's	Africa	Texas	75
Zebra, Damara	Africa	Texas	b
Zebra, Grant's	Africa	Texas	445
Zebra, Grevy's	Africa	Florida	11
		Texas	98
Zebra, mountain	Africa	Texas	c
Other zebras	Africa	Florida	9
		Texas	28
		Utah	1
Zebra-donkey hybrids	Hybrid	Florida	2
Tapirs			
Tapir, Malayan	Asia	Texas	b
Tapir, other	Central or S. America	Texas	b

Table 18–1 *Continued*

Animal	Origin	Major North American locations	Estimated number
Rhinoceroses			
Rhinoceros, black	Africa	Texas	13
Rhinoceros, white (squared-lipped rhinoceros)	Africa	Florida	5
		Texas	21
Other rhinoceroses		Texas	27
Hippopotamuses			
Hippopotamus	Africa	Texas	b
Hippopotamus, pygmy	Africa	Texas	2
Elephants			
Elephant	Asia or Africa	Texas	18
Other exotics (Chamois, suni, Malayan tapir and other tapirs combined for reporting privacy)		Texas	32
Totals	Canada	Canada	27,000
	U.S.A.	Alabama	200–300
		California	> 900
		Colorado	175
		Florida	16,487–19,542
		Georgia	500
		Hawaii	> 500–> 5235
		Kentucky	500–600
		Louisiana	1926–2126
		Maryland	4300
		Mississippi	457
		New Hampshire	> 1000
		New Mexico	6675
		New York	> 500
		North Carolina	700–900
		Oregon	18,200–18,400
		Tennessee	2300
		Utah	109
		Virginia	1200

[a]Some novel domestics are included.
[b]Source suppresses some Texas figures to safeguard owner privacy.
[c]In most, if not all, areas wild boar have interbred with feral swine. Feral swine are free ranging in at least 18 states, forming a population estimated at 500,000 to more than 1 million spread through the southeastern United States from Virginia south across Florida (36,000), west through Texas (nearly 121,700), into Arizona (200–400), and large concentrations in California (30,500) and Hawaii (> 100,000).

can create serious ecological problems. Mungall (2000) summarized the most common reasons for the opposition to exotic big game in North America (Decker 1978, Geist 1988).

1. Exotics can create problems if added to areas that are at carrying capacity. If at carrying capacity and new stock is added, the range will be abused if the animals use the same resources.
2. Native wildlife can be displaced by exotics. Exotics have the ability to outcompete native species (Feldhamer and Armstrong 1993, Demarais et al. 1998) and sika deer, axis, and fallow deer have outcompeted whitetails (Demarais et al. 1998). Bighorn sheep have also been outcompeted when exotics (i.e., barbary sheep) have been introduced into their habitat.
3. Exotics can disturb the ecological balance of a site away from its pristine or representative condition. Sites maintained as examples of certain types of habitat may be changed beyond recall with the introduction of exotics.
4. Managers should concentrate on indigenous species instead of trying to create biotic uniformity.
5. Money spent on exotics are resources eliminated from the management of indigenous species.
6. Exotics promote commercialization of animals normally considered wild, which erodes the basic pillars of wildlife conservation.
7. Exotics are difficult to control and at some time will escape (e.g., animals make holes, poachers cut wire, floods destroy fences). Some exotics can colonize rapidly and flourish even when they are not wanted (e.g., Reeve's muntjac spread throughout most of the South and the Midlands, England) (Harris and Duff 1970). In Texas, at least seven species of exotic ungulates have escaped from their enclosures and established wild populations (i.e., axis deer, nilgai antelope, blackbuck, aoudad, mouflon, sika deer, and fallow deer). Other mammals that were introduced to North America and established populations include the nutria, reindeer, and wild pigs (Moulton and Sanderson 1997).
8. Diseases and parasites can be introduced by exotics. There are serious regulations that have to be followed when importing exotics that have precluded many diseases, but these regulations have not applied to movement of exotics between private owners until recently. As a result, there is concern for disease transmission between exotics and native ungulates (Miller and Thorne 1993).
9. Exotics can alter the gene pool of native species by interbreeding and by creating hybrids. Hybridization has been a serious problem between mouflon sheep and wild sheep, red deer and wild elk, and wild boar and feral domestic hogs. Native populations are adapted to local selective pressures to function efficiently. Adding genes from other populations, species, or subspecies is not an advisable practice.
10. Exotics can alter the indigenous habitat of native species and can cause serious damage to forest crops and other forms of agriculture. Management of

exotics is difficult because there is no universal agreement as to how exotics should be managed in North America (White 1987). In most Canadian provinces exotics cannot be hunted but game farming is allowed (Haigh and Hudson 1993). In the United States some states have thriving deer farms (e.g., Washington, New York, Texas, Florida) or exotic hunting (e.g., Texas, New Mexico), but other states discourage any type of activity related to exotic ungulates (e.g., Arizona, Kentucky, Wyoming).

The management of exotics is a growing industry. Some of the issues associated with exotics are value judgments. However, if exotics are detrimental to native species and their habitats, the choices should be clearer. Mungall and Sheffield (1994) and Mungall (2000) provide more detail on exotics.

WILDLIFE-LIVESTOCK INTERACTIONS

Rangelands in the western United States provide habitat for more than 3000 species of mammals, birds, reptiles, fish, and amphibians. Rangelands are also shared with most of the cattle produced in the western United States. Managing livestock and wildlife on public lands in the West is one of the most challenging issues for managers of rangelands. To complicate issues further, the influences of livestock on flora and fauna are intuitively considered to be detrimental, but these relationships are not well understood (Krausman 1996). However, scientific organizations (e.g., Society for Range Management, The Wildlife Society, The Society for Conservation Biology) call for rangeland management that is governed by scientific inquiry and the application of those results to management problems. For example, The Wildlife Society Position Statement on Livestock Grazing on Federal Rangelands in the Western United States declares, "that properly functioning rangeland ecosystems supporting a wide diversity of native plant species are critically important to sustaining wildlife diversity and productivity in the American west. Scientifically sound management plans and practices are key to restoring lands degraded by many years of livestock grazing that damaged soils, water, and plant diversity." Most conservation organizations, agencies, and foundations have had to consider the influence of livestock on wildlife. The topic has been controversial since livestock were introduced to North America. Complete volumes (e.g., Krausman 1996) have been produced about the conflict and the issue is yet to be resolved to the satisfaction of everyone. To illustrate the concern, the following will briefly outline the history of livestock grazing in western North America and how livestock have influenced some wildlife on rangelands in North America (Table 18–2).

History of Livestock Grazing in Western North America

Christopher Columbus brought Moroccan cattle to the Carribean in the early 1500s, and some of these cattle were eventually transported to Mexico (Herndon 1984). In the mid-1500s, Spanish settlers brought a hardy breed of cattle to the new world,

Table 18–2   Area and percent of rangeland in federal ownership in the United States and adjacent parts of Canada and Mexico (Holechek et al. 1989)

	Area (ha) millions	Federal ownership (%)
Grasslands		
Tallgrass prairie	15	1
Southern mixed prairie	20	5
Northern mixed prairie	30	25
Shortgrass prairie	20	5
Palouse prairie	3	15
California annual grassland	3	6
Alpine grassland	4	99
Desert shrublands		
Sagebrush/grassland	39	65
Salt-desert shrub	35	85
Mojave	14	55
Sonoran	27	55
Chihuahuan	45	55
Woodlands		
Pinyon-juniper	17	65
Mountain shrub	13	70
Western coniferous forest	59	62
Southern pine forest	81	6
Eastern deciduous forest	100	7
Oak woodland	16	4

which grazed in the future state of Texas (Herndon 1984). The descendants of these cattle were the Texas longhorn. The cattle were used for hides and tallow because there was no easily accessible market for the meat (Ferguson and Ferguson 1983). Later, cattle were used to feed people on army posts and Indian reservations. During the gold rush of 1848–1849, western cattle fed many of the miners on their way to, and in, California; some miners also brought their own cattle from the East. The Civil War blocked cattle drives east from Texas and other western regions, so 4.5 million head of cattle accumulated in Texas, to be shipped after the war to stock other ranges (Wilkinson 1992). The first great United States investments in western cattle took place after the Civil War (Ferguson and Ferguson 1983, Wilkinson 1992). Wealthy investors from Europe and the eastern states invested in cattle. With the completion of the transcontinental railroad in 1865, large numbers of cattle were shipped to eastern markets from western rangelands. In the 1870s–1880s, cattle kingdoms with as many as 75,000 head of cattle were established to meet the demand for cattle products.

By the 1880s, problems had begun to develop on the range (Ferguson and Ferguson 1983). Cattlemen had originally expanded throughout the West with no restrictions on their use of the land. With the passage of the Homestead Act of 1862, other settlers began to move west, settling the rich bottomlands and planting crops (Wilkinson 1992). This restricted vast herds of cattle to the drier, less fertile soils of the western deserts and the Great Basin. In 1886–1887, the West experienced one of the worst winters in many ranchers' memory. Thousands of cattle died; 100,000 head were lost by one company alone. In Montana, 70% of all the cattle were lost; throughout the West one-third of all cattle died (Stuart 1925). When spring came, the carcasses of dead cattle were everywhere. Cattle had come into towns and attempted to eat the tarpaper off of houses, and anything else they could reach. One witness observed that he "could stand at one carcass and throw a rock to the next one" (Ferguson and Ferguson 1983). The few survivors of the winter were scattered and weak. In short, the winter of 1886–1887 was a disaster from which many western ranchers never fully recovered, but no changes were made in the way ranchers grazed their cattle on western rangeland. Overgrazing turned much of the rangeland into large fields of dust (Wilkinson 1992).

In the 1930s, drought and depression convinced livestock operators they would have to do something drastic to save their industry. In 1934, Congress passed the Taylor Grazing Act. The House of Representatives passed the Act on 11 April, but it took a storm blowing dust from the western rangelands all the way to Washington, D.C, to convince the Senate to pass the Act, which they did on 12 June (Sharp 1984). The purpose of the act was ". . . to stop injury to public grazing lands by preventing overgrazing and soil deterioration, to provide for the orderly use, improvement and development of rangelands, and to stabilize the livestock industry dependent on the public range." The Act was first administered under the Division of Grazing Control formed in the Department of the Interior (Wilkinson 1992). This became the Grazing Service in 1939, and the Division of Range Management in the Bureau of Land Management in 1940. The secretary of the interior was given permission to establish grazing districts and to specify the number of livestock and season when grazing would be permitted in each district. The range was overstocked, so some ranchers had to leave. Priority to stay was given to local residents and owners of the land and water who had used the land in the past. Advisory boards were set up to assist in the functioning of grazing districts (Sharp 1984).

Adjudication of grazing districts took until the 1960s to complete, for a variety of reasons (Sharp 1984). The Division of Grazing Control had a limited number of employees and a small budget. During World War II, the number of livestock was increased to feed troops. Finally, conflicts over raising grazing fees slowed the process. After the completion of adjudication, the Taylor Grazing Act focused on developing grazing plans such as rest rotation, deferred rotation, and others.

Since the inception of the Taylor Grazing Act, policy concerning livestock grazing on public lands has been a topic of great controversy. As the Act itself was debated, western cattlemen protested that they wanted to control the future of western rangelands, not leave it in the hands of eastern politicians (Boyd 1984). At present, about 70% of the land in the 11 western states (e.g., Washington, Oregon, California, Idaho, Colorado, Montana, Wyoming, New Mexico, Arizona, Utah, and Texas) is grazed by livestock, making it the most widespread influence on native ecosystems in North America (Wagner 1978) (including nongame upland game-birds, and wetland birds; rodents and lagomorphs; predators; endangered species; and elk and other ungulates).

Nongame Birds

Over 90% of the native birds in North America are nongame and birds are influenced by livestock, depending on their habitat requirements and life history. Aquatic species are not generally influenced by livestock, birds of prey are influenced because grazing can influence densities of their prey, wetland-associated species can be influenced by livestock altering nesting cover, and livestock can alter the food and cover of terrestrial species (Knoph 1996).

Nongame birds generally respond to the influences livestock have on native vegetation, especially for species that are dependent on herbaceous and shrub layers. The response of different species is highly variable between sites, years, and species. Although grazing has been criticized as detrimental to many nongame birds, the scientific data are lacking or flawed because of poor study design (e.g., no controls or replications), study sites selected to represent grazing that actually represent severely overgrazed areas, and investigator advocacy for a natural resources agency (Knoph 1996) or policy.

Upland Gamebirds

The literature about the influence of livestock on upland gamebirds has also been confounded with unsupported conjecture, mixed research results, and conflicting recommendations (Guthery 1996). To obtain an objective approach to the question of how livestock influences upland gamebirds, Guthery (1996) proposed the following theory.

> "The optimum approach to grazing in any environment is the approach that maximizes space-time by optimizing structure of herbaceous vegetation for the gamebird in question, given that site integrity is maintained. Because sites differ in primary productivity and vegetation structure, optimal grazing may range from no use to heavy use of forage by livestock" (Guthery 1996:64).

His theory is based on five broad principles.

1. Upland game birds are adapted through natural selection and evolution, to habitat with a certain structure (e.g., height, density, biomass).
2. Primary production influences structure of vegetation and primary production varies with climate, weather, and soils. In general, growth and biomass of herbaceous vegetation are greater in rich environments (e.g., longer growing seasons, higher precipitation, higher soil fertility) than in environments with shorter growing seasons, lower precipitation, or lower soil fertility.
3. Structure and composition of herbaceous vegetation are confounded with each other. Biomass generally increases with each seral stage (e.g., late seral

stage with tall grass would have more biomass than early seral stages of annual forbs).

4. Structure of vegetation, not composition, should serve as the keystone to grazing management decisions.
5. Within specific sites, gamebird abundance is positively correlated with usable space in time.

Application of these principles to the theory can guide livestock-gamebird habitat decisions (Table 18–3) that are beneficial to livestock and gamebirds with proper management.

Table 18–3 Grazing management issues related to major upland gamebirds in the United States and Canada (Guthery 1996)

Group Species	Grazing management issues
Grouse	
Sage grouse	Contextual[a]
Blue grouse	Peripheral[b]
Spruce grouse	Nongermane[c]
Ptarmigan	Nongermane
Greater prairie chicken	Contextual
Lesser prairie chicken	Contextual
Sharp-tailed grouse	Contextual
Ruffed grouse	Peripheral
New World quails	
Mountain quail	Peripheral
Scaled quail	Contextual
Gambel's quail	Contextual
California quail	Contextual
Bobwhite	Contextual
Harlequin quail	Contextual
Old World partridges and pheasants	
Gray partridge	Contextual
Chukar partridge	Contextual
Turkeys	
Wild turkey	Contextual
Pigeons and doves	
Band-tailed pigeon	Nongermane
Mourning dove	Contextual

[a]Grazing by cattle can be negative, neutral, or positive, depending upon the habitat context within which it is applied.
[b]Grazing by cattle is a minor management concern or irrelevant over most occupied range of the species in question, but grazing could have negative, neutral, or positive effects under local conditions.
[c]Because of the distribution of the species in question or the type of habitat it occupies, grazing by cattle is not a relevant habitat management consideration.

Figure 18–1 Some wildlife associated with wetlands.

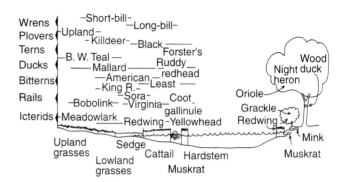

Wetland Birds

Wetlands are often small but important for avian diversity and biological productivity (Figure 18–1). Wetlands also provide forage and water for livestock, which have the potential to modify avian habitats. Damage to the habitat of wetland birds by livestock is most common and creates the majority of problems and conflicts when jointly managing wetlands for wildlife and livestock (Weller 1996). Braun et al. (1978) reviewed 55 studies that suggested livestock grazing on waterfowl was negative (e.g., nest trampling, lower nest density). Others have noted different influences of grazing on duck habitat for different species. Those birds that favor dense terrestrial vegetation during nesting would be influenced the most (e.g., dabbling ducks) (Weller 1996).

Even moderate grazing can reduce bird species richness, population size, and nest success of birds that depend on wetlands (Fig 18–1) and the habitats adjacent to them. Grazing of wetlands should be managed with concern, knowledge, planning, and regular monitoring to protect the integrity of wetland functions (Weller 1996).

Rodents and Lagomorphs

There are numerous rodents and lagomorphs that have been influenced by livestock grazing (Table 18–4) and in turn influence rangelands. Prairie dogs colonize sites that have been overgrazed, so livestock grazing can promote high prairie dog colonies, which expand under heavy grazing. Because prairie dogs depend on sight to avoid predators, they further clip vegetation around their burrows. They also dig extensive burrow systems, disturb soil, and promote growth of disturbance-oriented vegetation. Their activities alter the habitat toward short grasses and annual forbs instead of taller grasses and shrubs (Fagerstone and Ramey 1996). They also change rangelands by decreasing forb and grass cover in colonies, causing higher silicon concentrations in grasses found in areas they graze, and removal of plant biomass. Positive influences include increased plant species diversity and increased forage quality inside colonies, and they are a valuable prey item (e.g., black-footed ferret).

Ground squirrels also have positive effects on rangelands by deepening soils, causing soil mixing, and improving the water-holding capacity of soils. However, they are of-

Table 18–4   Numbers of species of some native lagomorphs and rodents that occur at various latitudes in the central part of North America (Jones and Manning 1996)

Orders	Manitoba	North Dakota	South Dakota	Nebraska	Kansas	Oklahoma	Texas
Lagomorpha	4	5	5	4	5	4	4
Rodentia	32	32	33	34	33	44	63

ten considered pests by livestock operators because they consume forage that could be used by livestock. In reality, there are little data available that support the negative influence of ground squirrels on livestock (Fagerstone and Ramey 1996).

Pocket gophers, on the other hand, can reduce the standing biomass by 20% or more by consumption and may alter plant composition and abundance. They also influence vegetation survival, growth, and biomass by creating soil mounds, feeding tunnels, and underground food caches. By creating tunnels and mound building, the pocket gophers can cause bare ground on 5 to 15% of the ground surface. This is partially compensated for by burying vegetation, which continually provides germination sites for early successional annual plant species (Fagerstone and Ramey 1996).

Lagomorphs have also been controversial when they share habitats with livestock. The niche for jackrabbits and cottontails is under study, as additional data are needed on the interactions between lagomorphs and livestock (Fagerstone and Ramey 1996). Their role in the environment is likely far more positive than any detrimental influences they have on livestock.

Predators

One of the biggest ongoing debates in wildlife management is the influence predators have on livestock. Predators kill approximately 2.5% of the adult sheep and 9.0% of the lambs annually in the western United States. Most adults and lambs (i.e., more than 74%) are killed by coyotes and others are killed by dogs, red foxes, mountain lions, black bears, grizzly bears, gray wolves, and bobcats (Andelt 1996). Some claim this is a small price to pay for using the habitat of predators. The livestock industry has a different view.

This controversy has led to an array of techniques to prevent predation (e.g., nonlethal: confinement, disposal of livestock carcasses, use of herders, fencing, guard dogs, frightening devices; and lethal: trapping, snaring, denning, shooting from the air and ground, and livestock protection collars) and an array of public acceptability (Figure 18–2) (Andelt 1996). "This problem likely will be minimized through effective public education programs, the adoption of nonlethal and more humane lethal control techniques, and the development of new and improved techniques. The trend away from predator population reduction method is likely to continue, especially when considering current public sentiment, predator population dynamics, costs, and environmental hazards of control techniques" (Andelt 1996:151).

Figure 18–2 Range of public acceptability for nonlethal and lethal predator control techniques (from Andelt 1996).

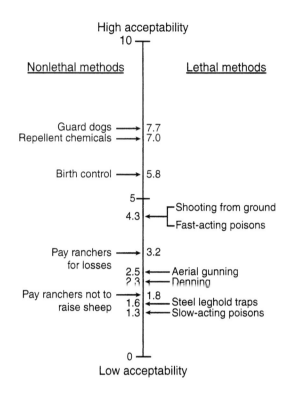

Endangered Species

The influence of livestock on endangered species was appropriately summarized by Carrier and Czech (1996:45–46).

> "The livestock industry has come as a shock to many American ecosystems, endangering species in the process. Species have suffered direct and indirect effects from livestock. Direct effects have included predator and rodent control, reduction of competitors, and accidental mortality. Indirect effects have included trophic linkage, disease and internal parasitism, external parasitism, and chemical contamination. Classic examples of species that have suffered the direct effects of livestock production are the wolf, prairie dog, and desert tortoise. Those that have suffered indirect effects include the black-footed ferret, willow flycatcher, and California condor. Some species suffer direct and indirect effects.
>
> The 'peaceful coexistence' of species in competition with each other is an oxymoronic ideal, as is the coexistence of domestic livestock and many native species. Ecosystems that appear relatively stable in the managerial time frame are actually dynamic in the evolutionary long term (Botkin 1990). The argument that endangered species will or should evolve to conform with modern conditions is unreasonable, because the molecular clock of evolutionary supply does not run at the speed of industrial demand. As Soule (1983:115) pointed out, 'we are dealing with the problem that the rate of environmental change . . . is several orders of magnitude higher than the rate at which genes are substituted in . . . vertebrates.'

Hopefully, the livestock industry will soon accept its share of the responsibility to perpetuate native ecosystems, including the full array of native species. After all, industry is a collective form of citizenship, and citizenship involves an ongoing effort to synthesize questions of 'what is best for the world' with 'what is best for me' (Landy 1993:20). The American public's desire for the conservation of endangered species is well documented in federal and state legislation. If this desire is not acknowledged by the livestock industry, the public will conclude that livestock interests view public lands as resources devoted to what is best for them; i.e., that the livestock industry is not a good citizen. This may turn the political tide in favor of species conservation, but by then, many more species may be extinct."

Elk

Present-day conflicts between domestic livestock and wild ungulates on North America's western rangelands encompass biological, cultural, and political spheres. For example, in Montana, elk migrate in winter from publicly owned lands at high elevations, to privately owned land at lower elevations that are already in use for agriculture, livestock grazing, and timber harvest (Cool 1992). Elk must then compete with humans and domestic livestock for use of the land. Resource managers and hunters advocate managing the land to maintain elk numbers; nonhunter urbanites also tend to support the presence of native wildlife on these lands. Ranchers and others who support themselves financially by raising livestock or harvesting other types of crops from the land, however, feel unfairly pressured to make economic sacrifices to preserve elk habitat (Little 1992). Thus, biological decisions made to benefit native wildlife, viewed by different cultures (e.g., urbanites, hunters, ranchers) have become a center for political controversy. Fee hunting, which has been suggested as a means of creating revenue to ranchers who have elk and other wildlife on their lands, has not been financially successful, and has opened up new debates about the ideology of letting private landowners manage wildlife (Benson 1992, Lacey et al. 1993).

My objectives in the remainder of this section are to bring perspective to the problem of livestock and wildlife interactions by describing elk-livestock interactions to discuss the conflicts involved, evaluate current methods of dealing with these conflicts, and make management recommendations that may minimize conflicts between wild ungulates and livestock in the future.

Grazing by livestock has been so prevalent since the mid-1800s that few accurate benchmarks remain to judge the influence livestock have had on western ecosystems (Fleischner 1994). The different ungulates in ecosystems are affected differently by grazing and the range of these effects is great. Because some of the biggest controversies swirl around the elk-livestock interactions, I will use elk as an example of conflicts between ungulates and livestock.

Elk and livestock are large, generalist herbivores that may compete for resources in areas where both species coexist (Vavra et al. 1989). Elk and livestock often make similar dietary choices and can obtain high nutritive value from grasses, but can also make use of a large variety of forbs and shrubs (Nelson and Leege 1982). In addition, they use similar habitat (Nelson and Leege 1982, Nelson 1982).

Simply occurring in the same area does not make two species competitors, even if they use the same resources (Birch 1957). Individuals of the same or different species must use a common resource that is in short supply (e.g., exploitive competition), or must use a common abundant resource and harm each other in the process (Nelson 1982) for competition to occur. The latter form of competition can occur as aggressive defense of a territory (e.g., *interference competition*) or as intimidation or avoidance of one species by another, which ends in one species leaving the area (e.g., *disturbance competition*). As a result of competition, one or both species reduces the other's population size to levels below what would exist without the other species' presence (Odum 1971).

It is difficult to determine the amount of competition that occurs between elk and livestock (Vavra et al. 1989). The potential for competition varies with habitat, forage availability, length of grazing season, grazing systems, time of year, feeding ecology of elk and livestock, population structure of elk and livestock, and spatial distribution of animals (Nelson 1982, Nelson and Leege 1982, Vavra 1992). The spatial distribution of elk and livestock may be particularly important because the two species have not coevolved (Odum 1071, Vavra et al. 1989). Coevolved species may partition their use of a habitat to avoid competition, but with an introduced exotic like livestock, this rarely occurs. A number of trends have been observed in elk-livestock interactions. Each of these will be discussed following a brief discussion of the history of elk in western North America.

Brief History of Elk in Western North America

Elk were among the most common, widely distributed ungulates before the arrival of European settlers. An estimated 10 million elk roamed North America from the Atlantic to the Pacific coasts (Seton 1929). Six subspecies of elk encompassed the species' range. Merriam elk were found in Arizona and New Mexico; Tule elk existed in southern California; Roosevelt elk dominated northern California, Oregon, and Washington; Rocky Mountain elk prevailed in eastern Washington and Oregon, and in Nevada, Utah, Idaho, western Montana, and Wyoming; Manitoban elk were found east almost to the Mississippi River; eastern elk inhabited eastern North America. Elk were absent in the driest portions of the Great Basin and desert southwest (McCabe 1982).

By 1900, as a result of uncontrolled hunting and competition with livestock and disease, the distribution of all elk subspecies was drastically reduced, and Merriam's elk had been exterminated (O'Neil 1985). From the turn of the century until 1940, elk were live-trapped and transported from the Yellowstone area for release in 36 states (Murie 1951). The present distribution of elk includes Rocky Mountain elk in western Montana, Wyoming, Utah, western Colorado, northeastern Arizona, and northwestern New Mexico; Roosevelt elk in northern California, western Oregon, and Washington; and pockets of Tule elk in central and southern California (Wisdom and Thomas 1996). Elk numbers were estimated at more than 700,000 in 1995 (Wisdom and Thomas 1996). In some states, there is now concern that there are too many elk, and steps are being taken to reduce elk numbers (Ogden and Mosley 1989).

Historically, elk made expansive, mostly elevational seasonal movements in response to weather changes and forage availability. Elk migrated to higher elevations
as summer progressed, where forage growth occurred later and quality of forage remained high into late summer. In winter, they migrated back to lower elevations to
escape snow and harsh weather (Vavra 1992). Ninety percent of the land traditionally used by elk in summer is public land (Wisdom and Thomas 1996). Most winter
ranges are privately owned and are devoted to grazing livestock, agriculture, or timber harvest (Lyon and Ward 1982, Vavra 1992).

Elk-Livestock Interactions

An indirect effect of elk and livestock using the same rangeland is a reduction of
plant vigor, which in turn affects the amount and quality of forage available (Mackie
1978). Reproductive plant parts, young plants, and rare but locally important plants
are also adversely affected by dual grazing (Mackie 1978). In eastern Oregon, the
frequency of grasses decreased, whereas the frequency of forbs increased under
heavy dual grazing by elk and livestock (Krueger and Winward 1974). Generally, the
impact on herbaceous vegetation was greatest when cattle and big game grazed a
common range.

The potential for competition is highest at all times on unproductive rangelands, especially in arid ecosystems (Fleischner 1994, Wisdom and Thomas 1996).
On productive rangelands, the potential for competition between elk and livestock is greatest on winter and spring/fall ranges where forage quality and/or
quantity is limited and where both ungulates share small ranges on low-elevation
bottomlands or foothills (Mackie 1978, Nelson 1982). Rocky Mountain elk in
northwestern Colorado avoid snow at high elevations by migrating to lowlands in
winter, where they compete with cattle for forage (Hobbs et al. 1996*a*). The supply of forage available to cattle declines with increasing elk numbers. This causes
a reduction in cattle production, but the magnitude of the effects is not propor-
tionate to elk density. If a sufficient amount of forage (greater than a certain
threshold) is available to cattle following elk grazing, elk populations do not harm
cattle production (Hobbs et al. 1996*b*).

Competition between elk and cattle is generally lowest on high-elevation summer ranges where forage of moderate-to-high quality is available and where animals
have a greater land base on which to graze (Wisdom and Thomas 1996). Summer
habitats must provide critical nutrition to allow elk to put on fat stores for winter. Females must create fat reserves to deal with reproduction and postnatal growth
(Baker and Hobbs 1982). Depending upon weather conditions, the summer period
of high-nutrition forage may be brief (Baker and Hobbs 1982). The potential for
competition increases during later summer and fall on high-elevation ranges, especially if there is seasonal drought or other weather abnormality. This is a critical time
for elk, which are preparing for a winter of less nutritious forage, and for livestock,
which are culled for meat production (Hobbs et al. 1979).

Competition between elk and livestock due to the similarity of their diets varies
throughout the year. In summer in Montana, elk diets are composed of 75% forbs,

10% grasses; cattle diets are composed of 33% forbs, 64–71% grasses (Stevens 1966, Mackie 1970). Competition for a few preferred species may be greater than implied by the above percentages (Stevens 1966). Forage quality is highest in spring and early summer and declines sharply as the quantity of forage nutrients is diluted by the production of structural carbohydrates as plants grow (Hobbs et al. 1979). In winter, diet quality continues to decline, as nutrients are leached out of dead grasses and forbs. Elk compensate by altering their diet (Hobbs et al. 1979) and shifting their choice of forage from forbs to grasses, which creates more dietary overlap with cattle (Nelson and Leege 1982). When grass becomes less available, elk switch to shrubs. As snow depth increases, elk eat conifers and lichens. Where browse or forbs are not available, however, elk and cattle may both concentrate on grasses, increasing competition (Nelson 1982).

Despite the great potential for competition between elk and livestock, competition is often minimized because elk and cattle distance themselves from each other spatially and temporally (Vavra et al. 1989). In southeastern Wyoming, cattle graze on lowland range sites, whereas elk prefer upland sites (Hart et al. 1991). Cattle in Riding Mountain National Park, in Manitoba, use only 10–15% of the park; elk use almost the entire park (Nelson 1982). Elk also make more use of steep slopes, forest edges, and areas far from water than do livestock (Nelson 1982). Cattle are less disturbed by roads than elk, which avoid roads (Ward et al. 1973, Nelson 1982). Finally, elk winter in foothills and lowlands, whereas livestock use these areas in spring and fall (Nelson 1982).

Although exploitive competition may be avoided by elk and cattle distancing themselves from one another, elk may suffer from disturbance competition as a result of cattle grazing. In many cases, elk assume a subordinate role to cattle; elk use of an area has been observed to decrease as cattle use increases (Skovlin et al. 1968, Wallace and Krausman 1987). Yeo et al. (1993) found that elk in central Idaho preferred rested pastures over pastures grazed, for all or part of the grazing season, by livestock. Elk gradually adjusted to the presence of cattle, making more use of pastures with livestock in later years of the study, but preferences for pastures without cattle continued. In a Montana study, only one of 29 radio-collared elk spent more than a few days among cattle (Lyon 1985). Other studies, however, have disputed the claim that elk suffer from being in proximity to cattle. In southeastern Wyoming, elk did not avoid cattle, even when to do so would only have involved moving a short distance (Ward 1973). Frisina (1992) found elk to be tolerant of cattle, and hypothesized that removal of vegetation by livestock had a greater impact on elk than social intolerance of elk for cattle. On winter ranges, cattle and elk are often found feeding together, leading to speculation that features of the habitat, season of the year, cattle stocking densities, breed of cattle and subspecies of elk may all influence the degree and type of social interactions between these two ungulates (Nelson 1982).

Current management for elk and livestock

The Great Basin and desert southwest have been historically overgrazed, and are largely in poor condition, so these ranges will only tolerate low stocking rates of elk and cattle (Holechek et al. 1989, Wisdom and Thomas 1996). On productive rangelands, a number of factors have been used to determine ecologically sound levels of grazing by elk and livestock. Equations have been developed that address the consumption equivalence, diet similarity, forage availability as influenced by potential competition, height of reach for browse, and distribution patterns in a habitat (Nelson 1982, Vavra 1992). Timing of range use and social interaction factors have received less mathematical attention, but are also used to determine the number of ungulates that can graze together on a range (Nelson 1982). The equations used are often simplistic, and do not deal effectively with the complexity involved in each factor (Vavra et al. 1989). To further confuse the issue, the allocation of ungulates to a range is often referred to in relative terms such as "light," "moderate," or "heavy" stocking rates (Fleischner 1994). The simplistic means by which stocking rates are reached and the ambiguous terms used to define these rates make determining appropriate rates difficult, but research has led to some guidelines.

A knowledge of the nutritional requirements of elk and livestock can help to determine the proper stocking rates and grazing systems for a particular area. Crude protein and dry matter digestibility are most often evaluated as indices of the nutritive value of forage consumed by ruminants (Mautz 1978). Cow elk require at least 5% crude protein in their diet all year; this increases to 11% during lactation in spring and summer (Nelson and Leege 1982). Elk requirements for digestible forage have not been studied at length, but 55% dry matter digestibility has been deemed adequate for adult elk, given that crude protein requirements are met (Wisdom and Thomas 1996). The nutritional status of available forage should be evaluated in conjunction with the quantity of available forage, because elk and livestock will disproportionately choose nutritious forage and if there is only a small amount of the nutritious forage available, evaluations based only on nutrient content of the forage are misleading (Baker and Hobbs 1982).

Different range management systems will have varying effects on forage quality. Light stocking rates of livestock have been found to have positive or neutral effects on elk. On productive Montana rangelands, light cattle grazing reduces litter cover, increasing the palatability of major forage species (Alt et al. 1992, Frisina 1992). Jourdonnais and Bedunah (1990) also found that cattle grazing removed litter, while at the same time maintaining a cover over the soil surface to prevent erosion. Hobbs et al. (1996*a*) found that elk grazing in northwestern Colorado increased the digestibility of perennial grasses available to lightly stocked cattle. Moderate grazing by cattle may have positive, neutral, or negative effects on elk, depending upon the timing of grazing and the productivity of the site (Wisdom and Thomas 1996). High levels of cattle grazing always create competition for forage, as do large numbers of elk. Many ungulates grazing any rangeland will lower the nutritive value of the forage and eliminate key forage species (Krueger and Winward 1974, Mackie 1978).

Since the late 1960s, the pros and cons of different grazing systems have been debated. Holistic resource management (Savory 1988) has gained worldwide attention, and rest-rotation grazing (Hormay 1970) has been advocated by range managers in western North America. Coordinated resource management planning is recommended by Anderson (Anderson and Scherzinger 1975, Anderson et al. 1990*a,b*) as a system wherein the time that high nutrient levels are present in forage is extended. None of these systems have been adequately tested; most studies of the effects of grazing have concentrated on grazed versus ungrazed systems.

In Montana, however, rest-rotation grazing has been implemented by a number of ranchers, with the cooperation of the Montana Department of Fish, Wildlife, and Parks. Frisina (1992) describes a prototypical Montana rest-rotation grazing system as being composed of three treatments. In the first treatment, pastures are grazed continually by livestock; in the second treatment, pastures are grazed by livestock only after seed-ripe (mid-October); no livestock are grazed in the pastures of the third treatment. All three treatments are always available to elk and other wild ungulates. The concept of this system is that treatments two and three provide vegetative rest. Treatment two also provides trampling action after seed-ripe, which allows microenvironments that are conducive to seed germination to be created. Preliminary results indicate that treatments one and two negatively affect elk use of pastures in the present season, but they establish higher-quality forage for the following spring. Under proper stocking rates of cattle and elk, treatment three can be used by elk to compensate for lower use of treatments one and two. Similar grazing systems in Montana have resulted in an increase in elk numbers (Frisina and Morin 1991) and an additional month of livestock grazing (Alt et al. 1992). In northeastern Oregon and southeastern Washington, grazing system and cattle stocking rate interacted to influence elk use of ranges. Under light cattle stocking, elk preferred season-long grazing, but under heavy cattle stocking, they preferred deferred rotation (Skovlin et al. 1968).

Prescribed burning has been used in conjunction with dual grazing of ranges by elk and livestock to reduce litter and improve the nutritive value of forage. Fire treatments in eastern Montana consumed heavy litter accumulation more effectively than cattle grazing, increasing early spring crude protein levels in forage plants (Jourdonnais and Bedunah 1985). Fall burning did expose soil surfaces to wind erosion, however. The long-term effects of fire may be minimal; nutritive value of burned forage declines two years after burning (Nelson and Leege 1982).

Water developments are used, especially on arid rangelands, to distribute cattle more evenly over an area (Holechek et al. 1989). Lack of water traditionally partitions elk and cattle use of an area; cattle stay closer to water than do elk. By extending the time span and area that cattle can graze, supplemental water may increase the potential for competition between elk and cattle (Mackie 1978). Water developments may favor cattle to the detriment of elk, because developments are often placed near roads, which are avoided by elk (Ward et al. 1973, Nelson 1982). In addition, elk often avoid cattle (Wallace and Krausman 1987, Hart et al. 1991, Yeo et al. 1993); if cattle are more evenly distributed, elk will be more limited in their choice of rangelands.

Suggestions for Future Management

The traditional paradigm of resource management, including the management of wildlife, has been to manage for multiple use and sustained yield of resources (Cortner and Moote 1994). Under this paradigm, habitat protection for a diversity of wildlife is often viewed as a constraint to realizing management goals. In recent years, a new paradigm, ecosystem management, which is concerned with preserving the complex interactions that drive ecological processes and individual species, has gained attention. Ecosystem management demands that interactions between live-stock and wild ungulates be viewed as a complex of interactions, in which the alteration of any one part greatly affects all others and the whole range (Mackie 1985). The dynamics of these interactions vary from location to location; rangeland prescriptions for the desert Southwest must necessarily differ from those for the Pacific Northwest. Therefore, each area must be treated as a unique situation with its own requirements (Mackie 1978).

Three important components of ecosystem management are public involvement; conservation partnerships among state and federal agencies, private landowners, and conservation organizations; and good working relationships among land managers and research scientists (Cortner and Moote 1992). The probability of conflicts among range users is high where wild ungulates and livestock share the range. Involved parties must build partnerships and work together to solve problems on private and federal rangelands (Cool 1992). This is happening in Plumas National Forest, in the California Sierra Mountains, where the forest service, Pacific Gas and Electric, California Department of Forest, and private landowners are working to restore Clark's Creek watershed (McLean 1994). They have collectively decided to make commodity outputs such as timber and cattle sales secondary to this goal. Projects such as this one may enable livestock and wildlife to coexist in perpetuity.

◼ TRANSLOCATIONS

Restoring native populations to historic range has been one of the success stories of wildlife conservation in North America for some species. However, the process is difficult, expensive, and often controversial (Nielsen 1988). Many early attempts to move animals were for reasons other than conservation, were not well thought out, and resulted in failure. Many of these early releases were *introductions,* (i.e., the release of free-ranging, wild animals to an area where the species does not or has not naturally occurred), or *reintroduction,* (i.e., the release of free-ranging, wild animals in an area different from which they come, and where previous attempts have been made to introduce the species). More recently, wildlife *translocations,* (i.e., the transport and release of free-ranging, wild animals from one location to another, but where the species presently occurs or naturally occurred in the past) are conducted for ecological or conservation efforts (Nielsen 1988).

One successful approach to the conservation of large mammals has been their translocation into former habitats. In the early 1900s, large mammals were at an all-time low in North America, as discussed earlier. The history of big game conservation has generally followed three stages. The first stage occurred when Europeans arrived in North America. There was little concern for the abundant wildlife, and exploitation was the norm. The second stage occurred when the human population realized that wanton exploitation was leading to the demise of large mammals, which led to efforts to protect the remaining stock. Finally, numerous conservation efforts led to the rise and evolution of efforts to restore and scientifically manage wildlife (Mackie 2000). This conservation effort included translocation, which is deeply rooted in the history of wildlife management.

In 1878 sportsmen translocated 18 white-tailed deer (*Odocoileus virginianus*) from New York to Vermont in the first restoration of big game in the United States (Mackie 2000). Several years later (1892) elk were translocated from Yellowstone National Park, beginning a program of trapping and translocating that helped reestablish elk populations over North America and other parts of the world. The translocation of other large mammals followed, but the scientific basis for translocations was not developed until the advent of wildlife management in the 1920s and 1930s. Likely reasons translocation evolved slowly are because translocations of large animals are time consuming, expensive, and logistically and politically challenging (Wolf et al. 1996, Dunham 1997, Fritts et al. 1997). Guidelines for the successful translocations of animals have been published (International Union for Conservation of Nature and Natural Resources 1995, Wolf, et al. 1996), yet the success and failures of many translocations are poorly documented (Short et al. 1992), translocation techniques are rarely tested (Morgant and Krausman 1981), and many projects are based partly or entirely on concepts that may or may not be correct (Hein 1997).

For example, translocations were a new approach when bighorn sheep restoration efforts began, and in the early years numerous errors were made (e.g., trying to capture sheep in padded steel leghold traps) (Krausman 2000). However, with the advent of the drop net, drugs, net gun, and use of helicopters, success was rapid. For example, from 1954 to 1978, 153 desert bighorn sheep were successfully trapped and translocated. Since then, over 2000 desert bighorn sheep have been successfully captured and relocated. Over 50% of all present-day populations of bighorn sheep stem from translocations (Bailey 1990).

Although efforts to restore populations of bighorn sheep have relied heavily on translocations (Bailey 1990, Jessup et al. 1995), most restoration programs have not been successful (Risenhoover et al. 1988). For example, only 53% of 87 translocation populations in nine western states were rated as successful in the 1980s (Leslie 1980), and more recently, only 41 out of 100 translocations in six western states from 1923 to 1997 were rated as successful (Singer et al. 2000). However, translocation programs have been successful in reestablishing populations in localized areas in the United States where they have been extirpated (Buechner 1960, Trefethen 1975). By 1991, 14 state game and fish agencies had transplanted sheep onto more than 200 historic ranges (Bailey and Klein 1997). Bighorn sheep provide only one example. In 1985 alone at least 29 states in the United States translocated mammals: 24 translocated native game species, five translocated nongame species, and five translocated endangered species. The most popular species to be translocated were bears, white tailed deer, elk, bighorn sheep, racoons, river otters, moose, mountain goats, beavers, and pronghorns (Boyer and Brown 1988).

Reasons To Move Wildlife

Nielsen (1988) listed three reasons for moving wild animals: conservation and ecology, commerce and recreation, and humanitarian concerns. Today, most animals are moved for conservation when they are or have been threatened locally by excessive hunting or predation, their habitat has been altered, or for any natural or other anthropogenic factor that caused the reduction or extinction of the original population. Translocations should not occur in this situation unless the limiting factor has been reduced or eliminated.

Translocations also occur to introduce genetic variability to isolated wild populations, to place animals into suitable but empty habitat, or to reduce the density of populations that are so high they are having a negative effect on their habitat. Finally, rare and endangered species are translocated to avoid local influences or disastrous events that may jeopardize them. Sometimes rare and endangered species are moved to areas that afford protection and the opportunity to reproduce in a secure environment.

Many animals have been introduced or translocated for commerce and recreation (e.g., hunting, trapping), or for any array of reasons (e.g., captive breeding, research, education), and released in parks, sanctuaries, or game farms. Finally, animals are moved for humanitarian concerns when conflicts arise between humans and animals. Sometimes translocating animals to different habitats may be more socially acceptable than extermination.

Potential Problems With Animal Translocations

Unless translocations are well thought out and properly conducted, problems will certainly arise. The most important and likely negative implications of poorly planned wildlife introductions include the following (South African Wildlife Management Association 1988).

1. Hybridization. The loss of genetic integrity of subspecies and ecotypes leading to a loss of genetic variability should be avoided. When genotypes from different areas are pooled, much of this variability can be lost.
2. Altered vegetation. Introduced species can often alter vegetation that native species depend on and can outcompete indigenous species.
3. Introduction of disease. If introduced animals have diseases and parasites, the indigenous species may be at risk.
4. Poor introduction sites. If the areas that animals are moved to are not adequate, they will not survive, resulting in a loss of rare and valuable animals.
5. Wasted resources. Besides the natural capital that can be lost due to poorly thought out or planned relocations, extensive amounts of human capital and money can also be wasted. Some relocations can cost $4000 per animal (Boyer and Brown 1988) or more. That expenditure deserves the best planning possible.

The Basics Of Translocations

Nielsen (1988) reviewed some of the basic issues of translocating wildlife that should be considered to maximize success.

1. Sufficient monetary and technical resources to meet objectives. Translocations require planning, time, professional expertise, technology, and other factors, all of which are costly. Because each translocation is different (e.g, different personnel, species, habitats), cookbook approaches are not successful. In addition, wild animal behavior is not consistent. When planning resources for translocations, always consider time and money for the unexpected. If the entire project cannot be conducted properly, the weak link may compromise the entire effort.

In order for proper resources to be allocated to the project, specific objectives should be established (e.g., to supplement a population of bighorn sheep with 20 females and four males over two years to increase the population to at least 100 animals within five years). Because of the diversity of species, environmental conditions, and peculiar circumstances for specific translocation, the results of each translocation need to be evaluated based on the established objective.

2. Conduct a feasibility study. As with any management project, an exhaustive feasibility study should be conducted. In the feasibility study items that should be considered include the following:

   a. The reason and justification of the study should be stated.
   b. The status of the wildlife to be translocated, including the ecology, biology, and ethology of the species, current size, movements and distribution on the present range, male:female ratio, recruitment, mortality, and other life history characteristics. The health and condition of the population should also be considered. If diseased animals are translocated, they can jeopardize the health of indigenous species. Also, animals that are in poor health do not have the full complement of resources to withstand the rigors of translocation. The genetic component of the populations being translocated should also be considered. As more technology is being developed, the genetic management of populations is gaining scientific stature. The advantages and disadvantages of mixing the gene pools of populations carry risks that may not be desirable. Furthermore, translocations of animals that have been bred and raised in captivity may not be viable or desirable because they may be products of limited genetic variability.

3. Site selection. Determine the status of the present range and the range animals where transplantation will occur, including location, size, ownership, land use, climate, topography, vegetation cover, water availability, estimated carrying capacity, predators, public access, and other habitat and environmental factors.

4. Translocation strategy. The translocation strategy should be well thought out and the strategy should be designed to meet the objectives of the operation. Some of the considerations include the number of animals and the age and sex ratios of animals to be translocated, number of releases, number of release sites, and time of translocation.

5. Capture techniques and release technology. There are numerous capture techniques that need to be carefully selected for the species being translocated. Whether a mechanical capture (e.g., nets, traps, capture corrals), chemical capture (e.g., drugs), or combination is used, it is important to have consulted with veterinarians or agency personnel that specialize in animal capture to assist with the actual capture, postcapture processing and handling, transportation to the release site, and the actual release into the new area. Each of these steps requires detailed planning to maximize success.

6. Monitoring the population. Postrelease surveys should be standard following translocations. Without adequate monitoring, little is learned about the success or failure of the operation. Considerable expense has gone into the program up to this point and it is poor practice not to monitor the release. Ideally, the translocated animals should be radio-tagged so their movements can be determined. Other data that should be collected (and collected often enough to be meaningful) include: use of the resources, mortalities and survival, productivity and recruitment, population health, competition with indigenous wildlife, and the influence of the translocation on vegetation. Other specific data may be collected, depending on the objective of the postrelease study. If specific objectives of the translocation were established, the follow-up studies will allow for the determination of whether or not the objectives have been met.

FROM GAME MANAGEMENT TO BIODIVERSITY MANAGEMENT

Since the development of the science of game management, there has been increased emphasis on understanding all life throughout the world. This emphasis has led to the concept of *biodiversity* (i.e., variety of life and ecological processes) (Wilcove and Samson 1987). Understanding and managing for biodiversity are important for at least three reasons (Wilson 1988).

1. An increasing number of humans are degrading the habitats animals depend on.
2. Science is attempting to learn how to understand biological diversity so it can relieve human suffering and environmental destruction.
3. The destruction of natural habitats for animals is causing extinctions.

Each of these clearly illustrates the importance of habitat management to maintain biodiversity that is critical to society. Without healthy ecosystems, society will fail (Payne and Bryant 1998). Yet, science has only described fewer than 2,000,000 of the 5 to 30 million species on earth (Wilson 1988, Alonso and Dallmeier 2000) (Table 18–5). To maintain this biodiversity and minimize its decline, land managers will have to shift their philosophy to biological diversity management from which wood, forage, and recreation are secondary products. "That philosophy will entail a change in the way forestlands and rangelands have been managed in the past. Forest and range management should be coordinated closely with wildlife habitat-management

Table 18–5 Numbers of described species of living organisms from Wilson 1988

Kingdom and major subdivision	Common name	Number of described species	Totals
Virus			
	Viruses	1000 (order of magnitude only)	1000
Monera			
Bacteria	Bacteria	3000	
Myxoplasma	Bacteria	60	
Cyanophycota	Blue-green algae	1700	4760
Fungi			
Zygomycota	Zygomycete fungi	665	
Ascomycota (including 18,000 lichen fungi)	Cup fungi	28,650	
Basidiomycota	Basidiomycete fungi	16,000	
Oomycota	Water molds	580	
Chytridiomycota	Chytrids	575	
Acrasiomycota	Cellular slime molds	13	
Myxomycota	Plasmodial slime molds	500	46,983
Algae			
Chlorophyta	Green algae	7000	
Phaeophyta	Brown algae	1500	
Rhodophyta	Red algae	4000	
Chrysophyta	Chrysophyte algae	12,500	
Pyrrophyta	Dinoflagellates	1100	
Euglenophyta	Euglenoids	800	26,900
Plantae			
Bryophyta	Mosses, liverworts, hornworts	16,600	
Psilophyta	Psilopsids	9	
Lycopodiophyta	Lycophytes	1275	
Equisetophyta	Horsetails	15	
Filicophyta	Ferns	10,000	
Gymnosperma	Gymnosperms	529	
Dicotolydonae	Dicots	170,000	
Monocotolydonae	Monocots	50,000	248,428
Protozoa			
	Protozoans: Sarcomastigophorans, ciliates, and smaller groups	30,800	30,800
Animalia			
Porifera	Sponges	5000	
Cnidaria, Ctenophora	Jellyfish, corals, comb jellies	9000	
Platyhelminthes	Flatworms	12,200	

Table 18–5 *Continued*

Kingdom and major subdivision	Common name	Number of described species	Totals
Nematoda	Nematodes (roundworms)	12,000	
Annelida	Annelids (earthworms and relatives)	12,000	
Mollusca	Mollusks	50,000	
Echinodermata	Echinoderms (starfish and relatives)	6100	
Arthropoda	Arthropods		
Insecta	Insects	751,000	
Other arthropods		123,161	
Minor invertebrate phyla		9300	989,761
Chordata			
Tunicata	Tunicates	1250	
Cephalochordata	Acorn worms	23	
Vertebrata	Vertebrates		
Agnatha	Lampreys and other jawless fishes	63	
Chrondrichthyes	Sharks and other cartilaginous fishes	843	
Osteichthyes	Bony fishes	18,150	
Amphibia	Amphibians	4184	
Reptila	Reptiles	6300	
Aves	Birds	9040	
Mammalia	Mammals	4000	43,853
TOTAL, all organisms			1,392,485

strategies to enhance biodiversity and sustain production of commercial timber and livestock along with that of wildlife for balanced ecosystems and for recreation and tourism" (Payne and Bryant 1998:3). Without a shift in philosophy, human populations will continue to increase with a corresponding per capita consumption that will not be able to minimize extinctions. For example, the causation of species endangerment in the United States is dominated by anthropogenic forces (Table 18–6). We need a new set of objectives if the maintenance of biodiversity is to be a priority. Payne and Bryant (1998) suggested five basic objectives for the management of biodiversity.

1. Land managers need to assess the cumulative impact of individual projects on regional populations and resources beyond their boundaries under their control. Biotic integrity should be emphasized and based on natural history, not politics. In other words, multispecies management should replace single-species management. Management plans should account for metapopulation structure, succession, cumulative and regional effects, and changing species

Table 18–6   Causes of endangerment for species classified as threatened or endangered by the U.S. Fish and Wildlife Service (Czech and Krausman 1997)

Cause	Number of species endangered by cause and rank of frequency[*]	Number of species endangered and rank of frequency[†]
Interactions with nonnative species	305—1	115—8
Urbanization	275—2	247—1
Agriculture	224—3	205—2
Outdoor recreation and tourism development	186—4	148—4
Domestic livestock and ranching activities	182—5	136—6
Reservoirs and other running water diversions	161—6	160—3
Modified fire regimes and silviculture	144—7	83—10
Pollution of water, air, or soil	144—8	143—5
Mineral gas, oil, and geothermal extraction or exploration	140—9	134—7
Industrial, institutional, and military activities	131—10	8—12
Harvest, intentional and incidental	120—11	101—9
Logging	109—12	79—13
Road presence, construction, and maintenance	94—13	83—11
Loss of genetic variability, inbreeding depression, or hybridization	92—14	33—16
Aquifer depletion, wetland draining or filling	77—15	73—15
Native species interactions, plant succession	77—16	74—14
Disease	19—17	7—18
Vandalism (destruction without harvest)	12—18	11—17

[*]Including Hawaiian and Puerto Rican species.
[†]Not including Hawaiian and Puerto Rican species.

composition. Practices that promote site-specific diversity (i.e., *alpha diversity*) should be replaced with practices that promote between-habitat diversity (i.e., *beta diversity*). Management decisions need to be broad based and progress in a top-down ecological hierarchy, beginning with regional diversity (i.e., *gamma diversity*) of ecosystems.

2. Enough habitat should be provided for all aspects of an animal's life history, including connectivity between habitat patches, and all stages of succession required by the animals being managed, and other habitat components to minimize fragmentation.

3. Problem species and ecosystems should be monitored by using guilds (i.e., a group of species that share a need for common resources in the environment; Hunter 1990) or other indices that will assist in evaluating the health of the entire ecosystem, not just a single species.

4. Biological diversity needs to be sustainable because it is critical to the environment and human society.

5. As with any management plan that is objective driven, land managers need to be kept informed to anticipate and correct problems.

As management for biodiversity increases, the data base for wildlife will also increase, providing more information from which informed decisions can be made. The increased knowledge will only enhance biodiversity.

MINIMUM VIABLE POPULATION

One of the current topics biologists are struggling with is the issue of how many animals are enough. What should the goals be when trying to increase populations? These questions are especially important when managing rare and endangered species, or others threatened with habitat fragmentation (Landres 1992). There is often no clear answer because of the complexity of the species, the limited area, and the different ways to calculate a minimum viable population size. For example, Reed et al. (1988) calculated the minimum viable population size of the red-cockaded woodpecker at 509 breeding pairs that require at least 25,450 hectares, which was larger than the estimates of population size and area recommended in the recovery plan (U.S. Fish and Wildlife Service 1985).

Determining a viable population is complex, requires data on most aspects of an animals' life history, and is often controversial because estimates have to be made for missing data. For example, some of the information used to calculate a minimum viable population includes the population growth rate, stochastic growth (or decline), the mean population size, variation, gene diversity, and probability of extinction (Gilpin and Soulé 1986).

Deterministic Rate of Increase

The *deterministic population growth rate* is a projection of the mean rate of growth of the population, calculated from average birth rates and death rates only. The population increases when the growth rate is more than zero, decreases when less than zero, and does not grow when the rate equals zero. In other words, the deterministic rate of increase is approximately the rate of growth or decline in the population each year. This rate can predict future population growth if the population has a stable age distribution, birth and death rates are constant, there is no inbreeding depression, females can always find mates, and there is no density dependence in mortality and mortality rates. Obviously, these assumptions are violated, collectively or individually, resulting in empirical population growth less than the deterministic growth rate.

Stochastic Rate of Increase

The mean rate of *stochastic population growth* incorporates environmental variation, catastrophes, or genetic instabilities into the model. The stochastic rate of increase will usually be less than the deterministic rate that is predicted from birth and death rates. It would be similar to the deterministic if population growth is steady and robust.

Mean Population Size

The mean population size is averaged across simulated populations that are not extinct.

Standard Deviation of the Mean Population Size

Determining the standard deviation across simulated populations indicates population stability. Standard deviations greater than approximately half the mean population size often indicate unstable population sizes. When the standard deviation is large (relative to the population size), the population is vulnerable to large random fluctuations and could go extinct. A small standard deviation usually indicates the population is growing steadily or declining rapidly.

Genetic Diversity

The genetic diversity (i.e., expected heterozygosity) is expressed as a percent of the initial gene diversity of the population. When gene diversity declines, fitness declines (Lacy 1993). Impacts of inbreeding in wild populations are poorly understood but may be severe (Jiménez et al. 1994).

The Probability of Extinction

The data available and data estimated are used to determine the probability of extinction. Most models will determine the proportion of iterations (e.g., 500–1000) within a given set of data that go extinct (i.e., the lack of males or females) in the simulation.

Obviously, small populations are more vulnerable to these factors than larger ones. When a population experiences average levels of fluctuation, it must maintain 1000 individuals to assure population viability (Thomas 1990). However, populations of birds and mammals may need 10,000 individuals to maintain minimum viable populations (Noss 1992).

When developing management plans that include minimum viable population sizes, one goal should be to maintain populations above minimum viability. However, because of the stochastic processes (i.e., genetic, demographic, environmental, catastrophic) that influence populations, it is impractical to determine specific levels of minimum population size. If data cannot be collected, some general guidelines should be established that would make management meaningful and measurable.

1. A certain level of probability of persisting (e.g., 90–99%) for
2. a specific time period (i.e., 100–150 years) within
3. a minimum area.

With these guidelines, managers can develop management plans to increase populations in a measurable way. The world is ripe with uncertainty and changes will have to be made, as they can, along the way. Also, as more data are available about minimum population size, minimum habitat requirements, landscape ecology, and ecosystem management, more informed decisions can be made about more species and communities.

SUMMARY

Nearly all wildlife management actions are controversial because humans have different opinions about how wildlife should be managed. This chapter has selected some of the controversial issues that cut across the goals of wildlife management and provided suggestions for management.

19

Wildlife Management in the Twenty-First Century[a]

"There is no end to the path—our present notions will as surely be outdated as those which we here outdate."

A. Leopold

FROM LEOPOLD TO THE NEXT CENTURY

There have been numerous advances and accomplishments in the field of wildlife management, yet the field is still young. The early progress was described by Leopold (1999) as a necessity to prevent game species from being reduced by "slick and clean" (i.e., farming every piece of land, leaving nothing for wildlife) agriculture. "The meaning is this: 'slick and clean' agriculture has removed the game food and cover, and thus increased the resistance which the environment offers to natural increase. Hence a lesser fraction of the breeding potential is realized, hence game is decreasing" (Leopold 1999a: 28–29).

"The technical process of adapting agriculture and game production to each other—this art of raising game as a wild by-product of the land" . . . is game management. "It offers an almost virgin field for the practical application of biological science. Its techniques are just beginning to be developed. To hasten their development, and start the training of a professional class competent to apply them, the ammunition industry has set up a series of research fellowships at several universities or agricultural colleges. One in Minnesota is studying the management of ruffed grouse, at Wisconsin the management of bobwhite quail, at Michigan of Hungarian partridge, at Arizona of Gambel's quail. Under the eye of the agricultural authorities, and with the advisory guidance of the U.S. Biological Survey, these

[a] From Krausman (2000a).

'Game Fellows' are amassing a body of skill on how to raise wild game on modern farms and in modern forests. The keynote of the whole venture is to find out what slight modifications in methods of managing the primary crop will decrease the environmental resistance to the increase of game" (Leopold 1999*a*: 30–32).

The approach to this new discipline was to follow the scientific method that has been the backbone of the acquisition of knowledge. As the field developed and experimentation enhanced the wildlife resources, Leopold recognized that science alone was not enough and "Something new must be done" (Leopold 1933:411). Leopold summarized the new trend as an American game policy based on three principles. First, America has adequate habitat for wildlife, the means to afford it, and the love of sport to assure that successful production will be rewarded. Second, there are conflicts in how the habitat, the funds for habitat, and the love of hunting can be developed into a productive relationship. How the land is used and what it is used for, and by whom, often are at odds. The best plan, however, is the one that most nearly satisfies the desires of the hunter, landowner, and the public. Third, there are not enough biological facts available on how to make the land produce game. "All factions, whatever their other differences, should unite to make available the known facts, to promote research to find the additional facts needed, and to promote training of experts qualified to apply them" (Leopold 1933:411–412). In the 1930s, the emphasis was on game species and their importance to society for recreation. It would not take much wordsmithing to adopt this policy to contemporary management. Nor would Leopold, had he been able to see his profession in the twenty-first century, be enlightened to see the advances we have made. Habitat still exists in North America and we certainly have conflicts, but we are only recently looking at the bigger picture of wildlife management and how it influences other land uses (i.e., ecosystem management). Unless humans maintain the "integrity of the parts" of the system we depend on, our future will be dim indeed (Leopold 1999*a*).

Perhaps the profession of wildlife management is too young to make predictions about our future and what the new century will bring. However, Leopold was extremely farsighted and identified many of the battles we are fighting today on the conservation front. Many of our ills are caused by overpopulation. Human impacts have caused the collapse of fisheries (Erickson 2000), over 40% of the earth's land surface has been transformed for human use, most accessible fresh water is in use, and most terrestrial nitrogen fixation is caused by humans (Vitousek et al. 1997). Furthermore, approximately 200,000 square kilometers of forests are being cut each year, and species extinction is 120,000 times higher than normal (i.e., one species lost every four years); species extinction is at least 30,000 species per year (Leaky and Lewin 1995). These staggering statistics force the question of "what is the future of wildlife?".

Over twenty-five years ago, Scheffer (1976) made at least twelve predictions about wildlife management and how it would change.

1. Biologists will not attempt to increase, decrease, or stabilize populations of wildlife without some estimation of how their actions influence other species. This is a philosophy preached by Leopold that has been reawakened with the resurgence of ecosystem management.

2. Life will be controlled with life or biocontrol through habitat modification.
3. Habitat modification for native wildlife will replace artificial feeding. Artificially maintained populations will lose favor with the public.
4. The introduction of exotics will be discouraged and banned.
5. There will be a movement away from killing.
6. There will be more emphasis placed on biodiversity and increased management for nongame.
7. Some standard wildlife management practices will be banned (i.e., leghold traps).
8. The names of state game departments will change from an emphasis on game to a broader emphasis on natural resources.
9. State game and fish commissioners will represent a broader base of the public (e.g., broader than the hunting community).
10. More money for wildlife will be allocated from general revenues and less from hunting license return.
11. Public land use decisions will be more representative of truly all-American decisions, compared with land use decisions in favor of hunters, trappers, and local stockmen.
12. The field of wildlife management has been weakened by inbreeding. The field needs a "whole-earth, web-of-life, one-people brand of philosophy" (Scheffer 1976:54).

He believed this philosophy would dominate the future of the wildlife profession. Looking back, he was not far off the mark. Others that have looked at the future of wildlife have concentrated on the negotiation skills that will be needed. Cutler (1982) argued that much of the wildlife habitat was in poor ecological health that abounded in urban sprawl at the expense of old-growth forests, swamps, uplands, hedgerows, windbreaks, shelterbelts, soils, streams, and other wetlands and wildlife habitat. In short, the natural world is becoming less complex, causing the job of the wildlife biologists to be moreso. Unless areas are set aside by law, they will be developed, which further complicates the issues. Cutler (1982), in describing the type of wildlife biologists needed in the 1980s, predicted that wildlifers would no longer be able to call their own shots but would be on interdisciplinary land and resource planning teams. There would be more people and less wildlife, resulting in increased competition for the use of wildlands. Increased competition would yield fewer areas designed for single species of wildlife, more birders and others interested in nongame wildlife, and fewer hunters. To keep up with the complex issues of contemporary wildlife management, regular updating and retraining will be as essential a part of the life of the professional biologists in the twenty-first century as it was in the twentieth century (Krausman 1979, Nelson 1980, Cutler 1982). Continuing education would allow an improvement of skills in ecology, economics, communications, and a broadening of skills in other related disciplines (e.g., agronomy, economics, geography). But just as important, if not more important, is the ability of future wildlifers to be able to *lead* and not only *respond* to situations. By being "leaders" instead of "responders," wildlifers would have to develop visions, set objectives, and obtain decision-making skills in integrated planning. It would force the profession to consider social needs and the development of successful negotiation skills. The wildlifer envisioned by Cutler (1982) should be one that is *educated*, not *trained.* An educated person would be an effective partner in dealing with the challenges to land use and the citizens that are interested in enhancing and maintaining wildlife and its habitat.

In subsequent discussions about future needs of wildlife biologists to successfully manage our wildlife resources, these same themes have been repeated with different but similar emphasis. For example, Wagner (1989) argued that the profession concentrated too much on consumptive use and the control of populations and needed to broaden horizons, embrace change, and work in a more scientific arena to manage and satisfy social values. Biologists still did not emphasize theory enough and were not sophisticated enough scientifically to make commitments or even recognize to the full range of values of wildlife resources that include the diverse and complex issues of endangered species, biodiversity, and international issues. To rectify these shortcomings, "we need to make a commitment to the full range of values which society assigns to wildlife resources, and we need to strengthen the teaching, research, and application of our science" (Wagner 1989:359).

As the need for broader backgrounds is emphasized, understanding the animal and its habitat is often overlooked. This is shortsighted, as emphasized by Peek (1989:364).

> "Conversely, agencies often downplay the biological or technical side of the education, in favor of courses in subjects as planning, policy, or environmental law. As planning becomes more comprehensive, better integration of resource management occurs, and refinements in population management are sought, we will need better information about wildlife populations and habitats. We will need to be able to predict consequences of management in terms of population and habitat response. We will need to be fully cognizant of the assumptions upon which our management activities rest, and we will be more fully critical of our information base. That means we will need to know more about the fundamental biology and ecology of the populations we manage. In short, biologists will have to use more skills and knowledge about wildlife biology and habitats than currently required. A wildlife biologist will have to design and execute plans for obtaining information, and be able to analyze the information using statistical procedures. This is inevitable. It should not be threatening. It simply means better management and conservation of the resources."

Keppie (1990) and Steidl et al.(2000) also viewed a better future for wildlife with more rigor in our studies, as that would create better thinkers with broader outlooks. That broader outlook will certainly involve the field of human dimensions (Thomas and Pletscher 2000) that evolved to encompass beliefs, values, customs, knowledge, and laws (Witter and John 1998). DeMillo et al. (1998:955) emphasize the human component of the wildlife management triad by stating that "it may become difficult to succeed in the wildlife profession without a fundamental understanding of human dimensions." This may be especially relevant as wildlife management occurs at the global level. When considering the human values of North Americans alone, wildlife management is complex enough, but as we become global citizens, science and technology will not do the job alone. The relationships between wildlife and people will have to be studied and understood before effective management can occur (Kessler et al. 1998).

So we as a wildlife profession have developed in the last century. As a new science, we are still struggling, as there are numerous problems, numerous ways to solve them, and numerous voices telling us where more problems lie. Most agree, however, that education plays a dominant role as the profession evolves. I would go further to state that the future of our wildlife resources is directly tied to solid education both in and out of the classroom, with wildlife, their habitats, and all of the anthropogenic forces that threaten the future of wildlife. These are not new ideas and were concisely stated by Leopold (1966:190) over 50 years ago.

> "The outstanding scientific discovery of the twentieth century is not television, or radio, but rather the complexity of the land organism. Only those who know the most about it can appreciate how little we know about it. The last word in ignorance is the man who says of an animal or plant: 'What good is it?' If the land mechanism as a whole is good, then every part is good, whether we understand it or not. If the biota, in the course of eons, has built something we like but do not understand, then who but a fool would disregard seemingly useless parts? To keep every cog and wheel is the first precaution of intelligent tinkering."

Intelligent Tinkering

It would be useful to have a magical glass ball to look into the future. Unfortunately, they do not exist and predictions of the future will have to be based elsewhere. The future is an unknown that I will not pretend to explain, but there are certain concepts that will help in creating a vision for the future. We do know that since the 1900s, when wildlife was at one of its lowest ebbs, the wildlife management experiment begun by Leopold (1933) has been successful for many species. It is also evident that as the year 2000 closes out the century, the face of wildlife has changed dramatically. There is more competition for land and wildlife habitat worldwide, species are becoming extinct faster than they have in the past, complex issues develop daily, and solutions may not come easily. Even when they do, society may have other alternatives that have to be considered. There is no doubt that being a wildlife biologist in the new century will be a cooperative, compromising, and often politically driven profession. There is also no doubt that even with the ups and downs of contemporary management and the complexities we have to deal with, it is our present activities that will drive the future success of wildlife management (Figure 19–1).

If I did have that crystal ball to peer into, I suspect I would continue to see successful wildlife management, but different than in the past. We cannot ignore the mass of humanity destroying or altering wildlife habitats throughout the world or the record pace of declining diversity. This is the present, and if the human spirit is to thrive, we will deal with these threats. The tools used are so highly technical that Klein (2000) believes we must become *technological naturalists*; biologists that spend

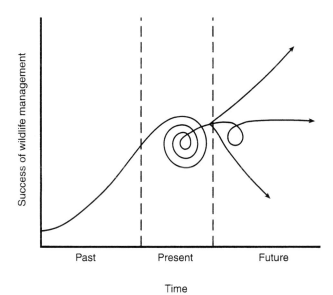

Figure 19–1 Since the 1900s, wildlife management has been successful. It is more complex as the new century begins, and our present activities will determine if we maintain the status quo, continue to succeed, or fail.

considerable time in the field but still take advantage of technology that will help us in our jobs. In my view, that will only happen with better and broader education as called for by Matter and Steidl (2000), Thomas and Pletscher (2000), and others. More emphasis will be required in the international arena (Kessler at al. 1998, Shaw 2000) and we will clearly incorporate more of the human element into management. However, the educational system will not be successful if that is done at the expense of understanding the basics of evolution, ecology, and life histories (Peek 1989, Bleich and Oehler 2000, Brown and Nielsen 2000). Furthermore, the education of future biologists will be incorporated into state, federal, and university cooperation (Bleich and Oehler 2000) similar to the model developed by the Cooperative Research Unit Program (Bissonette et al. 2000). There is no doubt that new techniques will be developed in this educational process, such as distance education (Edge and Leogering 2000). However, the key ingredient to the success of wildlife in the future will be the students who have the desire, drive, ambition, and dedication to become actively and passionately involved in the process of managing wildlife resources and the habitats they depend upon.

Appendix

Common and scientific names of animals mentioned

Alphabetical by common name		Alphabetical by scientific name	
Common name	**Scientific name**	**Scientific name**	**Common name**
Addax	*Addax nasomaculatus*	*Accipiter cooperii*	Cooper's hawk
African buffalo	*Syncerus caffer*	*Addax nasomaculatus*	Addax
Alpaca	*Lama pacos*	*Aepyceros melampus*	Impala
American antelope	*Antilocapra americana*	*Agelaius phoeniceus*	Red-winged blackbird
American crocodile	*Crocodylus acutus*	*Alcelaphus buselaphus*	Red hartebeest
American robin	*Turdus migratorius*	*Alcelaphus* spp.	Hartebeest
Antelope ground squirrel	*Ammospermophilus leucurus*	*Alces alces*	Moose
Aoudad	*Ammotragus lervia*	*Alectoris chukar*	Chukar partridge
Arabian oryx	*Oryx leucoryx*	*Ammospermophilus interpres*	Texas antelope squirrel
Argali	*Ovis ammon*	*Ammospermophilus leucurus*	Antelope ground squirrel
Axis (spotted) deer	*Axis axis*	*Ammotragus lervia*	Aoudad
Bactrian camel	*Camelus bactrianus*	*Ammotragus lervia*	Barbary sheep
Badger	*Taxidea taxus*	*Anas platyrhynchos*	Mallard
Band-tailed pigeon	*Columba fasciata*	*Antidorcas marsupialis*	Black springbok
Banteng	*Babalus javanicus*	*Antidorcas marsupialis*	Common springbok
Barasingha or swamp deer	*Cervus duvauceli*	*Antidorcas marsupialis*	Springbok
Barbados	*Ovis aries*	*Antilocapra americana*	American antelope
Barbary sheep	*Ammotragus lervia*	*Antilope cervicapra*	Blackbuck antelope
Barren ground caribou	*Rangifer arcticus*	*Aphelocoma coerulescens coerulescens*	Florida scrub jay
Beaver	*Castor canadensis*	*Axis axis*	Axis (spotted) deer
Bighorn sheep	*Ovis canadensis*	*Axis porcinus*	Hog deer
Black rhino	*Diceros bicornus*	*Babalus javanicus*	Banteng
Black springbok	*Antidorcas marsupialis*	*Bison bison*	Forest buffalo
Black wildebeest	*Connochaetes gnou*	*Bonasa umbellus*	Ruffed grouse

Alphabetical by common name		Alphabetical by scientific name	
Common name	Scientific name	Scientific name	Common name
Blackbird	*Euphagus* spp.	*Bos gaurus*	Gaur
Blackbuck antelope	*Antilope cervicapra*	*Bos grunniens*	Yak
Black-footed ferret	*Mustela nigripes*	*Bos taurus*	Watusi
Blesbok	*Damaliscus dorcas*	*Boselaphus tragocamelus*	Nilgai antelope
Blue grouse	*Dendragapus obscurus*	*Bubalus arnee*	Water buffalo
Blue wildebeest	*Connochaetes taurinus taurinus*	*Callipepla californica*	California quail
Bobcat	*Lynx rufus*	*Callipepla gambelii*	Gambel's quail
Bongo	*Tragelaphus eurycerus*	*Callipepla squamata*	Scaled quail
Bontebok	*Damaliscus dorcas*	*Camelus bactrianus*	Bactrian camel
Brown trout	*Salmo trutta*	*Camelus dromedarius*	Dromedary
Burro	*Equus asinus*	*Canis latrans*	Coyote
Bushbuck	*Tragelaphus scriptus*	*Canis lupus*	Gray wolf
California condor	*Gymnogyps californianus*	*Canis lupus*	Wolf
California quail	*Callipepla californica*	*Canis rufus*	Red wolf
Catalina goat	*Capra hircus*	*Capra aegagrus*	Persian ibex
Caucasian tur	*Capra* spp.	*Capra aegagrus*	Persian wild goat
Chamois	*Rupicapra rupicapra*	*Capra falconeri*	Markhor
Chapman's zebra	*Equus quagga antiquorum*	*Capra hircus*	Catalina goat
Chevrotain	Family Tragulidae	*Capra ibex*	Ibex
Chimney Swift	*Chaetura pelagica*	*Capra ibex sibirica*	Siberian ibex
Chinese water deer	*Hydropotes inermis*	*Capra* spp.	Caucasian tur
Chukar partridge	*Alectoris chukar*	*Capreolus capreolus*	Roe deer
Collared peccary	*Pecari tajacu*	*Castor canadensis*	Beaver
Common duiker	*Sylvicapra grimmia*	*Centrocercus urophasianus*	Sage grouse
Common loon	*Gavia immer*		
Common springbok	*Antidorcas marsupialis*	*Ceratotherium simum*	White rhino
Cooper's hawk	*Accipiter cooperii*	*Cervus duvauceli*	Barasingha or swamp deer
Cottontail rabbit	Genus *Sylvilagus*	*Cervus elaphus*	Elk
Coues white-tailed deer	*Odocoileus virginianus couesi*	*Cervus elaphus*	Red deer
Coyote	*Canis latrans*	*Cervus elaphus canadensis*	Elk, Eastern
Dall's sheep	*Ovis dalli*	*Cervus elaphus manitobensis*	Elk, Manitoban
Dama gazelle or Addra gazelle	*Gazella dama*	*Cervus elaphus merriami*	Elk, Merriam
Damara zebra	*Equus quagga antiquorum*	*Cervus elaphus nannodes*	Elk, Tule
Deer	*Odocoileus* spp.	*Cervus elaphus nelsoni*	Elk, Rocky Mountain
Deer mouse	*Peromyscus maniculatus*	*Cervus elaphus roosevelti*	Elk, Roosevelt

Alphabetical by common name		Alphabetical by scientific name	
Common name	**Scientific name**	**Scientific name**	**Common name**
Desert cottontail	*Sylvilagus audubonii*	*Cervus eldi*	Eld's deer
Desert mule deer	*Odocoileus hemionus eremicus*	*Cervus nippon*	Sika deer
Desert tortoise	*Gopherus agassizii*	*Cervus timorensis*	Rusa deer
Dik-dik	*Madoqua* spp.	*Cervus unicolor*	Sambar deer
Dorcas gazelle	*Gazella dorcas*	*Chaetura pelagica*	Chimney swift
Dromedary	*Camelus dromedarius*	*Choeropsis liberiensis*	Pigmy hippopotamus
East African oryx	*Oryx beisa*	*Colinus virginianus*	Northern bobwhite
Eland	*Taurotragus oryx*	*Columba fasciata*	Band-tailed pigeon
Eld's deer	*Cervus eldi*	*Connochaetes gnou*	Black wildebeest
Elephant	Family Elephantidae	*Connochaetes gnou*	White-tailed gnu
Elk	*Cervus elaphus*	*Connochaetes tarinus*	Wildebeest
		Connochaetes taurinus taurinus	Blue wildebeest
Elk, Eastern	*Cervus elaphus canadensis*	*Crocodylus acutus*	American crocodile
Elk, Manitoban	*Cervus elaphus manitobensis*	*Cygnus olor*	Mute swan
Elk, Merriam	*Cervus elaphus merriami*	*Cynomys* spp.	Prairie dog
Elk, Rocky Mountain	*Cervus elaphus nelsoni*	*Cyrtonyx montezumae*	Harlequin (Montezuma) quail
Elk, Roosevelt	*Cervus elaphus roosevelti*	*Dama dama*	Fallow deer
Elk, Tule	*Cervus elaphus nannodes*	*Damaliscus dorcas*	Blesbok
European boar	*Sus scrofa ferus*	*Damaliscus dorcas*	Bontebok
Fallow deer	*Dama dama*	*Dendragapus canadensis*	Spruce grouse
Florida panther	*Felis concolor coryi*	*Dendragapus obscurus*	Blue grouse
Florida scrub jay	*Aphelocoma coerulescens coerulescens*	*Dendragapus* spp.	Grouse
Forest buffalo	*Bison bison*	*Diceros bicornus*	Black rhino
Four-horned antelope	*Tetracerus quadricornis*	*Didelphis virginiana*	Opossum
		Diomedea epomophora	Royal albatross
Fox	*Vulpes* and *Urocyon* spp.	*Elaphurus davidianus*	Pere David's deer
Gambel's quail	*Callipepla gambelii*	*Empidonax trailii*	Willow flycatcher
Gaur	*Box gaurus*	*Equus asinus*	Burro
Gemsbok	*Oryx gazella*	*Equus grevyi*	Grevy's zebra
Gerenuk	*Litocranius walleri*	*Equus quagga antiquorum*	Chapman's zebra
Giraffe	*Giraffa camelopardalis*	*Equus quagga antiquorum*	Damara zebra
Grant's gazelle	*Gazella granti*	*Equus quagga bohmi*	Grant's zebra

Alphabetical by common name		Alphabetical by scientific name	
Common name	Scientific name	Scientific name	Common name
Grant's zebra	*Equus quagga bohmi*	*Equus* spp.	Zebra
Gray partridge	*Perdix perdix*	*Equus zebra*	Mountain zebra
Gray wolf	*Canis lupus*	*Euphagus* spp.	Blackbird
Greater prairie chicken	*Tympanuchus cupido*	*Falco peregrinus*	Peregrine falcon
Grevy's zebra	*Equus grevyi*	Family Elephantidae	Elephant
Grizzly bear	*Ursus arctos horribilis*	Family Tragulidae	Chevrotain
Grouse	*Dendragapus* spp.	*Felis concolor coryi*	Florida panther
Guanaco	*Lama guanicoe*	*Gavia immer*	Common loon
Harlequin (Montezuma) quail	*Cyrtonyx montezumae*	*Gazella dama*	Dama gazelle or Addra gazelle
Hartebeest	*Alcelaphus* spp.	*Gazella dorcas*	Dorcas gazelle
Hedgehog	Subfamily Erinaceinae	*Gazella granti*	Grant's gazelle
Himalayan thar	*Hemitragus jemlahicus*	*Gazella leptoceros*	Slender-horned gazelle
Hippopotamus	*Hippopotamus amphibius*	*Gazella thomsonii*	Thomson's gazelle
Hog deer	*Axis porcinus*	*Gazella* spp.	Persian gazelle
House sparrow	*Passer domesticus*		
Humpback whale	*Megaptera novaeangliae*	Genus Sylvilagus	Cottontail rabbit
		Giraffa camelopardalis	Giraffe
Iberian lynx	*Lynx pardinus*	*Glaucidium brasilianum*	Pygmy owl, ferruginous
Ibex	*Capra ibex*	*Gopherus agassizii*	Desert tortoise
Impala	*Aepyceros melampus*	*Grus americana*	Whooping crane
Kudu	*Taurotragus strepsiceros*	*Gymnogyps californianus*	California condor
Langur	*Presbytis* spp.		
Lechwe	*Kobus leche*	*Hemitragus jemlahicus*	Himalayan thar
Lesser prairie chicken	*Tympanuchus pallidicinctus*	*Hippopotamus amphibius*	Hippopotamus
Llama	*Lama glama*	*Hippotragus equinus*	Roan antelope
Lynx	*Lynx canadensis*	*Hippotragus niger*	Sable antelope
Mallard	*Anas platyrhynchos*	*Hydropotes inermis*	Chinese water deer
Markhor	*Capra falconeri*	*Kinosternon flavescens*	Yellow mud turtle
Merriam's turkey	*Meleagris gallopavo merriami*	*Kinosternon sonoriense*	Sonoran mud turtle
Moose	*Alces alces*	*Kobus ellipsiprymnus*	Waterbuck
Mouflon sheep	*Ovis musimon*	*Kobus leche*	Lechwe
Mountain goat	*Oreamnos americanus*	*Lagopus lagopus*	Red or willow grouse
Mountain lion	*Puma concolor*	*Lagopus leucurus*	White-tailed ptarmigan
Mountain quail	*Oreortyx pictus*	*Lagopus* spp.	Ptarmigan
Mountain zebra	*Equus zebra*	*Lama glama*	Llama
Mourning dove	*Zenaida macoura*	*Lama guanicoe*	Guanaco

Alphabetical by common name		*Alphabetical by scientific name*	
Common name	**Scientific name**	**Scientific name**	**Common name**
Mt. Graham red squirrel	*Tamiasciurus hudsonicus*	*Lama pacos*	Alpaca
Muntjac	*Muntiacus* spp.	*Lama vicugna*	Vicuna
Musk deer	*Moschus moschiferus*	*Leopardus pardalis*	Ocelot
Muskox	*Ovibos moschatus*	*Litocranius walleri*	Gerenuk
Muskrat	*Ondatra zibethicus*	*Lontra canadensis*	Otter
Mute swan	*Cygnus olor*	*Lontra canadensis*	River otter
Nilgai antelope	*Boselaphus tragocamelus*	*Lynx canadensis*	Lynx
Northern bobwhite	*Colinus virginianus*	*Lynx pardinus*	Iberian lynx
Nyala	*Tragelaphus angasi*	*Lynx rufus*	Bobcat
Ocelot	*Leopardus pardalis*	*Madoqua* spp.	Dik-dik
Okapi	*Okapia johnstoni*	*Mazama americana*	Red brocket deer
Opossum	*Didelphis virginiana*	*Megaptera novaeangliae*	Humpback whale
Ostrich	*Struthio camelus*	*Meleagris gallopavo*	Turkey
Otter	*Lontra canadensis*	*Meleagris gallopavo*	Wild turkey
Pere David's deer	*Elaphurus davidianus*	*Meleagris gallopavo merriami*	Merriam's turkey
Peregrine falcon	*Falco peregrinus*	*Melospiza melodia*	Song sparrow
Persian gazelle	*Gazella* spp.	*Mephitis* or *Spilogale* spp.	Skunk
Persian ibex	*Capra aegagrus*	*Moschus moschiferus*	Musk deer
Persian wild goat	*Capra aegagrus*	*Muntiacus* spp.	Muntjac
Pheasant	*Phasianus colchicus*	*Mustela erminea*	Stoat
Pigmy hippopotamus	*Choeropsis liberiensis*	*Mustela nigripes*	Black-footed ferret
Plethodontid salamander	*Plethodon* spp.	*Mustela* spp.	Weasel
Prairie chicken	*Tympanuchus* spp.	*Nesotragus moschatus*	Suni
Prairie dog	*Cynomys* spp.	*Odocoileus hemionus eremicus*	Desert mule deer
Prairie garter snake	*Thamnophis radix*	*Odocoileus* spp.	Deer
Psoroptic mite	*Psoroptic ovis*	*Odocoileus virginianus couesi*	Coues white-tailed deer
Ptarmigan	*Lagopus* spp.	*Okapia johnstoni*	Okapi
Pygmy owl, ferruginous	*Glaucidium brasilianum*	*Ondatra zibethicus*	Muskrat
Raccoon	*Procyon lotor*	*Oreamnos americanus*	Mountain goat
Red brocket deer	*Mazama americana*	*Oreortyx pictus*	Mountain quail
Red deer	*Cervus elaphus*	*Oryx beisa*	East African oryx
Red fox	*Vulpes vulpes*	*Oryx dammah*	Scimitar-horned oryx
Red hartebeest	*Alcelaphus buselaphus*	*Oryx gazella*	Gemsbok
Red or willow grouse	*Lagopus lagopus*	*Oryx gazella*	South African oryx (gemsbok)

Alphabetical by common name		Alphabetical by scientific name	
Common name	**Scientific name**	**Scientific name**	**Common name**
Red sheep	*Ovis orientalis*	*Oryx leucoryx*	Arabian oryx
Red wolf	*Canis rufus*	*Ovibos moschatus*	Muskox
Red-cockaded woodpecker	*Picoides borealis*	*Ovis ammon*	Argali
Red-winged blackbird	*Agelaius phoeniceus*	*Ovis aries*	Barbados
Reindeer	*Rangifer tarandus*	*Ovis canadensis*	Bighorn sheep
Ring-necked pheasant	*Phasianus colchicus*	*Ovis dalli*	Dall's sheep
River otter	*Lontra canadensis*	*Ovis dalli stonei*	Stone's sheep
Roan antelope	*Hippotragus equinus*	*Ovis musimon*	Mouflon sheep
Roe deer	*Capreolus capreolus*	*Ovis orientalis*	Red sheep
Royal albatross	*Diomedea epomophora*	*Ovis orientalis*	Urial
Ruffed grouse	*Bonasa umbellus*	*Passer domesticus*	House sparrow
Rusa deer	*Cervus timorensis*	*Pecari tajacu*	Collared peccary
Russian boar	*Sus Scrofa*	*Perdix perdix*	Gray partridge
Sable antelope	*Hippotragus niger*	*Permoyscus maniculatus*	Deer mouse
Sage grouse	*Centrocercus urophasianus*	*Phasianus colchicus*	Pheasant
Sambar deer	*Cervus unicolor*	*Phasianus colchicus*	Ring-necked pheasant
Scaled quail	*Callipepla squamata*	*Picoides borealis*	Red-cockaded woodpecker
Scimitar-horned oryx	*Oryx dammah*	*Plethodon* spp.	Plethodontid salamander
Sharp-tailed grouse	*Tympanuchus phasianellus*	*Presbytis* spp.	Langur
Siberian ibex	*Capra ibex sibirica*	*Procyon lotor*	Raccoon
Sika deer	*Cervus nippon*	*Psoroptic ovis*	Psoroptic mite
Sitatunga	*Tragelaphus spekei*	*Puma concolor*	Mountain lion
Skunk	*Mephitis* or *Spilogale* spp.	*Rangifer arcticus*	Barren ground caribou
Slender-horned gazelle	*Gazella leptoceros*	*Rangifer tarandus*	Reindeer
Song sparrow	*Melospiza melodia*	*Rupicapra rupicapra*	Chamois
Song thrush	*Turdus philomelos*	*Salmo trutta*	Brown trout
Sonoran mud turtle	*Kinosternon sonoriense*	*Scolopax minor*	Woodcock (American)
South African oryx (gemsbok)	*Oryx gazella*	*Struthio camelus*	Ostrich
Springbok	*Antidorcas marsupialis*	Subfamily Erinaceinae	Hedgehog
Spruce grouse	*Dendragapus canadensis*	*Sus scrofa*	Russian boar
Stoat	*Mustela erminea*	*Sus scrofa*	Wild board
Stone's sheep	*Ovis dalli stonei*	*Sus scrofa ferus*	European boar
Suni	*Nesotragus moschatus*	*Sylvicapra grimmia*	Common duiker

Alphabetical by common name		*Alphabetical by scientific name*	
Common name	**Scientific name**	**Scientific name**	**Common name**
Texas antelope squirrel	*Ammospermophilus interpres*	*Sylvilagus audubonii*	Desert cottontail
Thomson's gazelle	*Gazella thomsonii*	*Syncerus caffer*	African buffalo
Turkey	*Meleagris gallopavo*	*Tamiasciurus hudsonicus*	Mt. Graham red squirrel
Urial	*Ovis orientalis*	*Taurotragus oryx*	Eland
Vicuna	*Lama vicugna*	*Taurotragus strepsiceros*	Kudu
Water buffalo	*Bubalus arnee*	*Taxidea taxus*	Badger
Waterbuck	*Kobus ellipsiprymnus*	*Tetracerus quadricornis*	Four-horned antelope
Watusi	*Bos taurus*	*Thamnophis radix*	Prairie garter snake
Weasel	*Mustela* spp.	*Tragelaphus angasi*	Nyala
White rhino	*Ceratotherium simum*	*Tragelaphus eurycerus*	Bongo
White-tailed gnu	*Connochaetes gnou*	*Tragelaphus scriptus*	Bushbuck
White-tailed ptarmigan	*Lagopus leucurus*	*Tragelaphus spekei*	Sitatunga
White-winged dove	*Zenaida asiatica*	*Turdus migratorius*	American robin
Whooping crane	*Grus americana*	*Turdus philomelos*	Song thrush
Wild boar	*Sus scrofa*	*Tympamuchus cupido*	Greater prairie chicken
Wild turkey	*Meleagris gallopavo*	*Tympanuchus pallidicinctus*	Lesser prairie chicken
Wildebeest	*Connochaetes tarinus*	*Tympanuchus phasianellus*	Sharp-tailed grouse
Willow flycatcher	*Empidonax traillii*	*Tympanuchus* spp.	Prairie chicken
Wolf	*Canis lupus*	*Ursus arctos horribilis*	Grizzly bear
Woodcock (American)	*Scolopax minor*	*Vulpes* and *Urocyon* spp.	Fox
Yak	*Bos grunniens*	*Vulpes vulpes*	Red fox
Yellow mud turtle	*Kinosternon flavescens*	*Zenaida asiatica*	White-winged dove
Zebra	*Equus* spp.	*Zenaida macoura*	Mourning dove

Common and scientific names of plants mentioned

Alphabetical by common name		*Alphabetical by scientific name*	
Common name	**Scientific name**	**Scientific name**	**Common name**
Balsam fir	*Abies balsamea*	*Abies balsamea*	Balsam fir
Beargrass	*Nolina microcarpa*	*Acacia greggii*	Catclaw acacia
Burroweed	*Ambrosia dumosa*	*Acer* spp.	Maple
Catclaw acacia	*Acacia greggii*	*Agave lecheguilla*	Lecheguilla
Cocklebur	*Xanthium strumarium*	*Ambrosia dumosa*	Burroweed
Eucalyptus	*Eucalyptus* spp.	*Aristida* spp.	Three-awn grass
Hackberry	*Celtis occidentalis*	*Artemisia* spp.	Sagebrush
Hazelnut	*Corylus* spp.	*Baccharis salicifolia*	Seep willow
Hemlock	*Tsuga* spp.	*Betula papyrifera*	Paper birch
Hobblebush	*Viburnum alnifolium*	*Carpinus caroliniana*	Ironwood
Ironwood	*Carpinus caroliniana*	*Celtis occidentalis*	Hackberry

Alphabetical by common name		*Alphabetical by scientific name*	
Common name	**Scientific name**	**Scientific name**	**Common name**
Jojoba	*Simmondsia chinensis*	*Cereus giganteus*	Saguaro
Juniper	*Juniperus* spp.	*Corylus* spp.	Hazelnut
Larkspur	*Delphinium* spp.	*Delphinium* spp.	Larkspur
Lecheguilla	*Agave lecheguilla*	*Distichlis* spp.	Salt grass
Lehmann lovegrass	*Eragrostis lehmanniana*	*Eragrostis lehmanniana*	Lehmann lovegrass
Lupine	*Lupinus* spp.	*Eucalyptus* spp.	Eucalyptus
Maple	*Acer* spp.	*Foiquieria splendens*	Ocotillo
Mesquite	*Prosopis* spp.	*Fragaria* spp.	Strawberry
Mullein	*Verbascum thapsus*	*Gutierrezia sarothrae*	Snakeweed
Nightshade	*Solanum* spp.	*Juniperus* spp.	Juniper
Ocotillo	*Fouquieria splendens*	*Lupinus* spp.	Lupine
Paper birch	*Betula papyrifera*	*Nolina microcarpa*	Beargrass
Pinyon	*Pinus edulis*	*Opuntia engelmannii*	Prickly pear
Prickly pear	*Opuntia engelmannii*	*Picea glauca*	White spruce
Sagebrush	*Artemisia* spp.	*Pinus edulis*	Pinyon
Saguaro	*Cereus giganteus*	*Platanus occidentalis*	Sycamore
Salt grass	*Distichlis* spp.	*Prosopis* spp.	Mesquite
Seep willow	*Baccharis salicifolia*	*Salix* spp.	Willows
Snakeweed	*Gutierrezia sarothrae*	*Simmondsia chinensis*	Jojoba
Strawberry	*Fragaria* spp.	*Solanum* spp.	Nightshade
Sycamore	*Platanus occidentalis*	*Tsuga* spp.	Hemlock
Three-awn grass	*Aristida* spp.	*Verbascum thapsus*	Mullein
White spruce	*Picea glauca*	*Viburnum alnifolium*	Hobblebush
Willows	*Salix* spp.	*Xanthium strumarium*	Cocklebur

Glossary

Abiotic: The nonliving components of the environment (e.g., air, rocks, soil, water) (Thomas 1979*a*).

Abrupt edge: An edge between stands or communities that is regular (i.e., straight lines or gently sweeping curves), leading to relatively low amounts of edge per unit area (Thomas 1979*a*).

Abundance: The number of individuals or of a resource. Abundance usually should be evaluated concurrently with the availability of the resource to specific organisms (Bolen and Robinson 1995).

Accident: Death or injury from physical causes alone.

Accuracy: The nearness of a measurement to the actual value of the variable being measured.

Active management: Direct manipulation of animal populations (e.g., translocation, hunt).

Adapted: The suitability of an organism for a particular condition, usually habitat, which comes from the process by which an organism becomes better suited to its environment or particular functions (Thomas 1979*a*).

Adaptive management: The process of implementing a policy decision incrementally, so that changes can be made if the desired results are not being achieved. It is a process similar to a scientific experiment in that predictions and assumptions in management plans are tested, and experience and new scientific findings are used as the basis to improve resource management practices and future planning.

Additive mortality: A concept that the effect of one kind of mortality is added to those of other sources of mortality. For example, if predation takes 10% of a population and ice storm takes 20%, the total mortality for the year is 30%. If, in the next year, predation takes 20% and an ice storm takes 20% for a total mortality of 40%, the effects of the two factors are said to be additive. In this concept, hunting mortality adds to the total natural mortality rate of a population. See Compensatory mortality (Bolen and Robinson 1995).

Adult: An animal, or age class of animals, that has reached breeding age (Bolen and Robinson 1995).

Age ratio: The relative proportions of various age groups in a population. May be expressed in several ways: number of juveniles per adult, juveniles per adult female, juveniles per 100 females, or juveniles per pair of adults. Age ratios often are determined from examinations of hunter-killed animals (e.g., wings of woodcock, quail, or ducks) and may be used as a measure of breeding success in animal populations (Bolen and Robinson 1995).

Aggregation: Coming together of organisms into a group (Anderson 1991).

Allopatric: Different, usually used in reference to populations that occupy mutually exclusive (but usually adjacent) geographic areas (Anderson 1991).

Allotment: A designated area of land available for livestock grazing. Usually a grazing permit is issued designating a specified number and kind of livestock to be grazed according to direction found in an allotment management plan. It is the basic land unit used in the management of livestock on National Forest System lands, and associated lands administered by the Forest Service.

Alluvial soil: Soil deposited by running water, showing practically no horizon development (save A-1 information) or other modification (Anderson 1991).

Alpha diversity: Diversity within a community (Anderson 1991).

Ambient: Surrounding.

Angiosperm: Any class of plants that is identified by having their seeds enclosed in an ovary (Thomas 1979a).

Animal community: The species of animals supported by a combination of habitat niches (Thomas 1979a).

Animal unit (AU):   Considered to be one mature (455 kilogram) cow or the equivalent based upon average daily forage consumption of 11.8 kilograms of dry matter per day. Therefore, 1 AU = 7.7 white-tailed deer or 5.8 mule deer (Bolen and Robinson 1995).

Annual: In plants, a species that completes its life cycle in one growing season. Many common weeds are annuals (Bolen and Robinson 1995).

Annulus (plural, Annuli):   A mark, usually circular, indicating one year of growth, and therefore a useful measure of age. Accuracy increases or decreases in relation to the distinctiveness of seasonal growth patterns (i.e., north temperate versus tropical environments). Examples include rings on fish scales, trees, and horns (Bolen and Robinson 1995).

Antler: One of paired bony structures protruding from the skulls of deer, elk, moose, and caribou (Cervidae). Covered with velvet during development. Shed annually and usually branched; on males only, except caribou (Bolen and Robinson 1995).

Apical growth: Growth from the terminal shoot meristem (Thomas 1979a).

Aquaculture: The rearing of plants or animals in water under controlled conditions (Anderson 1991).

Aquatic habitat: Habitat that occurs in free water (Thomas 1979a).

Arboreal: Term referring to trees; for wildlife, a lifestyle adapted to trees. For example, squirrels are arboreal species (Bolen and Robinson 1995).

Area-kill: The annual kill per unit area.

Arthropod: Any invertebrate organism of the phylum Arthropoda, which includes the insects, crustaceans, arachnids, and myriapods, having a horny, segmented, external covering and jointed limbs (Thomas 1979a).

Artificial establishment: A planting of wildlife maintained only through renewed plantings or artificial propagation.

Asexual: Having no evident sex or sex organs; sexless; pertaining to or characterizing reproduction involving a single male or female gamete, such as binary fission or budding (Thomas 1979*a*).

Aspect: The direction toward which a slope faces (Thomas 1979*a*).

Assemblage: A group of species under study.

Association: A major unit in community ecology, characterized by essential uniformity of species composition (Anderson 1991).

Atrical: Born in a helpless state, so not able to move or support itself (Anderson 1991).

Autotrophic: Not requiring an exogenous factor for normal metabolism; refers to organisms, usually green plants, that are capable of converting solar energy to chemical energy (sugar) by photosynthesis (Anderson 1991).

Availability: With abundance, part of the ecological equation for measuring food and other resources. For example, earthworms often are abundant beneath logs, but under those conditions, such worms are not available as food for robins or woodcock. Similarly, twigs on tall shrubs and small trees, although abundant, may not be available as browse until deep snows bring the upper branches within the reach of deer and rabbits (Bolen and Robinson 1995).

Aversive conditioning: Learning that relies on the stimuli of undesirable experiences (i.e., negative reinforcement). For example, some experiments suggest that coyotes fed mutton treated with distasteful substances may avoid killing sheep (Bolen and Robinson 1995).

Avifauna: A term referring to all species of birds occupying a designated area or time. For example, about 487 species of birds make up the current avifauna of Texas (Bolen and Robinson 1995).

Backyard management: A part of urban wildlife management in suburban zones; installation of birdhouses, feeders, and layered vegetation are common techniques (Bolen and Robinson 1995).

Bag limit: Number of animals that can be taken in a unit of time, usually a day (Anderson 1991).

Band: A loose aggregation of wildlife, sometimes all of one sex.

Bare ground: All land surface not covered by vegetation, rock, or litter.

Basal area: The cross-sectional area of a tree as measured at breast height (approximately 1.37 meters) expressed in square feet per acre.

Basal metabolism: The measure of metabolism of the body in the resting state, determined by the amount of oxygen utilized or of heat produced (Thomas 1979*a*).

Bedding: The process of an animal lying down for a rest (Thomas 1979*a*).

Behavioral carrying capacity: The maximum population size a given area will support when intrinsic behavioral or physiological mechanisms are the primary force controlling the population.

Best estimate: The best possible approximation given a combination of available information and understanding of a situation (Thomas 1979*a*).

Beta diversity: Diversity comparison between similar communities (Anderson 1991).

Bias: The difference between the expected value of a population estimate and the actual population size; a systematic distortion away from true density.

Biennial: A plant living two years, usually flowering the second year; occurring every two years (Anderson 1991).

Big game: Large animals hunted, or potentially hunted, for sport—for example, elk, mule deer, big horn sheep, pronghorn, black bear (Anderson 1991).

Binomial: The latinized name of an organism consisting of two words, the first of which is the genus and the second the species (Thomas 1979a).

Biological amplification: The process in which organisms higher in the food chain accumulate and retain materials, such as organochlorines, from organisms lower in the food chain (Anderson 1991).

Biological clock: An internal mechanism that signals animals that it is time for some activity, such as migration or nesting (Anderson 1991).

Biological control: The use of organisms or viruses to control parasites, weeds, or other pests (Thomas 1979a).

Biological diversity (biodiversity):   The variety of life, typically expressed in terms of species richness, but also may be applied to genes and ecosystems. The preservation of biodiversity is the primary goal of conservation biology (Bolen and Robinson 1995).

Biological potential: The maximum production of a selected organism that can be attained under optimum management (Thomas 1979a).

Biomagnification: The accumulation of matter with each succeeding trophic level, bottom to top, in an ecosystem. Harmless levels of DDT applied at the lower trophic levels thereby were amplified to harmful levels in the bodies of bald eagles and other carnivores (Bolen and Robinson 1995).

Biomass: The total quantity of living organisms of one or more species per unit of space, which is called species biomass, or of all the species in a community, which is called community biomass (Thomas 1979a).

Biome: A complex of communities with a distinct type of vegetation (Anderson 1991).

Biosphere: The zone of life on Earth, a region usually considered to range from a few hundred meters above ground occupied by high-flying birds to a few meters below ground level where burrowing animals occur. Species living on the ocean floor extend the biosphere by several kilometers (Bolen and Robinson 1995).

Biosphere people: Society that is aware of and uses the biosphere.

Biota: All the plants and animals within an area or region (Anderson 1991).

Biotic: Life or the act of living (Thomas 1979a).

Biotic factors: Living entities—living plants and animals; opposite of *abiotic factors* (Anderson 1991).

Biotic potential: The number of births divided by the population size in a given area in a given time (Anderson 1991).

Biotrophic levels: The feeding levels in a food chain (Thomas 1979a).

Birth flow: A breeding system that pertains to populations whose rate of breeding is relatively constant throughout the year.

Birth pulse: A breeding system that pertains to populations whose rate of breeding occurs one time of the year.

Birth rate: Proportion of a population newly born in a unit of time (Anderson 1991).

Bog: An extremely wet, poorly drained area characterized by a floating spongy mat of vegetation often composed of sphagnum, sedges, and heaths (Anderson 1991).

Bottleneck: See Genetic bottleneck. Also used to describe periods, typically in winter, when food or other resources are limiting, usually to the point of markedly increasing mortality in a wildlife population (Bolen and Robinson 1995).

Breeding (or reproduction) potential: The maximum or unimpeded increase rate of a species in an "ideal" environment.

Brood: A family of young birds from a single mother; sometimes applied to fishes and reptiles (Bolen and Robinson 1995).

Brood parasite: See Nest parasite (Bolen and Robinson 1995).

Browse: Palatable twigs, shoots, leaves, and buds of woody plants. Term often used to describe a category of ungulate forage.

Brucellosis: A bacterial disease that affects mammals, often causing abortions in cattle and some ungulates (Anderson 1991).

Buffer: A species constituting food for predators and acting as a "buffer" to protect primary prey from predators.

Buffer strip: A strip of vegetation that is left or managed to reduce the impact of a treatment or action of one area on another (Thomas 1979*a*).

Burrow: A hole or tunnel dug in the ground by an animal.

Calving area: The areas, usually on spring-fall range, where females give birth to calves and maintain them during their first few days or weeks.

Cannibalism: A special case of carnivory where the prey and predator are the same species.

Canopy: A network of the uppermost branches of a forest that partially or fully covers the understory (Anderson 1991).

Canopy: The more or less continuous cover of branches and foliage formed collectively by the crowns of adjacent trees and other woody growth (Thomas 1979*a*).

Captive breeding: Breeding animals in a captive facility, usually done with endangered species for later release into the wild (Anderson 1991).

Carnivore: A flesh-eating animal (Anderson 1991).

Carrion: Dead animal flesh (Anderson 1991).

Carrying capacity: The maximum rate of animal stocking possible without inducing damage to vegetation or related resources; may vary from year to year because of fluctuating forage production (Thomas 1979*a*).

Catastrophic events: Events resulting from a great and sudden calamity or disaster. In the case of forest stands, such events include windstorms, wildlife, floods, snowslides, and insect outbreaks (Thomas 1979*a*).

Cave: A natural underground chamber that is open to the surface (Thomas 1979*a*).

Cavity: The hollow excavated in snags by birds; used for roosting and reproduction by many birds and mammals (Thomas 1979*a*).

Cavity nester: Wildlife species that nest in tree cavities—for example, woodpeckers (Anderson 1991).

Cecum (plural, ceca):   A dead-end outpouching of the digestive tract. Prevalent in fishes, birds, and mammals with high-fiber diets (e.g., spruce grouse, with a winter diet of evergreen needles) (Bolen and Robinson 1995).

Cementum annuli: Layers of the teeth of some animals that can be used for determining age (Anderson 1991).

Census: A complete enumeration of an entity.

Chaining: A vegetation-maintenance technique in which a heavy anchor chain is dragged between two tractors to break off or uproot plants (Anderson 1991).

Check-out system: Measuring the number of hunters or their kill by checking them in and out at points of entry and exit.

Chill factor: The increased chilling effect on an animal, attributable to wind velocity (Thomas 1979*a*).

Chlorophyll: A complex of mainly green pigments in the chloroplasts, characteristic of plants whose light-energy-transforming properties permit photosynthesis (Thomas 1979*a*).

Chloroplast: The protoplasm body or plastid in plant cells that contains chlorophyll (Thomas 1979*a*).

Circadian rhythm: The regular fluctuations in bodily functions (e.g., temperature) and behavior (e.g., sleeping) during a cycle approximating 24 hours (Bolen and Robinson 1995).

Clean farming: The elimination of diversity and interspersion on farmland and substitution of a monoculture. Hedgerows, roadside vegetation, and similar areas are eliminated. Instead, fields of single crops are cultivated from border to border without cover for wildlife (Bolen and Robinson 1995).

Clearcut: An area from which all trees have been removed by cutting (Thomas 1979*a*)

Climax: The culminating stage in plant succession for a given site where the vegetation has reached a highly stable condition (Thomas 1979*a*).

Climax community: The final or stable biotic community in a developmental series. It is self-perpetuating and in equilibrium with the physical habitat and environment. The presumed end point in succession.

Clumped distribution: An aggregated distribution pattern—for example, a herd of animals (Anderson 1991).

Cluster sampling: Simple random sampling applied to distinct groups of population numbers (Anderson 1991).

Clutch: The eggs laid by a bird or reptile in a single nesting attempt (Bolen and Robinson 1995).

Cohort: A group of individuals in a population born during a particular time period, such as a year (Anderson 1991).

Colonial nesters: Birds that nest in large groups (Anderson 1991).

Colony: A group of the same kind of animals or plants living together (Thomas 1979*a*)

Commensalism: The relation between two populations living together when only one receives a benefit; the other population is neither harmed nor benefited (Anderson 1991).

Commercial harvest: The cutting and marketing of trees for use (Thomas 1979*a*).

Commercial thinning: Any type of thinning that produces merchantable material at least equal to the value of the direct costs of harvesting (Thomas 1979*a*).

Commercial timber production: The process of growing wood products for sale or use (Thomas 1979*a*).

Community: A group of one or more populations of plants and animals in a common spatial arrangement; an ecological term used in a broad sense to include groups of various sizes and degrees of integration.

Community type: An aggregation of all plant communities with similar structure and floristic composition. A unit of vegetation within a classification with no particular successional status implied. A taxonomic unit of vegetation classification referencing existing vegetation.

Compartment: An organization unit or small subdivision of forest area for the purpose of orientation, administration, and silvicultural operations. An area defined by permanent boundaries, either of natural features or artificially marked, which is not necessarily coincident with stand boundaries.

Compensatory mortality: The concept that one kind of mortality largely replaces another kind of mortality in animal populations. In simple logic, an animal dying from one cause (e.g., hunting or disease) cannot die from another cause (e.g., predation or starvation), so one source of mortality compensates for the other. Therefore, the total mortality rate normally is not greatly influenced by changes for any single cause of death. For example, if predation takes 10% of a population and disease takes 20%, total natural mortality for the year is 30%. If, in the next year, predation takes 25% and disease takes 5%, the effect of predation is said to be compensatory. Of importance, the total annual mortality rate may remain essentially unchanged with or without legal hunting (Bolen and Robinson 1995).

Compensatory population growth: Equivalent to density-dependent population growth. That is, growth of a population is rapid when numbers are low and less rapid when numbers are high. Such growth may result from decreased mortality, increased natality, or both (Bolen and Robinson 1995).

Competition: The active demand by two or more organisms for a commonly required resource that is limited (Anderson 1991).

Competitive exclusion principal (Gause's hypothesis): No two species can occupy the same niche at the same time (Anderson 1991).

Compound 1080: Sodium monofluoroacetate, commonly used as a coyote poison before restrictions were imposed in the United States in 1973. Considered dangerous because of nontarget victims (Bolen and Robinson 1995).

Condition index: A measure of an animal's well-being, usually expressed in terms of fat content. Most condition indices are adjusted for size differences among individuals by dividing fat weight by some other physical feature (e.g., wing length for birds). In mammals, fat deposits in bone marrow or around kidneys are common condition indices (Bolen and Robinson 1995).

Conifer (coniferous): A cone-bearing plant (Anderson 1991).

Coniferous forest: A forest dominated by cone-bearing trees (Thomas 1979*a*).

Connectors: Strips or patches of vegetation used by wildlife to move between habitats (Thomas 1979*a*).

Conservation: Wise maintenance and use of natural resources (Anderson 1991).

Conservation biology: A field of many disciplines united with the common goal of preserving biodiversity. Genetics, physiology, geography, population biology, wildlife management, forestry, and veterinary science are among the basic and applied disciplines contributing to conservation biology; the professional staffs at many zoos also provide crucial expertise (Bolen and Robinson 1995).

Consumer: A member of the animal community. Consumers occupy the higher trophic levels in an ecosystem (Bolen and Robinson 1995).

Consumptive use: Use of resources that involves removal (e.g., hunting and fishing) (Anderson 1991).

Contour line: An imaginary line, or its representation on a contour map, joining points of equal elevation (Thomas 1979*a*).

Contrast: In wildlife management, the degree of difference in vegetative destructive structure along edges where plant communities meet or where successional stages or vegetative conditions within plant communities meet (Thomas 1979*a*).

Control: In wildlife management, the process of managing populations of a species to accomplish an objective; usually used in the sense of depressing population numbers of a pest species to prevent or decrease the impact of that species (Thomas 1979*a*).

Coordinated resource management (CRM): The process whereby various user groups are involved in discussion of alternative resource uses and collectively diagnose management problems, establish goals and objectives, and evaluate multiple-use resource management.

Coordinated timber-wildlife management: The melding of timber and wildlife management planning and action into one plan so that goals of both timber and wildlife are met (Thomas 1979*a*).

Coprophagy: The practice of eating feces. Coprophagy enables rabbits to recover nutrients from their droppings that escaped initial digestion (Bolen and Robinson 1995).

Corpora lutea: See Corpus luteum. (Bolen and Robinson 1995).

Corpus luteum (plural, Corpora lutea): A structure formed in the mammalian ovary from the follicle that once contained an ovum (egg); functions as an endocrine gland by secreting hormones to maintain pregnancy. Corpora lutea, which are visible when an ovary is sectioned, indicate the number of ova (eggs) shed by the ovary, thereby providing data on fertility (Bolen and Robinson 1995).

Corridor: A strip or block or habitat connecting otherwise isolated units of suitable habitat that allows the dispersal of organisms and the consequent mixing of genes (Bolen and Robinson 1995).

Cover: A general term used to describe vegetation and topography. Vegetative cover is divided into three categories: (1) the overstory of trees, (2) the midstory composed mainly of large shrubs and small trees and (3) the understory that includes small shrubs, grasses, and forbs. Cover requirements for various species of wildlife are discussed in the handbook. Types of cover include: brood cover (low-lying vegetation such as grasses and forbs that afford protection for young game birds, usually quail, turkeys, and grouse), escape cover (shrubby or herbaceous cover, hollow trees, water, rock crevices, or burrows that provide a means of getting away when harassed by predators), nesting cover (vegetation required by birds or animals to rear their young. Examples are tree dens for squirrels or grassy patches near openings for quail, turkey, grouse, or rabbits), roosting cover (cover required by game birds. It ranges from conifers for turkey to idle fields or sparse timberlands for quail) and wintering cover. All cover required by game to overwinter. (It varies from trees with squirrel dens, to brush cover for quail, and rock outcrops for bear).

Cover, canopy: The percentage of ground covered by a particular projection of the outermost perimeter of the natural spread of foliage of plants. Small openings within the canopy cover may exceed 100%.

Cover, ground: The percentage of material, other than bare ground, covering the soil surface. It may include organic material such as vegetation basal cover (live and standing dead), mosses and lichens, and litter, and inorganic material such as cobble, gravel, stones, and bedrock. Ground cover plus bare ground will total 100%.

Cover patch: A discrete area covered by vegetation that meets either the definition of hiding or thermal cover (Thomas 1979*a*).

Cover type: A taxonomic unit of vegetation classification referencing existing vegetation. Cover type is a broad talon based on existing plant species that dominate, usually within the tallest layer.

Covert: A geographic unit of game cover.

Covey: A small flock of birds which "lie."

Crash: The period of severe mortality following the peak of a cycle.

Critical area: An area which must be treated with special consideration because of inherent site factors, condition, values, or significant potential conflicts among uses.

Critical foods: Foods essential or necessary to the welfare of a game species; i.e, hard mast for gray squirrel.

Critical habitat: A legal term describing the physical or biological features essential to the conservation of a species.

Crop: Anatomically, an expandable part of a bird's esophagus used for food storage and, perhaps, a small amount of digestion. Analysis of crop contents helps determine the diets of birds because foods in the gizzard (stomach) often are ground beyond recognition. Crop also refers to the harvested part of a plant or animal population (Bolen and Robinson 1995).

Crop tree: Any tree forming, or selected to form, part of the final crop; generally a tree selected in a young stand for that purpose (Thomas 1979*a*).

Crown: The upper part of a tree or other woody plant, carrying the main branch system and foliage, and surmounting at the crown base a more or less clean stem (Thomas 1979*a*).

Cruising radius: The distance between locations at which an individual animal is found at various hours of the day or at various seasons, or during various years.

Culling: Removal of animals from a population, usually individuals at risk. For example, predators normally remove sick and injured individuals from a prey population. In some cases, culling may apply to the general thinning of a population, as when the density of foxes or raccoons is reduced for control of rabies (Bolen and Robinson 1995).

Cycle: A regular pattern, such as a repetitious change in population size (Anderson 1991).

DDT: Chemical abbreviation for dichloro-diphenyl-trichloroethane, perhaps the most familiar pesticide in the "family" of chlorinated hydrocarbons; banned in the United States in 1972, in part because of its harmful effects on wildlife (Bolen and Robinson 1995).

Dead and down woody material: All woody material, from whatever source, that is dead and lying on the forest floor (Thomas 1979*a*).

Debris: The scattered remains of something broken or destroyed; ruins, rubble; fragments (Thomas 1979*a*).

Deciduous: Pertaining to any plant organ or group of organs that is shed naturally; perennial plants that are leafless for some time during the year (Thomas 1979*a*).

Deciduous tree: A tree that drops its leaves each autumn. Usually broad-leaved trees of which maple, oak, birch, and hickory are examples (Bolen and Robinson 1995).

Decimating factors: Those that kill directly (e.g., hunting, predation, disease and parasites, accidents, starvation).

Decompose: To separate into component parts or elements; to decay or putrefy (Thomas 1979*a*).

Decomposer: One of the many organisms, principally bacteria, that reduce animal wastes and the carcasses of complex organisms into elemental components (Bolen and Robinson 1995).

Deer yard: An area of heavy cover where deer congregate in the winter for food and shelter (Anderson 1991).

Delphi technique: The process of combining expert opinions into a consensus; a method of making predictions (Thomas 1979*a*).

Den trees or dens: A rainproof, weather-tight cavity in a tree. Dens take eight to 30 years to develop following injury, disease, or natural pruning and are aided by periodic enlargement by squirrels or other wildlife.

Denning site: A place of shelter for an animal; also where an animal gives birth and raises young (Thomas 1979*a*).

Density: The number of individuals per unit area.

Density dependent: More severely affecting a population as the population size increases (Anderson 1991).

Density-dependent factor: A factor that acts in proportion to the density of animals. Some diseases are density dependent because a higher percentage of the population becomes infected as density increases. Natality and mortality often fluctuate with changes in density (Bolen and Robinson 1995).

Density independent: Having impact on a population not related to the population's size (Anderson 1991).

Density-independent factor: A factor that acts independently of population density. Weather is often considered density independent. A flood, for example, may kill an entire population regardless of density (Bolen and Robinson 1995).

Dependent variable: The variable in a relationship that is influenced by the independent variable (Thomas 1979*a*).

Desertion limit: The number of days after incubation starts when normal disturbances of the nest will not cause desertion.

Deterministic model: A mathematical model in which all the relationships are fixed and the concept of probability does not enter; a given input produces a predictable output (Anderson 1991).

Detritus food chain: A process in which dead organisms are decomposed by other organisms, such as worms, larvae, and bacteria (Anderson 1991).

Diameter breast high (d.b.h.):   The standard diameter measurement for standing trees, including bark, taken at 1.37 meters above the ground (Thomas 1979*a*).

Direct habitat improvement: Habitat manipulations primarily for the benefit of wildlife (Thomas 1979*a*).

Direct improvement: Land treatment measures or structures installed to benefit game, i.e., seeded wildlife openings, waterholes, or daylighting fruit-producing shrubs.

Dispersal: Movement of organisms into unfamiliar locations; behavior usually associated with younger animals upon leaving natal areas. *Pioneering* is a synonym (Bolen and Robinson 1995).

Dispersion (also see "law of dispersion"):   The pattern of distribution of individuals in an animal population; in the mathematical sense, dispersion describes the probability of occurrence of such individuals in particular places (Thomas 1979*a*).

Dispersion failure: A planting of wildlife followed by immediate dispersal and disappearance.

Distribution: The spread or scatter of an entity within its range.

Diversity: The total range of wildlife species, plant species, communities, and habitat features in an area (Anderson 1991).

Diversity index: A number that indicates the relative degree of diversity in habitat per unit area. It is expressed mathematically:

$$DI = \frac{TP}{2\sqrt{A \cdot \pi}}$$

where TP is the total perimeter of an area plus any edge within the area in meters or feet, A is the area in square meters, and π is 3.1416 (Thomas 1979*a*).

Dominant: Plant species or species groups that by means of their number, coverage, or size have considerable influence over control upon the conditions of existence of associated species. Also, individual animals that determine the behavior of one or more other animals, resulting in the establishment of a social hierarchy (Thomas 1979*a*).

Dragging: The use of a log or metal grate behind a tractor to loosen cattle manure in a field (Anderson 1991).

Drum: To make a reverberating sound by beating the wings rapidly as grouse do or by tapping on a suitable surface as woodpeckers do (Thomas 1979*a*).

Dusting: The process of rolling or exercising vigorously in dust or duff; in birds, has the function of aligning barbules and maintaining feathers (Thomas 1979*a*).

Dynamic: Characterized by or tending to produce continuous change or advance (Thomas 1979*a*).

Easement: An access area across another's land (Anderson 1991).

Ecological characteristics: The basic features of a species related to distribution, habitat, reproduction, growth characteristics and needs, and responses to habitat changes (Anderson 1991).

Ecological equivalents: Organisms occupying the same niche but living in different communities. Examples include bison and wildebeest occupying grazing niches in the plains of North America and Africa, respectively (Bolen and Robinson 1995).

Ecological longevity: The average length of life of individuals of a population under stated conditions (Anderson 1991).

Ecological niche: The role a particular organism plays in the environment (Thomas 1979*a*).

Ecological role: The part or influence of an organism in an ecosystem (Thomas 1979*a*).

Ecology: The study of interactions between organisms and their environment. Term coined by German zoologist Ernst Haeckel in 1866 based on the Greek root *oikos,* meaning "home," and *logos,* meaning "study" (Bolen and Robinson 1995).

Ecosystem: Living and nonliving components in an environment functioning together (Anderson 1991).

Ecosystem management: Ecosystem management means using an ecological approach to achieve the multiple-use management of national forests and grasslands by blending the needs of people and environmental values in such a way that national forests and grasslands represent diverse, healthy, productive, and sustainable ecosystems.

Ecosystem people: Humans that live directly off the land.

Ecotone: The community formed where two other communities meet, sometimes called an edge (Anderson 1991).

Edaphic: An adjective pertaining to soil (e.g., concerning texture, drainage, and fertility). Edaphic factors often are of direct importance to burrowing animals, but also exert major indirect influences on all wildlife because of ecological links with vegetation (Bolen and Robinson 1995).

Edge: The area where two communities meet (ecotones) (Anderson 1991).

Edge effect: The ecological result of increasing edges in homogeneous habitats, principally the increased abundance and diversity of species. A benefit in the management of some animals (bobwhite and other "edge species"), although increased predation or nest parasitism may result in other cases (Bolen and Robinson 1995).

Efficiency: Proportion of incoming solar energy converted to chemical energy (Anderson 1991).

Elk calving habitat: A habitat used by elk for calving; usually located on spring-fall range in areas of gentle slope; contains forage areas and hiding and thermal cover close to water (Thomas 1979*a*).

Emigration: Movement out of a given area (Anderson 1991).

Endangered species: A wildlife species officially designated by the U.S. Fish and Wildlife Service as having its continued existence threatened over its entire range because its habitat is threatened with destruction, drastic modification, or severe curtailment, or because of overexploitation, disease, predation, or other factors (Thomas 1979*a*).

Endemic: Native to a region (Anderson 1991).

Environment: The sum total of all the external conditions that may influence organisms (Thomas 1979*a*).

Environmental analysis report: A report on environmental effects of proposed federal actions that require an Environmental Impact Statement under section 102 of the National Environmental Policy Act (U.S. Laws, Statutes, etc. Public Law 91–190, 1970); an "in house" document that becomes the final document on projects whose effects are so minor as not to require a formal Environmental Impact Statement; though not formally required by the Act, this document is commonly used to determine if section 102 applies to the contemplated action (Thomas 1979*a*).

Environmental factor: Any influence on the combined plant and animal community (Thomas 1979*a*).

Environmental Impact Statement: The final version of the statement required under section 102 of the National Environmental Policy Act (U.S. Laws, Statutes, etc. Public Law 91–190, 1970) for major federal actions affecting the environment; a revision of the draft statement that includes public and governmental agency comments; a formal document meeting legal requirements and used as the basis for judicial decisions concerning compliance with the act; also refers to similar statements required by state and local laws patterned after the act (Thomas 1979*a*).

Environmental resistance: Factors that act to slow a population's growth (Anderson 1991).

Enzootic: The chronic level of disease frequency; that is, a low but constant occurrence of a disease in a population (Bolen and Robinson 1995).

Ephemeral streams: Streams that contain running water only for brief periods (Thomas 1979*a*).

Epizootic: An outbreak of disease. Large numbers of animals die in a short period (Bolen and Robinson 1995).

Epizootiology: The study of disease ecology. Addresses the "how" and "why" of diseases at either enzootic or epizootic levels (Bolen and Robinson 1995).

Equilibrium carrying capacity: The density of a population is at a constant equilibrium but the limiting variables are unknown.

Escape cover: Usually vegetation dense enough to hide an animal; used by animals to escape from potential enemies (Thomas 1979*a*).

Esophagus: A tube in the digestive system of vertebrates that connects the mouth with the stomach. Expands in some birds for temporary food storage (Bolen and Robinson 1995).

Estimate: A result of a statistical sample, often for determining population size. Used instead of a census or index. To obtain an estimate, animals or inanimate objects (e.g., nests or dens) may be counted on one or more sample areas known as plots (Bolen and Robinson 1995).

Ethology: The study of animal behavior (Bolen and Robinson 1995).

Etiology: The study of the cause of disease.

Eury-:   A prefix used to identify wide tolerances for specific components of the environment. White-tailed deer are eurythermal because of their wide tolerance of temperature extremes (Bolen and Robinson 1995).

Eurytopic: Able to withstand wide variations in environmental conditions (Anderson 1991).

Eutrophic: Term describing the enriched nature of some freshwater lakes and rivers. Organic materials and nutrients accumulate, leading to increased biological productivity. The enrichment process is known as eutrophication. Human activities may affect the process abnormally (Bolen and Robinson 1995).

Eutrophication: Process of lake succession by addition of nutrients (Anderson 1991).

Even-aged management: A system of forest management in which stands are produced or maintained with relatively minor differences in age (Thomas 1979*a*).

Evolution: The change in a population's genetic composition over time, leading to adaptations to the environment (Anderson 1991).

Exotic species: An organism introduced—intentionally or accidentally—from its native range into an area where the species did not previously occur. Asian animals such as pheasants, sika deer, and carp are exotic species in North America (Bolen and Robinson 1995).

Exponential growth: Population growth that exceeds the carrying capacity until population numbers saturate the habitat; growth characterized by a progressively increasing, nonlinear relation between population numbers and time (Anderson 1991).

Extinction: The complete loss—forever—of a unique constellation of genes known as a species (e.g., passenger pigeon) (Bolen and Robinson 1995).

Extirpation: The elimination of a species from one or more specific areas, but not from all areas. Extirpation should not be confused with extinction. Bison have been extirpated from most of their former range in North America, but many hundreds still exist on private ranches, zoos, and federal refuges (Bolen and Robinson 1995).

Factor: One of the forces reducing the numbers (decimating factor) or retarding the increase rate (welfare factor) of wildlife.

Fauna: A term for all animal life.

Fawning area: An area, usually on spring-fall range, where females give birth to fawns and maintain them in their first few days or weeks (Thomas 1979a).

Featured species: The selected wildlife species whose habitat requirements guide wildlife management including coordination, multiple-use planning, direct habitat improvements, and cooperative programs for a unit of land.

Featured-species management: A management policy keyed to a single species, perhaps at the expense of others. Often used for endangered species (e.g., forest management for red-cockaded woodpeckers) but also applied to more abundant species (Bolen and Robinson 1995).

Fecal material: Material discharged from the bowels; more generally, any discharge from the digestive tract of an organism (Thomas 1979a).

Fecundity: Actual number of organisms produced.

Feedback: The output of a given system that affects the state of that same system (Anderson 1991).

Feeding substrate: The surface on which an animal finds its food (Thomas 1979a).

Feral: Having reverted to a wild state after being domesticated; e.g., feral horses (Anderson 1991).

Fertility: The potential capability of an organism to produce young.

Fitness: The competence of an organism to pass on its genes to the next generation. All else being equal, some individuals within a population are of greater fitness than others. Fitness may be influenced by many factors, including physical, physiological, and social conditions (Bolen and Robinson 1995).

Flight limit: The maximum distance a bird can traverse at one continuous flight.

Flora: A term for all plant life.

Fluctuations: Irregular changes.

Food chain: The energy flow from green plants through consumer organisms at each trophic level (Anderson 1991).

Food habits: A term for the diet of wildlife. Thorough studies of food habits include feeding behavior and food availability as well as quantifying items in the diet (Bolen and Robinson 1995).

Food web: A complex food chain (Anderson 1991).

Forage: Browse and herbage, which is available and may provide food for grazing animals or be harvested for feeding.

Forage areas: Forest stands that do not qualify as either hiding or thermal cover and all natural and man-made openings (Thomas 1979a).

Forage medium: The environment in which feeding by a wildlife species occurs (Thomas 1979a).

Forb: Any herbaceous plant other than grasses or grasslike plants.

Forced trailing: Trails that compel animals, particularly big game and domestic livestock, to take a particular route of travel (Thomas 1979*a*).

Forest: Generally, an ecosystem characterized by tree cover; more particularly, a plant community predominantly of trees and other woody vegetation, growing closely together; an area managed for the production of timber and other forest produce, or maintained in forest cover for such indirect benefits as protection of catchment areas or recreation; an area of land proclaimed to be a forest under a forest act or ordinance (Thomas 1979*a*).

Forest management: The application of scientific, economic, and social principles to the management of a forest estate for specified objectives; the branch of forestry concerned with its overall administrative, economic, legal, social, scientific, and technical aspects—especially silviculture, protection, and forest regulation (Thomas 1979*a*).

Fossorial species: Animals adapted for living underground (Bolen and Robinson 1995).

Free water: Water that is not bound to any surface, particularly soil particles, and is available for transpiration of plants (Thomas 1979*a*).

Furbearer: A mammal commonly harvested for its hide (e.g., muskrat, mink) (Anderson 1991).

Gallinaceous bird: A bird of the order of Galliformes. These are chickenlike birds, including pheasants, quail, grouse, prairie chickens, ptarmigan, and turkeys (Bolen and Robinson 1995).

Game: Species of vertebrate wildlife hunted by man for sport (Thomas 1979*a*).

Gamma diversity: Diversity comparison of large, heterogenous areas (Anderson 1991).

Gene pool: Narrowly, all the genes of a localized interbreeding population; broadly all the genes of a species throughout its entire range (Anderson 1991).

Generality: The applicability of a model to appropriate situations (Anderson 1991).

Genetic bottleneck: The temporary reduction of a population to only a few individuals, thereby limiting the gene pool and increasing inbreeding. When such a population later increases in size, it will still have a limited gene pool (Bolen and Robinson 1995).

Genetic composition: The total genetic makeup of a population (Anderson 1991).

Genotype: The entire genetic constitution of an organism; contrast with *phenotype* (Anderson 1991).

Genus: One or a group of related species used in classification of organisms, the first word in binomial or scientific name (Thomas 1979*a*).

Geomorphic: Of or like the earth or the configuration or shape of the earth's surface (Thomas 1979*a*).

Gestation: The length of time from conception to birth (Anderson 1991).

Gizzard: The muscular stomach of birds that grinds food, usually with the aid of grit. Gizzards serve as the functional equivalent of teeth in mammals (Bolen and Robinson 1995).

Gleaning: A process of feeding, particularly in birds, in which food items are gathered from the surface of the foraging substrate, usually plants (Anderson 1991).

Gradient: The rise or fall of a ground surface expressed in degrees of slope (Thomas 1979*a*).

Grass: Any plant species that is a member of the family Gramineae.

Grass-like plant: A plant that vegetatively resembles a true grass of the Gramineae.

Gravid: Term describing females containing ripening eggs. Major physiological differences aside, the gravid condition in egg-laying species is somewhat analogous to pregnancy in mammals (Bolen and Robinson 1995).

Grazing food chain: The movement of energy from green plants to herbivores to carnivores, excluding composition (Anderson 1991).

Grazing system: A specialization of grazing management that defines systematically recurring periods of grazing and deferment for two or more pastures or management units.

Gross primary productivity: The total amount of energy available from the conversion of solar energy to chemical energy during photosynthesis by green plants (Anderson 1991).

Ground cover: The percentage of material, other than bare ground, covering the land surface. It may include live vegetation, standing dead vegetation, litter, cobble, gravel, stones, and bedrock. Ground cover plus bare ground would total 100%.

Guild: A group of species that exploits the same class of environment resources in a similar way.

Habitat: The resources and conditions present in an area that produce occupancy, including survival and reproduction by a given organism. Habitat is species specific.

Habitat availability: The accessibility and procurability of physical and biological components in a habitat.

Habitat avoidance: An oxymoron that should not be used; wherever an animal occurs defines its habitat.

Habitat block: An area of land covered by a relatively homogenous plant community, essentially a single successional stage or condition (Thomas 1979*a*).

Habitat component: A simple part, or a relatively complex entity regarded as a part, of an area or type of environment in which an organism or biological population normally lives or occurs (Thomas 1979*a*).

Habitat evaluation procedure (HEP): A method of the U.S. Fish and Wildlife Service that documents the quality and quantity of resources available for selected species of wildlife; involves estimation of a habitat suitability index (HSI) and the amount of such habitat available. HEP numerically predicts the effects of altering habitats by development or by natural events (Bolen and Robinson 1995).

Habitat niche: The peculiar arrangement of food, cover, and water that meets the requirements of a particular species (Thomas 1979*a*).

Habitat preference: Used to describe the relative use of different locations (habitats) by an individual or species.

Habitat quality: The ability of the area to provide conditions appropriate for individual and population persistence.

Habitat richness: The relative degree of ability of a habitat to produce numbers of species of either plants or animals; the more species produced the richer the habitat (Thomas 1979*a*).

Habitat selection: A hierarchical process involving a series of innate and learned behavioral decisions made by an animal about what habitat it would use at different scales of the environment.

Habitat suitability index (HSI):   Part of habitat evaluation procedure. A value ranging from 0.0 to 1.0, assigned on the basis of known food and cover requirements of a species. Habitat suitability index is multiplied by the area of habitat (in acres) to obtain the habitat units (HUs) available for each species (Bolen and Robinson 1995).

Habitat type: The aggregate of all areas that support, or can support, the same primary vegetative climax, a classification of environmental settings characterized by a single plant association; the expression through the plants present of the sum of the environmental factors that influence the nature of the climax (Thomas 1979*a*).

Habitat unit (HU):   Part of habitat evaluation procedure. The product resulting from Habitat Suitability Index (HSI) x the area available with that HSI. Used for assigning a comparative value for habitat of various species (Bolen and Robinson 1995).

Habitat use: The way an animal uses (or "consumes," in a generic sense) a collection of physical and biological entities in a habitat.

Habituation: A behavioral tendency wherein animals become accustomed to unnatural components in their environment. Deer feeding at the edges of highways often become habituated to traffic (Bolen and Robinson 1995).

Half-shrub: A perennial plant with a woody base; the annually produced stems die each year.

Hard snag: A snag composed primarily of sound wood, particularly sound sapwood; generally merchantable (Thomas 1979*a*).

Hardwood: The wood of broad-leaved trees, and the trees themselves, belonging to the botanical group Angiospermae; distinguished from softwoods by the presence of vessels (Thomas 1979*a*).

Harvest: Removal of animals from a population (Anderson 1991).

Hawking: The feeding behavior of birds wherein they capture food in flight (Thomas 1979*a*).

Herb: Any vascular plant except those developing persistent woody stems above ground.

Herbaceous: Term descriptive of nonwoody plants. Herbaceous vegetation includes forbs and grasses (except bamboo). Also the nonwoody parts of trees and shrubs (e.g., leaves) (Bolen and Robinson 1995).

Herbage: The aboveground material of any herbaceous plant.

Herbicide: A chemical substance used for killing plants (Thomas 1979*a*).

Herbivore: An animal that feeds on plants (Anderson 1991).

Herd: Any large aggregation, or detached unit, of hoofed mammals.

Herd behavior: The collective behavior exhibited by social groups of ungulates (Thomas 1979*a*).

Herptofauna: Term combining the amphibian and reptilian members of a community. *Herptiles* (or *herps*) likewise is a term combining amphibians and reptiles (Bolen and Robinson 1995).

Heterotroph: An organism that uses chemical energy supplied by autotrophic organisms (Anderson 1991).

Heterozygosity: The presence of both forms (known as *alleles*)—dominant and recessive—of a single gene in an individual or population. Heterozygosity generally conveys an advantage for coping with new situations. An individual with a genetic composition (genotype) represented as AaBbCc is heterozygous, whereas another individual of the same species with AabbCC or aaBBCC is homozygous. In a population, heterozygosity provides a variety of genotypes, thereby increasing the likelihood that some individuals are capable of surviving stresses acting on the population and thereby increasing fitness (Bolen and Robinson 1995).

Hibernacula: Habitat niches where certain animals overwinter (Thomas 1979*a*).

Hiding cover for deer: Vegetation capable of hiding 90% of a standing adult deer from the view of a human at a distance equal to or less than 61 meters; generally, any vegetation used by deer for security or to escape from danger (Thomas 1979*a*).

Hiding cover for elk: Vegetation capable of hiding 90% of a standing adult elk from the view of a human at a distance equal to or less than 61 meters; generally, any vegetation used by elk for security or to escape from danger (Thomas 1979*a*).

Hole nesters: Wildlife species that nest in cavities (Thomas 1979*a*).

Holistic: Emphasizing the importance of the whole and the interdependence of its parts (Thomas 1979*a*).

Home range: The area traversed by an animal during its activities in a specified period of time.

Homeostasis: A stable state or the tendency of a system to maintain a stable or balanced state (Anderson 1991).

Homotherm: A warm-blooded animal that can regulate its body temperature physiologically (Anderson 1991).

Homeothermic: Term indicating maintenance of stable body temperatures independently of environmental temperatures. Homeotherms—mammals and birds—can produce metabolic heat to meet the stress of low ambient temperatures. "Warm blooded" is a population description of homeothermic species (Bolen and Robinson 1995).

Homozygosity: The limited presence or total absence of either the dominant or recessive form (allele) of a gene in an individual or population. The genes of a homozygous individual may be represented as AabbCC or AABBcc. A population composed mainly of homozygous individuals lacks genetic diversity, thereby reducing the likelihood that some individuals are genetically more capable of surviving stresses than those in a heterozygous population. Therefore, a homozygous population is generally less able to survive over a long period of time than a heterozygous population (Bolen and Robinson 1995).

Horn: A structure protruding from the skulls of goats, sheep, and bovines (antelope, cows, and bison) consisting of a keratin sheath surrounding a core of bone. Usually paired, except in rhinoceros. There is no velvet covering during development. Rarely branched or shed, but pronghorns are notable exceptions. Horns occur in both sexes (Bolen and Robinson 1995).

Host: An organism that furnishes food, shelter, or other benefits to another species (Anderson 1991).

I-carrying capacity: The population density of the inflection point of the logistic curve.

Immature: An animal, or age class, incapable of breeding and not looking like an adult; usually young-of-the-year. Juvenile is a synonym (Bolen and Robinson 1995).

Immigration: Movement into a given area (Anderson 1991).

Impact (population): A change in a population's natality, growth, and/or survival caused by some disturbance (Anderson 1991).

Imprinting: Recognition fixed through a short-interval learning process in animals; young might imprint on parents or animals might imprint on a nesting habitat (Anderson 1991).

Inactive management: No direct action is allocated toward the manipulation of wildlife populations.

Inbreeding: Breeding among genetically similar individuals in a population leading to a reduced genetic variability (*homozygosity*) (Anderson 1991).

Inbreeding depression: The undesirable result of repeated matings within a small population of related individuals, typically reducing the gene pool and producing abnormalities and lessened fitness in the offspring (Bolen and Robinson 1995).

Inclusion: Areas of less than stand size that have an inherently different management type and productivity than that of the stand in which they lie. They can be treated differently than the stand if they are classified as key areas.

Independent variable: The variable in a relationship that is judged for its effect on the dependent variable (Thomas 1979*a*).

Index: A means of comparing the relative size of an animal population, usually from year to year, but it does not provide a census or estimate of actual numbers. Roadside drumming counts of ruffed grouse provide a useful population index (Bolen and Robinson 1995).

Indicator: A condition that is visible, and that denotes some other condition that is invisible.

Indicator species: Species that indicate certain environmental conditions, seral stages, or treatments.

Indicator species system: In wildlife, analagous to "featured species management"; in plant ecology, plant species used to indicate special environmental factors or plant community types (Thomas 1979*a*).

Indices: Indicators of population changes through repeated measurements; generally show population trends (Anderson 1991).

Indirect habitat improvement: Habitat manipulation done for purposes other than wildlife habitat improvement but exploited to accomplish wildlife management objectives (Thomas 1979a).

Induced diversity index: A number that indicates the relative degree of induced diversity in habitat per unit area produced by edges formed at the junction of sucessional stages, or vegetative conditions within plant communities; expressed mathematically:

$$\text{Induced DI} = \frac{\text{TE}_s}{2\sqrt{A \cdot \pi}}$$

where TE_s is the total length of edges between successional stages or conditions within plant communities, in meters or feet; A is the area expressed in square meters, and π is 3.1416 (Thomas 1979a).

Induced edge: An edge that results from the meeting of two successional stages or vegetative conditions within a plant community, can be controlled by management action (Thomas 1979a).

Influence: An environmental variable that influences a factor.

Inherent: In ethology, a behavioral characteristic that is not learned. Innate and instinctive are synonyms (Bolen and Robinson 1995).

Inherent diversity index: A number that indicates the relative degree of inherent diversity in habitat per unit area produced by plant community to plant community edges; expressed mathematically:

$$\text{Induced DI} = \frac{\text{TE}_c}{2\sqrt{A \cdot \pi}}$$

where TE_c is the total edge between plant communities within or on the perimeter of the area; A is the area and π is 3.1416 (Thomas 1979a).

Inherent edge: An edge that results from the meeting of two plant community types (Thomas 1979a).

Innate capacity for increase (r_m): A measure of the rate of increase of a population under "ideal" conditions (Anderson 1991).

Insectivorous: Insect eating (Anderson 1991).

Integrated pest management (IPM): A combination of chemical, cultural, and biological methods designed to minimize pest damage. For example, IPM recognizes the beneficial role of predaceous insects in the control of herbivorous insects (Bolen and Robinson 1995).

Interface: A surface forming a common boundary between two regions or between two things (Thomas 1979a).

Intermittent stream: A stream that ordinarily goes dry at one or more times during the year but sustains flows for some period (Thomas 1979a).

Interspecific competition: Competition between individuals of different species (Anderson 1991).

Interspersion: The degree to which environmental types are intermingled rangeland.

Intraspecific competition: Competition between members of the same species (Anderson 1991).

Intrinsic rate of increase: Difference between the birth and death rates in a population (Anderson 1991).

Inventory: A detailed list of things in possession; especially, a periodic survey of goods and material in stock; the process of survey; the items and the quantity of goods and materials listed (Thomas 1979a).

Invertebrate: Any animal without a vertebral column (i.e., backbone). Insects and crustaceans are particularly abundant groups of invertebrates (Bolen and Robinson 1995).

Irruption: A large, sudden, nonperiodic increase in density, often accompanied by an extension into hitherto unoccupied range.

Island biogeography: The study of natural communities on islands, with an emphasis on species diversity as related to an island's area and its distance from the mainland. Principles apply not only to actual islands, but also to fragmented forests and other patches of habitat (Bolen and Robinson 1995).

Juvenile: Immature (Bolen and Robinson 1995).

Juxtapose: To situate side by side; to place together (Thomas 1979a).

Juxtaposition: The act of arranging in space.

K-carrying capacity: The maximum number of animals of a given population that can be supported by available resources.

Key area: A portion of rangeland selected because of its location, grazing or browsing value, or use. It serves as a monitoring and evaluation point for range condition, trend, or degree of grazing use. Properly selected key areas reflect the overall acceptability of current grazing management over the rangeland. A key area guides the general management of the entire area of which it is a part.

Key species: 1. Forage species whose use serves as an indicator to the degree of use of associated species. In many cases, key species include indicator species and species traditionally referenced as increasers, decreasers, desirables, or intermediates. 2. Those species that must, because of their importance, be considered in the management program.

Key species management: In wildlife, analagous to "featured species management;" in range management, the most palatable and common plant species used by livestock; the plants on whose status livestock management decisions are based (Thomas 1979a).

Kill: The number of head killed per year from a unit of population.

Kill ratio: The proportion or percent of the game population that can be killed yearly without diminishing subsequent crops. The ratio of the yield to the population.

K-selected species: Species adapted to low birth rates, low death rates, and long life spans. Depressed populations of such species recover slowly and often have specialized habitat needs. As a result, overkill and other disasters easily jeopardize K-selected species. Whales, elephants, and whooping cranes are examples (Bolen and Robinson 1995).

Lagomorph: Any member of the mammalian order Lagomorpha, which includes rabbits (cottontails), hares, and pikas. Often mistaken for rodents, lagomorphs have two pairs of upper incisors, one small set directly behind the larger front pair (Bolen and Robinson 1995).

Land base: The amount of land with which the land manager has to work (Thomas 1979*a*).

Landform: Any physical, recognizable form or feature of the earth's surface having characteristic shape and produced by natural causes.

Landscape: A spatially heterogeneous area used to describe features (e.g., stand type, site, soil) of interest.

Landscape feature: Widespread or characteristic features within the landscape (e.g., stand type, site, soil, patch).

Landscape foreground: Areas managed particularly for their impact on the visual attributes of an area as viewed from a specified point (Thomas 1979*a*).

Law of dispersion: An ecological theory; the potential density of wildlife species with small home ranges that requires two or more types of habitat and is roughly proportional to the sum of the peripheries of those types (Thomas 1979*a*).

Law of interspersion: The number of resident wildlife species that requires two or more types of habitat depends on the degree of interspersions of numerous blocks of such types (Thomas 1979*a*).

Law of the minimum: An ecological axiom that states that any factor a population requires that is present in the smallest amount limits the population's growth accordingly (Anderson 1991).

Legume: Any of a large group of the pea family that has five pods enclosing the seeds (Anderson 1991).

Lek: A site where birds (primarily grouse) traditionally gather for sexual display and courtship (Anderson 1991).

Life form: A group of wildlife species whose requirements for habitat are satisfied by similar successional stages within given plant communities (Thomas 1979*a*).

Life-form association: A group of organisms whose requirements are satisfied by similar successional stages in the development of communities (Bolen and Robinson 1995).

Life tables: A table of population data based on a sample (often 1000 individuals) of the population showing the age at which each member died. A dynamic or **cohort** life table starts with a group of individuals all born during the same time period. A **static** or time-specific life table has a sample of individuals from each age class in the population (Anderson 1991).

Limiting factor: That factor or condition greater than or outweighing other factors in limiting wildlife population growth.

Lincoln index: A ratio based on banding and used for census.

Litter: 1. A family of young mammals from a single mother. 2. Uppermost layer of organic debris on the soil surface; essentially freshly fallen or slightly decomposed vegetative material (Bolen and Robinson 1995).

Livestock: Domestic animals, usually ungulates, raised for use, profit, or pleasure (Thomas 1979*a*).

Loafing cover: A place offering shade in summer or sun and wind protection in winter for idling.

Logistic growth: Growth of a population that approaches and remains near the carrying capacity (Anderson 1991).

Lopping: After felling, the chopping of small trees and branches and tops of large trees so that the resultant slash will lie close to the ground and decay more rapidly or be available for forage (Thomas 1979*a*).

Managed forest: A forest that has been brought under management to accomplish specified objectives.

Managed stand: A stand subject to silvicultural manipulation planned to give a desired result—usually increased wood production.

Management: Manipulation of populations or habitats to achieve desired goals by people (Anderson 1991).

Management by objectives: Planning by putting program objectives in priority order (Anderson 1991).

Management for species richness: A wildlife management strategy to produce a relatively high number of species per unit area (Thomas 1979*a*).

Marsh: A low, treeless, wet area, characterized by sedges, rushes, and cattails (Anderson 1991).

Mast: The fruit or nuts of such trees as oaks, beech, walnuts, and hickories. These are commonly referred to as hard mast. Soft mast is fruits and nuts of such plants as dogwood, viburnums, blueberry, huckleberry, crataegus, grape, raspberry, and blackberry.

Maximum population level: The greatest number of a wildlife species that can occur if the constraints of food, cover, and water are removed; the greatest number that can exist without losses caused by social strife (Thomas 1979*a*).

Maximum sustained yield: The largest number of fish or wildlife that can be removed without destroying a population's reproduction capability (Anderson 1991).

Mean: Average; the total of a series of measurements divided by the number of measurements (Thomas 1979*a*).

Mean annual mast yield: Total mast yield divided by the total age.

Mesic: A site characterized by a moderate amount of moisture.

Metapopulation: Strictly, a system of populations of a given species in a landscape linked by balanced rates of extinction and colonization. More loosely, the term is used for groups of populations of a species, some of which go extinct whereas others are established, but the entire system may not be in equilibrium.

Microclimate: The climatic conditions within a small or local habitat that is well defined (Thomas 1979*a*).

Micronutrient: One of several minerals necessary for plant and animal growth, but required in minute amounts. Primarily functions as a catalyst in enzyme systems. Zinc, copper, cobalt, and iron are examples (Bolen and Robinson 1995).

Migration: The periodic movement to and from a breeding area, but the term often is used loosely for other types of movements. Migration recurs each year in individuals of many species, but other species make only one round trip before dying (e.g., Pacific salmon). Other patterns occur (e.g., monarch butterflies) that complicate a universal definition. Altitudinal migration involves changes in elevation (e.g., some grouse), whereas latitudinal migration involves changes in latitude (e.g., the typical north-south migration of geese in North America) (Bolen and Robinson 1995).

Migration corridor: A belt, band, or stringer of vegetation that provides a completely or partially suitable habitat and that animals follow during migrations (Thomas 1979*a*).

Migration route: A travel route used routinely by wildlife in their seasonal movement from one habitat to another (Thomas 1979*a*).

Migrational homing: The return to a site of previous experience, usually a breeding area. Female ducks of many species, for example, return to the same breeding area where they nested in the previous year. The term is often shortened to *homing* (Bolen and Robinson 1995).

Mineral cycling: The cycling of minerals throughout an ecosystem (Anderson 1991).

Minimum-impact carrying capacity: The population density that minimizes impact on other wildlife or vegetation without eliminating the population.

Mitigation: The replacement, usually by substitution, of wildlife habitat lost to development (Bolen and Robinson 1995).

Mobility: The tendency of the individual animal to change location during the day, or between seasons or years.

Model: Any formal representation of the real world. A model may be conceptual, diagrammatic, mathematical, or computational.

Monitoring: The orderly collection, analysis, and interpretation of resource data to evaluate progress toward meeting management objectives.

Monoculture: A term describing unbroken expanses committed to a single crop, whether in forest or farmland. That is, interspersion is lacking (Bolen and Robinson 1995).

Monogamy (monogamous):   A type of mating behavior in which one male unites with one female, forming either seasonal (e.g., most ducks) or lifetime (e.g., geese and swans) pair bonds (Bolen and Robinson 1995).

Mortality: In wildlife management, the loss in a population from any cause, including hunter kill, poaching, predation, accidents, or diseases and parasites (Thomas 1979*a*)

Mortality rate: The proportion of a population dying in a unit of time (death rate) (Anderson 1991).

Mosaic: The intermingling of plant communities and their successional stages in such a manner as to give the impression of an interwoven design (Thomas 1979*a*).

Mosaic edge: An edge between stands or communities that is highly irregular, leading to a relatively large amount of edge per unit area (Thomas 1979*a*).

Multiple use: A concept of land management in which a number of products are deliberately produced from the same land base; in National Forests, the simultaneous provision of water, wood, recreation, wildlife, is required by Multiple-Use Sustained Yield Act (U.S. Laws, Statutes, etc. Public Law 86–517, 1960) (Thomas 1979*a*).

Multiple-use management: A concept of land management that integrates various activities and natural resources. For example, many national forests provide a combination of skiing, hunting, fishing, hiking, canoeing, camping, mining, and grazing in addition to lumbering and watershed protection. Wildlife represents a major component in multiple-use management (Bolen and Robinson 1995).

Multiple-use planning: The planning of management activities for a defined area to simultaneously accomplish goals for several distinct purposes, such as production of wood, water, wildlife, recreation, and grazing (U.S. Laws, Status, etc. Public Law 86–517, 1960).

Mutualism.   Mutually beneficial association of different kinds of organisms (Anderson 1991).

Natality: Birth rate. Expressed in several ways, but often as the number of offspring per female per year or per 100 females per year (Bolen and Robinson 1995).

Natality rate: The proportion of a population born in a unit of time (birth rate) (Anderson 1991).

National forest: In the United States, a Federal reservation, forest, range, or other wild land administered by the Forest Service of the U.S. Department of Agriculture under a program of multiple use and sustained yield.

Natural regulation: The regulation of a population through naturally occurring biological processes as opposed to regulation through anthropogenic practices.

Negative feedback: Feedback that inhibits or stops a system's progress (Anderson 1991).

Nest parasitism: The act or result of one species (the parasite) of bird laying its eggs in the nest of another (the host), after which the host raises the parasite's young at the expense of its own offspring. The brown-headed cowbird is a common nest parasite in North America (Bolen and Robinson 1995).

Nesting population level: The number of individuals or pairs of a species in an area during the breeding season.

Net primary productivity: Energy available in a plant following respiration. Gross primary productivity minus respiration equals net primary productivity (Anderson 1991).

Net productive rate: The average number of female offspring produced by the females in a population (Anderson 1991).

Niche: The functional role of an organism within its habitat (Anderson 1991).

Nitrogen fixation: The conversion of elemental, atmospheric nitrogen (N_2) to organic combinations or to forms readily usable in biological processes (Anderson 1991).

Nocturnal: Active at night (Anderson 1991).

Nonconsumptive use: Use of natural resources without removing them—for example, photography and watching wildlife (Anderson 1991).

Nongame wildlife: All wildlife not subject to harvest regulations (Anderson 1991).

Nonmanagement: See inactive management.

Nonpasserine: Term applied collectively to all birds other than those in the order Passeriformes. See Passeriformes (Bolen and Robinson 1995).

Noxious weeds: Those plant species designated as noxious weeds by federal or state law (it is a legal administrative declaration). Noxious weeds generally possess one or more of the characteristics of being aggressive and difficult to manage, parasitic, a carrier or host of serious insects or disease, and are nonnative, new to, or not common to the United States.

Nutrition: A process animals use to procure and process portions of their external chemical environment for the continued functioning of internal metabolism.

Objective: A clear, quantifiable statement of planned results to be achieved within a stated time period. An objective is achievable, quantifiable, and explicit. The completion of an objective must occur within a stated timeframe and the results must be documented.

Obligate: A plant or animal that occurs in a narrowly defined habitat (Thomas 1979*a*).

Odd area: A farm site where poor drainage or another shortcoming prevents tillage and crop production, thereby providing potential habitat for wildlife (Bolen and Robinson 1995).

Old growth stand: A stand that is past full maturity and showing decadence; the last stage in forest succession.

Omnivore: An animal that feeds on plants and animals.

Open canopy: A canopy condition that allows large amounts of direct sunlight to reach the ground (Thomas 1979*a*).

Opening: A break in the forest canopy; the existence of an area of essentially bare soil, grasses, forbs, or shrubs in an area dominated by trees (Thomas 1979*a*).

Optimum carrying capacity: Wildlife carrying capacity related to human values, which varies from place to place and may change as values change.

Optimum yield: The amount of material that, removed from a population, will maximize biomass (or numbers, or profit, or any other type of "optimum") on a sustained basis (Anderson 1991).

Organic matter in soil: Materials derived from plants or animals, much of it in an advanced state of decomposition (Thomas 1979*a*).

Organism: Any living individual of any plant or animal species (Thomas 1979*a*).

Outbreeding depression: The undesirable result of crossings between individuals from separate populations with incompatible traits; the offspring lack fitness (e.g., young ibex born in winter instead of spring; see text) (Bolen and Robinson 1995).

Overgrazing: 1. Severe and frequent grazing during active growth periods that impacts the recovery capability of a plant species or plant community. Generally this results in lower plant diversity and low plant community successional levels. 2. A continued overuse, usually by ungulates, that creates a deteriorated range condition (Anderson 1991).

Overstory: The upper canopy of plants. Usually refers to trees, tall shrubs, or vines.

Palatability: The relative degree of attractiveness of plants to animals as forage.

Parameter: Any variable or arbitrary constant appearing in a mathematical expression, the values of which restrict or determine the specific form of the expression (Thomas 1979*a*).

Parasite: An organism that benefits while feeding upon, securing shelter from, or otherwise injuring another organism (the host) (Anderson 1991).

Parasitic: Growing on and deriving nourishment from another organism (Anderson 1991).

Parasitism: The interaction of two individuals in which one, the host, serves as a food source for the other, the parasite (Anderson 1991).

Passeriformes: The largest order of birds, commonly known as *perching birds* or "songbirds," which includes such diverse groups as crows, wrens, swallows, thrushes, warblers, blackbirds, and sparrows (Bolen and Robinson 1995).

Passive management: See inactive management.

Patch: A recognizable area on the surface of the Earth that contrasts with adjacent areas and has definable boundaries.

Pathogen: A disease-causing agent; includes bacteria, viruses, and parasites.

Pathology: The anatomic or functional manifestations of disease (Thomas 1979*a*).

Pattern: Type of distribution (random, regular, or aggregate) (Anderson 1991).

Perennial: For plants, species that live two or more growing seasons. Trees and other woody plants are examples, but cattails, sod-forming grasses, and many other herbaceous species also are perennials (Bolen and Robinson 1995).

Perennial stream: A stream that ordinarily has running water on a year-round basis (Thomas 1979*a*).

Permafrost: Ground that is frozen a few inches below the surface all year (Anderson 1991).

Pesticide: Any of several chemicals that kill pests. Includes insecticides, herbicides, rodenticides, and fungicides. See Herbicide; (Bolen and Robinson 1995).

Phenology: 1. A branch of science dealing with relations between climate and periodic biological phenomena such as flowering, germination, and growth patterns. 2. Periodic biological phenomena that are correlated with climatic conditions.

Phenotype: Expression of the characteristics of an organism as determined by the interaction of its genetic constitution and the environment; contrast with genotype (Anderson 1991).

Pheromone: A naturally produced chemical secretion for olfactory communication (i.e., smell), usually between individuals of the same species. When synthesized, pheromones acting as sex attractants offer a means of luring insects into lethal traps, thereby reducing the need for toxic insecticides (Bolen and Robinson 1995).

Photo point: A permanently identified point from which photographs are taken at periodic intervals. Sometimes called a camera point.

Photoperiod: Day length (Anderson 1991).

Photoperiodism: Response of plants and animals to the relative duration of light and darkness—for example, some migration patterns are triggered by day length (Anderson 1991).

Photosynthesis: Formation of chemical bond (sugar) from solar energy by green plants (Anderson 1991).

Physiognomy: Vegetation classified according to shape and structure irrespective of the species included.

Physiological longevity: Maximum lifespan of individuals in a population under specified conditions; the organisms die of senescence (Anderson 1991).

Phytoplankton: Minute plants that float in an aquatic system; the plant community in marine and freshwater that floats free in the water and contains many species of algae and diatoms (Anderson 1991).

Pioneer: The first species or community in succession. On bare soil, annual weeds usually are the pioneer species, but some kinds of woody plants are also pioneers (e.g., jack pine after fires). Together with its associated animal life, such vegetation forms pioneer communities (Bolen and Robinson 1995).

Plant association: A potential natural plant community; definite floristic composition, and uniform appearance represented by stands occurring in places with similar environments. A taxonomic unit of vegetation classification.

Plant community: An assemblage of plants living and interacting together in a specific location. No particular ecological status is implied. Plant communities may include exotic or cultivated species.

Plant vigor: Plant health.

Plot: A sampling of an ecosystem or of a site.

Poikilotherm: A cold-blooded animal; the internal temperature remains similar to that of the environment. Some poikilotherms regulate their temperature behaviorally (Anderson 1991).

Poikilothermic: Term indicating the inability to maintain stable body temperatures when environmental temperatures change. Poikilotherms—fishes, amphibians, reptiles, and invertebrates—cannot generate metabolic heat to meet the stress of low ambient temperatures nor cool themselves at high ambient temperatures. "Cold blooded" is a popular description for poikilothermic species (Bolen and Robinson 1995).

Point: A map feature described by a single set of coordinates.

Point of resistance: The minimum population or density necessary for recovery of productivity.

Polyandry: Mating of a female with two or more males during mating season (Anderson 1991).

Polygamy: Mating of a male with two or more females during mating season (Anderson 1991).

Population: A group of organisms of a single species that interact and interbreed in a common place (Anderson 1991).

Population dynamics: The totality of changes in number, sex, and age that take place during the life of a population (Thomas 1979*a*).

Positive feedback: Return of output to a system that allows it to continue in its direction; feedback that enhances or promotes a system's progress (Anderson 1991).

Precipitation: Any form of atmospheric moisture reaching the Earth's surface, including rain, snow, fog, and hail (Bolen and Robinson 1995).

Precision: The closeness to each other of repeated measurements of the same quantity.

Precocial: Able to move about at an early age (Anderson 1991).

Predation: The act of predators capturing prey (Anderson 1991).

Predator: An organism that depends in total or part on killing another animal for its food. Bobcats, owls, and bass are well-known predators, but so are shrews, robins, bullfrogs, and dragonflies. A few species of plants are also predaceous (e.g., Venus flytrap) (Bolen and Robinson 1995).

Predator lane: A narrow strip of cover in which predators easily can find prey. For example, a predator lane is created when clean farming leaves only a small ring of vegetation around a pond edge; in this way skunks, raccoons, and other predators easily locate and destroy duck nests (Bolen and Robinson 1995).

Prescribed burning: The use of fire as a management tool for improving plant and animal habitats. *Controlled burning* is a synonym (Bolen and Robinson 1995).

Prescribed fire: Fire used as a management tool under specified conditions for burning a defined area (Thomas 1979*a*).

Prescription: In silvicultural terms, the formal written plan of action to carry out a silvicultural treatment of a forest stand to achieve specific objectives.

Prey: An organism killed and eaten by a predator (Bolen and Robinson 1995).

Prey switching: When two prey types are available and one is more abundant than the other, predators will concentrate on the prey that is more abundant until their numbers are low, at which time they switch to the other prey.

Primary association: The relationship between a wildlife species and a habitat condition that reflects a dependence on such habitat; a relationship that is strong and predictable (Thomas 1979*a*).

Primary cavity nesters: Wildlife species that excavate cavities in snags or trees.

Primary consumer: A trophic level; consists of herbivores (Bolen and Robinson 1995).

Primary excavator: A species that digs or chips out cavities in wood to provide itself or its mate with a site for nesting or roosting (Thomas 1979*a*).

Primary production: Production by green plants (Anderson 1991).

Primitive road: A one-lane unimproved forest road in fair to poor condition that is seldom or never maintained (Thomas 1979*a*).

Probability: The frequency, expressed as a proportion or percentage of the total occurrences, which, over a long series of trials, will produce a specified value for a variable (Anderson 1991).

Producer: A trophic level; consists of green plants (Bolen and Robinson 1995).

Production: Amount of energy (or material) formed by an individual, population, or community in a specified time period (Anderson 1991).

Productivity: The rate at which mature breeding stock produces other mature stock, or mature removable crop.

Promiscuous: Not restricted to one sexual partner (Anderson 1991).

Protein: Amino acids linked end to end in a specific order (Anderson 1991).

Proventriculus: The glandular prestomach of birds, located between the esophagus and gizzard where chemical digestion softens foods (Bolen and Robinson 1995).

Put and take: Planting of hatchery fish for removal by fishing enthusiasts or game-farm birds for removal by hunters (Anderson 1991).

Quantify: To determine or express the number of some item (Thomas 1979*a*).

Quantity: A number or amount of anything, either specific or indefinite (Thomas 1979*a*).

Race: A geographic variant of a species; often considered the same as subspecies (Anderson 1991).

Radiocollar: A collar containing a radio transmitter that is fastened on an animal. signals from the transmitter are received and used by wildlife biologists to gain information, usually about the position of the animal (Thomas 1979*a*).

Rain forest (tropical):   A multicanopied forest in humid tropical zones. Although lush, rain-forest vegetation actually is highly susceptible to human disturbances and, once cut, can seldom recover. More than half of the world's biota probably occur in tropical rain forests. Rain forests also occur in some temperate zones (e.g., Olympic Peninsula, Washington) (Bolen and Robinson 1995).

Random distribution: A distribution pattern in which an organism's position is independent of that of others (Anderson 1991).

Random sample: A sample in which the selection potential of each individual in the population is known; a sample free from selection bias (Anderson 1991).

Randomized sampling: A research method whereby each individual within a population has an equal chance of being selected as a sample. "Picking numbers from a hat" is a popular explanation (Bolen and Robinson 1995).

Range: The limits within which an entity operates or can be found.

Range of tolerance: The range of environmental conditions (e.g., temperature) that limits the abundance and distribution of organisms (Bolen and Robinson 1995).

Raptor: Any predatory bird—such as a falcon, hawk, eagle, or owl—that has feet with sharp talons or claws adapted for seizing prey and a hooked beak for tearing flesh (Thomas 1979*a*).

Recruitment: Increment of new individuals added to a wildlife population by natural reproduction, immigration, or stocking.

Reforestation: Reestablishment of a tree crop on forest sites (Thomas 1979*a*).

Refugium carrying capacity: The number of animals a habitat will support when the necessary welfare factors to alleviate predation are present in the proper amount.

Regenerate: To renew a tree crop through artificial or natural means (Thomas 1979*a*).

Regeneration: The renewal of the tree crop by natural or artificial means; also, the young crop (Thomas 1979*a*).

Regulate: To control or direct by rule, principle, or method; to adjust to some standard or requirement; to put in desired order (Thomas 1979*a*).

Regulating mechanisms: Factors that act to control the density of a population (Anderson 1991).

Relative density: The ranking of populations by density (e.g., area A has 40% more mice/ha than area B).

Release:kill ratio: The ratio of the number of head game annually released for restocking, and the number killed.

Relic: A surviving memorial of something past; an object having interest by reason of its age or association with the past; a surviving trace of something; remaining parts or fragments (Thomas 1979*a*).

Remise: A European term for an artificially established game-bird covert. Sometimes includes food as well.

Renesting: A nesting attempt which follows an earlier failure. (*Not* a second brood following an earlier success).

Replacement rate (net production rate) (R_o):   The average number of female offspring produced by the females in a population (Anderson 1991).

Resident species: The wildlife species commonly found in a specific area (Thomas 1979*a*).

Resolution: The smallest spatial scale at which we portray discontinuities in biotic and abiotic factors in map form.

Resource: Any biotic and abiotic factor directly used by an organism.

Resource abundance: The absolute amount (or size or volume) of an item in an explicitly defined area.

Resource availability: A measure of the amount of a resource actually available to the animal (i.e., the amount exploitable).

Resource preference: The likelihood that a resource will be used if offered on an equal basis with others.

Resource selection: The process by which an animal chooses a resource.

Resource use: A measure of the amount of resource taken directly (e.g., consumed, removed) from an explicitly defined area.

Respiration: The breakdown of sugar into usable energy by living organisms (Anderson 1991).

Response curve: A curve describing the response in potential use by elk or deer on a land type to changes in the cover-forage area ratio (Thomas 1979*a*).

Rest period: The period when an incubating hen normally leaves the nest for rest, food, or recreation.

Richness: A measure of the relative degree or number of plant or wildlife species or both associated with particular habitat conditions (Thomas 1979*a*).

Riparian area: Geographically delineable area with distinctive resource values and characteristics that are comprised of the aquatic and riparian ecosystems.

Riparian complex: Ecological units that support or may potentially support a specified pattern of riparian ecosystems, soils, landforms, and hydrologic characteristics.

Riparian ecosystem: A transition between the aquatic ecosystem and the adjacent terrestrial ecosystem: identified by soil characteristics or distinctive vegetation communities that require free or unbound water. Riparian ecosystems often occupy distinctive landforms, such as flood plains or alluvial benches.

Riparian zone: An area identified by the presence of vegetation that requires free or unbound water or conditions more moist than normally found in the area (Thomas 1979*a*).

Rookeries: In the United States, colonies of nesting herons, usually great blue (Anderson 1991).

Rotation: The planned number of years between the regeneration of a stand and its final cutting at a specified area (Thomas 1979*a*).

***r*-selected species:** Species adapted to high reproductive rates and short life spans, often with wide ranges of tolerance to environmental conditions. Bobwhites, mourning doves, and cottontails are examples (Bolen and Robinson 1995).

Rumen: The first of four chambers in the stomach of ruminants (e.g., deer). Food in the rumen undergoes fermentation, after which the contents are returned to the mouth for further chewing ("chewing the cud") (Bolen and Robinson 1995).

Ruminant: A cud-chewing animal with a complex stomach that is often divided into four chambers. Deer, goats, sheep, and bovines (antelopes, cattle, and bison) are examples (Bolen and Robinson 1995).

Runoff: Precipitation that is not retained on the site where it fell; natural drainage away from an area (Thomas 1979*a*).

Rut: Breeding season of some ungulates (Anderson 1991).

Sample: A subset of the total number of units in a population (Anderson 1991).

Sampling bias: An error in collecting or analyzing samples representing a population. Fish populations sampled with nets may not include the younger age classes if smaller individuals escape through the mesh, thereby introducing sampling bias based on equipment. Does and fawns usually are underrepresented in the deer harvest simply because of sampling bias introduced by hunter selection for mature bucks (Bolen and Robinson 1995).

Sampling theory: The assumption that samples reflect the same attributes (e.g., age or sex ratios, stomach contents, or parasite loads) as the remainder of the population from which the samples were collected (Bolen and Robinson 1995).

Saturation point: The maximum wild density common to widely separated optimum ranges.

Savanna: Lowland tropical and subtropical grassland with a scattering of trees and shrubs (Anderson 1991).

Scale: The resolution at which patterns are measured, perceived, or represented. Scale can be broken into several components, including grain and extent.

Scale of observation: The spatial and temporal scales at which observations are made. Scale of observation has two parts, extent and grain.

Scat: Animal fecal matter (Anderson 1991).

Scavenger: An organism feeding on dead organisms. Vultures and hyenas are examples, but many other species also scavenge at times (e.g., crows, golden eagles, and coyotes) (Bolen and Robinson 1995).

Scientific name: The binomial or two-word latinized name of an organism; the first word describes the genus, the second the species (Thomas 1979*a*).

Secondary cavity nester: Wildlife that occupies a cavity in a snag that was excavated by another species (Thomas 1979*a*).

Secondary consumer: A trophic level; consists of predators that feed on primary consumers. A bobcat is a secondary consumer (Bolen and Robinson 1995).

Secondary poisoning: Unintended contamination of organisms (nontarget species) feeding on previously poisoned animals. Vultures and other scavengers are particularly susceptible (e.g., feeding on the carcasses of poisoned coyotes) (Bolen and Robinson 1995).

Secondary species: Wildlife species that are less important than the featured species but will be present on a given unit of land. Management activities aimed at the primary species may sometimes benefit the secondary species. At times the secondary species may outnumber and provide more recreation than the primary wildlife species (such as deer on an area where bear are featured).

Sediment: Material suspended in liquid or air; the deposition of that material onto the surface underlying this liquid or air, usually the deposition of organic and inorganic soil materials by water (Thomas 1979*a*).

Selective grazing: The grazing of certain plant species on the range to the exclusion of others.

Self-regulation: The process of population regulation in which population increase is prevented by a deterioration in the quality of individuals that make up the population; population regulation by internal adjustments in behavior and physiology within the population rather than by external forces, such as predators (Anderson 1991).

Self-sustaining population: A wildlife population of sufficiently large size to assure its continued existence within the area of concern without introduction of other individuals from outside the area (Thomas 1979*a*).

Semiaquatic species: Wildlife species that spend part of their lives on land and part in water (Thomas 1979*a*).

Sensitive species: Those plants and animal species identified by a Regional Forester for which population viability is a concern, as evidenced by: (a) significant current or predicted downward trend in population numbers or density; or (b) significant current or predicted downward trend in habitat capability that will reduce species' existing distribution.

Seral community: Any community that is not at potential. A relatively transitory community that develops under ecological succession, toward or away from a potential natural community.

Seral stage: Successional plant communities are often classified into quantitative seral stages to depict the relative position on a classical successional pathway.

Sere: The stages that follow one another in an ecologic succession (Thomas 1979*a*).

Sex ratio: The ratio of males to females in animal populations, with the percentage of males expressed first (i.e., in a population with a 45:55 sex ratio, 45% are males). Sex ratios are subdivided by age: primary (fertilization); secondary (birth or hatching); tertiary (juvenile); and quaternary (adult) (Bolen and Robinson 1995).

Shade-intolerant plants: Plant species that do not germinate or grow well in shade (Thomas 1979*a*).

Shade-tolerant plants: Plants that grow well in shade (Thomas 1979*a*).

Sheet erosion: Loss or movement of soil in thin layers (Thomas 1979*a*).

Shelterbelt: A strip of trees or shrubs planted or left standing in prairie areas to help reduce wind and erosion of topsoil (Anderson 1991).

Shelterwood cutting: Any regeneration cutting designed to establish a new tree crop under the protection of remnants of the old stand (Thomas 1979*a*).

Shrub: A plant with persistent woody stems, relatively low growth habit, and generally several basal shoots instead of a single bole. It differs from a tree by its low stature and nonarborescent form (Thomas 1979*a*).

Sight barrier: Any object that serves to block the vision of an observer (Thomas 1979*a*).

Sight distance: The distance at which 90% or more of an adult elk or deer is hidden from the view of a human (Thomas 1979*a*).

Sigmoid curve: S-shaped curve—for example, the logistic curve (Anderson 1991).

Silvics: The study of the life history and general characteristics of forest trees and stands, with particular reference to site factors, in order to provide a basis for the practice of silviculture (Thomas 1979*a*).

Single-tree selection : A method of harvesting under an uneven-aged forest management system in which trees are individually selected for harvest (Thomas 1979*a*).

Single use: A concept of management in which a single management objective is paramount (Thomas 1979*a*).

Sink populations: In a landscape, a population or site that attracts colonists while not supplying migrants to other sites or populations.

Site: A single, specific point on the land. The sample point where data measurements are taken.

Site type: Classification of an area considered quantitatively in terms of how its environment determines the type, quality, and growth rate of the potential vegetation (Thomas 1979*a*).

Slash: The residue left on the ground after trees are felled or accumulated there as a result of storm, fire, or silvicultural treatment (Thomas 1979*a*).

Slope: The incline of the land surface measured in degrees from the horizontal or in percent as determined by the number of units change in elevation per 100 of the same measurement units; also characterized by the compass direction in which it faces (Thomas 1979*a*).

Snag: A standing dead tree from which the leaves and most of the limbs are missing due to environmental factors.

Source populations: In a landscape, a population or a site that supplies colonists to other patches.

Species: The basic taxonomic unit. A population whose members freely interbreed and share a common gene pool. Species are identified scientifically in

Latin with a two-part name known as a binomial, which includes both genus and species names (e.g., *Bubo virginianus* is the binomial for the great horned owl) (Bolen and Robinson 1995).

Species composition: The proportion of plant species or aggregations of species in relation to a total area. It may be expressed in terms of canopy cover, frequency, or weight.

Species diversity: A measure combining richness and evenness. Diversity increases when both the number of species increases and the number of individuals of each species are more evenly distributed. See Species evenness; Species richness (Bolen and Robinson 1995).

Species evenness: The relative abundance of individuals among those species present in a specified area and/or time. Some species, although present, may have only a few individuals, whereas others are abundant; such conditions diminish evenness. See Species richness (Bolen and Robinson 1995).

Species richness: An indicator of the number of species of plants or animals present in an area—the more species present, the higher the degree of species richness (Anderson 1991).

Specific natality rate: The number of individuals produced per unit of time per breeding female.

Spring-fall range: An area between summer range at high elevation and winter range at low elevation that is used by deer and elk during spring and fall as they move between summer and winter range (Thomas 1979*a*).

Stand: An uninterrupted unit of vegetation, homogenous in composition and of the same age. The vegetation can be of any physiognomic class.

Standard deviation: A statistical term; a measure of the dispersion about the mean of a population; i.e., the positive square root of the variance (Thomas 1979*a*).

Standard error: A statistical term; the standard deviation of a distribution of means or of any other statistic determined from samples; determined by dividing the standard deviation by the square root of the number of observations (Thomas 1979*a*).

Standardized sampling: A research method whereby an individual within a population is selected on a predetermined basis. For example, every twentieth oak tree along a line transect might be measured for acorn production. For comparison, see Randomized sampling (Bolen and Robinson 1995).

Stationary age distribution: The allocation of age classes in a population when the population does not change in size and the age structure is constant over time.

Stem: The principal axis of a plant, from which buds and shoots develop; with woody species, the term applies to all ages and thicknesses (Thomas 1979*a*).

Steno-:   A prefix used to identify narrow tolerances for specific components of the environment. Frogs are stenohaline because their eggs and larvae have almost no tolerance to salt water (Bolen and Robinson 1995).

Stenotopic: Having little ability to withstand modification of environmental conditions (Anderson 1991).

Steppe: An extensive area of natural, dry grassland; usually used in reference to grasslands in southwestern Asia and southeastern Europe; equivalent to *prairie* in North American usage (Anderson 1991).

Stochastic: Random or expected by chance. Wildfires, storms, and flooding are stochastic events, each with a statistical probability of happening. For example, rain may fall only 10 times per year in a desert, but the date of the next rainfall cannot be predicted accurately, whereas the *probability* of its occurrence on a given day can be calculated rather precisely. Rainfall therefore is a stochastic event. Some features of population ecology also may occur stochastically (e.g., chance variation in a sex ratio or birth rate) (Bolen and Robinson 1995).

Stochastic model: A mathematical model based on probabilities; the prediction of the model is not a single fixed number but a range of possible numbers; opposite of *deterministic model* (Anderson 1991).

Stock: A group of fish or other aquatic animal that can be treated as a single unit for management purposes (Anderson 1991).

Straggling failure: A translocation followed by initial thrift but ultimate dwindling and disappearance.

Strain: Type of fish having genetic adaptability to a specific set of physical conditions (e.g., temperature) (Anderson 1991).

Strata: See "vegetative strata."

Stratified sampling: Sampling by groups: a precision-increasing method (Anderson 1991).

Stream management unit: A management zone alongside a stream where the management objective is the protection of the riparian zone or of water quality or quantity or both (Thomas 1979*a*).

Stream protection zone: A zone in which management activity is excluded or modified to protect the riparian zone or quality and quantity of water or both (Thomas 1979*a*).

Stringer: Vegetation arranged in a long, thin, linear fashion (Thomas 1979*a*).

Structure (community):   Physical makeup of vegetation in a community (Anderson 1991).

Study area: An arbitrary spatial extent chosen by the investigator within which to conduct a study (contrast with site and scale).

Stump: The woody base of a tree left in the ground after felling (Thomas 1979*a*).

Subadult: An animal, or age class, that resembles an adult but does not breed because of behavioral and/or sexual immaturity (Bolen and Robinson 1995).

Subspecies: Geographic variant of a species (Anderson 1991).

Substitutable area: An area that may serve in the place of another area for a particular purpose selected by the forest manager (Thomas 1979*a*).

Substrate: Supporting material; in biology usually refers to soil or soil-like material, as *community substrate* (Anderson 1991).

Success ratio: The ratio of number of hunters to number of game killed.

Succession: The process of vegetative and ecological development whereby an area becomes successively occupied by different plant communities.

Successional stage: A stage or recognizable condition of a plant community that occurs during its development from bare ground to climax (Thomas 1979*a*).

Sucker: A shoot arising from below ground level, either from a rhizome or from a root (Thomas 1979*a*).

Summer range: A range, usually at higher elevation, used by deer and elk during the summer; a summer range is usually much more extensive than a winter range (Thomas 1979*a*).

Sunning: The process of an animal methodically exposing itself to direct sunlight; also called sunbathing (Thomas 1979*a*).

Surrogate species: A close relative of a rare or endangered species; used to determine capture and breeding techniques, rearing procedures, and in physiological tests for endangered species (Anderson 1991).

Survivorship curve: Data from column l_x in a life table (individuals alive at the beginning of each interval) plotted on semilog paper (Anderson 1991).

Suspended sediment: Soil and other material suspended in water, usually disturbed or moving water; this material drops out of suspension when water movement slows or ceases (Thomas 1979*a*).

Sustained yield: Number of animals or plants that can be removed from a population year after year without jeopardizing future yields (Anderson 1991).

Swamp: A wet area that usually has standing trees (Anderson 1991).

Symbiosis: A relationship between two or more kinds of living organisms where all benefit; sometimes obligatory to one or more of the organisms in the relationship (Thomas 1979*a*).

Sympatric: Similar; usually used in reference to populations that occupy the same geographic region (Anderson 1991).

Taiga: Boreal forests of the north (Anderson 1991).

Talus: The accumulation of broken rocks that occurs at the base of cliffs or other steep slopes (Thomas 1979*a*).

Taxon (plural taxa):   Any level of the hierarchy of taxonomy. Species, genus, and family are each a taxon (Bolen and Robinson 1995).

Taxonomy: The classification of organisms into units known as taxa (singular, taxon). Carolus Linnaeus, a Swedish biologist, developed the currently used taxonomic system in 1753 (for plants) and 1758 (for animals). The species is the basic taxonomic unit. After the species level, taxa progress into larger units: genus, family, order, class, and phylum. For some plants and animals, scientists identify a smaller taxon, the subspecies (Bolen and Robinson 1995).

Terrestrial vertebrates: Animals with backbones that dwell primarily on land (Thomas 1979*a*).

Territory: Any area defended by an individual against intrusion by others of the same species; apparently assures spacing in keeping with availability of food and other resources (the classical explanation). Typical territorial behavior (by males) simultaneously attracts females while warning other males (Bolen and Robinson 1995).

Thermal cover: Cover used by animals to ameliorate effects of weather (Thomas 1979*a*).

Thermal-neutral zone: An area where the ambient conditions do not trigger a metabolic response on the part of the occupying animal (Thomas 1979*a*).

Threatened species: Any species that is likely to become an endangered species within the foreseeable future throughout all or a significant portion of its range and that the appropriate secretary has designated as a threatened species. (Some states also have declared certain species as threatened through their regulations or statutes).

Tolerance: The capacity of wildlife to withstand or adjust to any disturbed conditions in the environment.

Transect: A linear plot, usually represented by a line, along which are often placed regularly spaced quadrates (plot frames), loops, or other devices.

Transition zone: An area of land where two forest types blend. It occurs between upland and lowland. This band usually contains high-quality wildlife food, including soft and hard mast, forage, and fruiting shrubs.

Translocate: Move from one place to another within current or historic habitat (Anderson 1991).

Treatment: In experimentation, a stimulus applied in order to observe its effect on an experimental situation or to compare its effect with the effects of other treatments; in practice, may refer to anything capable of controlled application according to experimental requirements; the act or manner of treating something (Thomas 1979a).

Trends: Indications of changes in populations over time (Anderson 1991).

Tribe: A subdivision or subfamily based on structure, plumage, habits, or courtship behavior (Anderson 1991).

Trophic level: A structure of producer and consumer organisms superimposed on food chains to trace the flow of energy (Anderson 1991).

Tundra: Northern, high-elevation biome with few sizable wood plants because of a short growing season (Anderson 1991).

Understory: 1. Foliage, consisting of seedlings, shrubs, and herbs, that lies beneath and is shaded by canopy or taller plants (Anderson 1991). 2. In silviculture, trees growing under the canopy formed by taller trees; in range management, herbaceous and shrub vegetation under a brushwood or tree canopy (Thomas 1979a).

Uneven-aged management: A forestry practice in which individuals or small groups of trees are cut (e.g., single-tree selection cut; group selection cut), thereby producing a forest with trees of different ages and heights as cutting and regrowth continue. For comparison, see Even-aged management (Bolen and Robinson 1995).

Uneven-aged stand: A forest stand managed to maintain an intermingling of trees that differ markedly in age; stands continuously or periodically regenerated, tended, and harvested with no real beginning or end (Thomas 1979a).

Ungulate: A hooved animal (Anderson 1991).

Unit area: A parcel of land in which wildlife needs for food, cover, water, and reproduction must be met.

Urbanization: The process or degree of human concentration in relatively small areas (cities), where huge amounts of basic materials are consumed from stocks produced elsewhere (e.g., food from farms, ranches, and oceans). At its extreme, urbanization covers the soil with a monoculture of buildings and pavement (Bolen and Robinson 1995).

Utilization: The available forage by weight consumed or trampled through livestock grazing. Usually expressed as a percent.

Variable: Generally, any quantity that varies; more precisely, a quantity that may take any one of a set of values; a single influence or one of several measurable influences acting on a particular process (Thomas 1979*a*).

Variance: A statistical term; a measure of variability within a finite population or sample; the total of the squared deviations of each observation from the arithmetical mean divided by one less than the total number of observations (Thomas 1979*a*).

Vector: An organism that transmits a disease within and between populations, commonly an insect (e.g., mosquito) or other arthropod (e.g., tick). Fleas are vectors for sylvatic plague (Bolen and Robinson 1995).

Vegetation classification: The process of analyzing vegetation community data and defining hierarchical entities based on that data. There are two branches of vegetation classification: potential natural and existing. The potential natural vegetation classification hierarchy includes series, subseries, plant associations, and plant association phases. The existing vegetation classification hierarchy covers types and community types.

Vegetation management status: The relative degree to which kinds, proportions, and amounts of vegetation in the present plant community resemble the desired plant community chosen for an ecological site.

Vegetation strata: The layers of vegetation that may be discerned in a plant community (Thomas 1979*a*).

Vegetation structure: The form or appearance of a stand; the arrangement of the canopy; the volume of vegetation in tiers or layers (Thomas 1979*a*).

Vegetation type: A kind of existing plant community with distinguishable characteristics described in terms of the present vegetation that dominates the aspect or physiognomy of the area.

Velvet: The soft, highly vascularized tissues that cover developing antlers; normally rubbed and sloughed off when antlers reach full growth and harden (Bolen and Robinson 1995).

Ventriculus: The muscular, grit-bearing stomach of birds, designed for grinding food; also called the gizzard. May be modified in some species (e.g., hummingbirds) (Bolen and Robinson 1995).

Versatile: Capable of or adapted for survival in several plant communities or successional stages or both (Thomas 1979*a*).

Vertebrate: Any animal with a backbone, including cartilaginous fishes (e.g., sharks). A taxon known scientifically as Vertebrata includes fishes, amphibians, reptiles, birds, and mammals. Among the vertebrates, birds and mammals are most often considered in the narrow view of "wildlife," although such a limited view is not in keeping with modern wildlife management (Bolen and Robinson 1995).

Vertical diversity: The diversity in an area that results from the complexity of the aboveground structure of the vegetation; the more tiers of vegetation or the more diverse the species makeup or both, the higher the degree of vertical diversity (Thomas 1979*a*).

Viability: Strictly, the ability to live or grow. In conservation biology, the probability of survival of a population for an extended period of time (Thomas 1979*a*).

Viable population: A population large enough and with adequate habitat to perpetuate itself (Anderson 1991).

Vigor: The relative robustness of a plant in comparison to other individuals of the same species. It is reflected primarily by the size of a plant and its parts in relation to its age and the environment in which it is growing.

Visual management zone: An area in which the overriding management concern is for an esthetically pleasing appearance; in most forested areas, such zones are managed to give the impression of mature and relatively unbroken forest (Thomas 1979*a*).

Vitamins: Organic compounds, which occur in food in minute amounts and cannot be synthesized by animals, and are essential for normal life and functioning.

Water-holding capacity: A measure of the ability of soil to soak up and hold water (Thomas 1979*a*).

Water quality: Determined by a series of standard parameters: turbidity, temperature, bacterial content, pH, and dissolved oxygen (Thomas 1979*a*).

Water quantity: The amount of water coming from a watershed or drainage (Thomas 1979*a*).

Welfare factors: Necessary aspects for survival (e.g., food, water, cover, special factors).

Wetland(s):   1. Any area where the water table is near or above the surface of the land during a considerable part of the year (Anderson 1991). 2. Those areas that are inundated by surface or ground water with a frequency sufficient to support, and under normal circumstances do or would support, a prevalence of vegetation or aquatic life that requires saturated or seasonally saturated soil conditions for growth and reproduction. Generally includes swamps, marshes, bogs, and similar areas such as sloughs, potholes, wet meadows, river overflows, mud flats, and natural ponds.

Wilderness: Lands designated by law as wilderness; no road building or timber management is allowed on such lands; they are intentionally managed to maintain their primitive character (Thomas 1979*a*).

Wildfire: An unplanned fire requiring suppression action, as contrasted with a prescribed fire burning within prepared lines enclosing a designated area under prescribed conditions; a free-burning fire unaffected by fire suppression measures (Thomas 1979*a*).

Wildlife: All undomesticated animals in a natural environment (Anderson 1991).

Wildlife, consumptive use: Unharvested game or nongame species harvested for sport, food, study, or commerce.

Wildlife habitat management: The manipulation or maintenance of vegetation to yield desired results in terms of habitat suitable for designated wildlife species or groups of species (Thomas 1979a).

Wildlife logs: Logs left in place on the forest floor for wildlife habitat (Thomas 1979a).

Wildlife management: The scientifically based art of manipulating habitats to produce some level of a desired species or manipulating animal populations to achieve a desired end (Thomas 1979a).

Wildlife openings: Openings maintained to meet various food or cover needs for wildlife. They may contain native vegetation or planted crops and can be maintained by burning, discing, mowing, planting, fertilizing, grazing, or herbicides.

Wildlife range: A range, usually at lower elevation, used by migratory deer and elk during the winter months; usually better defined and smaller than summer ranges (Thomas 1979a).

Wolf tree: A tree of dominant size and position that usurps light and space from smaller understory, preventing its growth (Anderson 1991).

Xeric: 1. Deficient in moisture for the support of life—said of a desert environment (Anderson 1991). 2. A site low or deficient in moisture that is available for the support of plant life.

Xerophyte: Plant that can grow in dry places (Anderson 1991).

Yard: A wintering ground used by deer during deep snow. Paths are trampled down to afford access to browse food.

Yield: The sustained kill per unit of area or population.

Zooplankton: Animal portion of the plankton; the animal community in marine and freshwater situations that floats free in the water, independent of the shore and the bottom, moving passively with the currents (Anderson 1991).

Literature Cited

Abdellatif, E. M., K. B. Armitage, M. S. Gaines, and M. L. Johnson. 1982. The effect of watering on a prairie vole population. *Acta Theriologica* 27:243–255.

Acord, B. R., C. A. Ramey, and R. W. Werge. 1994. Charting a future: process and promise. Proceedings of the Vertebrate Pest Conference 16:5–8.

Adams, D. A. 1993. *Renewable resource policy.* Island Press, Washington, D.C., USA.

Adams, L. G., B. W. Dale, and L. D. Mech. 1996. Wolf predation of caribou calves in Denali National Park, Alaska. Pages 245–260 *in* L. N. Carbyn, S. H. Fritts, and D. R. Seip, editors. *Ecology and conservation of wolves in a changing world.* Canadian Circumpolar Institute, University of Alberta, Edmonton, Alberta, Canada.

Albert, S. K., and P. R. Krausman. 1993. Desert mule deer and forage resources in southwest Arizona. *Southwestern Naturalist* 38:198–205.

Alderman, J. A., P. R. Krausman, and B. D. Leopold. 1989. Diel activity of female desert bighorn sheep in western Arizona. *Journal of Wildlife Management* 53:264–271.

Alexander, R. D. 1974. The evaluation of social behavior. *Annual Review of Ecology and Systematics* 5:325–383.

Alkon, P. U., B. Pinshaw, and A. A. Degen. 1982. Seasonal water turnover rates and body water volumes in desert chukars. *Condor* 84:332–337.

Allen, D. L. 1954. *Our wildlife legacy.* Funk and Wagnalls, New York, New York, USA.

Allen, R. W. 1980. Natural mortality and debility. Pages 172–185 *in* G. Monson and L. Sumner, editors. *The desert bighorn.* The University of Arizona Press, Tucson, Arizona, USA.

Allen S. H., and A. B. Sargent. 1975. A rural mail-carrier index of North Dakota red foxes. *Wildlife Society Bulletin* 3:74–77.

Alonso, A., and F. Dallmeier. 2000. Working for biodiversity. Monitoring and assessment of biodiversity program. Smithsonian Institution, Washington, D.C., USA.

Alt, K. L., M. R. Frisina, and F. J. King. 1992. Coordinated management of elk and cattle, a perspective—Wall Creek Wildlife Management Area. *Rangelands* 14:12–15.

Andelt, W. F. 1996. Carnivores. Pages 133–155 *in* P. R. Krausman, editor. *Rangeland wildlife.* Society for Range Management, Denver, Colorado, USA.

Andelt, W. F., R. L. Phillips, R. H. Schmidt, and R. B. Gill. 1999. Trapping furbearers: an overview of the biological and social issues surrounding a public policy controversy. *Wildlife Society Bulletin* 27:53–64.

Andersen, A. W. 1949. Early summer foods and movements of the mule deer (*Odocoileus hemionus*) in the Sierra Vieja range of southwestern Texas. *Texas Journal of Science* 1:45–50.

Anderson, A. E., D. E. Medin, and D. C. Bowden. 1972. Indices of carcass fat in a Colorado mule deer population. *Journal of Wildlife Management* 36:579–594.

Anderson, A. E., D. E. Medin, and D. C. Bowden. 1974. Growth and morphometry of the carcass, selected bones, organs, and glands of mule deer. *Wildlife Monograph* 39.

Anderson, D. R., and K. P. Burnham. 1976. Population ecology of the mallard: VI. The effect of exploitation on survival. *U.S. Fish and Wildlife Service Resource Publication* 128.

Anderson, D. R. J. L. Loake, B. R. Crain, and K. P. Burnham. 1979. Guidelines for line transect sampling of biological populations. *Journal of Wildlife Management* 43:70–78.

Anderson, D. R., K. P. Burnham, and G. C. White. 1982. The effect of exploitation on annual survival of mallard ducks: an ultra-structural model. Proceedings of the International Biometry Conference 11:33–39.

Anderson, E. W., and R. J. Scherzinger. 1975. Improving quality of winter forage for elk by cattle grazing. *Journal of Range Management* 28:120–125.

Anderson, E. W., D. L. Franzen, and J. E. Melland. 1990*a*. Prescribed grazing to benefit watershed-wildlife livestock. *Rangelands* 12:105–111.

Anderson, E. W., D. L. Franzen, and J. E. Melland. 1990*b*. Forage quality as influenced by prescribed grazing. Pages 56–70 *in* K. E. Severson, editor. *Can livestock be used as a tool to enhance wildlife habitat?* U.S. Department of Agriculture Forest Service General Technical Report RM-194.

Anderson, J. K. 1985. *Hunting in the ancient world.* University of California Press, Berkeley, California, USA.

Anderson, S. H. 1991. *Managing our wildlife resources.* 2nd ed. Prentice Hall, Englewood Cliffs, New Jersey, USA.

Andrawantha, H. G., and L. C. Birch. 1954. *The distribution and abundance of animals.* University of Chicago Press, Chicago, Illinois, USA.

Aney, W. W. 1974. Estimating fish and wildlife harvest, a survey of methods used. Proceedings of the Western Association of Game and Fish Commissioners 54:70–79.

Arizona Game and Fish Department. 2000. 2000–2001 Arizona hunting regulations. Arizona Game and Fish Department. Phoenix, Arizona, USA.

Arney, J. D. 1985. *User's guide for the Stand Projection System (SPS).* Report 1. Applied Biometrics, Spokane, Washington, USA.

Askins, R. A., M. J. Philbrick, and D. S. Sugeno. 1987. Relationships between the regional abundance of forest and the composition of forest bird communities. *Biological Conservation* 39:129–152.

Atkinson, B., editor. 1965. *Walden and other writings of Henry David Thoreau.* The Modern Library, New York, New York, USA.

Atwell, G. C. 1962. Moose investigations: mortality studies, south central Alaska. Alaska Department of Fish and Game,

Federal Aid in Wildlife Restoration Project W-6–R-3, Job 2.

Bahnak, B. R., J. C. Holland, L. J. Verme, and J. J. Ozoga. 1979. Seasonal and nutritional effects on serum nitrogen constituents in white-tailed deer. *Journal of Wildlife Management* 43:454–460.

Bailey, J. A. 1968. A weight-length relationship for evaluating physical conditions of cottontails. *Journal of Wildlife Management* 32:835–841.

Bailey, J. A. 1984. *Principles of wildlife management.* John Wiley and Sons, New York, New York, USA.

Bailey, J. A. 1990. Management of Rocky Mountain bighorn sheep herds in Colorado, Colorado Division of Wildlife, Fort Collins, Colorado. Special Report 66.

Bailey, J. A., and D. R. Klein. 1997. United States of America. Pages 307–316 *in* D. M. Shackleton, editor. Wild sheep and their relatives: status survey and conservation action plan for Caprinae. International Union for the Conservation of Nature and Natural Resources, Gland, Switzerland.

Baker, D. L., and N. T. Hobbs. 1982. Composition and quality of elk summer diets in Colorado. *Journal of Wildlife Management* 46:694–703.

Ballard, W. B., and P. S. Gipson. 2000. Wolf. Pages 321–346 *in* S. Demarais and P. R. Krausman, editors. *Ecology and management of large mammals in North America.* Prentice Hall, Upper Saddle River, New Jersey, USA.

Ballard, W. B., and V. Van Ballenberghe. 1997. Predator/prey relationships. Pages 247–273 *in* A. W. Franzmann and C. C. Schwartz, editors. *Ecology and management of the North American moose.* Smithsonian Institution Press, Washington, D.C., USA.

Ballard, W. B., D. Lutz, T. W. Kreegan, L. H. Carpenter, and J. C. DeVos, Jr. In press. Deer-predator relationships: a review of recent North American studies with emphasis on mule and black-tailed deer. *Wildlife Society Bulletin.*

Ballard W. B., J. S. Whitman, and C. L. Gardner. 1987. Ecology of an exploited wolf population in south-central Alaska. *Wildlife Monograph* 98.

Ballard, W. B., L. A. Ayres, P. R. Krausman, D. J. Reed, and S. G. Fancy. 1997. Ecology of wolves in relation to a migratory caribou herd in northwest Alaska. *Wildlife Monograph* 135.

Banci, V., and A. Harestad. 1988. Reproduction and natality of wolverine (*Gulo gulo*) in Yukon. *Annuals of Zoology Fennici* 25:265–270.

Bangs, E. E., T. N. Bailey, and M. F. Portner. 1989. Survival rates of adult female moose on the Kenai Peninsula, Alaska. *Journal of Wildlife Management* 53:557–563.

Banks, R. C. 1979. Human related mortality of birds in the United States. U.S. Fish and Wildlife Service Special Scientific Report 215.

Baskett, T.S., M.S. Armbruster, and M.W. Sayre. 1978. Biological perspectives for the mourning dove call-count survey. Transactions of the North American Wildlife and Natural Resources Conference 43:163–180.

Bartholomew, G. A. 1972. The water economy of seed-eating birds that survive without drinking. Proceedings of the International Ornithological Congress 15:237–254.

Bartholomew, G. A., and T. J. Code. 1963. The water economy of land birds. *Auk* 80:504–539.

Bayless, S. R. 1969. Winter food habits, range use, and home range of antelope in Montana. *Journal of Wildlife Management* 33:538–551.

Beale D. M., and D. A. Smith. 1970. Forage use, water consumption, and productivity of pronghorn antelope in western Utah. *Journal of Wildlife Management* 34:570–582.

Bean, M. J. 1983. *The evolution of national wildlife law.* Praeger, New York, New York, USA.

Belanger, D. O. 1988. *Managing American wildlife: A history of the International Association of Fish and Wildlife Agencies.* University of Massachusetts Press, Amherst, Massachusetts, USA.

Belcher, D. M. 1982. TWIGS: the woodman's ideal growth projection system. Pages 70–95 *in* J. W. Moser, Jr., editor. *Microcomputers: a new tool for foresters.* Purdue University Press, West Lafayette, Indiana, USA.

Belcher, D. M., M. R. Holdaway, and G. J. Brand. 1986. A description of STEMS—the stand and tree evaluation modeling system. U.S. Forest Service General Technical Report NC-79.

Bellantoni, E. S., and P. R. Krausman. 1991. Movements of desert mule deer and collared peccary in Saguaro National Monument. Pages 181–193 *in* C. P. Stone and E. S. Bellantoni, editors. Proceedings of the Symposium on Research in Saguaro National Monument. Rincon Institute and Southwest Parks and Monuments Association, Tucson, Arizona, USA.

Bellrose, F. C. 1959. Lead poisoning as a mortality factor in waterfowl populations. *Bulletin of the Illinois Natural History Survey* 27:235–288.

Bender, L. C., and R. D. Spencer. 1999. Estimating elk population size by reconstruction from harvest data and herd ratios. *Wildlife Society Bulletin* 27:636.

Benolkin, P. 1990. Strategic placement of artificial watering devices for use by chukar partridge. Pages 59–62 *in* G. K. Tsukamoto and S. J. Stiver, editors. Proceedings of the Wildlife Water Development Symposium. Nevada Chapter of The Wildlife Society, U.S. Bureau of Land Management, and Nevada Department of Wildlife, Las Vegas, Nevada, USA.

Benson, D. E. 1991. Values and management of wildlife and recreation on private land in South Africa. *Wildlife Society Bulletin* 19:497–510.

Benson, D. E. 1992. Commercialization of wildlife: a value-added incentive for conservation. Pages 539–553 *in* R. D. Brown, editor. *The biology of deer.* Springer-Verlag, New York, New York, USA.

Benson, G. L., and W. F. Laudenslayer, Jr. 1986. DYNAST: simulating wildlife responses to forest-management strategies. Pages 351–355 *in* J. Verner, M. L. Morrison, and C. J. Ralph, editors. *Wildlife 2000: modeling habitat relationships of terrestrial vertebrates.* University of Wisconsin Press, Madison, Wisconsin, USA.

Berger, J. 1992. Facilitation of reproductive synchrony by gestation adjustment in gregarious mammals: a new hypothesis. *Ecology* 73:323–329.

Bergerud, A. T. 1971. The population dynamics of Newfoundland caribou. *Wildlife Monograph* 25.

Bergerud, A. T. 1978. Caribou. Pages 83–101 *in* J. L. Schmidt and D. L. Gilbert, editors. *Big game of North America.* Stackpole Books, Harrisburg, Pennsylvania, USA.

Bergerud, A. T. 2000. Caribou. Pages 658–693 *in* S. Demarais and P. R. Krausman, editors. *Ecology and management of large mammals in North America.* Prentice Hall, Upper Saddle River, New Jersey, USA.

Bergerud, A. T., and F. Manuel. 1969. Aerial census of moose in central Newfoundland. *Journal of Wildlife Management* 33:910–916.

Bergerud, A. T., and W. B. Ballard. 1989. Wolf predation on Nelchina caribou: a reply. *Journal of Wildlife Management* 53:251–259.

Berryman, J. H. 1994. Blurred images: and the future of wildlife damage management. Proceedings of the Vertebrate Pest Conference 16:2–4.

Bickle, T. S. 1969. Water developments and their effects on mule deer on the Fort Stanton Range. Thesis, New Mexico State University, Las Cruces, New Mexico, USA.

Birch, L. C. 1957. The meaning of competiton. *American Naturalist* 91:5–18.

Bissonette, J. A., C. S. Loftin, D. M. Leslie, Jr., L. A. Nordstrom, and W. J. Fleming. 2000. The Cooperative Research Unit Program and wildlife education: historic development, future challenges. *Wildlife Society Bulletin* 28:534–541.

Biswell, H. H., and A. M. Schultz. 1958. Effects of vegetation removal on spring flow. California Fish and Game 44:211–230.

Blackmore, H. L. 1971. *Hunting weapons.* Walker and Company, New York, New York, USA.

Blacksell, M., and A. W. Gilg. 1981. *The countryside: planning and change.* George Allen & University, London, United Kingdom.

Blair, W. F. 1941. Techniques for the study of mammal populations. *Journal of Mammalogy* 22:149–157.

Blankenship, L. H., A. B. Humphrey, and D. MacDonald. 1971. A new stratification for mourning dove call-count routes. *Journal of Wildlife Management* 35:319–326.

Bleich, V. C., and M. W. Oehler. 2000. Wildlife education in the United States: thoughts from some agency biologists. *Wildlife Society Bulletin* 28:542–545.

Bleich, V. C., J. Coombes, and J. H. Davis. 1982. Horizontal wells as a wildlife habitat improvement technique. *Wildlife Society Bulletin* 10:324–328.

Bleich, V. C., R. T. Bowyer, and J. D. Wehausen. 1997. Sexual segregation in mountain sheep: resources or predation. *Wildlife Monograph* 134:1–50.

Block, W. M., and L. A. Brennan. 1993. The habitat concept in ornithology: theory and applications. Pages 35–91 *in* D. M. Power, editor. *Current Ornithology.* Volume 11. Plenum Press, New York, New York, USA.

Blong, B., and W. Pollard. 1968. Summer water requirements of desert bighorn in the Santa Rosa Mountains, California, in 1965. *California Fish and Game Department* 54:289–296.

Blunden, J., and N. Curry, editors. 1985. *The changing countryside.* Croom Helm, London, United Kingdom.

Bock, C. E. 1987. Distribution-abundance relationships of some Arizona landbirds: a matter of scale? *Ecology* 68:124–129.

Bock, C. E., and J. H. Bock. 1999. Response of winter birds to drought and short-duration grazing in southeastern Arizona. *Conservation Biology* 13:1117–1123.

Bodick, J. E., and M. Peffer, editors 1993. Science and Technology Desk Reference. Gale Research, Incorporated, Washington, D.C., USA.

Boinski, S. 1987. Birth synchrony in squirrel monkeys (*Saimiri oerstedi*): a strategy to reduce neonatal predation. *Behavioral Ecology and Sociobiology* 21:393–400.

Bolen, E. G., and W. L. Robinson. 1995. *Wildlife ecology and management.* 3rd ed. Prentice Hall, Englewood Cliffs, New Jersey, USA.

Bookhout, T. A., editor. 1994. *Research and management techniques for wildlife and habitats.* 5th ed. The Wildlife Society, Bethesda, Maryland, USA.

Botkin, D. B. 1990. *Discordant harmonies.* Oxford University Press, Oxford, United Kingdom.

Bowsky, W. M. editor. 1971. *The black death: A turning point in history?* Holt, Rinehart and Winston, New York, New York, USA.

Bowyer, R. T. 1986. Habitat selection by southern mule deer. *California Fish and Game Journal* 72:153–169.

Boyce, S. 1980. Management of forests for optimal benefits (DV-NAST-OB). U.S. Forest Service Research Paper SE-204.

Boyce, W. M., D. A. Jessup, and R. K. Clark. 1991. Serodiagnostic antibody response to *Psoroptes* infestations in bighorn sheep. *Journal of Wildlife Disease* 27:10–15.

Boyd, R. J., A. Y. Cooperrider, P. C. Lent, and J. A. Bailey. 1986. Ungulates. Pages 519–564 *in* A. Y. Cooperrider, R. J. Boyd, and H. R. Steward, editors. Inventory and monitoring of wildlife habitat. U.S. Bureau of Land Management, Denver, Colorado, USA.

Boyd, S. 1984. Excerpts from the Idaho wool-growers association. Pages 90–95 *in* The Taylor Grazing Act: 50 years of progress. U.S. Department of the Interior, Washington, D.C., USA.

Boyer, D. A., and R. D. Brown. 1988. A survey of translocations of mammals in the United States in 1985. Pages 1–11 *in* L. Nielsen and R. D. Brown, editors. Translocation of wild animals. Wisconsin Humane Society, Milwaukee, Wisconsin, and Caesar Kleberg Wildlife Research Institute, Kingsville, Texas, USA.

Boyle, K. J., R. L. Dressler, A. G. Clark, and M. F. Teisl. 1993. Moose hunter preferences and setting season timings. *Wildlife Society Bulletin* 21:498–504.

Bracey, H. E. 1972. *People and the countryside.* Routledge and Kegan Paul, London, United Kingdom.

Bradford, D. F. 1975. The effects of an artificial water supply on free living *Peromyscus true. Journal of Mammalogy* 56:705–707.

Bradley, W. G. 1963. Water metabolism in desert mammals with special reference to desert bighorn sheep. Desert Bighorn Council Transactions 7:26–39.

Brand, C. J. 1987. Duck plague. Pages 117–127 *in* M. Friend, editor. Field guide to wildlife diseases. U.S. Fish and Wildlife Service Resource Publication 167.

Brand, G. J., S. R. Shifley, and L. F. Ohmann. 1986. Linking wildlife and vegetation models to forecast the effects of management. Pages 383–387 *in* J. Verner, M. L. Morrison, and C. J. Ralph, editors. *Wildlife 2000: modeling habitat relationships of terrestrial vertebrates.*

University of Wisconsin Press, Madison, Wisconsin, USA.

Braun, C. E., K. W. Harmon, J. A. Jackson, and C. D. Littlefield. 1978. Management of national wildlife refuges in the United States: its impact on birds. *Wilson Bulletin* 90:309–321.

Braun, C. E., R. K. Schmidt, Jr., and G. E. Rogers. 1971. Census of Colorado white-tailed ptarmigan with tape-recorded calls. *Journal of Wildlife Management* 37:90–93.

Briese, L. A., and M. H. Smith. 1974. Season abundance and movement of nine species of small mammals. *Journal of Mammalogy* 55:615–629.

Brody, S. 1945. *Bioenergetics and growth.* Haufner, New York, New York, USA.

Bromley, D. W., editor. 1998. *The handbook of environmental economics.* Blackwell Publishers, Malden, Massachusetts, USA.

Brown, B. T., and R. R. Johnson. 1983. The distribution of bedrock depressions (*tinajas*) as sources of surface water in Organ Pipe Cactus National Monument, Arizona. *Journal of the Arizona-Nevada Academy of Science* 18:61–68.

Brown, D. 1966. Deer management information. Arizona Game and Fish Department, Federal Aid in Wildlife Restoration Report W-53-R-16, WP2-J3:55.

Brown, D. E. 1998. Water for wildlife: belief before science. Pages 9–16 *in* Environmental, economic, and legal issues related to rangeland water developments: proceedings of a symposium. Arizona State University, College of Law, Tempe, Arizona, USA.

Brown, R. D. 1984. The use of physical and physiological indices to predict the nutritional condition of deer—a review. Pages 52–63 *in* P. R. Krausman and N. S. Smith, editors. Deer in the Southwest: a workshop. Arizona Cooperative Wildlife Research Unit and the School of Renewable Natural Resources, The University of Arizona, Tucson, Arizona, USA.

Brown, R. D., and L. A. Nielsen. 2000. Leading wildlife academic programs into the new millennium. *Wildlife Society Bulletin* 28:495–502.

Brown, T. L., and D. J. Decker. 1979. Incorporating farmers' attitudes into management of white-tailed deer in New York. *Journal of Wildlife Management* 43:236–239.

Brownlee, S. L. 1979. Water development for desert mule deer. Booklet 7000-32. Texas Parks and Wildlife Department, Austin, Texas, USA.

Broyles, B. 1995. Desert wildlife water developments: questioning use in the Southwest. *Wildlife Society Bulletin* 23:663–675.

Broyles, B., and T. L. Cutler. 1999. Effect of surface water on desert bighorn sheep in the Cabeza Prieta National Wildlife Refuge, southwestern Arizona. *Wildlife Society Bulletin* 27:1082–1088.

Bubenik, A. B. 1982. Physiology. Pages 125–179 *in* J. W. Thomas and D. E. Toweill, editors. *Elk of North America, ecology and management.* Stackpole Books, Harrisburg, Pennsylvania, USA.

Bubenik, A. (Tony) B. 1991. The wildlife resource—mine, ours, or whose? Pages 143–151 *in* L. A. Renecker and R. J. Hudson, editors. Wildlife production: conservation and sustainable development. University of Alaska, Fairbanks, Alaska. Agricultural and Forestry Experiment Station miscellaneous publication 91-6.

Buckland, S. T., D. R. Anderson, K. P. Burnham, and J. L. Laake. 1993. *Distance sampling: estimating abundance of biological populations.* Chapman and Hall, London, United Kingdom.

Buechner, H. K. 1960. The bighorn sheep in the United States: its past, present, and future. *Wildlife Monograph* 4.

Bull, E. L. 1981. Indirect estimates of abundance of birds. Pages 76–80, *in* C. J. Ralph and J. M. Scott, editors. Estimating the numbers of terrestrial birds. *Studies in Avian Biology* 6.

Burkett, D. W., and B. C. Thompson. 1994. Wildlife association with human-altered water sources in semiarid vegetation communities. *Conservation Biology* 8:682–690.

Burnham, K. P., D. R. Anderson, and J. L. Laake. 1980. Estimation of density from line transect data. *Journal of Wildlife Management* 43:992–996.

Burnham, K. P., G. C. White, and D. R. Anderson. 1984. Estimating the effect of hunting on annual survival rates of adult mallards. *Journal of Wildlife Management* 48:350–361.

Busch, D. E., J. C. Rorabaugh, and K. R. Rautenstrauch. 1984. Deer entrapment in canals of the western United States: a review of the problem and attempted solutions. Pages 95–100 *in* P. R. Krausman and N. S. Smith, editors. Deer in the Southwest: a workshop. Arizona Cooperative Wildlife Research Unit and the School of Renewable Natural Resources, The University of Arizona, Tucson, Arizona, USA.

Buss, I. O., and C. V. Swanson. 1950. Some effects of weather on pheasant reproduction in southeastern Washington. Transactions of the North American Wildlife Conference 15:364–378.

Byers, J. A. 1997. *American pronghorn: social adaptations and the ghosts of predators past.* The University of Chicago Press, Chicago, Illinois, USA.

Cain, S. A., J. A. Kadlee, D. L. Allen, R. A. Cooley, M. C. Hornocker, A. S. Leopold, and F. H. Wagner. 1972. Predator control. 1971. Report to Council on Environmental Quality and the U.S. Department of the Interior. University of Michigan Press, Ann Arbor, Michigan, USA.

Calder, W. A. III. 1981. Diuresis on the desert? Effects of fruit—and nectar—feeding on the house finch and other species. *Condor* 83:267–268.

Calhoun, J. B. 1950. North American census of small mammals. Rosco B. Jackson Memorial Lab, Bar Harbor, Maine. Release 3:1–90.

Calhoun, J. B. 1952. The social aspects of population dynamics. *Journal of Mammalogy* 33:139–159.

Callan, R. J., T. D. Bunch, G. W. Workman, and R. E. Mock. 1991. Development of pneumonia in desert bighorn sheep after exposure to a flock of exotic wild and domestic sheep. *Journal of the American Veterinary Medical Association* 198:1052–1056.

Campbell, B. H., and R. Remington. 1979. Bighorn use of artificial water sources in the Buckskin Mountains, Arizona. Desert Bighorn Council Transactions 23:50–56.

Campbell, B. H., and R. Remington. 1981. Influence of construction activities on water-use patterns of desert bighorn sheep. *Wildlife Society Bulletin* 9:63–65.

Campbell, H. 1961. An evaluation of gallinaceous guzzlers for quail in New Mexico. *Journal of Wildlife Management* 24:21–26.

Campbell, H., D. K. Martin, P. E. Ferkovich, and B. K. Harris. 1973. Effects of hunting and some other environmental factors on scaled quail in New Mexico. *Wildlife Monographs* 34.

Capen, D. E., editor. 1981. The use of multivariate statistics in studies of wildlife habitat. U.S. Forest Service General Technical Report RM-87.

Carpenter, L. H. 2000. Harvest management goals. Pages 192–213 *in* S. Demarais and P. R. Krausman, editors. *Ecology and management of large mammals in North America*. Prentice Hall, Upper Saddle River, New Jersey, USA.

Carrier, W. D., and B. Czech. 1996. Threatened and endangered wildlife and livestock interactions. Pages 39–47 *in* P. R. Krausman, editor, *Rangeland Wildlife*. The Society for Range Management, Denver, Colorado, USA.

Castro, A. E., and W. P. Heuschele. 1992. *Veterinary diagnostic virology*. Mosby-Year Book, St. Louis, Missouri, USA.

Caughley, G. 1966. Mortality patterns in mammals. *Ecology* 47:906–918.

Caughley, G. 1970. Eruption of ungulate population, with emphasis on Himalayan thar in New Zealand. *Ecology* 51:51–72.

Caughley, G. 1974. Bias in aerial survey. *Journal of Wildlife Management* 38:921–933.

Caughley, G. 1976. Wildlife management and the dynamics of ungulate populations. Pages 183–246 *in* T. H. Coaker, editor. *Applied Biology*. Volume 1. Academic Press, New York, New York, USA.

Caughley, G. 1977. *Analysis of vertebrate populations*. John Wiley, United Kingdom.

Caughley, G. 1979. What is this thing called carrying capacity? Pages 2–8 *in* M. S. Boyce and L. D. Hayden-Wing, editors. *North American elk: ecology, behavior and management*. University of Wyoming Press, Laramie, Wyoming, USA.

Caughley, G., and A. R. E. Sinclair. 1994. *Wildlife ecology and management*. Blackwell Scientific Publications, Boston, Massachusetts, USA.

Channing, C. H. 1958. Highway casualties of birds and animals for one year period. *Murrelet* 39:41.

Chapman, D. G. 1951. Some properties of the hypergeometric distribution with applications to zoological sample censuses. University of California Publication Statistics 1:131–160.

Chapman, J. 1987. The extent and nature of parliamentary enclosure. *Agricultural History Review* 35:25–35.

Cheatum, E. L. 1949. The use of corpora lutea for determining ovulation incidence and variation in the fertility of white-tailed deer. *Cornell Veterinarian* 39:282–291.

Child, K. N. 1998. Incidental mortality. Pages 275–301 *in* A. W. Franzmann and C. L. Schwartz, editors. *Ecology and management of North American moose*. Smithsonian Institution Press, Washington, D.C., USA.

Chizek, J. T. 1985. Car-deer collisions. *Wisconsin Natural Resources* 9:41–42.

Christensen, G. C. 1954. The chukar partridge in Nevada. Nevada Fish and Game Commission Biological Bulletin 1.

Christian, D. P. 1979. Comparative demography of three Namib Desert rodents: responses to the provision of supplementary water. *Journal of Mammalogy* 60:679–690.

Church, D. C. 1988. *The ruminant animal*. Prentice Hall, Englewood Cliffs, New Jersey, USA.

Clark, D. E., 1952. A study of the behavior and movements of the Tucson Mountain mule deer. Thesis, University of Arizona, Tucson, Arizona, USA.

Clark, P. J., and F. C. Evans. 1954. Distance to nearest neighbor as a measure of spatial relationships in populations. *Ecology* 4:445–453.

Clutton-Brock, T. H., F. E. Guinners, and S. D. Albon. 1982. *Red deer: behavior and ecology of two sexes*. University of Chicago Press, Chicago, Illinois, USA.

Clutton-Brock, T. H., M. Major, S. D. Albon, and F. E. Guinness. 1987. Early development and population dynamics in red deer. Density-dependent effects on juvenile survival. *Journal of Animal Ecology* 56:53–67.

Cole, L. C. 1954. The population consequences of life history phenomena. *Quarterly Review of Biology* 29:103–137.

Collins, J. P. 1998. Adjudicating nature: science, law, and policy in a southern Arizona grassland. Pages 226–234 *in* Environmental, economic, and legal issues related to rangeland water developments: proceedings of a symposium. Arizona State University, College of Law, Tempe, Arizona, USA.

Committee on Merchant Marine and Fisheries. 1975. A compilation of federal laws relating to conservation and development of our nation's fish and wildlife resources, environmental quality and oceanography. U.S. Government Printing Office, Washington, D.C., USA.

Conover, M. R., W. C. Pitt, K. K. Kessler, T. J. DuBow, and W. A. Sanborn. 1995. Review of human injuries, illnesses, and economic losses caused by wildlife in the United States. *Wildlife Society Bulletin* 23:407–414.

Conner, M. C., R. A. Lancia, and K. H. Pollock. 1986. Precision of the change-in-ratio technique for deer population management. *Journal of Wildlife Management* 50:125–129.

Connolly, G. E. 1981. Assessing populations. Pages 287–345 *in* O. C. Wallmo, editor. *Mule and black-tailed deer of North America*. University of Nebraska Press, Lincoln, Nebraska, USA.

Connolly, G. E. 1981. Limiting factors and population regulation. Pages 245–285 *in* O. C. Wallmo, editor. *Mule and black-tailed deer of North America*. University of Nebraska Press, Lincoln, Nebraska, USA.

Connolly, G. E. 1978. Predators and predator control. Pages 369–394 *in* J. L. Schmidt and D. L. Gilbert, editors. *Big game of North America: ecology and management*. Stackpole Books, Harrisburg, Pennsylvania, USA.

Cook, J. G., L. J. Quinlan, L. L. Irwin, L. D. Bryant, R. A. Riggs, and J. W. Thomas. 1996. Nutrition-growth relations of elk calves during late summer and fall. *Journal of Wildlife Management* 60:528–541.

Cook, J. G., L. L. Irwin, L. D. Bryant, R. A. Riggs, and J. W. Thomas. 1998. Relations of forest cover and condition of elk: a test of the thermal cover hypothesis in summer and winter. *Wildlife Monograph* 141.

Cooke, B. D. 1982. Shortage of water in natural pastures as a factor limiting a population of rabbits, *Oryctolagus cuniculus* (L.) in arid northeastern south Australia. *Australian Wildlife Research* 9:465–476.

Cool, K. L. 1992. Seeking common ground on western rangelands. *Rangelands* 14:90–92.

Cortner, H. J., and M. A. Moote. 1992. Sustainability and ecosystem management forces shaping political agendas and public policy. Proceedings of the Society of American Foresters 1992 Convention.

Cortner, H. J., and M. A. Moote. 1994. Trends and issues in land and water resources management: setting the agenda for change. *Environmental Management* 18:167–173.

Costanza, R., editor. 1991. *Ecological economics: the science and management of sustainability*. Columbia University, New York, New York, USA.

Cottam, C., and J. B. Trefethen. 1968. *Whitewings*. D. Van Nostrand Company, Princeton, New Jersey, USA.

Coupland, R. T., editor. 1992. *Ecosystems of the world 8A: natural grasslands—introduction and Western Hemisphere*. Elsevier, New York, New York, USA.

Cowan, I. McT. 1956. Life and times of the coast black-tailed deer. Pages 523–617 *in* W. P. Taylor, editor. *The deer of North America*. Stackpole Books, Harrisburg, Pennsylvania, USA.

Craighead, J. J., J. S. Sumner, and J. A. Mitchell. 1995. *The grizzly bears of Yellowstone: their ecology in the Yellowstone ecosystem, 1959–1992*. Island Press, Washington, D.C., USA.

Creswell, P., S. Harris, R. G. M. Bunce, and D. J. Jeffferies. 1989. The badger (*Meles meles*) in Britain: present status and future population changes. *Biological Journal Linnean Society (London)* 38:91–101.

Crete, M. 1989. Approximation of K-carrying capacity for moose in eastern Quebec. *Canadian Journal of Zoology* 67:373–380.

Croon, G. W., D. R. McCullough, C. E. Olson, Jr., and L. M. Queal. 1968. Infrared scanning techniques for big game censusing. *Journal of Wildlife Management* 32:751–759.

Crow, L. Z. 1964. A field survey of water requirements of desert bighorn sheep. *Desert Bighorn Council Transactions* 8:77–83.

Curtis, R. O., G. W. Clendenen, and D. J. DeMars. 1981. A new stand simulator for coast Douglas fir: DFSIM user's guide. U.S. Forest Service General Technical Report PNW-128.

Cutler, M. R. 1982. What kind of wildlifers will be needed in the 1980s? *Wildlife Society Bulletin* 10:75–79.

Cutler, P. L. 1996. Wildlife use of two artificial water developments on the Cabeza Prieta National Wildlife Refuge, southwestern Arizona. Thesis, University of Arizona, Tucson, Arizona, USA.

Cutler, T. L., and D. E. Swann. 1999. Using remote photography in wildlife ecology: a review. *Wildlife Society Bulletin* 27:571–581.

Czech, B. 1995. American Indians and wildlife conservation. *Wildlife Society Bulletin* 23:568–573.

Czech, B. 1999. Big game management on tribal lands. Pages 277–289 *in* S. Demaris and P. R. Krausman, editors. *Ecology and management of large mammals in North America.* Prentice Hall, Upper Saddle River, New Jersey, USA.

Czech, B. 2000*a*. Economic growth as the limiting factor for wildlife conservation. *Wildlife Society Bulletin* 28:4–15.

Czech, B. 2000*b*. *Shoveling fuel for a runaway train: errant economists, shameful spenders, and a plan to stop them all.* University of California Press, Berkeley, California, USA.

Czech, B., and L. A. Tarango. 1998. Wildlife as an economic staple; an example from the San Carlos Apache Reservation. Pages 209–215 *in* G. J. Gottfried, C. B. Edminster, and M. C. Dillon, compilers. Cross border waters: fragile treasures for the 21st century. Ninth U.S./Mexico conference in recreation, parks, and wildlife. U.S. Forest Service, Rocky Mountain Research Station.

Czech, B., and P. R. Krausman. 1997. Distribution and causation of species endangerment in the United States. *Science* 277:1116–1117.

Czech, B., and P. R. Krausman. 1997. Implications of an ecosystem management literature review. *Wildlife Society Bulletin* 25:667–675.

Czech, B., and P. R. Krausman. 1999. Public opinion on endangered species conservation and policy. *Society and Natural Resources* 12:469–479.

Czech, B., P. R. Krausman, and P. K. Devers. 2000. Economic associations among causes of species endangerment in the United States. *BioScience* 50:593–601.

Czech, B., and P. R. Krausman. 2001. *The Endangered Species Act: history, conservation biology, and public policy.* Johns Hopkins University Press, Baltimore, Maryland, USA.

Dale, V. H., and M. Hemstrom. 1984. CLIMACS: a computer model of forest stand development for western Oregon and Washington. U.S. Forest Service Research Paper PNW-327.

Dalke, P. D., D. B. Pyrah, D. C. Stanton, J. E. Crawford, and E. F. Schlatterer. 1963. Ecology, productivity, and management of sagegrouse in Idaho. *Journal of Wildlife Management* 27:811–841.

Dalrymple, G. H., and N. G. Reichenbach. 1984. Management of an endangered species of snake in Ohio, USA. *Biological Conservation* 30:195–200.

Darby, H. C. 1973. *A new historical geography of England.* University Press, Cambridge, United Kingdom.

Dark, J., N. G. Forger, and I. Zucker. 1986. Regulation and function of lipid mass during the annual cycle of the golden-mantled ground squirrel. Pages 445–451 *in* H. C. Heller, X. J. Musacchia, and L. C. H. Wang, editors. *Living in the cold: physiological and biochemical adaptations.* Elsevier Scientific, New York, New York, USA.

Dasmann, R. F. 1945. A method for estimating carrying capacity of rangelands. *Journal of Forestry* 43:400–402.

Dasmann, R. F. 1964. *Wildlife biology.* John Wiley & Sons, New York, New York, USA.

Dasmann, R. F. 1981. *Wildlife biology.* 2nd ed. John Wiley & Sons, New York, New York, USA.

Daubenmire, R. 1968. *Plant communities: a textbook of plant synecology.* Harper and Row, New York, New York, USA.

Daubenmire, R. 1976. The uses of vegetation in assessing the productivity of forest lands. *Botanical Review* 42:115–143.

Dauphine, T. C., and R. L. McClure. 1974. Synchronous mating in Canadian barren-ground caribou. *Journal of Wildlife Management* 38:54–66.

Davidson, W. R., F. A. Hayes, V. F. Nettles, and F. E. Kellogg. 1981. *Diseases and parasites of white-tailed deer.* Heritage Printers, Charlotte, North Carolina, USA.

Davies, S. J. J. F. 1982. Behavioral adaptations of birds to environments where evaporation is high and water is in short supply. *Comparative Biochemistry and Physiology* 71A:557–566.

Davis, C. A., P. E. Sawyer, J. P. Griffing, and B. D. Borden. 1974. Bird populations in a shrub-grassland area, southeastern New Mexico. New Mexico State University, Agricultural Experiment Station Bulletin 619.

Davis, D. E. 1963. Estimating the numbers of game populations. Pages 89–118 *in* H. S. Mosley, editor. *Wildlife investigational techniques.* The Wildlife Society, Washington, D.C., USA.

Davis, D. E., editor. 1983. *CRC handbook of census methods for terrestrial vertebrates.* CRC Press, Boca Raton, Florida, USA.

Davis, J. W., L. H. Karstad, and D. O. Trainer. 1981*a. Infectious diseases of wild mammals.* 2nd ed. Iowa State University Press, Ames, Iowa, USA.

Davis, J. W., L. H. Karstad, and D. O. Trainer. 1981*b. Parasitic diseases of wild mammals.* 2nd ed. Iowa State University Press, Ames, Iowa, USA.

Davis, L. S. 1980. Strategy for building a location-specific, multi-purpose information system for wildland management. *Journal of Forestry* 78:402–408.

Davis, L. S., and L. I. DeLain. 1986. Linking wildlife-habitat analysis to forest planning with ECOSYM. Pages 361–369 *in* J. Verner, M. L. Morrison, and C. J. Ralph, editors. *Wildlife 2000: modeling habitat relationships of terrestrial vertebrates.* University of Wisconsin Press, Madison, Wisconsin, USA.

Dawson, J. W., and R. W. Mannan. 1991. The role of territoriality in the social organization of Harris' hawks. *Auk* 108:661–672.

Dawson, W. R., C. Carey, C. S. Adkisson, and R. D. Ohmant. 1979. Responses of Brewer's and chipping sparrows to water restriction. *Physiological Zoology* 52:529–541.

Deblinger, R. D., and A. W. Alldredge. 1991. Influence of fresh water on pronghorn distribution in a sagebrush/steppe grassland. *Wildlife Society Bulletin* 19:321–326.

DeCalestra, D. S., and S. L. Stout. 1997. Relative deer density and sustainability: a conceptual framework for integrating deer management with ecosystem management. *Wildlife Society Bulletin* 25:252–258.

Decker, D. J., and K. G. Purdy. 1988. Toward a concept of wildlife acceptance capacity in wildlife management. *Wildlife Society Bulletin* 16:53–57.

Decker, E. 1978. Exotics. Pages 249–256 *in* J. L. Schmidt and D. L. Gilbert, editors. *Big game of North America: ecology and management.* Stackpole Books, Harrisburg, Pennsylvania, USA.

Deevey, E. S., Jr. 1947. Life tables for natural populations of animals. *Quarterly Review of Biology* 22:283–314.

Degen, A. A., M. Kam, A. Hazan, and K. A. Nagy. 1986. Energy expenditure and water flux in three sympatric desert rodents. *Journal of Animal Ecology* 55:421–429.

Degen, A. A., B. Pinshow, and P. J. Shaw. 1984. Must desert chukars (*Alectoris chukar sinaica*) drink water? Water influx and body mass changes in response to dietary water content. *Auk* 101:47–52.

DeGraaf, R. M., N. G. Tilghman, and S. H. Anderson. 1985. Foraging guilds of North American birds. *Environmental Management* 9:493–536.

DelGiudice, G. D., and U. S. Seal. 1988. Classifying winter undernutrition in deer via serum and urinary urea nitrogen. *Wildlife Society Bulletin* 16:27–32.

DelGiudice, G. D., L. D. Mech, and U.S. Seal. 1990. Effects of winter undernutrition on body composition and physiological profiles of white-tailed deer. *Journal of Wildlife Management* 54:539–550.

del Hoyo, J., A. Elliot, and J. Sargatal, editors. 1992. *Handbook of birds of the world.* Vol. 1 Lynx edicions, Barcelona, Spain.

Demarais, S., J. T. Baccus, and M. S. Traweek, Jr. 1998. Nonindigenous ungulates in Texas: long term population trends and possible competitive mechanisms. Transactions of the North American Wildlife and Natural Resources Conference 63:49–55.

DeMillo, G. E., R. J. Warren, and J. P. Stowe, Jr. 1998. Is the non-thesis graduate degree an appropriate complement to the education of tomorrow's wildlife professional? *Wildlife Society Bulletin* 26:954–960.

Denney, R. N. 1978. Managing the harvest. Pages 395–408 *in* J. L. Schmidt and D. L. Gilbert, editors. *Big game of North America: ecology and management*. Stackpole Books, Harrisburg, Pennsylvania, USA.

DeStefano, S., C. J. Brand, and M. D. Samuel. 1995. Seasonal ingestion of toxic and nontoxic shot by Canada geese. *Wildlife Society Bulletin* 23:502–506.

de Steiguer, J. E. 1989. Forestry sector environmental effects. Pages 251–262 *in* P. V. Ellefson, editor. *Forest resource economics and policy*. Westview Press, Boulder, Colorado, USA.

de Steiguer, J. E. 1995. Three theories from economics about the environment. *BioScience* 45:552–557.

Dimbleby, G. W. 1984. Anthropogenic changes from neolithic through medieval times. *New Phytologist* 98:57–72.

DiSilvestro, R. L. 1985. The federal animal damage control program. Pages 130–148 *in The Audubon Wildlife Report*. National Audubon Society, New York, New York, USA.

Dodgshon, R. A. 1978. The early Middle Ages, 1066–1350. Pages 81–117 *in* R. A. Dodgshon and R. A. Butlin, editors. *A historical geography of England and Wales*. Academic Press, London, United Kingdom.

Dolton, D. D. 1993. The call-count survey: historic development and current procedures. Pages 233–252 *in* T. S. Baskett, M. W. Sayre, R. E. Tomlinson, and R. E. Mirarchi, editors. *Ecology and management of the mourning dove*. Stackpole Books, Harrisburg, Pennsylvania, USA.

Douglas, C. L., and D. M. Leslie, Jr. 1986. Influence of weather and density on lamb survival of desert bighorn sheep. *Journal of Wildlife Management* 50:153–156.

Downing, R. L., W. H. Moore, and J. Knight. 1965. Comparison of deer census techniques applied to a known population in a Georgia enclosure. Proceedings of the Annual Conference of the Southeastern Association of Fish and Wildlife Agencies 19:26–30.

Dozier, H. L., M. H. Markley, and L. M. Llewellyn. 1948. Muskrat investigations on the Blackwater National Wildlife Refuge, Maryland, 1941–1945. *Journal of Wildlife Management* 12:177–190.

Drew, M. L., D. A. Jessup, A. A. Burr, and C. E. Franti. 1992. Serologic survey for brucellosis in feral swine, wild ruminants, and black bear of California 1977 to 1989. *Journal of Wildlife Diseases* 28:355–363.

Dublin, L. I., and A. J. Lotka. 1925. On the true rate of natural increase as exemplified by the population of the United States, 1920. *Journal of the American Statistical Association* 20:305–339.

Dudzinski, M. L., and G. W. Arnold. 1967. Aerial photography and statistical analysis for studying behavior patterns of grazing animals. *Journal of Range Management* 20:77–83.

Duerr, A. E., and S. DeStefano. 1999. Using a metal detector to determine lead sinker abundance in waterbird habitat. *Wildlife Society Bulletin* 27:952–958.

Dueser, R. D., and J. H. Porter. 1986. Habitat use by insular small mammals: relative effects of competition and habitat structure. *Ecology* 67:195–201.

Duke, G. E. 1966. Reliability of census of singing male woodcocks. *Journal of Wildlife Management* 30:697–707.

Dunforth, A. A., and F. P. Errington. 1964. Casualties among birds along a selected road in Wiltshire. *Bird Study* 11:168–182.

Dunham, K. M. 1997. Population growth of mountain gazelles, *Gazella gazella*, reintroduced to central Arabia. *Biological Conservation* 81:205–214.

Dunlap, T. R. 1988. *Saving America's wildlife*. Princeton University Press, Princeton, New Jersey, USA.

Eberhardt, L., and R. C. Van Etten. 1956. Evaluation of the pellet group count as a deer census method. *Journal of Wildlife Management* 20:70–74.

Eberhardt, L., and T. J. Peterle, and R. Schofield. 1963. Problems in a rabbit population study. *Wildlife Monograph* 10.

Eberhardt, L. L., D. G. Chapman, and J. R. Gilbert. 1979. A review of marine mammal census methods. *Wildlife Monograph* 63.

Edeburn, R. M. 1973. Great horned owl impaled on barbed wire. *Wilson Bulletin* 85:478.

Edge, W. D., and J. P. Loegering. 2000. Distance education: expanding learning opportunities. *Wildlife Society Bulletin* 28:522–533.

Edge, W. D., S. L. Olson-Edge, and L. L. Irwin. 1990. Planning for wildlife in national forests: elk and mule deer habitats as an example. *Wildlife Society Bulletin* 18:87–98.

Edlin, H. L. 1972. *Trees, woods, and man.* Collins, London, United Kingdom.

Edwards, R. Y., and C. D. Fowle. 1955. The concept of carrying capacity. Transactions of the North American Wildlife Conference 20:587–602.

Egerton III, F. N. 1968. Ancient sources for animal demography. *Isis* 59:175–189.

Ek, A. R., and R. A. Monserud. 1974. FOREST: a computer model for simulating the growth and reproduction of mixed species forest stands. University of Wisconsin, School of Natural Resources Research Report R2635.

Elder, J. B. 1954. Notes on summer water consumption by desert mule deer. *Journal of Wildlife Management* 18:540–541.

Ellis, J. A., R. L. Westemeier, K. P. Thomas, and H. W. Norton. 1969. Spatial relationships among quail coveys. *Journal of Wildlife Management* 33:249–254.

Ellison, L. N. 1991. Shooting and compensatory mortality in tetaonids. *Ornis Scandinavica* 22:229–240.

Emsley, C. 1987. *Crime and society in England, 1750–1900.* Longman Group, Harlow, United Kingdom.

Erickson, J. D. 2000. Endangering the economics of extinction. *The Wildlife Society Bulletin* 28:34–41.

Errington, P. 1934. Vulnerability of bobwhite populations to predation. *Ecology* 15:110–127.

Errington, P. 1956. Factors limiting higher vertebrate populations. *Science* 124:304–307.

Errington, P. L., and F. N. Hamerstrom, Jr. 1935. Bobwhite winter survival on experimentally shot and unshot areas. Iowa State College. *Journal of Science* 9:625–639.

Estes, R. D. 1976. The significance of breeding synchrony in the wildebeest. *East African Wildlife Journal* 14:135–152.

Etchberger, R. C., and P. R. Krausman. 1999. Frequency of birth and lambing sites of a small population of mountain sheep. *The Southwestern Naturalist* 44:354–360.

Evans, C. D., W. A. Troyer, and C. J. Lensink. 1966. Aerial census of moose by quadrate sampling units. *Journal of Wildlife Management* 30:767–776.

Evans, W. 1969. Analysis of the mule deer population on Fort Stanton, New Mexico. Thesis, New Mexico State University, Las Cruces, New Mexico, USA.

Evenari, M. L., L. Shanan, N. Tadmor, and Y. Aharoni. 1961. Ancient agriculture in the Negev. *Science* 133:979–996.

Evenden, F. G. 1971. Animal road kills. *Atlantic National* 26:36–37.

Evenden, F. G. n.d. Wildlife biology and management. Pages 106–153 *in* H. C. Leppir, editor. 2nd ed. *Careers in conservation.* John Wiley & Sons, New York, New York, USA.

Fagerstone, K. A., and C. A. Ramey. 1996. Rodents and lagomorphs. Pages 83–132 *in* P. R. Krausman, editor. *Rangeland wildlife.* The Society for Range Management, Denver, Colorado, USA.

Fahrig, L., J. H. Pedlar, S. E. Pope, P. D. Taylor, and J. F. Wegner. 1995. Effect of road traffic on amphibian density. *Biological Conservation* 73:177–182.

Farner, 1955. *Birdbanding in the study of population dynamics.* University of Illinois Press, Urbana, Illinois, USA.

Favre, D. S. 1983. *Wildlife: cases, laws, and policy.* Associated Faculty Press, Tarrytown, New York, USA.

Feldhamer, G. A., and W. E. Armstrong. 1993. Interspecific competition between four exotic species and native artiodactyls in the United States. Transactions of the North

American and Natural Resources Conference 58:468–478.

Feldman, J. W. 1996. The politics of predator control. Thesis, Utah State University, Logan, Utah, USA.

Ferguson, D., and N. Ferguson. 1983. *Sacred cows and the public trough.* Maverick Publishers, Bend, Oregon, USA.

Ferreras, P., J. J. Aldoma, J. F. Beltron, and M. Delibes. 1992. Rates and causes of mortality in a fragmented population of Iberian lynx *Felis pardina* Temminck, 1824. *Biological Conservation* 61:197–202.

Ferrier, G. J., and W. G. Bradley. 1970. Bighorn habitat evaluation in the Highland Range of southern Nevada. *Desert Bighorn Council Transactions* 14:66–93.

Fight, R. D., J. M. Chittester, and G. W. Clendenen. 1984. DFSIM with economics: a financial analysis option for the DFSIM Douglas fir simulator. U.S. Forest Service General Technical Report PNW-175.

Finger, S. E., I. L. Brisbin, Jr., M. H. Smith, and D. F. Urbston. 1981. Kidney fat as a predictor of body condition in white-tailed deer. *Journal of Wildlife Management* 45:964–968.

Fink, D. H., K. R. Cooley, and G. W. Frasier. 1973. Wax-treated soils for harvesting water. *Journal of Range Management* 26:396–398.

Finnis, R. G. 1960. Road casualties among birds. *Bird Study* 7:21–32.

Flader, S. 1974. *Thinking like a mountain: Aldo Leopold and the evaluation of an ecological attitude toward deer, wolves, and forests.* University of Nebraska Press, Lincoln, Nebraska, USA.

Fleischner, T. L. 1994. Ecological costs of livestock grazing in western North America. *Conservation Biology* 8:629–644.

Flinders, J. T., and R. M. Hansen. 1973. Abundance and dispersion of leporids within a shortgrass ecosystem. *Journal of Mammalogy* 54:287–291.

Flinders, J. T., and R. M. Hansen. 1975. Spring population responses of cottontails and jackrabbits to cattle grazing shortgrass prairie. *Journal of Range Management* 28:290–293.

Flower, S. S. 1938. Further notes on the duration of life in animals. IV. Birds. Proceedings of the Zoological Society of London 108(A):195–235.

Flower, S. S. 1947. Further notes on the duration of life in mammals. V. The alleged and actual ages to which elephants live. Proceedings of the Zoological Society of London 117:680–688.

Flueck, W. J. 1989. The effect of selenium on reproduction of black-tailed deer (*Odocoileus hemionus columbianus*) in Shasta County, California. Dissertation, University of California, Davis, California, USA.

Foote, L. E., H. S. Peters, and A. L. Finkner. 1958. Design tests for mourning dove call-count sampling in seven southeastern states. *Journal of Wildlife Management* 22:402–408.

Foreyt, W. J., and D. A. Jessup. 1982. Fatal pneumonia in bighorn sheep following direct association with domestic sheep. *Journal of Wildlife Disease* 18:163–168.

Forman, R. T. 2000. Estimate of the area affected ecologically by the road system in the United States. *Conservation Biology* 14:31–35.

Fox, K. B., and P. R. Krausman. 1994. Fawning habitat of desert mule deer. *Southwestern Naturalist* 39:269–275.

Fox, L. M. 1997. Nutritional content of forage in Sonoran pronghorn habitat. Thesis, The University of Arizona, Tucson, Arizona, USA.

Fox, L. M., and P. R. Krausman. In press. Water and nutrient content of forage in Sonoran pronghorn habitat, Arizona. *California Fish and Game Journal.*

Francis, W. J. 1973. Accuracy of census methods of territorial red-winged blackbirds. *Journal of Wildlife Management* 37:98–102.

Franzmann, A. W. 2000. Moose. Pages 578–600 *in* S. Demarais and P. R. Krausman, editors. *Ecology and management of large mammals in North America.* Prentice Hall, Upper Saddle River, New Jersey, USA.

Frasier, G. W. 1980. Harvesting water for agricultural, wildlife, and domestic uses. *Journal of Soil and Water Conservation.* May–June: 125–128.

Frasier, G. W., K. R. Cooley, and J. R. Griggs. 1978. Performance evaluation of water harvesting catchments. *Journal of Range Management* 32:453–456.

Freddy, D. J., and R. B. Gill. 1977. A life table for managing deer populations. Colorado Department of Natural Resources. Outdoor Facts 104.

Freese, C. H., and D. L. Trauger. 2000. Wildlife markets and biodiversity conservation in North America. *Wildlife Society Bulletin* 28:42–51.

Friend, M. 1987*a*. Avian cholera. Pages 69–82 *in* M. Friend, editor. Field guide to wildlife diseases. U.S. Fish and Wildlife Service Resource Publication 167.

Friend, M. 1987*b*. Lead poisoning. Pages 175–190 *in* M. Friend, editor. Field guide to wildlife diseases. U.S. Fish and Wildlife Service Resource Publication 167.

Friggens, M. T. In press. Carnivory on desert cottontails by Texas antelope ground squirrels. *Southwestern Naturalist*.

Frisinia, M. R. 1992. Elk habitat use within a rest-rotation grazing system. *Rangelands* 14:9–96.

Frisinia, M. R., and F. G. Morin. 1991. Grazing private and public land to improve the Fleecer elk winter range. *Rangelands* 13:291–294.

Fritts, S. H., E. E. Bangs, J. A. Fontaine, M. R. Johnson, M. K. Phillips, E. D. Koch, J. R. Gunson. 1997. Planning and implementing a reintroduction of wolves to Yellowstone National Park and central Idaho. *Restoration Ecology* 5:7–27.

Frye, O. E., Jr. 1961. A review of bobwhite quail management in eastern North America. Transactions of the North American and Natural Resources Conference 26:273–281.

Fryxell, J. M., and A. R. E. Sinclair. 1988. Causes and consequences of migration by large herbivores. *Trends in Ecology and Evolution* 3:237–241.

Fuller, M. R., and J. A. Mosher. 1981. Methods of detecting and counting raptors: a review. Pages 235–246 *in* C. J. Ralph and J. M. Scott, editors. Estimating numbers of terrestrial birds. *Studies in Avian Biology* 6.

Fuller, T. 1989. Population dynamics of wolves in north-central Minnesota. *Wildlife Monograph* 105.

Fuller, T. K. 1991. Do pellet counts index white-tailed deer numbers and population change? *Journal of Wildlife Management* 55:393–396.

Furniss, R. 1991. Marketing surplus game-ranched animals to big game hunters. Pages 311–320 *in* L. A. Renecker and R. J. Hudson, editors. Wildlife production: conservation and sustainable development. University of Alaska, Fairbanks, USA. Agricultural and Forestry Experiment Station Miscellaneous Publication 91–6.

Gasaway, W. C., R. O. Boertje, D. V. Grongaard, D. J. Kelleyhouse, R. O. Stephenson, and D. G. Larsen. 1992. The role of predation in limiting moose at low densities in Alaska and Yukon and implications for conservation. *Wildlife Monograph* 120.

Gates, C. E. 1969. Simulation study of estimators for the line transect method. *Biometrics* 25:317–328.

Gates, C. E. 1979. Line transects and related issues. Pages 71–154 *in* R. M. McCormick, P. Patil, and D. S. Robson, editors. Sampling biological populations. *Statistical Ecology* Series 5: International Cooperation Publication House, Fairland, Maryland, USA.

Gates, C. E., W. H. Marshall, and D. P. Olson. 1968. Line-transect method of estimating grouse population densities. *Biometrics* 24:135–145.

Gates, C. E. 1980. Linetran, a general computer program for analyzing line-transect data. *Journal of Wildlife Management* 44:658–661.

Gates, C. E., and W. B. Smith. 1972. Estimation of density of mourning doves from aerial estimation. *Biometrics* 28:345–359.

Gavin, T. A., L. H. Suring, P. A. Vohs, and E. C. Meslow. 1984. Population characteristics, spatial organization, and natural mortality in the Columbian white-tailed deer. *Wildlife Monograph* 91.

Geissler, P. H. 1990. Confidence intervals for the federal waterfowl harvest surveys. *Journal of Wildlife Management* 54:201–205.

Geissler, P. H., and J. R. Sauer. 1990. Topics in route-regression analysis. Pages 54–57 *in* J. R. Sauer and S. Droege, editors. Survey designs and statistical methods for the estimation of avian population trends. U.S. Fish and Wildlife Service Biological Report 90.

Geist, V. 1988. How markets in wildlife meats and parts, and the sale of hunting privileges, jeopardize wildlife conservation. *Conservation Biology* 2:15–26.

Geluso, K. N. 1978. Urine concentrating ability and renal structure of insectivorous bats. *Journal of Mammalogy* 59:312–323.

George, W. G. 1974. Domestic cats as predators and factors in winter shortages of raptor prey. *Wilson Bulletin* 86:384–396.

Ghobrial, L. I. 1970. The water relations of the desert antelope *Gazella dorcas dorcas*. *Physiological Zoology* 43:249–256.

Gibbens, R. P., and A. M. Schultz. 1962. Manipulation of shrub form and browse production in game range improvement. *California Fish and Game* 48:49–63.

Gilbert, F. F., and D. G. Dodds. 1987. *The philosophy and practice of wildlife management*. Robert E. Kreiger Publishing, Malabar, Florida, USA.

Giles, R. H., Jr. 1978. *Wildlife management*. W. H. Freeman, San Francisco, California, USA.

Gilpin, M. E., and M. E. Soulé. 1986. Minimum viable populations: processes of species extinction. Pages 19–34 *in* M. E. Soulé, editor. *Conservation biology: the science of scarcity and diversity*. Sinauer, Sunderland, Massachusetts, USA.

Gionfriddo, J. P., and P. R. Krausman. 1986. Summer habitat use by mountain sheep. *Journal of Wildlife Management* 50:331–336.

Girouard, M. 1987. *A Country House Companion*. Yale University Press, New Haven, Connecticut, USA.

Glading, B. 1943. A self-filling quail watering device. *California Fish and Game Journal* 29:157–164.

Glading, B. 1947. Game watering devices for the arid southwest. Transactions of the North American Wildlife Conference 12:286–292.

Glover, F. A. 1969. Birds in grassland ecosystems. Pages 279–289 *in* R. L. Dix and R. G. Beidleman, editors. The grassland ecosystem: a preliminary synthesis. *Grassland Biome, International Biome Project*. Colorado

State University, Range Science Department Science Series 2.

Goldstein, D. L., and K. A. Nagy. 1985. Resource utilization by desert quail: time and energy, food and water. *Ecology* 66:378–387.

Golightly, R. T., Jr., and R. D. Ohmart. 1984. Water economy of two desert canids: coyote and kit fox. *Journal of Mammalogy* 65:51–58.

Golley, F. B. 1961. Energy values of ecological materials. *Ecology* 42:581–584.

Goodwin, J. G., Jr., and C. R. Hungerford. 1977. Habitat use by native Gambel's and scaled quail and released masked bobwhite quail in southern Arizona. U.S. Forest Service Research Paper RM-197.

Gowdy, J. M. 2000. Terms and concepts in ecological economics. *Wildlife Society Bulletin* 28:26–33.

Graff, W. 1980. Habitat protection and improvement. Pages 310–319 *in* G. Monson and L. Sumner, editors. *The desert bighorn*. University of Arizona Press, Tucson, Arizona, USA.

Graves, B. D. 1961. Waterhole observations of bighorn sheep. Desert Bighorn Council Transactions 5:27–29.

Gray, R. S. 1974. Lasting waters for bighorn. Desert Bighorn Council Transactions 18:25–27.

Gregg, M. A., M. Bray, K. M. Kilbride, and M. R. Dunbar. In press. Importance of birth synchrony to survival of pronghorn fawns in the Northern Great Basin. *Journal of Wildlife Management*.

Grinnel, J. 1917. The niche-relationship of the California thrasher. *Auk* 34:427–433.

Gross, J. E. 1960. Investigation of seasonal sheep and deer habitat factors. New Mexico Department of Game and Fish, Federal Aid Project W-100-R-1.

Gross, J. E., L. C. Stoddard, and F. H. Wagner. 1974. Demographic analysis of a northern Utah jackrabbit population. *Wildlife Monograph* 40.

Gross, J. E., N. T. Hobbs, and B. A. Wunder. 1993. Independent variables for predicting intake rate of mammalian herbivores: biomass density, plant density, or bite size? *Oikos* 68:75–81.

Gullett, P. A. 1987. Oil taxicoses. Pages 191–196 *in* M. Friend, editor. Field guide to wildlife diseases. U.S. Fish and Wildlife Service Resource Publication 167.

Gullion, G. W. 1966. A viewpoint concerning the significance of studies of game bird food habits. *Condor* 68:372–376.

Gullion, G. W. 1977. The use of drumming behavior in ruffed grouse population studies. *Journal of Wildlife Management* 30:717–729.

Guthery, F. 1996. Upland gamebirds. Pages 59–69 *in* P. R. Krausman, editor. *Rangeland wildlife.* Society for Range Management, Denver, Colorado, USA.

Guthery, F. S., and R. L. Bingham. 1992. On Leopold's principle of edge. *Wildlife Society Bulletin* 20:340–344.

Hahn, H. C., Jr. 1949. A method of censusing deer and its application in the Edwards Plateau region of Texas. Texas Game, Fish, and Oyster Commission, Austin, Texas, USA.

Haigh, J. C., and R. J. Hudson. 1993. *Farming wapiti and red deer.* Mosley, Chicago, Illinois, USA.

Hailey, T. L. 1967. Reproduction and water utilization of Texas transplanted desert bighorn sheep. Desert Bighorn Council Transactions 11:53–58.

Hairston, N. G., F. E. Smith, and L. B. Slobodkin. 1960. Community structure, population control and competition. *The American Naturalist* 44:421–425.

Hall, C. A. S., P. W. Jones, T. M. Donovan, and J. P. Gibbs. 2000. The implications of mainstream economics for wildlife conservation. *Wildlife Society Bulletin* 28:16–25.

Hall, F. C., L. W. Brewer, J. F. Franklin, and R. L. Werner. 1985. Plant communities and stand conditions. Pages 17–31 *in* E. R. Brown, editor. *Management of wildlife and fish habitats in forests of western Oregon and Washington.* U.S. Forest Service, Portland, Oregon, USA.

Hall, L. S., P. R. Krausman, and M. L. Morrison. 1997. The habitat concept and a plea for standard terminology. *Wildlife Society Bulletin* 25:173–182.

Hall-Martin, A. J. 1986. Recruitment in a small black rhino population. *Pachyderm* 7:6–8.

Halloran, A. F. 1949. Desert bighorn management. North American Wildlife and Natural Resource Conference Transactions 14:527–537.

Halloran, A. F., and O. V. Deming. 1958. Water development for desert bighorn sheep. *Journal of Wildlife Management* 22:1–9.

Halls, L. K. 1978. White-tailed deer. Pages 46–66 *in* J. L. Schmidt, and D. L. Gilbert, editors. *Big game of North America: ecology and management.* Stackpole Books, Harrisburg, Pennsylvania, USA.

Hamilton, W. D. 1971. Geometry of the selfish herd. *Journal of Theoretical Biology* 31:295–311.

Hamlin, K. L., S. J. Riley, D. Pyrah, A. R. W. Dood, and R. J. Mackie. 1984. Relationships among mule deer fawn mortality, coyotes, and alternate prey species. *Journal of Wildlife Management* 48:489–499.

Hanks, J. 1981. Characterization of population condition. Pages 47–73 *in* C. W. Fowler and T. D. Smith, editors. *Dynamics of large mammal populations.* John Wiley and Sons, New York, New York, USA.

Hansen, M. C. 1996. Foraging ecology of female Dalls' sheep in the Brooks Range, Alaska. Dissertation, University of Alaska Fairbanks, Fairbanks, Alaska, USA.

Hanson, W. R., and C. Y. McCulloch. 1955. Factors influencing mule deer in Arizona brushlands. Transactions of the North American Wildlife Conference 20:568–588.

Haramis, G. M., J. R. Goldsberry, D. G. McAuley, and E. L. Derleth. 1985. An aerial photographic census of Chesapeake Bay and North Carolina canvasbacks. *Journal of Wildlife Management* 49:449–454.

Hardan, A. 1975. Discussion: session 1. Page 60 *in* Proceedings of a Water Harvesting Symposium. ARS W-22. U.S. Department of Agriculture, Phoenix, Arizona, USA.

Harkin, J. M. 1973. Lignin. Pages 323–373 *in* G. W. Butler and R. W. Bailey, editors. *Chemistry and biochemistry of herbage,* Volume 1. Academic Press, New York, New York, USA.

Harlow, R. F. 1984. Habitat evaluation. Pages 601–628 *in* L. K. Halls, editor. *White-tailed deer ecology and management.* Stackpole Books, Harrisburg, Pennsylvania, USA.

Harris, R. A., and K. R. Duff. 1970. *Wild deer in Britain.* Taplinger, New York, New York, USA.

Hart, J. S., and M. Berger. 1972. Energetics, water economy and temperature regulation during flight. Proceedings of the International Ornithological Congress 15:189–199.

Hart, R. H., K. W. Hepworth, M. A. Smith, and J. W. Waggoner, Jr. 1991. Cattle grazing behavior on a foothill elk winter range in southeastern Wyoming. *Journal of Range Management* 44:262–266.

Haugen, A. O., and D. L. Trauger. 1962. Ovarian analysis for data on corpora lutea changes in white-tailed deer. *Iowa Academy of Sciences* 69:230–238.

Hawkes, J., and C. Hawkes. 1958. *Prehistoric Britain.* Penguin Books, Harmondsworth, United Kingdom.

Hawthorne, D. W., G. A. Nunley, and V. Prothro. 1999. A history of the Wildlife Services program. Probe Newsletter 196–197.

Hay, K. G. 1958. Beaver census methods in the Rocky Mountain region. *Journal of Wildlife Management* 22:395–402.

Hazam, J. E., and P. R. Krausman. 1988. Measuring water consumption of desert mule deer. *Journal of Wildlife Management* 52:528–534.

Heady, H. F., and J. Bartolome. 1977. The Vale rangeland rehabilitation program: the desert repaired in southeastern Oregon. U.S. Forest Service Pacific Northwest Forest and Range Experiment Station Resource Bulletin PNW-70.

Hearel, D. C. 1992. The effect of wolf predation and snow cover on musk-ox group size. *American Naturalist* 139:190–204.

Hediger, H. 1950. *Wild animals in captivity.* Butterworths Scientific Publications, London, United Kingdom.

Hein, E. W. 1997. Improving translocation programs. *Conservation Biology* 11:1270–1274.

Hepworth, W. G. 1965. Investigation of pronghorn antelope in Wyoming Antelope States Workshop Proceedings 1:1–12.

Herbeck, L. A., and D. A. Larsen. 1999. Plethodontid salamander response to silvicultural practices in Missouri Ozark Forests. *Conservation Biology* 13:623–632.

Herndon, P. 1984. History of grazing on the public lands. Pages 1–6 *in The Taylor Grazing Act: 50 years of progress.* U.S. Department of the Interior, Washington, D.C., USA.

Hervert, J. J., and P. R. Krausman. 1986. Desert mule deer use of water developments in Arizona. *Journal of Wildlife Management* 50:670–676.

Hickey, J. J. 1966. Birds and pesticides. Pages 318–329 *in* A. Stefferud, editor. *Birds in our lives.* U.S. Bureau of Sport Fish and Wildlife, Washington, D.C., USA.

Hilden, O. 1965. Habitat selection in birds. *Annales Zoologici Fennici* 2:53–75.

Hines, W. W., and J. C. Lemos. 1979. Reproductive performance by two age classes of mule Roosevelt elk in southwestern Oregon. Oregon Department of Fish and Wildlife, Wildlife Research Report 8.

Hirst, S. M. 1969. Road-strip census techniques for wild ungulates in African woodland. *Journal of Wildlife Management* 33:40–48.

Hissa, R. H., Rinfamaki, P. Linden, and V. Vihko. 1990. Energy reserves of the capercallie *Tetrao urogullus* in Finland. *Comparative Biochemical Physiology* 97A:345–351.

Hobbs, N. T. 1988. Estimating habitat carrying capacity: an approach for planning reclamation and mitigation for wild ungulates. *Issues and Technology in Management of Impacted Wildlife* 3:3–7.

Hobbs, N. T., and T. A. Hanley. 1990. Habitat evaluation: do use/availability data reflect carrying capacity? *Journal of Wildlife Management* 54:515–522.

Hobbs, N. T., D. L. Baker, G. D. Bear, and D. C. Bowden. 1996*a.* Ungulate grazing in sagebrush grassland: mechanisms of resource competiton. *Ecological Applications* 6:200–217.

Hobbs, N. T., D. L. Baker, G. D. Bear, and D. C. Bowden. 1996*b.* Ungulate grazing in sagebrush grassland: effects of resource competiton on secondary production. *Ecological Applications* 6:218–227.

Hobbs, N. T., D. L. Baker, J. E. Ellis, and D. M. Swift. 1979. Composition and quality of elk diets during winter and summer: a preliminary review. Pages 47–52 *in* M. S. Boyace and L. D. Hayden-Wing, editors. *North American elk: ecology, behavior, and management.* University of Wyoming, Laramie, Wyoming, USA.

Hobbs, N. T., D. L. Baker, J. E. Ellis, D. M. Swift, and R. A. Green. 1982. Energy- and nitrogen-based estimates of elk winter-range carrying capacity. *Journal of Wildlife Management* 46:12–21.

Hobon, P. M. 1990. A review of desert bighorn sheep in the San Andres Mountains, New Mexico. Desert Bighorn Council Transactions 34:14–22.

Hodder, I. 1978. The human geography of Roman Britain. Page 29–55 *in* R. A. Dodgshon and R. A. Butlin, editors. *A historical geography of England and Wales.* Academic Press, London, United Kingdom.

Hodgdon, H. E. 1988. Employment of 1986 wildlife graduates. *Wildlife Society Bulletin* 16:333–338.

Hodgdon, H. E. 1991. Wildlife damage management policy and professional considerations. Proceedings of the Eastern Wildlife Damage Control Conference 5:28–30.

Hodson, N. L. 1962. Some notes on the causes of bird road casualties. *Bird Study* 9:168–173.

Hodson, N. L., and D. W. Snow. 1965. The road deaths inquiry, 1960–61. *Bird Study* 12:90–99.

Hoffman, R. A., and P. F. Robinson. 1966. Changes in some endocrine glands of white-tailed deer as affected by season, sex, and age. *Journal of Mammalogy* 47:266–280.

Hoffman, R. W., H. G. Shaw, M. A. Rumble, B. F. Wakeling, C. M. Molloham, S. D. Schemnitz, R. Engel-Wilson, and D. A. Hengel. 1993. Management guidelines for Merriam's wild turkeys. Colorado Division of Wildlife Report 18.

Holechek, J. L., H. Wofford, D. Arthun, M. L. Galyean, and J. D. Wallace. 1986. Evaluation of total fecal collection for measuring cattle forage intake. *Journal of Range Management* 39:2–4.

Holechek, J. L., R. D. Piper, and C. H. Herbel. 1989. *Range management principles and practices.* Prentice Hall, Englewood Cliffs, New Jersey, USA.

Holleman, D. F., J. R. Luick, and R. G. White. 1979. Lichen intake estimate for reindeer and caribou during winter. *Journal of Wildlife Management* 43:192–201.

Holling, C. S. 1959. The components of predation as revealed by a study of small-mammal predation of the European pine sawfly. *Canadian Entomologist* 91:293–320.

Holloran, A. F. 1949. Desert bighorn management. North American Wildlife Conference Transactions 14:527–537.

Holte, A. E., and M. A. Houck. 2000. Juvenile greater roadrunner (Cuculidae) killed by choking on a Texas horned lizard (Phrynosomatidae). *The Southwestern Naturalist* 45:74–76.

Holthausen, R. S. 1986. Use of vegetation projection models for management problems. Pages 371–375 *in* J. Verner, M. L. Morrison, and C. J. Ralph, editors. *Wildlife 2000: modeling habitat relationships of terrestrial vertebrates.* University of Wisconsin Press, Madison, Wisconsin, USA.

Hopps, H. C. 1964. *Principles of pathology.* 2nd ed. Appleton-Century-Crofts, New York, New York, USA.

Hormay, A. L. 1970. Principles of rest-rotation grazing and multiple-use land management. U.S. Forest Service Training Text 4. Washington, D.C., USA.

Hornaday, W. T. 1889. *The extermination of the American bison.* U.S. Government Printing Office, Washington, D.C., USA.

Hornocker, M. G. 1970. An analysis of mountain lion predation upon mule deer and elk in the Idaho Primitive Area. *Wildlife Monograph* 21.

Horst, R. 1971. Observations on the kidney of the desert bighorn sheep. Desert Bighorn Council Transactions 15:24–37.

Horst, R., and M. Langworthy. 1971. Observations on the kidney of the desert bighorn sheep. *Anatomical Record* 169:343.

Hourdequim, M. 2000. Ecological effects of roads. *Conservation Biology* 14:16–17.

Houston, D. B. 1968. The Shires moose in Jackson Hole, Wyoming. Grand Teton National History Association, Technical Bulletin 1.

Houston, D. B. 1982. *The northern Yellowstone elk: ecology and management.* Macmillan, New York, New York, USA.

Howard, W. E. 1960. Innate and environmental dispersal of individual vertebrates. *American Midland Naturalist* 63:152–161.

Howard, W. E. 1974. The biology of predator control. *Module in Biology 11.* Addison-Wesley, Reading, Pennsylvania, USA.

Howard, W. E., and H. E. Childs, Jr. 1959. Ecology of pocket gophers with emphasis on *Thomomys bottae mewa. Hilgardia* 29:277–858.

Howe, F. P., and L. D. Flake. 1988. Mourning dove movements during the reproductive season in southeastern Idaho. *Journal of Wildlife Management* 52:477–480.

Howell, J. C. 1951. The roadside census as a method of measuring bird populations. *Auk* 68:334–357.

Hrdy, S. B. 1977. *The langurs of Abu.* Harvard University Press, Cambridge, Massachusetts, USA.

Hudson, J. W. 1962. The role of water in the biology of the antelope ground squirrel, *Citellus lecurus. University of California Publications in Zoology* 64:1–56.

Hudson, R., and W. Watkins. 1986. Foraging rates of wapiti on green and cured pastures. *Canadian Journal of Zoology* 64:1705–1708.

Hughes, K. S. 1991. Sonoran pronghorn use of habitat in southwest Arizona. Thesis, The University of Arizona, Tucson, Arizona, USA.

Hughes, M. R., J. R. Roberts, and B. R. Thomas. 1987. Total body water and its turnover in free-living nestling glaucous-winged gulls with a comparison of body water and water flux in avian species with and without salt glands. *Physiological Zoology* 60:481–491.

Humbert, J., and W. P. Dasmann. 1945. Range management—a restatement of definition and objectives. *Journal of Forestry* 43:263–264.

Hungerford, C. R. 1960. Water requirements of Gambel's quail. Transactions of the North American Wildlife and Natural Resource Conference 25:231–240.

Hungerford, C. R. 1962. Adaptation shown in selection of food by Gambel's quail. *Condor* 64:213–219.

Hunter, M. L., Jr. 1990. *Wildlife, forests, and forestry: principles of managing forests for biological diversity.* Prentice Hall, Englewood Cliffs, New Jersey, USA.

Hutchinson, G. E. 1957. Concluding remarks. Cold Spring Harbor Symposium on Quantitative Biology. 22:415–427.

Hutto, R. L. 1985. Habitat selection by nonbreeding migratory land birds. Pages 455–476 *in* M. L. Cody, editor. *Habitat selection in birds.* Academic Press, Orlando, Florida, USA.

Hyde, O. D. 1986. Public lands, private lands and oranges. *Ambit* March:22–24.

Illius, A. W. 1989. Allometry of food intake and grazing behavior with body size in cattle. *Journal of Agricultural Science* 113:259–266.

Illius, A. W., and I. J. Gordon. 1990. Variation in foraging behavior in red deer and the consequences for population demography. *Journal of Animal Ecology* 59:89–101.

International Association of Fish and Wildlife Agencies. 1997. The economic importance of hunting. U.S. Fish and Wildlife Service Cooperative Grant Agreement 14-48-98210-97-GO47.

International Union for the Conservation of Nature and Natural Resources. 1995. Guidelines for reintroduction. Reintroduction Specialists Group. International Union for the Conservation of Nature and Natural Resources, Gland, Switzerland.

Irvine, C. A. 1969. The desert bighorn sheep of southeastern Utah. Thesis, Utah State University, Logan, Utah, USA.

Jameson, D. S. 1963. Responses of individual plants to harvesting. *Botanical Review* 29:532–594.

Jenks, J. A., D. M. Leslie, Jr., R. L. Lochmiller, and M. A. Melchiors. 1994. Variation in gastrointestinal characteristics of male and female white-tailed deer: implications for resource partitioning. *Journal of Mammalogy* 75:1045–1053.

Jensen, M., and P. R. Krausman. 1993. *Conservation Biology's* literature: new wine or just a new bottle? *Wildlife Society Bulletin* 21:199–203.

Jessup, D. A. 1993. Translocation of wildlife. Pages 493–499 *in* M. E. Fowler, editor. *Zoo and wild animal medicine.* 3rd ed. W. B. Saunders Company, Philadelphia, Pennsylvania, USA.

Jessup, D. A., and W. M. Boyce. 1996. Diseases of wild ungulates and livestock. Pages 395–412 *in* P. R. Krausman, editor. *Rangeland wildlife.* Society for Range Management, Denver, Colorado, USA

Jessup, D. A., E. A. Thorne, M. W. Miller, and D. L. Hunter. 1995. Health implications in the translocation of wildlife. Pages 381–385 *in* J. A. Bissonette and P. R. Krausman, editors. *Integrating people and wildlife for a sustainable future.* The Wildlife Society, Bethesda, Maryland, USA.

Jessup, D. A., J. DeForge, and S. Sandberg. 1991. Biobullet vaccination of captive and free-ranging bighorn sheep. Pages 428–434 *in* L. A. Renecker and R. J. Hudson, editors. *Wildlife production: conservation and sustainable development.* University of Alaska, Fairbanks, Alaska, USA.

Jiménez, J. A., K. A. Hughes, G. Alaks, L. Graham, and R. C. Lacy. 1994. An experimental study of inbreeding depression in a natural habitat. *Science* 266:271–273.

Johnson, D. H. 1980. The comparison of usage and availability measurements for evaluating resource preference. *Ecology* 61:65–71.

Johnson, D. H. 1994. Population analysis. Pages 419–444 *in* T. A. Bookhout, editor. *Research and management techniques for wildlife and habitats.* 5th ed. The Wildlife Society, Bethesda, Maryland, USA.

Johnson, D. H., G. L. Krapu, K. J. Reinceke, and D. G. Jorde. 1985. An evaluation of condition indices for birds. *Journal of Wildlife Management* 49:569–575.

Johnson, J. F. 1962. Wildlife water development, maintenance and evaluation. New Mexico Department of Game and Fish, Federal Aid in Wildlife Restoration Report W-78-0-08, Job 1.

Johnston, D. W., and T. P. Haines. 1957. Analysis of mass bird mortality in October, 1954. *Auk* 74:447–458.

Jones, C., and R. W. Manning. 1996. The mammals. Pages 29–38 *in* P. R. Krausman, editor. *Rangeland wildlife.* Society for Range Management, Denver, Colorado, USA.

Jones, G. R. J. 1978. Celts, Saxons, and Scandinavians. Pages 57–79 *in* R. A. Dodgshon and R. A. Butlin, editors. *A historical geography of England and Wales.* Academic Press, London, United Kingdom.

Jones, R. L., and H. C. Hanson. 1985. *Mineral licks, geophagy, and bioecochemistry of North American ungulates.* The Iowa State University Press, Ames, Iowa, USA.

Jordan, L. A., and J. P. Workman. 1989. Economics and management of fee hunting for deer and elk in Utah. *Wildlife Society Bulletin* 17:482–487.

Jorgensen, P. 1974. Vehicle use at a desert bighorn watering area. Desert Bighorn Council Transactions 18:18–24.

Joule, J., and G. N. Cameron. 1974. Field estimation of demographic parameters: influence of *Sigmodon hispidus* population structure. *Journal of Mammalogy* 55:309–318.

Jourdonnais, C., and D. J. Bedunah. 1985. Improving elk forage: range research along the Sun River. *Western Wildlands* 11:20–24.

Jourdonnais, C., and D. J. Bedunah. 1990. Prescribed fire and cattle grazing on an elk winter range in Montana. *Wildlife Society Bulletin* 18:232–240.

Kanuk, L., and C. Bevenson. 1975. Mail surveys and response rates: a literature review. *Journal of Market Research* 12:440–453.

Karasor, W. H. 1983. Water flux and water requirement in free-living antelope ground squirrels *Ammospermophilus leucurus. Physiological Zoology* 56:94–105.

Keen, W. H. 1982. Habitat selection and interspecific competition in two species of plethodontid salamanders. *Ecology* 63:94–102.

Keith, L. B. 1963. *Wildlife's ten-year cycle.* The University of Wisconsin Press, Madison, Wisconsin, USA.

Kelker, G. 1952. Yield tables for big game herds. *Journal of Forestry* 50:206–207.

Kellert, S. R. 1980. Contemporary values of wildlife in American society. Pages 31–60 *in* W. W. Shaw and E. H. Zube, editors. Wildlife values. U.S. Department of Agriculture Forestry Service, Rocky Mountain Forest and Range Experiment Station. Center for assessment of noncommodity natural resource values. Report 1.

Kennedy, C. 1958. Water developments on the Kofa and Cabeza Prieta Game Ranges. Desert Bighorn Council Transactions 2:28–31.

Kennedy, P. M., and G. E. Heinsohn. 1974. Water metabolism of two marsupials—the brushtailed possum, *Trichosurus vulpecula,* and the rock wallaby, *Petrogale inornata,* in the wild. *Comparative Biochemistry and Physiology* 47A:829–834.

Keppie, D. M. 1990. To improve graduate student research in wildlife education. *Wildlife Society Bulletin* 18:453–458.

Kerwin, M. L., and G. J. Mitchell. 1971. The validity of the wear-age technique for Alberta pronghorns. *Journal of Wildlife Management* 35:743–747.

Kessler, W. B., S. Csányi, and R. Field. 1998. International trends in university education for wildlife conservation and management. *Wildlife Society Bulletin* 26:927–936.

Kie, J. G. 1978. Femur marrow fat in white-tailed deer carcasses. *Journal of Wildlife Management* 42:661–703.

Kimmins, J. P. 1987. *Forest ecology.* Macmillan, New York, New York, USA.

King, R. T. 1937. Ruffed grouse management. *Journal of Forestry* 35:523–532.

Kirby, C., 1941. English game law reform. Pages 345–380 *in Essays in modern English history, in honor of Wilbur Cortez Abbott.* Kennikat Press, Port Washington, Wisconsin, USA.

Kirby, J., S. Delany, and J. Quinn. 1994. Mute swans in Great Britain: a review, current status and long-term trends. *Hydrobiologia* 279/280:467–482.

Kirkpatrick, R. L. 1980. Physiological indices in wildlife management. Pages 99–112 *in* S. D. Schmenitz, editor. *Wildlife techniques manual.*

The Wildlife Society, Washington, D.C., USA.

Kistner, T., and J. Bone. 1973. Diseases and parasites commonly found in big game. *Oregon State Game Bulletin* 28:3–12.

Kistner, T. P., and J. F. Bone. 1993. Diseases of game animals and birds and their relationship to humans. *Oregon State Game Commission Bulletin* 28:3–12.

Kistner, T. P., C. E. Trainer, and N. A. Hartman. 1980. A field technique for evaluating physical condition of deer. *Wildlife Society Bulletin* 8:11–17.

Klein, D. K. 1992. Comparative ecological and behavioral adaptation of *Ovibos moschatus* and *Rangifer tarandus. Rangifer* 12:47–55.

Klein, D. R. 1968. The introduction, increase, and crash of reindeer on St. Matthew Island. *Journal of Wildlife Management* 32:350–367.

Klein, D. R. 1995. The introduction, increase, and demise of wolves on Coronation Island, Alaska. Pages 275–280 *in* L. N. Carbyn, S. H. Fritts, and D. R. Seip, editors. *Ecology and conservation of wolves in a changing world.* Canadian Circumpolar Institute, University of Alberta, Edmonton, Alberta, Canada.

Klein, D. R. 2000. The muskox. Pages 545–558 *in* S. Demarais and P. R. Krausman, editors. *Ecology and management of large mammals in North America.* Prentice Hall, Upper Saddle River, New Jersey, USA.

Klein, D. R. 2000*a.* Walking in the footsteps of Aldo Leopold. *Wildlife Society Bulletin* 28:464–467.

Klimkiewicz, M. K., R. B. Clapp, and A. G. Futcher. 1983. Longevity records of North American birds: Remizidae through Parulinae. *Journal of Field Ornithology* 54:287–294.

Kline, P. D. 1965. Factors influencing roadside counts of cottontails. *Journal of Wildlife Management* 29:665–671.

Knoph, F. 1996. Perspectives on grazing nongame bird habitats. Pages 51–58 *in* P. R. Krausman, editor. *Rangeland wildlife.* Society for Range Management, Denver, Colorado, USA.

Knudsen, M. F. 1963. A summer waterhole study at Carrizo Spring, Santa Rosa Mountains of southern California. Desert Bighorn Council Transactions 7:185–192.

Koenen, K. K. G., and P. R. Krausman. In press. Desert mule deer habitat use in a semidesert grassland. *The Southwestern Naturalist.*

Koerth, N. E., and F. S. Guthery. 1990. Water requirements of captive northern bobwhites under subtropical seasons. *Journal of Wildlife Management* 54:667–672.

Krannich, R. S., and T. L. Teel. 1999. Attitudes and opinions about wildlife resource conditions and management in Utah: results of a 1998 statewide general public and license purchaser survey. Institute for Social Science Research on Natural Resources, Utah State University, Logan, Utah, USA.

Krausman, P. R. 1978. Forage relationships between two deer species in Big Bend National Park, Texas. *Journal of Wildlife Management* 42:101–107.

Krausman, P. R. 1979. Continuing education for wildlife administrators. *Wildlife Society Bulletin* 7:57–58.

Krausman, P. R. 1996*a*. Preface. Page vii *in* P. R. Krausman, editor. *Rangeland wildlife.* Society for Range Management, Denver, Colorado, USA.

Krausman, P. R. 1996*b*. Problems facing bighorn sheep in and near domestic sheep allotments. Pages 59–64 *in* W. D. Edge, editor. Sustaining Rangeland Ecosystems Symposium, Oregon State University, Corvallis, Oregon SR953.

Krausman, P. R. 1996*c*. The exit of the last wild mountain sheep. Pages 242–250 *in* G. P. Nabham, editor. *Counting sheep.* The University of Arizona Press, Tucson, Arizona, USA.

Krausman, P. R. 1997. The influence of scale on the management of desert bighorn sheep. Pages 349–367 *in* J. A. Bissonette, editor. *Landscape ecology and wildlife management.* Springer-Verlag, New York, New York, USA.

Krausman, P. R. 2000. An introduction to the restoration of bighorn sheep. *Restoration Ecology:* in press.

Krausman, P. R. 2000*a*. Wildlife management in the twenty-first century: educated predictions. *The Wildlife Society Bulletin* 28:490–495.

Krausman, P. R., and B. Czech. 1998. Water developments and desert ungulates. Pages 138–154 *in Environmental, Economic, and Legal Issues Related to Rangeland Water Development.* College of Law, Arizona State University, Tempe, Arizona, USA.

Krausman, P. R., and E. D. Ables. 1981. Ecology of the Carmen Mountains white-tailed deer. National Park Service Scientific Monograph Series 15.

Krausman, P. R., and R. C. Etchberger. 1993. Effectiveness of mitigation features for desert ungulates along the Central Arizona Project. Final report, U.S. Bureau of Reclamation, 9-CS-32-00350.

Krausman, P. R., and R. C. Etchberger. 1995. Response of desert ungulates to a water project in Arizona. *Journal of Wildlife Management* 59:292–300.

Krausman, P. R., and R. C. Etchberger. 1996. Desert bighorn sheep and water: a bibliography. U.S. Geological Survey, Cooperative Park Studies Unit, The University of Arizona, special report 13.

Krausman, P. R., and W. W. Shaw. 1984. Providing water for wildlife in arid lands. *Egyptian Journal of Wildlife and Natural Resources* 5:11–52.

Krausman, P. R., A. J. Kuenzi, R. C. Etchberger, K R. Rautenstrauch, L. L. Ordway, and J. J. Hervert. 1997. Diets of desert mule deer. *Journal of Range Management* 50:513–522.

Krausman, P. R., B. D. Leopold, R. F. Seegmiller, and S. G. Torres. 1989. Relationships between desert bighorn sheep and habitat in western Arizona. *Wildlife Monograph* 102.

Krausman, P. R., S. Torres, L. L. Ordway, J. J. Hervert, and M. Brown. 1985. Diel activity of ewes in the Little Harquahala Mountains, Arizona. Desert Bighorn Council Transactions 29:24–26.

Krebs, C. J. 1972. *Ecology.* Harper and Row, New York, New York, USA.

Krebs, C. J. 1978. *Ecology: the experimental analysis of distribution and abundance.* 2nd ed. Harper and Row, New York, New York, USA.

Krebs, C. J., S. Boutin, R. Boonstra, A. R. E. Sinclair, J. N. M. Smith, M. R. T. Dale, K. Martin, and R. Turkington. 1995. Impact of food and predation on the snowshoe hare cycle. *Science* 269:1112–1115.

Krueger, W. C., and A. H. Winward. 1974. Influence of cattle and big game grazing on understory structure of a douglas fir - ponderosa pine - Kentucky bluegrass community. *Journal of Range Management* 27:450–453.

Kushlan, J. A. 1988. Conservation and management of the American crocodile. *Environmental Management* 12:777–790.

Laake, J. L., K. P. Burnham, and D. R. Anderson. 1979. *User's manual for program TRANSECT.* Utah State University Press, Logan, Utah, USA.

Lack, D. 1954. *The natural regulation of animal numbers.* Clarendon Press, Oxford, United Kingdom.

Lacy, J. R., K. Jamtgaard, L. Riggle, and T. Hayes. 1993. Impacts of big game on private land in southwestern Montana: landowner perceptions. *Journal of Range Management* 46:31–37.

Lacy, R. C. 1993. Impacts of inbreeding in natural and captive populations of vertebrates: implications for conservation. *Perspectives in Biology and Medicine* 36:480–496.

Lamprey, H. F. 1964. Estimation of the large mammal densities, biomass and energy exchange in the Tarangire game reserve and the Masai Steepe in Tanganyika. *East African Wildlife Journal* 2:1–46.

Lancia, R. A., C. S. Rosenberry, and M. C. Conner. 2000. Population parameters and their estimation. Pages 64–83 *in* S. Demarais and P. R. Krausman, editors. *Ecology and management of large mammals in North America.* Prentice Hall, Upper Saddle River, New Jersey, USA.

Lancia, R. A., J. D. Nichols, and K. H. Pollock. 1994. Estimating the number of animals in wildlife populations. Pages 215–253 *in* T. A. Bookhout, editor. *Research and management techniques for wildlife and habitats.* The Wildlife Society, Bethesda, Maryland, USA.

Landres, P. B. 1992. Temporal scale perspectives in managing biological diversity.

Transactions of the North American Wildlife and Natural Resources Conference 57:292–307.

Landy, M. 1993. Public policy and citizenship. Pages 19–44 *in* H. Ingram and S. R. Smith, editors. *Public policy for democracy.* The Brookings Institute, Washington, D.C., USA.

Lautier, J. K., T. V. Dailey, and R. D. Brown. 1988. Effect of water restriction on feed intake of white-tailed deer. *Journal of Wildlife Management* 52:602–606.

Laycock, G. 1973. Saving western eagles from traps and zaps. *Audubon* 75:133.

Leaky, R., and R. Lewin. 1995. *The sixth extinction: patterns of life and the future of humankind.* Doubleday, New York, New York, USA.

Lehnert, M. E., and J. A. Bissonette. 1997. Effectiveness of highway crosswalk structures at reducing deer-vehicle collisions. *Wildlife Society Bulletin* 25:809–818.

Lent, P. C. 1978. Musk-ox. Pages 125–147 *in* J. L. Schmidt and D. L. Gilbert, editors. *Big game of North America.* Stackpole Books, Harrisburg, Pennsylvania, USA.

Leonard, R. M., and E. B. Fish. 1974. An aerial photographic technique for censusing lesser sandhill cranes. *Wildlife Society Bulletin* 2:191–195.

Leopold, A. 1933. *Game management.* Charles Scribner's Sons, New York, New York, USA.

Leopold, A. 1949. *A Sand County almanac and sketches here and there.* Oxford University Press, New York, New York, USA.

Leopold, A. 1966. *A Sand County almanac with other essays on conservation from Round River.* Oxford University Press, New York, New York, USA.

Leopold, A. 1999. The land-health concept and conservation. Pages 218–226 *in* J. B. Callicott and E. T. Freyfogle, editors. *For the health of the land.* Island Press, Washington, D.C., USA.

Leopold, A. 1999a. Game management. A new field for science. Pages 27–32 *in* J. B. Callicott and E. T. Freyfogle, editors. *For the health of the land.* Island Press, Washington, D.C., USA.

Leopold, A. S. 1977. *The California quail.* University of California Press, Berkeley, California, USA.

Leopold, A. S., S. A. Cain, C. M. Cottam, I. N. Gabrielson, and T. E. Kimbal. 1964. Predator and rodent control in the U.S. Transactions of the North American Wildlife and Natural Resources Conference 29:27–49.

Leopold, B. D. 2000. Changes, changes, but then again . . . *Wildlife Society Bulletin* 28:301.

Leopold, B. D., and P. R. Krausman. 1986. Diets of three predators in Big Bend National Park, Texas. *Journal of Wildlife Management* 50:290–295.

LeResche, R. E. 1968. Spring-fall calf mortality in an Alaska moose population. *Journal of Wildlife Management* 32:953–956.

LeResche, R. E., and R. A. Rausch. 1974. Accuracy and precision of aerial moose censusing. *Journal of Wildlife Management* 38:175–182.

Leslie, D. M., Jr. 1978. Differential utilization of water sources by desert bighorn sheep in the River Mountains, Nevada. Desert Bighorn Council Transactions 22:23–26.

Leslie, D. M., Jr. 1980. Remnant populations of desert bighorn sheep as a source for transplantation. Desert Bighorn Council Transactions 24:36–44.

Leslie, D. M., Jr., and C. L. Douglas. 1979. Desert bighorn sheep of the River Mountains, Nevada. *Wildlife Monograph* 66.

Leslie, D. M., Jr., and C. L. Douglas. 1980. Human disturbance at water sources of desert bighorn sheep. *Wildlife Society Bulletin* 8:284–290.

Levin, D. A. 1976. The chemical defenses of plants to pathogens and herbivores. *Annual Review of Ecological Systematics* 7:121–159.

Lewis, J. C., and E. Legler, Jr. 1968. Lead shot ingestion by mourning doves and incidence in soil. *Journal of Wildlife Management* 32:476–482.

Lewis, W. D. 1919. *The Life of Theodore Roosevelt.* United Publishers, New York, New York, USA.

Lincoln, F. C. 1930. Calculating waterfowl abundance on the basis of band returns. U.S. Department of Agriculture Circular 118.

Lincoln, F. C. 1931. Some causes of mortality among birds. *Auk* 48:538–546.

Lindstedt, S. L., and M. S. Boyce. 1985. Seasonality, fasting endurance and body size in mammals. *American Naturalist* 125:873–878.

Lindzey, F. G. 1981. Denning dates and hunting seasons for black bears. *Wildlife Society Bulletin* 9:212–216.

Linhart, S. G., and F. F. Knowlton. 1975. Determining the relative abundance of coyotes by scent station lines. *Wildlife Society Bulletin* 3:119–124.

Little, J. A. 1992. Historical livestock grazing perspective. *Rangelands* 14:88–90.

Litvaitis, J. A., K. Titus, and E. M. Anderson. 1994. Measuring vertebrate use of territorial habitat and foods. Pages 254–274 *in* T. A. Bookhout, editor. *Research and management techniques for wildlife and habitats.* 5th ed. The Wildlife Society, Bethesda, Maryland, USA.

Locke, L. N. 1987*a*. Aspergillosis. Pages 145–150 *in* M. Friend, editor. Field guide to wildlife diseases. U.S. Fish and Wildlife Service Resource Publication 167.

Locke, L. N. 1987*b*. Chlamydiosis. Pages 107–113 *in* M. Friend, editor. Field guide to wildlife diseases. U.S. Fish and Wildlife Service Resource Publication 167.

Locke, L. N., and M. Friend. 1987. Avian botulism. Pages 83–93 *in* M. Friend, editor. Field guide to wildlife diseases. U.S. Fish and Wildlife Service Resource Publication 167.

Locker, C. A., and D. J. Decker. 1989. Colorado black bear hunting referendum: what was behind the vote. *Wildlife Society Bulletin* 23:370–376.

Lode, T. 1993. The decline of otter *Lutra lutra* populations in the region of the Pays de Loire, western France. *Biological Conservation* 65:9–13.

Long, J. L. 1981. *Introduced birds of the world.* David and Charles, London, United Kingdom.

Lord, R. D. 1959. Comparison of early morning and spotlight roadside censuses for cottontails. *Journal of Wildlife Management* 23:458–460.

Lord, R. D., A. M. Vilches, J. I. Maiztegui, and C. A. Soldini. 1970. The tracking board: a relative census technique for studying rodents. *Journal of Mammalogy* 51:828–829.

Lotka, A. J. 1907. Studies on the mode of growth of material aggregates. *American Journal of Science* 24:199–216.

Lotka, A. J. 1922. The stability of the normal age distribution. Proceedings of the National Academy of Science 8:339–345.

Lovell, C. D., and R. A. Dolbeer. 1999. Validation of the United States Air Force bird avoidance model. *Wildlife Society Bulletin* 27:167–171.

Lowe, P. D. 1983. Values and institutions in the history of British nature conservation. Pages 329–352 *in* A. Warren and F. B. Goldsmith, editors. *Conservation in perspective.* John Wiley & Sons, Chichester, United Kingdom.

Lowe, V. P. W. 1969. Population dynamics of deer (*Cervus elaphus* L.) on Rhum. *Journal of Animal Ecology* 38:425–457.

Luxmoore, R. 1991. Monitoring trade in wildlife products, and the impact of ranching on wildlife conservation. Pages 49–54 *in* L. A. Renecker and R. J. Hudson, editors. Wildlife production conservation and sustainable development. University of Alaska Fairbanks. Agricultural and Forestry Experiment Station Miscellaneous Publication 91-6.

Lyon, L. J. 1985. Elk and cattle on the National Forests: a simple question of allocation or a complex management problem? *Western Wildlands* 11:16–19.

Lyon, L. J., and A. L. Ward. 1982. Elk and land management. Pages 443–477 *in* J. W. Thomas and D. E. Toweill, editors. *Elk of North America, ecology and management.* Stackpole Books, Harrisburg, Pennsylvania, USA.

MacDonald, D., editor. 1984. *The encyclopedia of mammals.* Facts on File, Incorporated. New York, New York, USA.

MacDonald, D., and E. G. Dillman. 1968. Techniques for estimating nonstatistical bias in big game harvest surveys. *Journal of Wildlife Management* 32:119–129.

MacFarlane, W. V., and B. Howard. 1972. Comparative water and energy economy of wild and domestic mammals. Symposium of the Zoological Society of London 31:261–296.

MacFarlane, W. V., B. Howard, and B. D. Seibert. 1969. Tritiated water in the measurement of milk intake and tissue growth of ruminants in the field. *Nature* (London) 221:578–579.

MacGregor, W. G. 1950. An evaluation of California quail management. Proceedings of the Annual Conference of the Western Association of State Game and Fish Commission 33:157–160.

MacIntyre, D. 1952. Trapping and vermin. Pages 136–157 *in* E. Parker, editor. *The Lonsdale keeper's book.* Seeley Service & Company, London, United Kingdom.

Mackie, R. J. 1970. Range ecology and relations of mule deer, elk, and cattle in the Missouri River breaks, Montana. *Wildlife Monograph* 20.

Mackie, R. J. 1978. Impacts of livestock grazing on wild ungulates. North American Wildlife Conference 43:462–476.

Mackie, R. J. 1985. The elk-deer-livestock triangle. Pages 51–56 *in* G. W. Workman, editor. *Western elk management: a symposium.* Utah State University, Logan, Utah, USA.

Mackie, R. J. 2000. History of management of large mammals in North America. Pages 292–320 *in* S. Demarais and P. R. Krausman, editors. *Ecology and management of large mammals in North America.* Prentice Hall, Upper Saddle River, New Jersey, USA.

Mackie, R. J., and H. K. Buechner. 1963. The reproductive cycle of the chukar. *Journal of Wildlife Management* 27:246–260.

Mackie, R. J., D. F. Pac, K. L. Hamlin, and G. L. Dusek. 1998. Ecology and management of mule deer and white-tailed deer in Montana. Montana Fish, Wildlife, and Parks. Federal Aid in Wildlife Restoration Project W-120-R.

Macnab, J. 1985. Carrying capacity and other slippery shibboleths. *Wildlife Society Bulletin* 13:403–410.

Maghini, M. T., and N. S. Smith. 1990. Water use and diurnal seasonal ranges of Coves white-tailed deer. Pages 21–34 *in* P. R. Krausman and N. S. Smith, editors. *Managing wildlife in the southwest: a symposium.* Arizona Cooperative Wildlife Research Unit and School of Renewable Natural Resources, University of Arizona, Tucson, Arizona. USA.

Maharaj, V., and J. Carpenter. 1996. *The economic impact of sport fishing in the United States.* American Sportfishing Association, Fairfax, Virginia, USA.

Mahon, C. L. 1971. Water development for desert bighorn sheep in southeastern Utah. Desert Bighorn Council Transactions 15:74–77.

Mann, W. M. 1934. *Wild animals in and out of the zoo, VI.* Smithsonian Scientific Series, New York, New York, USA.

Mannan, R. W., R. N. Conner, B. Marcot, and J. M. Peek. 1994. Managing forestlands for wildlife. Pages 689–721 *in* T. A. Bookhout, editor. *Research and management techniques for wildlife and habitats,* Fifth edition. The Wildlife Society, Bethesda, Maryland, USA.

Marby, T. J., and J. E. Gill. 1979. Sesquiterpene lactones and other terpenoids. Pages 501–537 *in* G. A. Rosenthal and D. H. Janzen, editors. *Herbivores: their interaction with secondary plant metabolites.* Academic Press, New York, New York, USA.

Marcot, B. G. 1988. 1st class expert systems: 1st class. *AI Expert* 3:77–80.

Marcot, B. G. 1986. Use of expert systems in wildlife-habitat modeling. Pages 145–150 *in* J. Verner, M. L. Morrison, and C. J. Ralph, editors. *Wildlife 2000: modeling habitat relationships of terrestrial vertebrates.* University of Wisconsin Press, Madison, Wisconsin, USA.

Mares, M. A. 1983. Desert rodent adaptation and community structure. *Great Basin Naturalist Memoirs* 7:30–43.

Margalef, R. 1969. Diversity and stability: a practical proposal and a model of interdependence. Diversity and stability in ecological systems. Brookhaven National Laboratories, Upton, New York. *Brookhaven Symposium of Biology* 22:25–37.

Marsh, G. P. 1965. *Man and nature.* Harvard University Press, Cambridge, Massachusetts, USA.

Marsh, G. P. 1907. *The earth as modified by human action.* 3rd ed. Charles Scribner's Sons, New York, New York, USA.

Martin, J. T., and B. E. Juniper. 1970. *The cuticles of plants.* Saint Martin's Press, New York, New York, USA.

Martin, M., H. Radtke, B. Eleveld, and S. D. Nafziger. 1988. The impacts of the Conservation Reserve Program on rural communities: the case of three Oregon counties. *Western Journal of Agricultural Economics* 13:225–232.

Matter, W. J., and R. J. Steidl. 2000. University undergraduate curriculum in wildlife: beyond 2000. *Wildlife Society Bulletin* 28:503–507.

Matthiessen, P. 1959. *Wildlife in America.* Viking Press, New York, New York, USA.

Mautz, W. W. 1978. Nutrition and carrying capacity. Pages 321–348 *in* J. W. Thomas and D. E. Toweill, editors. *Elk of North America, ecology and management.* Stackpole Books, Harrisburg, Pennsylvania, USA.

Mautz, W. W., H. Silver, J. B. Holter, H. H. Hayes, and W. E. Urban, Jr. 1976. Digestibility and related nutritional data for seven northern deer browse species. *Journal of Wildlife Management* 40:630–638.

Mayhew, W. W. 1968. Biology of desert amphibians and reptiles. Pages 195–365 *in* G. W. Brown, editor. *Desert biology.* Academic Press, New York, New York, USA.

Maynard, L. A., and J. K. Loosli. 1969. *Animal nutrition.* McGraw-Hill, New York, New York, USA.

Mazaika, R., P. R. Krausman, and R. C. Etchberger. 1992. Forage availability for mountain sheep in Pusch Ridge Wilderness, Arizona. *Southwestern Naturalist* 37:372–378.

McCabe, R. E. 1982. Elk and Indians: historical values and perspectives. Pages 321–348 *in* J. W. Thomas and D. E. Toweill, editors. *Elk of North America, ecology and management.* Stackpole Books, Harrisburg, Pennsylvania, USA.

McCall, T. C., R. D. Brown, and L. C. Brender. 1997. Comparison of techniques for determining the nutritional carrying capacity for white-tailed deer. *Journal of Range Management* 50:33–38.

McCarthy, C. W., and J. A. Bailey. 1994. Habitat requirements of desert bighorn sheep. Colorado Division of Wildlife Special Report 69.

McCarthy, T. 1973. Ocular impalement of a great horned owl. *Wilson Bulletin* 85:477–478.

McClure, H. E. 1951. An analysis of animal victims on Nebraska's highways. *Journal of Wildlife Management* 15:410–420.

McCullouch, C. Y., Jr. 1962. Population and range effects of rodents on the sand sagebrush grasslands of western Oklahoma. Oklahoma State University Arts and Science Studies, Biological Studies Series 9.

McCullough, D. R. 1974. *Status of larger mammals in Taiwan.* Tourism Bureau, Taipai, Taiwan.

McCullough, D. R. 1979. *The George Reserve deer herd—population ecology of a K-selected species.* University of Michigan Press, Ann Arbor, Michigan, USA.

McCullough, D. R. 1984. Lessons from the George Reserve, Michigan. Pages 211–242 *in* L. K. Halls, editor. *White-tailed deer: ecology and management.* Stackpole Books, Harrisburg, Pennsylvania, USA.

McCullough, D. R. 1987. The theory and management of *Odocoileus* populations. Pages 535–549 *in* C. M. Wemmer, editor. *Biology and management of the Cervidae.* Smithsonian Institution Press, Washington, D.C., USA.

McCullough, D. R. 1990. Detecting density dependence: filtering the baby from the bath water. Transactions of the North American Wildlife and Natural Resources Conference 55:534–543.

McCullough, D. R. 1992. Concepts of large herbivore population dynamics. Pages 967–984 *in* D. R. McCullough and R. H. Burnett, editors. *Wildlife 2001: populations.* Elsevier Science Publishers, London, United Kingdom.

McCullough, D. R. 1997. Irruptive behavior in ungulates. Pages 69–98 *in* W. J. McShea, H. B. Underwood, and J. H. Rappole, editors. *The science of overabundance.* Smithsonian Institution Press, Washington, D.C., USA.

McCutchen, H. E. 1990. A technique to visually assess physical condition of bighorn sheep. Desert Bighorn Council Transactions 29:27–28.

McGeary, M. N. 1960. *Gifford Pinchot, forester-politician.* Princeton University Press, Princeton, New Jersey.

McHugh, T. 1958. Social behavior of the American buffalo (*Bison bison bison*). *Zoologica* 43:1–40.

McIntosh, M. 1979. Menagerie of names . . . *The Conservationist* (Missouri) 4:3–5.

McKelvie, C. L. 1985. *A future for game?* George Allen & Unwin, London, United Kingdom.

McLead, M. N. 1974. Plant tannins—their role in forage quality. *Nutrition Abstract Review* 44:803–815.

McLead, M. N., P. M. Kennedy, and D. J. Minson. 1990. Resistance of leaf and stem fractions of tropical forage to chewing and passage in cattle. *British Journal of Nutrition* 63:105–119, April: 17-21.

McLean, H. E. 1994. Come back, cool stream. *American Forests,* March/April: 17–21.

McLellon, B. N. 1989. Dynamics of a grizzly bear population during a period of industrial resource extraction. III. Natality and rate of increase. *Canadian Journal of Zoology* 67:1865–1868.

McLeod, S. R. 1997. Is the concept of carrying capacity useful in variable environments? *Oikos* 79:529–542.

McNab, B. K. 1963. Bioenergetics and the determination of home range size. *American Naturalist* 97:133–140.

McNab, B. K. 1986. The influence of food habits on the energetics of eutherian mammals. *Ecological Monograph* 56:1–9.

McNabb, F.M.A. 1969. A comparative study of water balance in three species of quail. I. Water turnover in the absence of temperature stress. *Comparative Biochemistry and Physiology* 28:1045–1058.

Meagher, M. M. 1978. Bison. Pages 123–133 *in* J. L. Schmidt and D. L. Gilbert, editors. *Big game of North America.* Stackpole Books, Harrisburg, Pennsylvania, USA.

Meagher, M. M. 1973. The bison of Yellowstone National Park. National Park Service *Scientific Monograph* Series 1.

Mech, D. L. 1970. *The wolf: the ecology and behavior of an endangered species.* Natural History Press, Garden City, New York, New York, USA.

Mech, L. D., and G. D. DelGiudice. 1985. Limitations of the marrow-fat technique as an indicator of body condition. *Wildlife Society Bulletin* 13:204–206.

Mech, L. D., and P. D. Karns. 1977. Role of the wolf in a deer decline in the Superior National Forest. U.S. Forest Service Research Paper, North Central Forest Experiment Station, Saint Paul, Minnesota, NC-148.

Mellanby, K. 1967. *Pesticides and pollution.* Collins, London, United Kingdom.

Menasco, K. A. 1986. Stocktanks: an underutilized resource. Transactions of the North American Wildlife and Natural Resource Conference 51:304–309.

Mensch, J. L. 1969. Desert bighorn (*Ovis canadensis* nelsoni) losses in a natural trap tank. *California Fish and Game* 55:237–238.

Mentis, M. T. 1977. Stocking rates and carrying capacities for ungulates on African rangelands. *South African Journal of Wildlife Research* 7:89–98.

Merritt, M. F. 1974. Measurement of utilization of bighorn sheep habitat in the Santa Rosa Mountains. *Desert Bighorn Council Transactions* 18:4–17.

Messier, F. 1985. Social organization, spatial distribution, and population density of wolves in relation to moose density. *Canadian Journal of Zoology* 63:1068–1077.

Messier, F. 1991. The significance of limiting and regulating factors on the demography of moose and white-tailed deer. *Journal of Animal Ecology* 60:377–393.

Messmer, T. A., M. W. Brunson, D. Reiter, and D. G. Hewitt. 1999. United States public attitudes regarding predators and their management to enhance avian recruitment. *Wildlife Society Bulletin* 27:75–85.

Miller, D. G., S. P. Bratton, and J. Hadidian. 1992. Impacts of white-tailed deer on endangered plants. *Natural Areas Journal* 12:67–74.

Miller, F. L. 1974. Biology of the Kaminuriak population of barren-ground caribou, Part 2, Deviation as an indicator of age and sex; composition of socialization of the population. Canadian Wildlife Service Report Series 31.

Miller, J. E. 1995. The professional evolution of wildlife damage management. Eastern Wildlife Damage Management Conference 7:1–6.

Miller, K. V., and J. M. Wentworth 2000. Carrying capacity. Pages 140–155 *in* S. Demarais and P. R. Krausman, editors. *Ecology and management of large mammals in North America.* Prentice Hall, Upper Saddle River, New Jersey, USA.

Miller, M. W., and E. T. Thorne. 1993. Captive cervids as potential sources of disease for North Americas wild cervid population: avenues, implications, and preventive management. Transactions of the North American Wildlife and Natural Resources Conference 58:460–467.

Miller, M. W., J. M. Williams, T. J. Schiefer, and J. W. Seidel. 1991. Bovine tuberculosis in a captive herd in Colorado; epizoology, diagnosis, and management. *U.S. Animal Health Association* 95:533–542.

Miller, S. A. 1984. Estimation of animal production numbers for national assessments and appraisals. U.S. Forest Service General Technical Report, RM-105.

Mitchell, P. C. 1911. *Proceedings of the Zoological Society of London* 1911: 425.

Mohler, L. L., and D. E. Toweill. 1982. Regulated elk populations and hunter harvests. Pages 561–597 *in* J. W. Thomas and D. E. Toweill, editors. *Elk of North America.* Stackpole Books, Harrisburg, Pennsylvania, USA.

Monson, G. 1958. Water requirement. Desert Bighorn Council Transactions 2:64–66.

Morgart, J. R., and P. R. Krausman. 1981. The status of a transplanted bighorn population in Arizona using an enclosure. *Transactions of the Desert Bighorn Council* 25:46–49.

Morris, J. G. 1985. Nutritional and metabolic responses to argive deficiency in carnivores. *Journal Nutrition* 115:524–531.

Morrison, M. L., B. G. Marcot, and R. W. Mannan. 1992. *Wildlife-habitat relationships: concepts and applications.* University of Wisconsin Press, Madison, Wisconsin, USA.

Morrison, M. L., B. G. Marcot, and R. W. Mannan. 1998. *Wildlife-habitat relationships: concepts and applications.* 2nd ed. The

University of Wisconsin Press, Madison, Wisconsin, USA.

Morrison, M. L., I. C. Timossi, K, A. With, and P. N. Manley. 1985. Use of tree species by forest birds during winter and summer. *Journal of Wildlife Management* 49:1098–1102.

Morton, S. R. 1980. Field and laboratory studies of water metabolism in *Sminthopsis crassicaudata* (Marsupialia: Dasyuridae). *Australian Journal of Zoology* 28:213–227.

Moulton, C. R. 1923. Age and chemical development in mammals. *Journal of Biological Chemistry* 57:79–97.

Moulton, M. P., and J. Sanderson. 1997. *Wildlife issues in a changing world*. St. Lucie Press, Delray Beach, Florida, USA.

Mouton, R. J., and R. M. Lee. 1992. Actual costs of bighorn sheep water developments. *Desert Bighorn Council Transactions* 36:49–50.

Muir, R., and N. Muir. 1987. *Hedgerows: their history and wildlife*. Michael Joseph, London, United Kingdom.

Mullen, R. K. 1971. Energy metabolism and body water turnover rates of two species of free-living kangaroo rats, *Dipodomys merriami* and *Dipodomys microps*. *Comparative Biochemistry and Physiology* 39A:379–380.

Mumme, R. L., S. J. Schoech, G. E. Woolfenden, and J. W. Fitzpatrick. 2000. Life and death in the fast lane: demographic consequences of road mortality in the Florida scrub jay. *Conservation Biology* 14:501–512.

Mungall, E. C. 2000. Exotics. Pages 736–764 *in* S. Demarais and P. R. Krausman, editors. *Ecology and management of large mammals in North America*. Prentice Hall, Upper Saddle River, New Jersey, USA.

Mungall, E. C., and W. J. Sheffield. 1994. *Exotics on the range: the Texas example*. Texas A & M University, College Station, Texas, USA.

Munsche, P. B. 1981. *Gentlemen and poachers: The English game laws 1671–1831*. University Press, Cambridge, United Kingdom.

Murie, A. 1944. The wolves of Mount McKinley. *National Park Service Fauna* Series 5.

Murie, O. J. 1951. *The elk of North America*. Stackpole Books, Harrisburg, Pennsylvania, USA.

Murphy, D. A. 1965. Effects of various opening days on deer harvest and hunting pressure. Proceedings of the Annual Conference of the Southeastern Association of Game and Fish Commissioners 19:141–146.

Murphy, D. D., and B. D. Noon. 1991. Coping with uncertainty in wildlife biology. *Journal of Wildlife Management* 55:773–782.

Murton, R. K. 1972. *Man and birds*. Taplinger Publication Company, New York, New York, USA.

Myers, L. E. 1975. Water harvesting—2000 B.C. to 1074 A.D. Pages 1–7 *in* Proceedings of a Water Harvesting Symposium. ARS W-22. U.S. Department of Agriculture, Phoenix, Arizona, USA.

Myers, L. E., and G. W. Frasier. 1974. Asphalt-fiberglass for precipitation catchments. *Journal of Range Management* 27:12–15.

Nagy, K. A. 1987. How do desert animals get enough water? Pages 89–98 *in* L. Berkofsky and M. G. Wustele, editors. *Progress in desert research*. Rowman and Littlefield, Totowa, New Jersey, USA.

National Research Council. 1978*a*. Nutrient requirements of laboratory animals. National Academy of Sciences Publication 2767.

National Research Council. 1978*b*. Nutrient requirements of nonhuman primates. National Academy of Sciences Publication 2768.

National Research Council. 1982. *Nutrient requirements of mink and foxes*. National Academy of Sciences, Washington, D.C., USA.

National Research Council. 1984. *Nutrient requirements of poultry*. National Academy of Sciences, Washington, D.C., USA.

National Research Council. 1997. *Toward a sustainable future: addressing the long-term effects of motor vehicle transportation on climate and ecology*. National Academy Press, Washington, D.C., USA.

Neale, E. 1986. *The natural history of badgers*. Croom Helm, London, United Kingdom.

Neff, D. J. 1968. The pellet-group count technique for big game trend, census, and distribution: a review. *Journal of Wildlife Management* 32:597–614.

Nelson, J. R. 1982. Relationships of elk and other large herbivores. Pages 415–441 *in* J. W. Thomas and D. E. Toweill, editors. *Elk of North America, ecology and management.* Stackpole Books, Harrisburg, Pennsylvania, USA.

Nelson, J. R., R. M. Koes, W. H. Miller, and B. B. Davitt. 1982. Big game habitat management on a nutritional basis—a new approach. Western elk workshop. Flagstaff, Arizona 22–23 February.

Nelson, J. R., and T. A. Leege. 1982. Nutrition requirements and food habits. Pages 323–367 *in* J. W. Thomas and D. E. Toweill, editors. *Elk of North America, ecology and management.* Stackpole Books, Harrisburg, Pennsylvania, USA.

Nelson, L., Jr. 1980. Final report on the pilot program of continuing education in wildlife ecology and management. *The Wildlifer* 182:43–44.

Newcombe, C. P., and J. O. T. Jensen. 1996. Channel suspended sediment and fisheries: a synthesis for quantitative assessment of risk. *North American Journal of Fisheries Management* 16:693–727.

Newman, D. E. 1959. Factors influencing the winter roadside count of cottontails. *Journal of Wildlife Management* 23:290–294.

Newton, I., I. Wyllie, and A. Asher. 1991. Mortality causes in British barn owls *Tyto alba*, with a discussion of aldrin-dieldrin poisoning. *Ibis* 133:162–169.

Nichols, J. D. 1986. On the use of enumeration estimators for interspecific comparisons, with comments on a 'trappability' estimator. *Journal of Mammalogy* 67:590–593.

Nichols, J. D., M. J. Conray, D. R. Anderson, and K. P. Burnham. 1984. Compensatory mortality in waterfowl populations: a review of evidence and implications for research and management. Transactions of the North American Wildlife and Natural Resources Conference 49:535–554.

Nichols, J. D., and J. E. Hines. 1983. The relationship between harvest and survival rate of mallards: a straightforward approach with partitioned data sets. *Journal of Wildlife Management* 47:334–348.

Nichols, J. D., and K. H. Pollock. 1983. Estimation methodology in contemporary small mammal capture-recapture studies. *Journal of Mammalogy* 64:253–260.

Nichols, J. D., R. E. Tomlinson, and G. Waggerman. 1986. Estimating nest detection probabilities for white-winged dove nest transects in Tamaulipas, Mexico. *Auk* 103:825–828.

Nichols, R. G., and M. R. Pelton. 1972. Variation in fat levels of mandibular cavity tissue in white-tailed deer (*Odocoileus virginianus*) in Tennessee. Proceedings of the Southeastern Association of Game and Fish Commissioners 26:57–68.

Nielsen, D. B., and D. D. Lytle. 1985. Who gains (or loses) when big-game uses private lands? *Utah Science,* Summer:48–50.

Nielsen, L. 1988. Definitions, considerations, and guidelines for translocation of wild animals. Pages 12–49 *in* L. Nielsen and R. D. Brown, editors. *Translocation of wild animals.* Wisconsin Humane Society, Milwaukee, Wisconsin, and Caesar Kleberg Wildlife Research Institute, Kingsville, Texas, USA.

Nish, D. 1964. An evaluation of artificial water catchment basins on Gambel's quail populations in southern Utah. Thesis, Utah State University, Logan, Utah, USA.

Noon, B. R. 1986. Summary: biometric approaches to modeling—the researcher's viewpoint. Pages 197–201 *in* J. Verner, M. L. Morrison, and C. J. Ralph, editors. *Wildlife 2000.* University of Wisconsin Press, Madison, Wisconsin, USA.

Noss, R. F. 1992. The wildlands project land conservation strategy. *Wild Earth.* Special Issue:10–25.

Noy-Meir, I. 1975. Stability of grazing systems: an application of predator-prey graphs. *Journal of Ecology* 63:459–481.

Ockenfels, R. A., D. E. Brooks, and C. H. Lewis. 1991. General ecology of Coves white-tailed deer in the Santa Rita Mountains. Arizona Game and Fish Department Technical Report Number 6.

O'Connell, M. 2000. Saving urban hawks. *Arizona Daily Star,* 9 April B1, B5.

Odum, E. P. 1971. *Fundamentals of Ecology.* 3rd ed. W. E. Saunders, Philadelphia, Pennsylvania, USA.

Odum, E. P., and E. J. Kuenzler. 1955. Measurement of territory and home range size in birds. *Auk* 72:128–137.

Ogden, P. R., and J. C. Mosley. 1989. *Summary of the elk-livestock information workshops.* Cooperataive Extension Service, University of Arizona, Tucson, Arizona, USA.

Olech, L. A. 1979. Summer activity rhythms of peninsular bighorn sheep in Anza-Borrego Desert State Park, San Diego County, California. Desert Bighorn Council Transactions 23:33–36.

Oliver, C. D., and B. C. Larson. 1990. *Forest stand dynamics.* McGraw-Hill, New York, New York, USA.

Oliver, W. H. 1969. Riparian lands–icy habitat for upland birds. *Washington Department of Game Bulletin* 21:3–5.

Onderka, D. K., S. A. Rawluk, and W. D. Wishart. 1988. Susceptibility of Rocky Mountain bighorn sheep to pneumonia induced by bighorn and domestic livestock strains of *Pasturella haemolytica. Canadian Journal of Veterinary Research* 52:439.

Onderka, D. K., and W. D. Wishart. 1988. Experimental contact transmission of *Pasteurella haemolytica* from clinically normal domestic sheep causing pneumonia in Rocky Mountain bighorn sheep. *Journal of Wildlife Disease* 24:663–667.

O'Neil, J. 1985. Management strategies and future research on elk in Arizona. Pages 35–38 *in* G. W. Workman, editor. *Western elk management: a symposium.* Utah State University, Logan, Utah, USA.

Ordway, L. L. 1985. Habitat use by desert mule deer. Thesis, University of Arizona, Tucson, Arizona, USA.

Ordway, L. L., and P. R. Krausman. 1986. Habitat use by desert mule deer. *Journal of Wildlife Management* 50:677–683.

Overton, W. S., and D. E. Davis. 1969. Estimating the numbers of animals in wildlife populations. Pages 403–455 *in* R. H. Giles, editor. *Wildlife management*

techniques. The Wildlife Society, Washington, D.C., USA.

Owen, R. B., Jr. 1969. Heart rate, a measure of metabolism in blue-winged teal. *Comparative Biochemistry and Physiology* 31:431–436.

Ozoga, J. J. 1969. Same longevity records for female white-tailed deer in northern Michigan. *Journal of Wildlife Management* 33:1027.

Parra, R. 1978. Comparison of fore gut and hind gut fermentation of herbivores. Pages 205–229 *in* G. G. Montgomery, editor. *The ecology of arboreal folivores.* Smithsonian Institution Press, Washington, D.C., USA.

Parry, P. L. 1972. Development of permanent wildlife water supplies Joshua Tree National Monument. Desert Bighorn Council Transactions 16:92–96.

Pasitschniak-Arts, M., and F. Messier. 1998. Effects of edges and habitats on small mammals in a prairie ecosystem. *Canadian Journal of Zoology* 76:2020–2025.

Pasitschniak-Arts, M., and F. Messier. 2000. Brown (grizzly) and polar bears. Pages 409–428 *in* S. Demarais and P. R. Krausman, editors. *Ecology and management of large mammals in North America.* Prentice Hall, Upper Saddle River, New Jersey, USA.

Patterson, I. J. 1965. Timing and spacing of broods in the black-headed gull, *Larus ridibundus. Ibis* 107:433–459.

Patterson, R. L. 1952. *The sage grouse in Wyoming.* Wyoming Game and Fish Commission and Sage Books, Denver, Colorado, USA.

Patton, D. R. 1975. A diversity index for quantifying habitat "edge." *Wildlife Society Bulletin* 3:171–173.

Patton, D. R. 1992. *Wildlife habitat relationships in forested ecosystems.* Timber Press, Portland, Oregon, USA.

Paulik, G. J., and D. S. Robson. 1969. Statistical calculations for change-in-ratio estimators of population parameters. *Journal of Wildlife Management* 33:1–27.

Payne, N. F., and F. C. Bryant. 1998. *Wildlife habitat management of forestlands, rangelands, and farmlands.* Krieger Publishing Company, Malabar, Florida, USA.

Payne, N. F., and F. Copes. 1988. *Wildlife and fisheries habitat improvement handbook.* U.S. Forest Service, Washington, D.C., USA.

Pearce, D. W. 1992. *The MIT dictionary of modern economics.* 4th ed. MIT Press, Cambridge, Massachusetts, USA.

Pearl, R., and J. R. Miner. 1935. Experimental studies on the duration of life. XIV. The comparative mortality of certain lower organisms. *Quarterly Review of Biology* 10:60–79.

Peek, J. M. 1989. A look at wildlife education in the United States. *Wildlife Society Bulletin* 17:361–365.

Peek, J. M., and P. R. Krausman. 1996. Grazing and mule deer. Pages 183–192 *in* P. R. Krausman, editor. *Rangeland wildlife.* Society for Range Management, Denver, Colorado, USA.

Pelton, M. R. 2000. Black bear. Pages 389–408 *in* S. Demarais and P. R. Krausman, editors. *Ecology and management of large mammals in North America.* Prentice Hall, Upper Saddle River, New Jersey, USA.

Pendleton, G. W. 1992. Nonresponse patterns in the federal waterfowl hunter questionnaire survey. *Journal of Wildlife Management* 56:344–348.

Perrins, C. M. 1970. The timing of birds' breeding seasons. *Ibis* 112:242–255.

Perry, T. W. 1984. *Animal life-cycle feeding and nutrition.* Academic Press, New York, New York, USA.

Peterken, G. F. 1983. Woodland conservation in Britain. Pages 83–100 *in* A. Warren and F. B. Goldsmith, editors. *Conservation in perspective.* John Wiley & Sons, Chichester, United Kingdom.

Peterson, R. O., J. D. Woolington, and T. N. Bailey. 1984. Wolves of the Kenai Peninsula, Alaska. *Wildlife Monograph* 88.

Peterson, S. R. 1975. Ecological distribution of breeding birds. Pages 22–38 *in* Symposium on management of forest and range habitats for nongame birds. U.S. Forest Service General Technical Report WO-1.

Pickett, S. T. A., and P. S. White. 1985. *The ecology of natural disturbance and patch dynamics.* Academic Press, Orlando, Florida, USA.

Pierce, B. M., R. T. Bowyer, and V. C. Bleich. In Press. Habitat selection by mule deer: forage benefits or risk of predation by mountain lions? *Journal of Mammalogy.*

Pierce, B. M., V. C. Bleich, and R. T. Bowyer. In Press. Social organization of mountain lions: does a land-tenure system regulate population size? *Ecology.*

Pimlott, D. H., and P. W. Joslin. 1968. The status and distribution of the red wolf. Transactions of the North American Wildlife and Natural Resources Conference 33:373–389.

Player, I. 1972. *The white rhino saga.* William Collins Sons & Company, London, United Kingdom.

Pokras, M. A., and R. Chafel. 1992. Lead toxicosis from ingested fishing sinkers in adult common loons (*Gavia immer*) in New England. *Journal of Zoo and Wildlife Medicine* 23:92–97.

Pollock, K. H., J. D. Nichols, C. Brownie, and J. E. Hines. 1990. Statistical inference for capture-recapture experiments. *Wildlife Monograph* 107.

Porter, R. D. 1950. The Hungarian partridge in Utah. *Journal of Wildlife Management* 19:93–109.

Porter, W. F. 1992. Burgeoning ungulate populations in national parks: is intervention warranted. Pages 304–312 *in* D. R. McCullough and R. H. Barnett, editors. *Wildlife 2001: populations.* Elsevier Science Publishers, London, United Kingdom.

Porter, W. F., and G. A. Baldassarre. 2000. Future directions for the graduate curriculum in wildlife biology: building on our strengths. *Wildlife Society Bulletin* 28:508–513.

Potter, D. R. 1982. Recreational use of elk. Page 529 *in* J. W. Thomas and D. E. Towlill, editors. *Elk of North America.* Stackpole Books, Harrisburg, Pennsylvania, USA.

Potvin, F., and J. Huot. 1983. Estimating carrying capacity of a white-tailed deer wintering area in Quebec. *Journal of Wildlife Management* 47:463–475.

Prasad, N. L. N. S., and F. S. Guthery. 1986. Wildlife use of livestock water under short duration and contiguous grazing. *Wildlife Society Bulletin* 14:450–454.

Pritchard, G. T., and C. T. Robbins. 1990. Digestive and metabolic efficiencies of grizzly and black bears. *Canadian Journal of Zoology* 68:1645–1651.

Pulling, A. V. S. 1945. Nonbreeding in bighorn sheep. *Journal of Wildlife Management* 9:155–156.

Purdy, K. G., and D. J. Decker. 1989. Applying wildlife values information in management: the wildlife attitude and values scale. *Wildlife Society Bulletin* 17:494–500.

Rackham, O. 1976. *Trees and woodland in the British landscape*. J. M. Dent & Sons, London, United Kingdom.

Raedeke, K. J., and J. F. Lehmkuhl. 1986. A simulation procedure for modeling the relationships between wildlife and forest management. Pages 377–381 *in* J. Verner, M. L. Morrison, and C. J. Ralph, editors. *Wildlife 2000: modeling habitat relationships of terrestrial vertebrates*. University of Wisconsin Press, Madison, Wisconsin, USA.

Rahm, N. M. 1938. Quail range extension in the San Bernardino National Forest Progress Report. *California Fish and Game* 24:133–158.

Raphael, M. G., and R. H. Barrett. 1981. Methodologies for a comprehensive wildlife survey and habitat analysis in old-growth douglas-fir forests. *Cal-Neva Wildlife Transactions* 1981:106–121.

Rasker, R. 1989. Agriculture and wildlife: an economic analysis of waterfowl habitat management on farms in western Oregon. Dissertation, College of Forestry, Oregon State University, Corvallis, Oregon, USA.

Rasker, R., M. V. Martin, and R. L. Johnson. 1992. Economics: theory versus practice in wildlife management. *Conservation Biology* 6:338–349.

Ratcliffe, D. A. 1984. Postmedieval and recent changes in British vegetation: the culmination of human influence. *New Phytologist* 98:73–100.

Rathore, G. S. 1995. Conflicts between human economic development and endangered species. Pages 82–84 *in* J. A. Bissonette and P. R. Krausman, editors. Integrating people and wildlife for a sustainable future. Proceedings of the first International Wildlife Management Congress. The Wildlife Society, Bethesda, Maryland, USA.

Rautenstrauch, K. R., and P. R. Krausman. 1989. Influence of water availability and rainfall on movements of desert mule deer. *Journal of Mammalogy* 70:197–201.

Rautenstrauch, K. R., and P. R. Krausman. 1986*a*. Preventing desert mule deer drownings in the Mohawk Canal, Arizona. U.S. Bureau of Reclamation Final Report Contract 9-07-30-X069, Phoenix, Arizona, USA.

Rautenstrauch, K. R., and P. R. Krausman. 1989*b*. Preventing mule deer drowning in the Mohawk Canal, Arizona, *Wildlife Society Bulletin* 17:280–286.

Raveling, D. G. 1979. The annual cycle of body composition of Canada geese with special reference to control of reproduction. *Auk* 96:234–252.

Reed, J. M., P. D. Doerr, and J. R. Walters. 1988. Minimum viable population size of the red-cockaded woodpecker. *Journal of Wildlife Management* 52:385–391.

Reed, N. P., and D. Drabelle. 1984. *The United States Fish and Wildlife Service*. Westview Press, Boulder, Colorado, USA.

Reese, J. B., and H. Haines. 1978. Effects of dehydration on metabolic rate and fluid distribution in the jackrabbit, *Lepus californicus*. *Physiological Zoology* 51:155–165.

Reffalt, W. C. 1963. Some watering characteristics of two penned bighorn sheep on the Desert Game Range, Nevada. Desert Bighorn Council Transactions 7:156–166.

Reichardt, P. B., J. P. Bryant, T. P. Clausen, and G. D. Wieland. 1984. Defense of winter dormant Alaska paper birch against snowshoe hares. *Oceologia* 65:58–69.

Reiter, J., K. Parker, and B. J. LeBoeuf. 1981. Female competition and reproductive success in northern elephant seals. *Animal Behavior* 29:670–687.

Repetto, R. 1988. The forest for the trees? *Government policies and the misuse of forest resources*. World Resources Institute, Washington, D.C., USA.

Responsive Management. 1994. Illinois residents' opinions and attitudes regarding trapping, fur hunting, and furbearer management. Responsive Management, Harrisonburg, Virginia, USA.

Rhoades, D. F., and R. G. Cates. 1976. Toward a general theory of plant antiherbivore chemistry. Pages 168–213 *in* J. W. Wallace and R. L. Mansell, editors. Recent advances in phytochemistry, Volume 10. *Biochemical interaction between plants and insects*. Plenum, New York, New York, USA.

Rich, T. 1986. Habitat and nest-site selection by burrowing owls in the sagebrush steppe of Idaho. *Journal of Wildlife Management* 50:548–555.

Richardson, W. J. 1994. Serious birdstrike-related accidents to military aircraft of ten countries: preliminary analysis of circumstances. Proceedings of Bird Strike Committee Europe 21:129–152.

Richmond, C. R., W. H. Langham, and T. T. Trujillo. 1962. Comparative metabolism of titrated water by mammals. *Journal of Cell and Comparative Physiology* 59:45–53.

Ricklefs, R. E. 1979. *Ecology*. 2nd ed. Chiron Press, New York, New York, USA.

Rideout, C. B. 1978. Mountain goat. Pages 149–159 *in* J. L. Schmidt and D. L. Gilbert, editors. *Big game of North America*. Stackpole Books, Harrisburg, Pennsylvania, USA.

Riedman, M. L., J. A. Estes, M. M. Staedler, A. A. Giles, and D. R. Carlson. 1994. Breeding patterns and reproductive success of California sea otters. *Journal of Wildlife Management* 58:391–399.

Riney, T. 1982. *Study and management of large mummals*. John Wiley & Sons, New York, New York, USA.

Ringelman, J. K., and M. R. Szymczak. 1985. A physiological condition index for wintering mallards. *Journal of Wildlife Management* 49:564–568.

Risenhoover, K. L., J. A. Bailey, and L. A. Wakelyn. 1988. Assessing the Rocky Mountain bighorn sheep management problems. *Wildlife Society Bulletin* 16:346–352.

Robbins, C., and A. Moen. 1975. Uterine composition and growth in pregnant white-tailed deer. *Journal of Wildlife Management* 39:684–691.

Robbins, C. S., D. Bystrak, and P. H. Geissler. 1986. The breeding bird survey: its first fifteen years, 1965–1979. U.S. Fish and Wildlife Service Resource Publication 157.

Robbins, C. T. 1983. *Wildlife feeding and nutrition*. Academic Press, Orlando, Florida, USA.

Robbins, C. T. 1993. *Wildlife feeding and nutrition*. 2nd ed. Academic Press, New York, New York, USA.

Robbins, C. T., A. N. Moen, and J. T. Reid. 1974. Body composition of white-tailed deer. *Journal of Animal Science* 38:871–876.

Robbins, C. T., S. M. Parish, and B. L. Robbins. 1985. Selenium and glutathione peroxidase activity in mountain goats. *Canadian Journal of Zoology* 63:1544–1547.

Roberts, R. F. 1977. Big game guzzlers. *Rangeman's Journal* 4:80–82.

Robertson, R. J. 1976. Optimal niche space of the redwinged blackbird: spatial and temporal patterns of nesting activity and success. *Ecology* 54:1085–1093.

Robinette, W. L., N. V. Hancock, and D. A. Jones. 1977. The Oak Creek mule deer herd in Utah. Utah State Division of Wildlife Resources Publication 77-15:1–148.

Robinson, J. R. 1957. Functions of water in the body. Proceedings of the Nutritional Society 16:108–112.

Rodgers, L. L. 1987. Seasonal changes in defecation rates of free-ranging white-tailed deer. *Journal of Wildlife Management* 51:330–333.

Rodgers, III., A. D. 1951. *Bernard Eduard Fernow— a story of North American forestry*. Princeton University Press, Princeton, New Jersey, USA.

Roelke, M. E., J. S. Martenson, and S. J. O'Brien. 1993. The consequences of demographic reduction and genetic depletion in the endangered Florida panther. *Current Biology* 3:340–350.

Roffe, T. J. 1987. Avian tuberculosis. Pages 95–99 *in* M. Friend, editor. Field guide to wildlife diseases. U.S. Fish and Wildlife Service Publication 167.

Rogers, A., J. Blunder, and N. Curry. 1985. *The countryside handbook*. Croom Helm, London, United Kingdom.

Rogers, G. E. 1953. Function and operation of big game check stations in Colorado. *Journal of Wildlife Management* 17:256–267.

Rogers, J. P., J. D. Nichols, F. W. Martin, C. F. Kimball, and R. S. Pospahala. 1979. Can ducks be managed by regulation? An examination of harvest and survival rate of ducks in relation to hunting. Transactions of the North American Wildlife and Natural Resource Conference 44:114–126.

Romin, L. A., and J. A. Bissonette. 1996. Deer-vehicle collisions: status of state monitoring activities and mitigation efforts. *Wildlife Society Bulletin* 24:276–283.

Rominger, E. M., and C. T. Robbins. 1996. Winter foraging ecology of woodland caribou in northwestern Washington. *Journal of Wildlife Management* 60:719–728.

Roosevelt, T. 1926. *Theodore Roosevelt: an autobiography.* Charles Scribner's Sons, New York, New York, USA.

Roseberry, J. L. 1979. Bobwhite population responses to exploitation: real and simulated. *Journal of Wildlife Management* 43:285–305.

Rosen, P. C., and C. H. Lowe. 1994. Highway mortality of snakes in the Sonoran Desert of southern Arizona. *Biological Conservation* 68:143–148.

Rosen, P. C., C. R. Schwalbe, D. A. Parizek, Jr., P. A. Holm, and C. H. Lowe. 1995. Introduced aquatic vertebrates in the Chiricahua region: effects on declining native ranid frogs. Pages 251–261 *in* L. F. DeBano, G. J. Gottfried, R. H. Hamre, C. B. Edminster, P. F. Ffolliott, and A. Ortega-Rubio, technical coordinators. Biodiversity of the Madrean Archipelago: the Sky Islands of southwestern United States and northwestern Mexico. U.S. Forest Service General Technical Report Rm-264.

Rosenstock, S. S., W. B. Ballard, and J. C. DeVoz, Jr. 1999. Viewpoint: benefits and impacts of wildlife water developments. *Journal of Range Management* 52:302–311.

Rosenzweig, M. L. 1981. A theory of habitat selection. *Ecology* 62:327–335.

Ross, R. C. 1934. Age and fecundity of mule deer (*Odocoileus hemionus hemionus*). *Journal of Mammalogy* 15:72.

Ross, W. D. 1964. *Aristotle: a complete exposition of his works and thought.* 5th ed. Barnes and Noble, New York, New York, USA.

Rumble, M. A., and L. D. Flake. 1983. Management considerations to enhance use of stock ponds by waterfowl broods. *Journal of Range Management* 36:691–694.

Russo, J. P. 1964. The Kaibab North deer herd: its history, problems and management. Arizona Game and Fish Department Wildlife Bulletin 7.

Rutberg, A. T. 1987. Adaptive hypothesis of birth synchrony in ruminants: an interspecific test. *The American Naturalist* 130:692–710.

Sadleir, R. M. F. S. 1969. *The ecology of reproduction in wild and domestic mammals.* Methuen, United Kingdom.

Samson, J., J. T. Jorgenson, and W. D. Wishort. 1989. Glutathione peroxidase activity and selenium levels in Rocky Mountain bighorn sheep and mountain goats. *Canadian Journal of Zoology* 67:2493–2496.

Sanchez, J. E., and M. K. Haderlie. 1990. Water management on Cabeza Prieta and Kofa National Wildlife Refuges. Pages 73–77 *in* G. K. Tsukamoto and S. J. Stiver, editors. Proceedings of the Wildlife Water Development Symposium. Nevada Chapter of The Wildlife Society, U.S. Bureau of Land Management, and Nevada Department of Wildlife, Las Vegas, Nevada, USA.

Sarbello, W., and L. W. Jackson. 1985. Deer mortality in the town of Malone. *New York Fish and Game Journal* 32:141–157.

Sargent, A. B., and J. E. Forbes. 1973. Mortality among birds, mammals and certain snakes on 17 miles of Minnesota roads. *Loon* 45:4–7.

Sarrazin, J. P. R., and J. R. Bider. 1973. Activity: a neglected parameter in population estimates—the development of a new technique. *Journal of Mammalogy* 54:369–382.

Sauer, J. R., and M. S. Boyce. 1979. Time series analysis of the National Elk Refuge census. Pages 9–12 *in* M. S. Boyce and L. O. Hayden-Wing, editors. *North American elk: ecology, behavior, and management.* University of Wyoming, Laramie, Wyoming, USA.

Savory, A. 1988. *Holistic resource management.* Island Press, Washington, D.C., USA.

Scolario, J. A. 1990. On a longevity record of the Magellanie penguin. Journal of field Ornithology 61:377–484.

Schadle, D. 1958. Arizona's catchment then and now. Desert Bighorn council Transactions 2:32–35.

Schaller, G. B. 1972. *The Serengeti lion; a study of predator-prey relations.* University of Chicago Press, Chicago, Illinois, USA.

Schamberger, M., A. H. Farmer, and J. W. Terrell. 1982. Habitat suitability index models: introduction. U.S. Fish and Wildlife Service FWS/OBS/82/10.

Scheffer, V. B. 1976. The future of wildlife management. *Wildlife Society Bulletin* 4:51–54.

Schemnitz, S. D. 1961. Ecology of the scaled quail in the Oklahoma panhandle. *Wildlife Monograph* 8.

Schemnitz, S. E. 1994. Scaled quail. *Birds of North America* 106:1–14.

Schmidt, S. L., and D. C. Dalton. 1995. Bats of the Madrean Archipelago (Sky Islands): current knowledge, future directions. Pages 274–287 *in* L. F. DeBano, G. J. Gottfried, R. H. Hamre, C. B. Edminster, P. F. Ffolliott, and A. Ortega-Rubio, technical coordinators. Biodiversity of the Madrean Archipelago: the Sky Islands of southwestern United States and northwestern Mexico. U.S. Forest Service General Technical Report RM-264.

Schmidt, S. L., and S. DeStefano. 1996. Impact of artificial water developments of nongame wildlife in the Sonoran Desert of southern Arizona: 1996 annual report. Arizona Cooperative Fish and Wildlife Research Unit, University of Arizona, Tucson, Arizona, USA.

Schmidt-Nielsen, K. 1981. *Animal physiology: adaptation and environment.* Cambridge University Press, London, United Kingdom.

Schmitz, O. J., and A. R. E. Sinclair 1997. Rethinking the role of deer in forest ecosystem dynamics. Pages 201–223 *in* W. J. McShea, H. B. Underwood, and J. H. Rappole, editors. *The science of overabundance.* Smithsonian Institution Press, Washington, D.C., USA.

Schreiner, K. M. 1976. Critical habitat: what it is and is not. *Endangered Species Technical Bulletin* 1:1–4.

Schroeder, R. L. 1987. Community models for wildlife impact assessment: a review of concepts and approaches. U.S. Fish and Wildlife Service Biological Report 87(2).

Schwartz, C. C., and B. Bartley. 1991. Reducing incidental moose mortality: considerations for management. *Alces* 27:227–231.

Schwartz, C. C., W. L. Regelin, and J. C. Nagy. 1980. Deer preference for juniper forage and volatile oil treated foods. *Journal of Wildlife Management* 44:114–120.

Schwartz, O. A., V. C. Bleich, and S. A. Hall. 1986. Genetics and the conservation of mountain sheep *Ovis canadensis nelsoni. Biological Conservation* 37:179–190.

Schweitzer, A. 1966. *Reverence for life.* Harper and Row, New York, New York, USA.

Scott, J. E. 1998. Do livestock waters help wildlife? Pages 493–507 *in* Environmental, economic, and legal issues related to rangeland water developments: proceedings of a symposium. Arizona State University College of Law, Tempe, Arizona, USA.

Scott, T. G. 1938. Wildlife mortality on Iowa highways. *American Midland Naturalist* 20:527–539.

Scott, V. E., and E. L. Boeker. 1972. An evaluation of wild turkey call counts in Arizona. *Journal of Wildlife Management* 36:628–630.

Seal, U. S., L. J. Verme, and J. J. Ozoga. 1978. Dietary protein and energy effects on deer fawn metabolic patterns. *Journal of Wildlife Management* 42:776–790.

Seal, U. S., L. J. Verme, J. J. Ozoga, and A. W. Erickson. 1972. Nutritional effects on thyroid activity and blood of white-tailed deer. *Journal of Wildlife Management* 39:1041–1052.

Sears, J. 1988. Regional and seasonal variations in lead poisoning in the mute swan *Cygnus olar* in relation to the distribution of lead and lead weights, in the Thames area, England. *Biological Conservation* 46:115–134.

Seber, G. A. F. 1982. *The estimation of animal abundance and related parameters.* 2nd ed. MacMillian, New York, New York, USA.

Seip, D. R. 1992. Factors limiting woodland caribou populations and their interrelationships with wolves and moose in southeastern British Columbia, *Canadian Journal of Zoology* 70:1494–1503.

Seton, E. T. 1929. *Lives of game animals*. Doubleday, Doran and Co., Garden City, New York, USA.

Severinghaus, C. W. 1955. Deer weights on an index of range conditions on two wilderness areas in the Adirondack region. *New York Fish and Game Journal* 2:154–160.

Severinghaus, W. D. 1981. Guild theory development as a mechanism for assessing environmental impact. *Environmental Management* 5:187–190.

Severson, K. E., and A. L. Medina. 1983. Deer and elk habitat management in the Southwest. *Journal of Range Management Monograph* 2.

Shackley, M. 1992. Manatees and tourism in southern Florida: opportunity or threat? *Journal of Environmental Management* 34:257–265.

Shaeil, J. 1971. *Rabbits and their history*. David and Charles Newton, Abbot, United Kingdom.

Shaffer, M. L. 1981. Minimum population sizes for species conservation. *BioScience* 1:131–134.

Sharp, L. 1984. Overview of the Taylor Grazing Act. Pages 9–10 *in The Taylor Grazing Act: 50 years of progress*. U.S. Department of the Interior, Washington, D.C., USA.

Sharpe, F. R., and A. J. Lotka. 1911. A problem in age distribution. *Philadelphia Magazine* 21:435–438.

Shaw, G. E. 1973. Electrocution of a caribou herd caused by lightning in central Alaska. *Journal of Wildlife Disease* 9:311–313.

Shaw, H. G. 1977. Impact of mountain lions on mule deer and cattle in northwestern Arizona. Pages 17–32 *in* R. L. Phillips and C. Jonkel, editors. Proceedings of the 1975 Predator Symposium. Forest and Conservation Experiment Station, University of Montana, Missoula, Montana, USA.

Shaw, H. G., and C. Mollohan. 1992. Merriam's turkey. Pages 331–349 *in* J. G. Dickson, editor. *The wild turkey: biology and its management*. Stackpole Books, Harrisburg, Pennsylvania, USA.

Shaw, W. W. 2000. Graduate education in wildlife management: major trends and opportunities for serving international students. *Wildlife Management Bulletin* 28:514–517.

Scheffer, V. B. 1976. The future of wildlife management. *Wildlife Society Bulletin* 4:51–54.

Sheffield, W. J., Jr., E. D. Ables, and B. A. Fall. 1971. Geographic and ecologic distribution of nilgai antelope in Texas. *Journal of Wildlife Management* 35:250–257.

Sherman, P. W. 1976. Natural selection among same group-living organisms. Dissertation, University of Michigan, Ann Arbor, Michigan, USA.

Shipley, L. A., J. E. Gross, D. E. Spalinger, N. T. Hobbs, and B. A. Wunder. 1994. The scaling of intake rate in mammalian herbivores. *The American Naturalist* 143:1055–1082.

Short, H. L. 1977. Food habits of mule deer in a semidesert grass-shrub habitat. *Journal of Range Management* 30:206–209.

Short, H. L. 1981. Nutrition and metabolism. Pages 99–127 *in* O. C. Wallmo, editor. *Mule and black-tailed deer of North America*. University of Nebraska Press, Lincoln, Nebraska, USA.

Short, H. L. 1983. Wildlife guilds in Arizona desert habitats. U.S. Bureau of Land Management Technical Note 362.

Short, J., S. D. Bradshaw, J. Giles, R. I. T. Prince, and G. R. Wilson. 1992. Reintroduction of macropods (Marsupialia: Macrophodoidea) in Australia: a review. *Biological Conservation* 62:189–204.

Shugart, H. H. 1984. *A theory of forest dynamics*. Springer-Verlag, New York, New York USA.

Shugart, H. H., and S. W. Seagle. 1985. Modeling forest landscapes and the role of disturbance in ecosystems and communities. Pages 353–368 *in The ecology of natural disturbance and patch dynamics*. Academic Press, New York, New York, USA.

Silflow, R. M., W. J. Foreyt, S. M. Taylor, W. W. Laegreid, D. H. Liggitt, and R. W. Lied. 1989. Comparison of pulmonary defense mechanisms in Rocky Mountain bighorn (*Ovis canadensis canadensis*) and domestic sheep. *Journal of Wildlife Disease* 25:514.

Simmons, N. M. 1969. Heat stress and bighorn behavior in the Cabeza Prieta Game Range,

Arizona. *Desert Bighorn Council Transactions* 13:55–63.

Simmons, T. R., and U. P. Kreuter. 1989. Herd mentality: banning ivory sales is no way to save the elephant. *Policy Review* Fall 46:49.

Simpson, V. R., A. E. Hunt, and M. C. French. 1979. Chronic lead poisoning in a herd of mute swans. *Environmental Pollution* 18:187–202.

Singer, F. J., C. M. Papouchis, and K. K. Symonds. 2000. Translocations as a tool for restoring populations of bighorn sheep. *Restoration Ecology* 8:in press.

Siniff, D. B., and R. O. Skogg. 1964. Aerial censusing of caribou using stratified random sampling. *Journal of Wildlife Management* 28:391–401.

Skovlin, J. M. 1982. Habitat requirements and evaluations. Pages 369–413 *in* J. V. Thomas and D. E. Toweill, editors. *Elk of North America: ecology and management.* Stackpole Books, Harrisburg, Pennsylvania, USA.

Skovlin, J. M., P. J. Edgerton, and R. W. Harris. 1968. The influence of cattle management on deer and elk. *North American Wildlife Conference* 33:169–181.

Slobodkin, L. B. 1980. *Growth and regulation of animal populations.* 2nd ed. Dover Publications, New York, New York, USA.

Small, S. J. 1990. *Preserving family lands: a landowner's introduction to tax issues and other considerations.* Powers and Hall Professional Corporation, Boston, Massachusetts, USA.

Smith, A. D. 1964. Defecation rates of mule deer. *Journal of Wildlife Management* 28:435–444.

Smith, C. A., E. L. Young, C. R. Land, and K. P. Bovee. 1987. Predator-induced limitations on deer population growth in southeast Alaska. Alaska Department of Fish and Game, Federal Aid to Wildlife Restoration, Final Report Project W-22-4, W-22-5, and W-22-6.

Smith, G. W., and R. E. Reynolds. 1992. Hunting and mallard survival, 1979–88. *Journal of Wildlife Management* 56:306–316.

Smith, H. A., T. C. Jones, and R. D. Hunt. 1972. *Veterinary pathology.* 4th ed. Lin and Febiger, Philadelphia, Pennsylvania, USA.

Smith, L. M., I. L. Brisbin, Jr., and G. C. White. 1984. An evaluation of total trapline captures as estimates of furbearer abundance. *Journal of Wildlife Management* 48:1452–1455.

Smith, N. S. 1970. Appraisal of condition estimation methods for East African ungulates. *East African Wildlife Journal* 8:123–129.

Smith, N. S., and R. S. Henry. 1985. Short-term effects of artificial oases on wildlife. Final report, U.S. Bureau of Reclamation and Arizona Cooperative Wildlife Research Unit, University of Arizona, Tucson, Arizona, USA.

Smith, R. H., and A. LeCount. 1979. Some factors affecting survival of desert mule deer fawns. *Journal of Wildlife Management* 43:657–665.

Smith, R. H., and S. Gallizioli. 1965. Predicting hunter success by means of a spring call count of Gambel quail. *Journal of Wildlife Management* 29:806–813.

Smith, R. L. 1977. *Elements of ecology and field biology.* Harper and Row, New York, New York, USA.

Smith, R. L., and B. Read. 1992. Management parameters affecting the reproductive potential of captive, female black rhinoceros, *Diceros bicornis. Zoo Biology* 11:375–383.

Smith, T. M. 1986. Habitat-simulation models: integrating habitat-classification and forest simulation models. Pages 389–393 *in* J. Verner, M. L. Morrison, and C. J. Ralph, editors. *Wildlife 2000: modeling relationships of terrestrial vertebrates.* University of Wisconsin Press, Madison, Wisconsin, USA.

Smith, T. M., H. H. Shugart, and D. C. West. 1981. Use of forest simulation models to integrate timber harvest and nongame bird management. Transactions of the North American Wildlife and Natural Resource Conference 46:501–510.

Snyder, W. D. 1967. Experimental habitat improvement for scaled quail. Colorado Department of Game, Fish, and Parks Technical Publication 19.

Solmon, V. E. F. 1978. Gulls and aircraft. *Environmental Conservation* 5:277–280.

Solmon, V. E. F. 1974. Aircraft and wildlife. Pages 137–141 *in* J. H. Noyes and D. R. Progulske, editors. *Wildlife in an urbanizing environment.*

Cooperative Extension Service, University of Massachusetts, Amherst, Massachusetts, USA.

Solomon, M. E. 1949. The natural control of animal populations. *Journal of Animal Ecology* 18:1–35.

Soule, M. E. 1983. What do we really know about extinction? Pages 111–124 *in* C. M. Schoenwald-Cox, S. M. Chambers, B. MacBryde, and W. L. Thomas, editors. *Genetics and conservation: a reference for managing wild animal and plant populations.* Benjamin/Cummings Publishing Company, Inc., London, United Kingdom.

South African Wildlife Management Association. 1988. Towards a policy for the introduction and translocation of game animals. Pages 93–98 *in* L. Nielsen and R. D. Brown, editors. *Translocation of wild animals.* Wisconsin Humane Society, Inc. Milwaukee, Wisconsin, USA, and Caesar Kleberg Wildlife Research Institute, Kingsville, Texas, USA.

Sowell, B. F., B. H. Koerth, and F. C. Bryant. 1985. Seasonal nutrient estimates of mule deer diets in the Texas panhandle. *Journal of Range Management* 38:163–167.

Sowls, L. K. 1961. Hunter-checking stations for collecting data on the collared peccary (Peccary tajacu). Transactions of the North American Wildlife and Natural Resource Conference 26:497–505.

Spalinger, D. E. 2000. Nutritional ecology. Pages 108–139 *in* S. Demarais and P. R. Krausman, editors. *Ecology and management of large mammals in North America.* Prentice Hall, Upper Saddle River, New Jersey, USA.

Spalinger, D. E., T. Hanley, and C. Robbins. 1988. Analysis of the functional response in foraging in the Sitka black-tailed deer. *Ecology* 69:1166–1175.

Spinage, C. A. 1973. The role of photoperiodism in the seasonal breeding of tropical African ungulates. *Mammal Review* 3:71–84.

Squibb, R. C. 1985. Mating success of yearling and older bull elk. *Journal of Wildlife Management* 49:744–750.

Sredl, M. J., and J. M. Howland. 1995. Conservation and management of Madrean populations of the Chiricahua leopard frog. Pages 379–385 *in* L. F. DeBano, G. J. Gottfried, R. H. Hamre,

C. B. Edminster, P., F. Ffolliott, and A. Ortega-Rubio, technical coordinators. Biodiversity of the Madrean Archipelago: the Sky Islands of southwestern United States and northwestern Mexico. U.S. Forest Service General Technical Report RM-264.

Staaland, H., and R. G. White. 1991. Influence of foraging ecology on alimentary tract size and function of Svalbard reindeer. *Canadian Journal of Zoology* 69:1326–1334.

Stamps, J. A., M. Buechner, and V. V. Krishnan. 1987. The effects of edge permeability and habitat geometry on emigration from patches of habitat. *American Naturalist* 129:533–552.

Stegner, W. E. 1953. *Beyond the hundredth meridian.* University of Nebraska Press, Lincoln, Nebraska, USA.

Steidl, R. J., S. DeStefano, and W. J. Matter. 2000. On increasing the quality, reliability, and rigor of wildlife science. *Wildlife Society Bulletin* 28:518–521.

Steinert, S. F., H. D. Riffel, and G. C. White. 1994. Comparisons of big game harvest estimates from check station and telephone surveys. *Journal of Wildlife Management* 58:335–340.

Stephenson, R. O., W. B. Ballard, C. A. Smith, and K. Richardson. 1995. Wolf biology and management in Alaska, 1981–1992. Pages 43–54 *in* L. N. Carbyn, S. H. Fritts, and D. R. Seip, editors. *Ecology and conservation of wolves in a changing world.* Canadian Circumpolar Institute, University of Alberta, Edmonton, Alberta, Canada.

Stetson, L. 1994. *The wild Muir.* Yosemite Association, Yosemite National Park, California, USA.

Stevens, C. E., and I. D. Hume. 1995. *Comparative physiology of the vertebrate digestive system.* Cambridge University Press, Cambridge, United Kingdom.

Stevens, D. R. 1966. Range relationships of elk and livestock, Crow Creek drainage, Montana. *Journal of Wildlife Management* 30:349–363.

Steward, A. J. A., and A. N. Lance. 1983. Moor-draining: a review of impacts on land use. *Journal of Environmental Management* 17:81–99.

Steward-Oaten, A., J. R. Bence, and C. W. Osenberg. 1992. Assessing effects of unreplicated perturbations: no simple solutions. *Ecology* 73:1396–1404.

Steward-Oaten, A., W. W. Murdoch, and K. R. Parker. 1986. Environmental impact assessment: "pseudoreplication" in time? *Ecology* 67:929–940.

Stewart, P. A. 1973. Electrocution of birds by an electric fence. *Wilson Bulletin* 85:476–477.

Stickel, L. F. 1954. A comparison of certain methods of measuring ranges of small mammals. *Journal of Mammalogy* 35:1–15.

Stoddard, H. L. 1931. *The bob-white quail: its habits, preservation, and increase.* C. Scribner's Sons, New York, New York, USA.

Stoddart, L. A., A. D. Smith, and T. Box. 1975. *Range management.* 3rd ed. McGraw-Hill, New York, New York, USA.

Stout, J., and G. W. Cornwell. 1976. Nonhunting mortality of fledged North American waterfowl. *Journal of Wildlife Management* 40:681–693.

Strickland, D. M., H. J. Harjes, K. R. McCaffery, H. W. Miller, L. M. Smith, and R. J. Stall. 1994. Harvest management. Pages 445–473 *in* T. A. Bookhout, editor. *Research and management techniques for wildlife and habitats.* 5th ed. The Wildlife Society, Bethesda, Maryland, USA.

Stromayer, K. A., and R. J. Warren. 1997. Are overabundant deer herds in the eastern Untied States creating alternate stable states in forest plant communities? *Wildlife Society Bulletin* 25:227–234.

Strong, L. L., D. S. Gilmer, and J. A. Brass. 1991. Inventory of wintering geese with a multispectral scanner. *Journal of Wildlife Management* 55:250–259.

Stroud, R. K., and M. Friend. 1987. Avian salmonellosis. Pages 101–106 *in* M. Friend, editor. Field guide to wildlife diseases. U.S. Fish and Wildlife Service Resource Publication 167.

Stuart, D. G. 1925. *Forty years on the frontier.* Arthur H. Clark Col, Cleveland, Ohio, USA.

Sundstrom, C. 1968. Water consumption by pronghorn antelope and distribution related to water in Wyoming's Red Desert.

Proceedings of the Biennial Pronghorn Antelope States Workshop 3:39–46.

Swank, W. G. 1958. The mule deer in Arizona chaparral. Arizona Game and Fish Department, Phoenix, Arizona. *Wildlife Bulletin* 3.

Swank, W. G., R. M. Watson, G. H. Freeman, and T. Jones. 1968. Proceedings of the workshop on the use of light aircraft in wildlife management in East Africa. *East African Agricultural and Forestry Journal* 34:1–111.

Sweeney, J. M. 1986. Refinement of DYNAST's forest structure simulation. Pages 357–360 *in* J. Verner, M. L. Morrison, and C. J. Ralph, editors. *Wildlife 2000: modeling habitat relationships of terrestrial vertebrates.* University of Wisconsin Press, Madison, Wisconsin, USA.

Talbot, L. M., and M. H. Talbot. 1963. The wildebeest in western Masai land, East Africa. *Wildlife Monograph* 12.

Tappeiner, J. C., J. C. Gourley, and W. H. Emmingham. 1985. A user's guide for on-site determinations of stand density and growth with a programmable calculator. Oregon State University, Forest Research Laboratory, Special Publication 11.

Tarshis, I. B. 1971. An unusual fatality of a yearling Canada goose. *Jack-Pine Warbler* 49:128.

Taylor, C. R. 1972. The desert gazelle: a paradox resolved. *Symposium of Zoological Society of London* 31:215–227.

Taylor, C. R. 1969. The eland and the oryx. *Scientific American* 220:88–95.

Taylor, C. R., and C. P. Lyman. 1967. A comparative study of the envornmental physiology of an East African antelope, the eland, and the hereford steer. *Physiological Zoology* 40:280–295.

Taylor, D., and W. Neal. 1984. Management implications of size-class frequency distribution in Louisiana alligator populations. *Wildlife Society Bulletin* 12:312–319.

Teal, E. W. 1952. John Burroughs: Disciple of nature. *Coronet* 31:90–94.

Teer, J. A. 1991. Nonnative large ungulates in North America. Pages 55–66 *in* L. A.

Renecker and R. J. Hudson, editors. Wildlife production: conservation and sustainable development. University of Alaska Fairbanks. Agricultural and Forestry Experiment Station Miscellaneous Publication 91-6.

Teer, J. G., D. L. Drawe, T. L. Blankenship, W. F. Andelt, R. S. Cook, J. G. Kie, F. F. Knowlton, and M. White. 1991. Deer and coyotes: the Welder experiments. Transactions of the North American Wildlife and Natural Resources Conference 56:550–560.

Teer, J. G., J. W. Thomas, and E. A. Walker. 1965. Ecology and management of white-tailed deer in the Llano Basin of Texas. *Wildlife Monograph* 15.

The Wildlife Society. 1996. *Universities and colleges offering curricula in wildlife conservation.* The Wildlife Society, Bethesda, Maryland, USA.

The Wildlife Society. 1999. *Program for certification of professional wildlife biologists.* The Wildlife Society, Bethesda, Maryland, USA.

Thomas, C. D. 1990. What do real population dynamics tell us about minimum population sizes? *Conservation Biology* 4:324–327.

Thomas, J. W. 1979. Wildlife habitats in managed forests: the Blue Mountains of Oregon and Washington. *U.S. Forest Service Handbook* 553.

Thomas, J. W. 1979*a.* Glossary. Pages 470–494 *in* J. W. Thomas, technical editor. Wildlife habitats in managed forests. *U.S. Department of Agriculture, Forest Service, Agriculture Handbook* 553.

Thomas, J. W. 1979*b.* Introduction. Pages 10–21 *in* J. W. Thomas, technical editor. Wildlife habitats in managed forests. *U.S. Department of Agriculture, Forest Service, Agriculture Handbook* 553.

Thomas, J. W., and D. H. Pletscher. 2000. The convergence of ecology, conservation biology, and wildlife biology—necessary or redundant? *Wildlife Society Bulletin* 28:546–549.

Thomas, J. W., R. J. Miller, C. Maser, R. G. Anderson, and B. E. Carter. 1979. Plant communities and successional states. Pages 22–39 *in* J. W. Thomas, technical editor. Wildlife habitats in managed forests: the Blue Mountains of Oregon and Washington. *U.S. Forest Service Agricultural Handbook* 533.

Thomas, V. G. 1988. Body condition, ovarian hierarchies, and their relation to egg formation in Anseriform and Galliform species. Proceedings of the International Ornithological Congress 19:353–363.

Thompson, J. R., V. C. Bleich, S. G. Torres, and G. P. Mulchahy. 2000. Comparison of two mountain sheep translocation techniques. *Southwestern Naturalist:* in press.

Thoreau, H. D. 1854. *Walden, or life in the woods and civil disobedience.* New American Library, New York, New York, USA.

Thorne, E. T., M. W. Miller, D. A. Jessup, and D. L. Hunter. 1992. Disease as a consideration in translocation and reintroducing wild animals: western state wildlife management agency perspectives. Pages 18–25 *in* Proceedings of the joint conference of the American Association of Zoo Veterinarians, the American Association of Wildlife Veterinarians, and the American Association of Zoo Veterinarians.

Thorne, E. T., N. Kingston, W. R. Jolley, and R. C. Bergstrom. 1982. *Diseases of Wyoming wildlife.* 2nd ed. Wyoming Game and Fish Department, Cheyenne, Wyoming, USA.

Thorne, E. T., R. E. Dean, and W. G. Hepworth. 1976. Nutrition during gestation in relation to successful reproduction in elk. *Journal of Wildlife Management* 40:330–335.

Thorpe, J. 1996. Fatalities and destroyed civil aircraft due to bird strikes, 1912–1995. Proceedings of Bird Strike Committee Europe 23:17–31.

Threlfall, W. 1995. Conservation and wildlife management in Britain. Pages 27–74 *in* V. Geist and I. McTaggart-Cowan, editors. *Wildlife conservation policy.* Detselig Enterprises, Calgary, Alberta, Canada.

Tidemann, S. C., B. Green, and K. Newgrain. 1989. Water turnover and estimated food consumption in the three species of fairy-wren (Molurus spp.). *Australian Wildlife Research* 16:187–194.

Tietenberg, T. 1988. *Environmental and natural resource economics.* 2nd ed. Scott, Foresman and Company, Glenview, Illinois, USA.

Tilghman, N. G. 1989. Impacts of white-tailed deer on forest regeneration in northwestern

Pennsylvania. *Journal of Wildlife Management* 53:424–453.

Tinbergen, N., M. Impekoven, and D. Franck. 1967. An experiment on spacing-out as a defense against predation. *Behavior* 28:307–321.

Torre-Bueno, J. R. 1978. Evaporative cooling and water balance during flight in birds. *Journal of Experimental Biology* 75:231–236.

Trefethen, J. B. 1975. *An American crusade for wildlife.* Winchester Press and the Boone and Crockett Club, New York, New York, USA.

Trefethen, J. B. 1975. *The wild sheep of modern North America.* The Boone and Crockett Club and Winchester Press, New York, New York, USA.

Trombulak, S. C., and C. A. Frissell. 2000. Review of ecological effects of roads on terrestrial and aquatic communities. *Conservation Biology* 14:18–30.

Trost, R. E. 1987. Mallard survival and harvest rate: a reexamination of relationships. Transactions of the North American Wildlife and Natural Resources Conference 52:264–284.

True, G. H., Jr. 1933. The quail replenishment program. *California Fish and Game* 19:20–23.

Trzcinski, M. K., L. Fahrig, and G. Merriam. 1999. Independent effects of forest cover and fragmentation on the distribution of forest breeding birds. *Ecological Applications* 9:586–593.

Tsukamoto, G. K., and S. J. Stiver, editors. 1990. Wildlife water developments. Proceedings of the Wildlife and Water Development Symposium, Las Vegas, Nevada, USA.

Tubbs, C. 1986. *The new forest.* Collins, London, United Kingdom.

Tuggle, B. N. 1987a. Gizzard worms. Pages 159–163 *in* M. Friend, editor. Field guide to wildlife diseases. U.S. Fish and Wildlife Service Resource Publication 167.

Tull, C. E., P. Germain, and A. W. May. 1972. Mortality of thick-billed murres in the West Greenland salmon fishery. *Nature* 237:42–44.

Tullar, B. F., Jr. 1983. A long lived white-tailed deer. *New York Fish and Game Journal* 30:119.

Tuovila, V. R. 1999. Bobcat movements and survival near U.S. Highway 281 in southern

Texas. Thesis, Texas A & M University—Kingsville, Kingsville, Texas, USA.

Turner, J. C. 1970. Water consumption of desert bighorn sheep. Desert Bighorn Council Transactions 14:189–197.

Turner, J. C. 1973. Water, energy and electrolyte balance in the desert bighorn sheep, *Ovis canadensis.* Dissertation, University of California, Riverside, USA.

Turner, M. 1984. The landscape of parliamentary enclosure. Pages 132–166 *in* M. Reed, editor. *Discovering past landscapes.* Croom Helm, London, United Kingdom.

Tyler, N. J. C. 1987. Natural limitation of the abundance of the high Arctic Svalbard reindeer. Dissertation, University of Cambridge, Cambridge, United Kingdom.

Tyson, E. L. 1959. A deer drive vs. track census. Transactions of the North American Wildlife Conference 24:457–464.

Uhlig, H. G. 1956. The gray squirrel in West Virginia. West Virginia Conservation Commission, Division of Game Management, Bulletin 3.

United States Department of the Interior. 1952. Investigations of methods of appraising the abundance of mourning doves. Fish and Wildlife Service Special Science Report. Wildlife 17.

United States Department of Transportation. 1996. Highway statistics 1996. FHWA-PL-98-003. U.S. Department of Transportation, Office of Highway Information Management, Washington, D.C., USA.

United States Fish and Wildlife Service. 1980. Habitat evaluation procedures (HEP). Division of Ecology Services. ESM 102.

United States Fish and Wildlife Service. 1985. Red-cockaded woodpecker recovery plan. U.S. Fish and Wildlife Service, Atlanta, Georgia, USA.

United States Fish and Wildlife Service. 1988. Endangered Species Act of 1973, as amended through the 100th Congress. U.S. Department of the Interior, Washington, D.C., USA.

United States Fish and Wildlife Service. 1996. The 1996 national survey of fishing, hunting, and wildlife-associated recreation.

U.S. Fish and Wildlife Service, Washington, D.C., USA.

United States Forest Service. 1979. A generalized forest growth projection system applied to the Lakes States region. U.S. Forest Service General Technical Report NC-49.

Unsworth, J. W., D. F. Pac, G. C. White, and R. M. Bartman. 1999. Muledeer survival in Colorado, Idaho, and Montana. *Journal of Wildlife Management* 63:315–326.

Van Ballenberghe, V. 1985. Wolf predation on caribou: the Nelchina case history. *Journal of Wildlife Management* 49:711–720.

Van Ballenberghe, V., and W. B. Ballard. 1994. Limitation and regulation of moose populations: the role of predation. *Canadian Journal of Zoology* 72:2071–2077.

Van Ballenberghe, V., and W. B. Ballard. 1997. Population dynamics. Pages 223–245 *in* A. W. Franzmann and C. C. Schwartz, editors. *Ecology and management of the North American moose*. Smithsonian Institute Press, Washington, D.C., USA.

Van Etten, R. C., and C. L. Bennett, Jr. 1965. Some sources of error in using pellet-group counts for censusing deer. *Journal of Wildlife Management* 29:723–729.

Van Horne, B. 1983. Density as a misleading indicator of habitat quality. *Journal of Wildlife Management* 47:894–901.

Van Loben Sels, R. C., J. D. Congdon, and J. T. Austin. 1995. Aspects of the life history and ecology of the Sonoran mud turtle in southeastern Arizona. Pages 262–266 *in* L. F. De Bano, G. J. Gottfried, R. H. Hamre, C. B. Edminster, P. F. Ffolliott, and A. Ortega-Rubio, technical coordinators. Biodiversity of the Madre Archipelago: the Sky Islands of southwestern United States and northwestern Mexico. USDA Forest Service General Technical Van Soest, P. J. N utrtional ecology of the rominant. O & B Books, Corvallis, Oregon USA.Report RM-26.

Van Soest, P. J. 1982. *Nutritional ecology of the Dominant*. O&B Books, Corvallis, Oregon USA.

Van Soest, P. J. 1994. *Nutritional ecology of the ruminant*. 2nd ed. Cornell University Press, Ithaca, New York, USA.

Vavra, M. 1992. Livestock and big game forage relationships. *Rangelands* 14:57–59.

Vavra, M., M. McInnis, and D. Sheehy. 1989. Implications of dietary overlap to management of free-ranging large herbivores. Proceedings of the Western Section of the American Society of Animal Science 40:489–495.

Verner, J. 1985. Assessment of counting techniques. *Current Ornithology* 2:247–302.

Verner, J., and A. S. Boss, technical coordinators. 1980. California wildlife and their habitats: western Sierra Nevada. U.S. Forest Service General Technical Report PSW-37.

Verner, J., M. L. Morrison, and C. J. Ralph, editors. 1986. *Wildlife 2000: modeling habitat relationships of terestrial vertebrates*. University of Wisconsin Press, Madison, Wisconsin, USA.

Verner, J., and K. A. Miline. 1990. Analyst and observer variables in density estimates from spot mapping. *Condor* 92:313–325.

Vesey-Fitzgerald, B. 1946. *British game*. Collins, London, United Kingdom.

Vietmeyer, N. B. 1991. Opportunities for commercial utilization of exotic species. Pages 3–7 *in* L. A. Renecker and R. J. Hudson, editors. Wildlife production: conservation and sustainable development. University of Alaska Fairbanks, Alaska. Agricultural and Forestry Experiment Station Miscellaneous Publication 91-6.

Vitousek, P. M., H. A. Mooney, J. Lubchenco, and J. M. Melillo. 1997. *Human domination of Earth's ecosystem. Science* 277:494–499.

Wade, D. A. 1987. Economics of wildlife production and damage control on private lands. Pages 154–163 *in* J. D. Decker and G. R. Goff, editors. *Valuing wildlife: economic and social perspectives*. Westview Press, Boulder, Colorado, USA.

Wagner, F. H. 1978. Livestock grazing and industry. Pages 121–145 *in* H. P. Browkau, editor. *Wildlife and America*. Council on Environmental Quality, Washington, D.C., USA.

Wagner, F. H. 1989. American wildlife management at the crossroads. *Wildlife Society Bulletin* 17:354–360.

Wagner, F. H. 1988. *Predator control and the sheep industry: the role of science in policy formation.* Regina Books, Claremont, California, USA.

Wallace, M. C., and P. R. Krausman. 1987. Elk, mule deer, and cattle habitats in central Arizona. *Journal of Range Management* 40:80–83.

Waller, D. M., and W. S. Alverson. 1997. The white-tailed deer: a keystone herbivore. *Wildlife Society Bulletin* 25:217–226.

Wallmo, O. C., L. C. Carpenter, W. L. Reglin, R. B. Gill, and D. L. Baker. 1977. Evaluation of deer habitat on a nutritional basis. *Journal of Range Management* 30:122–127.

Walters, C. J. 1986. *Adaptive management of renewable resources.* MacMillan, New York, New York, USA.

Walters, C. J., and R. Hilborn. 1978. Ecological optimization and adaptive management. *Annual Review of Ecology and Systematics.* 9:157–188.

Ward, A. L., J. J. Cupal, A. L. Lea, C. A. Oakley, and R. W. Weeks. 1973. Elk behavior in relation to cattle grazing, forest recreation, and traffic. *Transactions of the North American Wildlife Conference* 28:327–337.

Warner, R. E., and S. J. Brady. 1994. Managing farmlands for wildlife. Pages 648–662 *in* T. A. Bookhout, editor. *Research and management techniques for wildlife and habitats.* The Wildlife Society, Bethesda, Maryland, USA.

Warren, R. J. 1991. Ecological justification for controlling deer populations in eastern national parks. *Transactions of the North American Wildlife and Natural Resources Conference* 56:56–66.

Warren, R. J., and R. L. Kirkpatrick. 1982. Evaluation of the nutritional status of white-tailed deer using fat indices. Proceedings of the Southeastern Association of Fish and Wildlife Agencies 36:463–472.

Warrick, G. D., and P. R. Krausman. 1989. Barrel cacti consumption by desert bighorn sheep. *The Southwestern Naturalist* 34:483–486.

Watkins, C. 1987. The future of woodlands in the rural landscape. Pages 71–96 *in* D. Lockhart and B. Ilberg, editors. *The future of the British Rural Landscape.* Geo Books, Norwich, Ontario, Canada.

Watson, R. M., and A. D. Graham. 1969. A census of the large mammals of Loliondo controlled area, northern Tanzania. *East African Wildlife Journal* 7:43–59.

Watson, R. M., I. S. C. Parker, and T. Allan. 1969. A census of elephant and other large mammals in the Mkomazi region of northern Tanzania and southern Kenya. *East African Wildlife Journal* 7:11–26.

Watts, T. J. 1979. Status of the Big Hatchet desert sheep population, New Mexico. Desert Bighorn Council Transactions 23:92–94.

Weaver, J., R. Escano, D. Mattson, T. Puchlerz, and D. Despain. 1985. A cumulative effects model for grizzly bear management in the Yellowstone ecosystem. Pages 234–246 *in* G. P. Contreras and K. E. Evans, compilers. Proceedings of the grizzly bear habitat symposium. U.S. Forest Service General Technical Report INT-207.

Weaver, R. A. 1959. Effects of burro on desert water supplies. *Desert Bighorn Council Transactions* 3:1–3.

Weaver, R. A. 1973. Burro versus bighorn. Desert Bighorn Council Transactions 17:90–97.

Weaver, R. A., F. Vernoy, and B. Craig. 1958 or 1959. Game water development on the desert. Desert Bighorn Council Transactions 2:21–27.

Wecker, S. C. 1964. Habitat selection. *Scientific American* 211:109–116.

Welch, B.L. 1962. Adrenal of deer as indicators of population conditions for purposes of management. *Proceedings of the National White-tailed Deer Symposium* 1:94–108.

Weller, M. W. 1996. Birds of rangeland wetlands. Pages 71–82 *in* P. R. Krausman, editor. *Rangeland wildlife.* Society for Range Management, Denver, Colorado, USA.

Welles, R. E., and F. B. Welles. 1961. The bighorn of Death Valley. U.S. National Park Service, Fauna Series 6.

Werner, W. E. 1984. Bighorn sheep water development in southwestern Arizona. Desert Bighorn Council Transactions 28:12–13.

Werner, W. E. 1985. Philosophies of water development for bighorn sheep in southwestern Arizona. Desert Bighorn Council Transactions 29:13–14.

Westerskov, K. A. J. 1957. Training for the wildlife profession. *New Zealand Science Review* 15:67–75.

Westman, W. E. 1990. Managing for biodiversity: unresolved science and policy questions. *Bioscience* 40:26–33.

Whitcomb, R. F., C. S. Robbins, J. F. Lynch, B. L. Whitcomb, M. K. Klimbiewicz, and D. Bystrak. 1981. Effects of forest fragmentation on ave found on the eastern deciduous forest. Pages 125–205 *in* R. L. Burgess and D. M. Sharpe, editors. *Forest island dynamics in man-dominated landscapes.* Springer-Verlag, New York, New York, USA.

White, G. C. 2000. Modeling population dynamics. Pages 84–107 *in* S. Demarais and P. R. Krausman, editors. *Ecology and management of large mammals in North America.* Prentice Hall, Upper Saddle River, New Jersey, USA.

White, R. J. 1987. *Big game ranching in the United States.* Wild Sheep and Goat International, Mesilla, New Mexico, USA.

Whittaker, M. E., and V. G. Thomas. 1983. Seasonal levels of fat and protein reserves of snowshoe hares in Ontario. *Canadian Journal of Zoology* 61:1339–1345.

Whyte, R. J., and E. G. Bolen. 1985. Variation in mallard digestive organs during winter. *Journal of Wildlife Management* 49:1037–1040.

Wiener, J. G., and M. H. Smith. 1972. Relative efficiencies of four small mammal traps. *Journal of Mammalogy* 53:868–873.

Wiens, J. A. 1984. Resource systems, populations, and communities. Pages 397–436 *in* P. W. Price, C. N. Slobodchikoff, and W. S. Gaud, editors. *A new ecology: novel approaches to interactive systems.* John Wiley and Sons, New York, New York, USA.

Wilcove, D. S., and F. B. Samson. 1987. Innovative wildlife management: listening to Leopold. Transactions of the North American Wildlife and Natural Resource Conference 52:321–329.

Wilkinson, C. F. 1992. *Crossing the next meridian: land, water, and the future of the west.* Island Press, Washington, D. C., USA.

Williams, A. C., A. J. T. Johnsingh, and P. R. Krausman. In press. Elephant-human conflict in Rajai National Park, Northwest India. *Wildlife Society Bulletin.*

Williams, G. L., K. R. Russell, and W. K. Seitz. 1977. Pattern recognition as a tool in the ecological analysis of habitat. Pages 521–531 *in* Classification, inventory, and analysis of fish and wildlife habitat. U.S. Fish and Wildlife Service, FWS/OBS-78/76.

Williams, P. L., and W. D. Koenig. 1980. Water dependence of birds in a temperate oak woodland. *Auk* 97:339–350.

Williamson, J. F. 1983. Woodplan: microcomputer programs for forest management. Pages 128–130 *in* Proceedings of the national workshop on computer uses in fish and wildlife programs. Virginia Polytechnic Institute State University, Blacksburg, Virginia, USA.

Williamson, K. 1972. The conservation of bird life in the new coniferous forests. *Forestry* 45:87–100.

Wilson, E. O. 1988. *Biodiversity.* National Academy Press, Washington, D.C., USA.

Wilson, E. O. 1988. The current state of biological diversity. Pages 3–18 *in* E. O. Wilson, editor. *Biodiversity.* National Academy Press, Washington, D.C., USA.

Wilson, L. O. 1971. The effects of free water on desert bighorn home range. Desert Bighorn Council Transactions 15:82–89.

Wilson, L. O., and D. Hannans. 1977. Guidelines and recommendations for design and modification of livestock watering developments to facilitate safe use by wildlife. Bureau of Land Management Technical Note 305.

Wisdom, M. J., and J. G. Cook. 2000. North American elk. Pages 694–735 *in* S. Demarais and P. R. Krausman, editors. *Ecology and management of large mammals in North America.* Prentice Hall, Upper Saddle River, New Jersey, USA.

Wisdom, M. J., and J. W. Thomas. 1966. Elk. Pages 157–181 *in* P. R. Krausman, editor. *Rangeland wildlife.* The Society for Range Management, Denver, Colorado, USA.

Witter, D. J., and L. R. John. 1998. Emergence of human dimensions in wildlife management.

Transactions of the North American Wildlife and Natural Resources Conference 63:200–214.

Wolf, G. M., B. Griffith, C. Reed, and S. A. Temple. 1996. Avian and mammalian translocations: update and reanalysis of 1987 survey data. *Conservation Biology* 10:1142–1154.

Wolfe, L. M. 1951. *Son of the wildnerness, the life of John Muir.* Alfred A. Knopf, New York, New York, USA.

Woodell, S. R. J., editor. 1985. The English landscape: past, present, and future. *Wolfson college lectures 1983.* Oxford University Press, Oxford, United Kingdom.

Woods, J. E., T. S. Bickle, W. Evans, J. C. Germany, and V. W. Howard, Jr. 1970. The Fort Stockton mule deer herd: some ecological and life history characteristics with special emphasis on the use of water. New Mexico State University Agricultural Experiment Station Bulletin 567.

Woodward, A. R., and W. R. Marion. 1978. An evaluation of factors affecting night-light counts of alligators. Proceedings of the Annual Conference of the Southeasern Association of Fish and Wildlife Agencies 32:291–302.

Wooley, J. B., Jr., and R. B. Owen, Jr. 1978. Energy costs of activity and daily energy expenditure in the black duck. *Journal of Wildlife Management* 42:739–745.

Woolf, A. 1976. Pathology and necropsy techniques. Pages 425–436 *in* J. L. Schmidt and D. L. Gilbert, editors. *Big game of North America.* Stackpole Books, Harrisburg, Pennsylvania, USA.

Wright, B. A., and D. R. Fesenmaier. 1988. Modeling rural landowners' hunting access policies in east Texas, U.S.A. *Environmental Management* 12:229–236.

Wright, J. T. 1959. Desert wildlife. Arizona Game and Fish Department Wildlife Bulletin 6. Phoenix, Arizona, USA.

Wyatt, C. L., M. Trivedi, and D. R. Anderson. 1980. Statistical evaluation of remotely sensed thermal data for deer census. *Journal of Wildlife Management* 44:397–402.

Wykoff, W. R., N. L. Crookston, and A. R. Stage. 1982. User's guide to the stand prognosis model. U.S. Forest Service General Technical Report INT-133.

Wynne-Edwards, V. C. 1965. Self-regulating system in populations of animals. *Science* 147:1543–1548.

Yeo, J. J., J. M. Peek, W. T. Wittinger, and C. T. Kvale. 1993. Influence of rest-rotation cattle grazing on mule deer and elk habitat use in east-central Idaho. *Journal of Range Management* 46:245–250.

Yoakum, J. D. 1978. Pronghorn. Pages 103–121 *in* J. L. Schmidt and D. L. Gilbert, editors. *Big game of North America.* Stackpole Books, Harrisburg, Pennsylvania, USA.

Yoakum, J. D. 1994. Water requirements for pronghorn. Pronghorn Antelope Workshop Proceedings 16.

Yoakum, J. D., and B. W. O'Gara. 2000. Pronghorn. Pages 559–577 *in* S. Demarais and P. R. Krausman, editors. *Ecology and management of large mammals in North America.* Prentice Hall, Upper Saddle River, New Jersey, USA.

Yosef, M. K., H. D. Johnson, W. G. Bradley, and S. M. Seif. 1974. Tritiated water-turnover rate in rodents: desert and mountain. *Physiological Zoology* 47:153–162.

Young, S. P. 1944. Their history, life habits, economic status, and control. Part I. Pages 1–386 *in* S. P. Young and E. A. Goldman, editors. *The wolves of North America.* Dover Press, New York, New York, USA.

Zervanos, S. M., and G. I. Day. 1977. Water and energy requirements of captive and free-living collared peccaries. *Journal of Wildlife Management* 41:527–532.

Zimmerman, D. A. 1954. Bird mortality on Michigan highways. *Jack-Pine Warbler* 32:60–66.

Index